S0-BVO-690

Statistical Decision Theory and Related Topics II

Academic Press Rapid Manuscript Reproduction

Statistical Decision Theory and Related Topics II

Edited by

Shanti S. Gupta and David S. Moore
Department of Statistics
Purdue University
West Lafayette, Indiana

Proceedings of a Symposium Held at Purdue University
May 17–19, 1976

Academic Press New York San Francisco London 1977
A Subsidiary of Harcourt Brace Jovanovich, Publishers

ACADEMIC PRESS, INC.
111 Fifth Avenue, New York, New York 10003

United Kingdom Edition published by
ACADEMIC PRESS, INC. (LONDON) LTD.
24/28 Oval Road, London NW1

Library of Congress Cataloging in Publication Data

Main entry under title:

Statistical decision theory and related topics II.

Bibliography: p.

1. Statistical decision–Congresses. I. Gupta, Shanti Swarup, Date II. Moore, David S.
III. Purdue University, Lafayette, Ind.
QA279.4.S74 519.5'4 76-55357
ISBN 0–12–307560–2

PRINTED IN THE UNITED STATES OF AMERICA

math

CONTENTS

Numbers in parentheses refer to AMS (MOS) 1970 subject classifications.

CONTRIBUTORS TO THE SYMPOSIUM

Numbers in parentheses indicate the pages on which the authors' contributions begin.

Robert E. Bechhofer, School of Operations Research, Cornell University, Ithaca, New York 14853 (1)

James O. Berger, Department of Statistics, Purdue University, West Lafayette, Indiana 47907 (19)

P. J. Bickel, Department of Statistics, University of California, Berkeley, California 94720 (37)

M. E. Bock, Department of Statistics, Purdue University, West Lafayette, Indiana 47907 (19)

Lawrence D. Brown, Statistics Center, Rutgers University, New Brunswick, New Jersey 08903 (57)

Herman Chernoff, Department of Mathematics, Massachusetts Institute of Technology, Cambridge, Massachusetts 02139 (93)

A. P. Dempster, Department of Statistics, Harvard University, Cambridge, Massachusetts 02138 (121)

Shanti S. Gupta, Department of Statistics, Purdue University, West Lafayette, Indiana 47907 (139)

Wassily Hoeffding, Department of Statistics, The University of North Carolina, Chapel Hill, North Carolina 27514 (157)

Deng-Yuan Huang, Institute of Mathematics, Academia Sinica, Nankang, Taipei, Taiwan, Republic of China (139)

Peter J. Huber, Fachgruppe für Statistik, Eidgenöessische Technische Hochschule, 8006 Zürich, Switzerland (165)

J. Kiefer, Department of Mathematics, Cornell University, Ithaca, New York 14853 (193)

T. L. Lai, Department of Mathematical Statistics, Columbia University, New York, New York 10027 (213)

L. Le Cam, Department of Statistics, University of California, Berkeley, California 94720 (223)

Michael B. Marcus, Department of Mathematics, Northwestern University, Evanston, Illinois 60201 (245)

David S. Moore, Department of Statistics, Purdue University, West Lafayette, Indiana 47907 (269)

S. E. Nevius, Department of Statistics, The Florida State University, Tallahassee, Florida 32306 (281)

Jerzy Neyman, Department of Statistics, University of California, Berkeley, California 94720 (297)

I. Olkin, Department of Statistics, Stanford University, Stanford, California 94305 (313)

F. Proschan, Department of Statistics, The Florida State University, Tallahassee, Florida 32306 (281)

C. Radhakrishna Rao, Indian Statistical Institute, New Delhi 110029, India (327)

Herbert Robbins, Department of Mathematical Statistics, Columbia University, New York, New York 10027 (213)

Herman Rubin, Department of Statistics, Purdue University, West Lafayette, Indiana 47907 (351)

Jerome Sacks, Department of Mathematics, Northwestern University, Evanston, Illinois 60201 (245)

Thomas J. Santner, School of Operations Research, Cornell University, Ithaca, New York 14853 (1)

J. B. Selliah, University of Sri Lanka, Pavillion View, Vaddukoddai, Sri Lanka (313)

J. Sethuraman, Department of Statistics, The Florida State University, Tallahassee, Florida 32306 (281)

D. Siegmund, Department of Mathematical Statistics, Columbia University, New York, New York 10027 (213)

Milton Sobel, Department of Mathematics, University of California, Santa Barbara, California 93106 (357)

J. N. Srivastava, Department of Statistics, Colorado State University, Fort Collins, Colorado 80523 (375)

W. J. Studden, Department of Statistics, Purdue University, West Lafayette, Indiana 47907 (411)

Bruce W. Turnbull, School of Operations Research, Cornell University, Ithaca, New York 14853 (1)

W. R. van Zwet, Rijksuniversiteit de Leiden, Postbus 2060, Leiden, The Netherlands (421)

Lionel Weiss, School of Operations Research, Cornell University, Ithaca, New York 14853 (439)

R. A. Wijsman, Department of Mathematics, University of Illinois, Urbana, Illinois 61801 (451)

J. Wolfowitz, Department of Mathematics, University of Illinois, Urbana, Illinois 61801 (193)

H. P. Wynn, Mathematics Department, Imperial College, Queens Gate, London SW7, England (471)

James Yackel, Department of Statistics, Purdue University, West Lafayette, Indiana 47907 (269)

Joseph A. Yahav, Department of Statistics, Tel-Aviv University, Ramat Aviv, Israel (37, 93)

PREFACE

Decision theory and areas of research related to it have been at the heart of advances in mathematical statistics during the past generation. This volume contains the invited papers presented at an international symposium on Statistical Decision Theory and Related Topics held at Purdue University in May, 1976. Decision theory was broadly interpreted to include related areas that have been the scene of rapid progress since the preceding Purdue symposium in 1970. This symposium featured sessions on general decision theory, multiple decision theory, optimal experimental design, and robustness. The researchers invited to participate, and to author papers for this volume, are among the leaders in these fields.

We are especially grateful to Professor Felix Haas, Executive Vice President and Provost, and to Professor Allan Clark, Dean of the School of Science, for the encouragement and financial support provided by Purdue. The symposium was also supported by the National Science Foundation under grant MPS75-23196, by the Air Force Office of Scientific Research under grant AFOSR 76-2969, and by the Office of Naval Research under contract N00014-75-C-0455. We wish to thank these agencies for their assistance, and in particular Dr. I. N. Shimi of the Air Force Office of Scientific Research and Drs. Bruce McDonald and Robert Lundegard of the Office of Naval Research.

Many individuals contributed to the success of the symposium and to the preparation of this volume. The program for the symposium was developed under the guidance of an advisory committee composed of S. S. Gupta, Chairman, Purdue University, R. R. Bahadur, University of Chicago, L. Le Cam, University of California at Berkeley, J. C. Kiefer, Cornell University, H. E. Robbins, Columbia University, and J. Wolfowitz, University of Illinois. The excellence of the program was due in large part to the efforts of these colleagues. Local arrangements were coordinated by G. P. McCabe, assisted by other faculty and students in the Purdue Department of Statistics.

Numerous colleagues at Purdue and elsewhere served as referees for the papers presented here. In many cases, their comments helped to strengthen the papers. We are happy to acknowledge the encouragement and assistance of Academic Press in preparing this volume for publication. Finally, the burden of typing the entire contents accurately and attractively was borne with great skill by Norma Lucas.

The papers presented here bear witness to the vigor of research in statistical theory. We are pleased to present them to the statistical community.

SELECTING THE LARGEST INTERACTION IN A TWO-FACTOR EXPERIMENT

By Robert E. Bechhofer, Thomas J. Santner and Bruce W. Turnbull
Cornell University

1. *Introduction.* The research described in the present paper was motivated by consideration of the following type of problem: Suppose that a medical research worker wishes to plan an experiment to study the effect of several (c) different methods of treatment on a physiological response of male and female subjects (who are otherwise matched with respect to other factors which might affect the response). It is assumed known that the effect of the treatment on the mean response is different for men than for women, and also that it varies from treatment to treatment. It is suspected that there may be a large interaction between sex and method of treatment, and it is desired to identify the sex-treatment combination for which this interaction is largest in the hope that such information might provide some clue as to the mechanism underlying the effectiveness of the methods of treatment. The statistical problem is to design the experiment on such a scale that this largest interaction can, if it is sufficiently large to be of practical importance to the experimenter, be detected with preassigned probability. The setup described above consists of 2c sex-treatment combinations, and is one of the general class of 2-factor experiments involving $r \geq 2$ levels of one qualitative factor and $c \geq 2$ levels of a second qualitative factor. More generally, one might consider multifactor experiments involving three or more qualitative factors. It should be noted that our present goal of selecting the combination associated with the

*Research supported by U. S. Army Research Office Contract DAHC04-73-C-0008, Office of Naval Research Contract N00014-75-C-0586 and National Science Foundation Contract ENG75-10487.

largest interaction is quite different from the classical one of selecting the combination associated with the largest response, the later goal having been treated in [1].

The present paper considers 2-factor experiments and concentrates on the $2 \times c$ case. It is organized as follows: In Section 2 we give the model and statistical assumptions that we adopt. The formulation of our selection problem for $r \times c$ factorial experiments is proposed in Section 3; this involves the statement of a goal and an associated probability requirement. A single-stage selection procedure is proposed in Section 4; the criterion for choosing sample size is introduced in this section, and this leads to a statement of the basic statistical problem that we seek to solve.

In Section 5 we derive an exact expression for the probability of a correct selection (PCS) in the $r \times c$ case when the single-stage selection procedure is used; special cases involving particular choices of r and c are studied in detail. Section 6 contains the main result of the paper. It concerns the explicit determination of the so-called least favorable (LF) configuration of the interactions (γ_{ij}) in the $2 \times c$ case; this result is stated as Theorem 6.1, the proof of which is given following the theorem. Directions of future research are indicated in Section 7.

2. *Model and statistical assumptions*. We consider a 2-factor experiment, both factors qualitative, the first factor being studied at r levels and the second at c levels. We assume the usual fixed-effects linear model with observations $Y_{ijk}(1 \leqq i \leqq r, 1 \leqq j \leqq c; 1 \leqq k \leqq n)$ which are normal and independent with $E\{Y_{ijk}\} = \mu_{ij} = \mu + \alpha_i + \beta_j + \gamma_{ij}$ $(\sum_{i=1}^{r} \alpha_i = \sum_{j=1}^{c} \beta_j = \sum_{i=1}^{r} \gamma_{ij} = \sum_{j=1}^{c} \gamma_{ij} = 0)$, $\mathrm{Var}\{Y_{ijk}\} = \sigma^2$. We further assume that μ, the α_i, β_j, and γ_{ij} are unknown, and that σ^2 is known. Our interest is in the γ_{ij}.

If $\gamma_{ij} \equiv 0$ (all i,j) then $\mu_{ij} = \mu + \alpha_i + \beta_j$, and the "effects" of the two factors on the mean are said to be strictly additive for all factor-level combinations; if $\gamma_{ij} \neq 0$ (all i,j) then interaction is said to be present. It is with this latter situation that we shall be particularly concerned in this paper.

Let $\gamma_{[1]} \leq \cdots \leq \gamma_{[rc]}$ denote the ranked values of the γ_{ij} $(1 \leq i \leq r,\ 1 \leq j \leq c)$. (Note that $\gamma_{[1]} \leq 0 \leq \gamma_{[rc]}$, and that for $r = c = 2$ we have $\gamma_{[1]} = \gamma_{[2]} = -\gamma_{[3]} = -\gamma_{[4]}$.) It is assumed that the experimenter has no prior knowledge concerning the pairing of the $\gamma_{[s]}$ $(1 \leq s \leq rc)$ with the levels of either of the factors.

3. *Goal and probability requirement.* We shall consider a particular goal which would appear to be appropriate in these situations. In our formulation of the decision problem we adopt a new variant of the so-called indifference-zone approach (Bechhofer [1]). The goal and formulation are given below.

(3.1) GOAL I: "To select the factor-level combination associated with $\gamma_{[rc]}$."

Equivalently we might be concerned with selecting the factor-level combination associated with $\gamma_{[1]}$. Other meaningful goals could also be posed. (See Section 7.)

The experimenter restricts consideration to selection procedures which guarantee the following

PROBABILITY REQUIREMENT

$$P\{\text{Correct Selection}\} \geq P^* \quad \text{whenever } \gamma_{[rc]} \geq \Delta^* \text{ and } \gamma_{[rc]} - \gamma_{[rc-1]} \geq \delta^*. \tag{3.2}$$

The three quantities $\{\Delta^*, \delta^*, P^*\}$ $\left(0<\Delta^*<\infty,\ 0<\delta^*< \frac{(r-1)(c-1)-1}{(r-1)(c-1)}\Delta^*,\ \frac{1}{rc} < P^* < 1\right)$ are to be specified by the experimenter prior to the start of experimentation, the choice of their values depending on economic and cost considerations. The event "Correct Selection" (CS) in (3.2) means the selection of the levels of the two factors associated with $\gamma_{[rc]}$ when $\gamma_{[rc]} - \gamma_{[rc-1]} > 0$.

REMARK 3.1. Values of δ^*/Δ^* greater than $[(r-1)(c-1)-1]/(r-1)(c-1)$ are inappropriate. For suppose $\gamma_{[rc]} = \gamma_{11}$ and $\gamma_{11} - \gamma_{ij} \geqq \delta^*$; then $\gamma_{11} = \sum_{a=2}^{r} \sum_{b=2}^{c} \gamma_{ab} \leqq (r-1)(c-1)(\gamma_{11}-\delta^*)$. Hence $\delta^* \leqq [(r-1)(c-1)-1]\gamma_{11}/(r-1)(c-1)$ which must be true for <u>all</u> $\gamma_{11} \geqq \Delta^*$.

REMARK 3.2. If the traditional indifference-zone approach were used, the inequality $\gamma_{[rc]} \geqq \Delta^*$ would not be present in (3.2). However, in the present situation in which our interest is in the largest <u>positive</u> interaction, we are interested in this interaction only when it is sufficiently large relative to <u>zero</u> which is the standard which defines <u>no</u> interaction. The situation here is similar to the one considered in Bechhofer-Turnbull [2].

4. *Selection procedure, and the choice of sample size.* We shall employ the following "natural" single-stage selection procedure for (3.1).

P: Take n independent observations Y_{ijk} $(1 \leqq k \leqq n)$ for each (i,j)-combination $(1 \leqq i \leqq r,\ 1 \leqq j \leqq c)$. Compute $\hat{\gamma}_{ij} = \bar{Y}_{ij.} - \bar{Y}_{i..} - \bar{Y}_{.j.} + \bar{Y}_{...}$ where $\bar{Y}_{ij.} = \sum_{k=1}^{n} Y_{ijk}/n$,

$$\bar{Y}_{i..} = \sum_{j=1}^{c} \bar{Y}_{ij.}/c,\quad \bar{Y}_{.j.} = \sum_{i=1}^{r} \bar{Y}_{ij.}/r, \text{ and} \tag{4.1}$$

$\bar{Y}_{...} = \sum_{i=1}^{r} \sum_{j=1}^{c} \bar{Y}_{ij.}/rc$. Select the factor-level combination which produced $\hat{\gamma}_{[rc]} = \max\{\hat{\gamma}_{ij} \mid 1 \leqq i \leqq r,\ 1 \leqq j \leqq c\}$ as the one associated with $\gamma_{[rc]}$.

Procedure *P* is completely defined with the exception of the common sample size n which is under the control of the experimenter. This leads to the central problem that we address in the present paper:

(4.2) PROBLEM. For given $c \geq 3$, $c \geq r \geq 2$ and specified $\{\Delta^*, \delta^*, P^*\}$, find the smallest value of n which will guarantee (3.2) when P is used.

The first stage in solving (4.2) is that of deriving an exact expression for the PCS when P is used.

5. *Expression for the PCS.* We assume that $\gamma_{[rc]} > 0$, i.e., $\gamma_{ij} \neq 0$, and that for $c \geq 3$, $c \geq r \geq 2$ there is a unique largest γ_{ij}; without loss of generality we assume $\gamma_{11} = \gamma_{[rc]}$. Then the PCS using P can be written as

$$P\left\{\hat{\gamma}_{11} > \max_{\substack{2\leq a\leq r\\ 2\leq b\leq c}} \hat{\gamma}_{ab},\ \hat{\gamma}_{11} > \max_{2\leq a\leq r} \hat{\gamma}_{a1},\ \hat{\gamma}_{11} > \max_{2\leq b\leq c} \hat{\gamma}_{1b}\right\}$$

$$= P\{\sum_{i=2}^{r}\sum_{j=2}^{c}(X_{ij}+\gamma_{ij})-(X_{ab}+\gamma_{ab})>0,\ (2 \leq a \leq r,\ 2 \leq b \leq c);$$

$$\sum_{i=2}^{r}\sum_{j=2}^{c}(X_{ij}+\gamma_{ij}) + \sum_{j=2}^{c}(X_{aj}+\gamma_{aj}) > 0,\ (2 \leq a \leq r); \tag{5.1}$$

$$\sum_{i=2}^{r}\sum_{j=2}^{c}(X_{ij}+\gamma_{ij}) + \sum_{i=2}^{r}(X_{ib}+\gamma_{ib}) > 0,\ (2 \leq b \leq c)\}$$

where the $X_{ij} = \hat{\gamma}_{ij} - \gamma_{ij}$ have an $(r-1)(c-1)$-variate normal distribution with $E\{X_{ij}\} = 0$ $(2 \leq i \leq r,\ 2 \leq j \leq c)$,

(5.2)

$$\mathrm{Var}\{X_{ij}\} = \frac{(r-1)(c-1)}{rcn}\sigma^2$$

$$(2 \leq i \leq r,\ 2 \leq j \leq c)$$

$$\mathrm{Cov}\{X_{i_1,j_1}, X_{i_2,j_2}\} = \frac{1}{rcn}\sigma^2$$

$$(i_1 \neq i_2,\ j_1 \neq j_2;\ 2 \leq i_1,i_2 \leq r,\ 2 \leq j_1,j_2 \leq c)$$

$$\mathrm{Cov}\{X_{i,j_1}, X_{i,j_2}\} = -\frac{r-1}{rcn}\sigma^2$$

$$(j_1 \neq j_2;\ 2 \leq i \leq r,\ 2 \leq j_1,j_2 \leq c)$$

$$\mathrm{Cov}\{X_{i_1,j}, X_{i_2,j}\} = -\frac{c-1}{rcn}\sigma^2$$

$(i_1 \neq i_2;\ 2 \leq i_1, i_2 \leq r,\ 2 \leq j \leq c)$.

Thus the exact PCS can be expressed in terms of rc-1 linear stochastic inequalities involving (r-1)(c-1) correlated normal variates.

In the remainder of the present paper we shall concentrate our attention on the $r = 2$, $c \geq 3$ case, although in (5.4), Remark 6.4 and the Appendix we shall also refer to the $r = c = 3$ case. Expressions for the exact PCS for these cases are given below.

2×c *case*. When $r = 2$ the PCS of (5.1) reduces to

$$P\{\sum_{j=2}^{c}(X_{2j}+\gamma_{2j}) - (X_{2b}+\gamma_{2b}) > 0,\ (2 \leq b \leq c);$$

(5.3)

$$\sum_{j=2}^{c}(X_{2j}+\gamma_{2j}) + (X_{2b}+\gamma_{2b}) > 0,\ (2 \leq b \leq c)\}$$

which involves only 2(c-1) inequalities.

3×3 *case*. When $r = c = 3$ the PCS of (5.1) reduces to

$$(5.4)\qquad P\{\sum_{i=2}^{3}\sum_{j=2}^{3}(X_{ij}+\gamma_{ij}) - (X_{ab}+\gamma_{ab}) > 0,\ (2 \leq a, b \leq 3)\}$$

which involves only 4 inequalities.

6. LF-*configuration of the* γ_{ij} *for the* 2×c *case*. The most difficult step in solving (4.2) is that of determining a least favorable (LF) configuration of the γ_{ij}, i.e., a set of γ_{ij} $(1 \leq i \leq r,\ 1 \leq j \leq c)$ which minimizes (5.1) for any $n \geq 1$, subject to $\gamma_{[rc]} \geq \Delta^*$ and $\gamma_{[rc]} - \gamma_{[rc-1]} \geq \delta^*$ where $0 < \Delta^* < \infty$ and $0 < \delta^* < \frac{(r-1)(c-1)-1}{(r-1)(c-1)}\Delta^*$. Our result for the 2×c case is presented in Theorem 6.1 below.

We introduce the following notation. Let $\underset{\sim}{\gamma} = (\gamma_2,\ldots,\gamma_c)$ denote a (c-1)-vector where $\gamma_b = \gamma_{2b}$ $(2 \leq b \leq c)$. For each $\gamma_{11} \geq \Delta^*$, let $E(\gamma_{11}) = \{\underset{\sim}{\gamma} \mid \sum_{j=2}^{c}\gamma_j = \gamma_{11};\ \delta^* - \gamma_{11} \leq \gamma_b \leq \gamma_{11} - \delta^*,$ $(2 \leq b \leq c)\}$ be the set of configurations in the preference zone for a CS on the plane $\sum_{j=2}^{c} \gamma_j = \gamma_{11}$; thus $E = \bigcup_{\gamma_{11} \geqq \Delta^*} E(\gamma_{11})$ is the entire preference zone.

A basic lemma. Our first result describes the infimum of the PCS over the section $E(\gamma_{11})$. Let $p = (c-3)/2$ or $(c-4)/2$ according as c is odd or even, and let j be an integer defined according to the rule

$$(6.1) \quad j = \begin{cases} p & , \text{ if } 0 < \delta^* < \dfrac{c-2-2p}{c-1-2p}\gamma_{11} \\ t \ (0 \leqq t < p), & \text{if } \dfrac{c-4-2t}{c-3-2t}\gamma_{11} \leqq \delta^* < \dfrac{c-2-2t}{c-1-2t}\gamma_{11}. \end{cases}$$

Clearly j always exists and is uniquely defined since $0 < \delta^* < (c-2)\Delta^*/(c-1)$.

LEMMA 6.1. For the 2×c case ($c \geqq 3$) and for fixed $\gamma_{11} \geqq \Delta^*$ the LF-configuration of the $\{\gamma_j\}$ over the section $E(\gamma_{11})$ is given by

$$(6.2) \quad \underset{\sim}{\gamma}(j,\gamma_{11}) = (\gamma_{11}-\delta^*,\ldots,\gamma_{11}-\delta^*,d_j,\delta^*-\gamma_{11},\ldots,\delta^*-\gamma_{11})$$

where $d_j = \delta^*(c-2-2j) - \gamma_{11}(c-3-2j)$, and there are $c-2-j$ and j elements $\gamma_{11}-\delta^*$ and $\delta^*-\gamma_{11}$, respectively.

REMARK 6.1. $\underset{\sim}{\gamma}(j,\gamma_{11})$ completely defines the γ_{ab} ($1 \leqq a \leqq 2$, $1 \leqq b \leqq c$) matrix since $\sum_{a=1}^{2} \gamma_{ab} = \sum_{b=1}^{c} \gamma_{ab} = 0$; all matrices obtained from $\{\gamma_{ab}\}_{2\times c}$ by permuting rows and/or columns have the same associated PCS.

We next note that representation (5.3) of the PCS can equivalently be expressed as

$$(6.3) \quad f(\underset{\sim}{\gamma}) = P\{\underset{\sim}{X} + \underset{\sim}{\gamma} \in A\}$$

where $\underset{\sim}{X} = (X_{22},\ldots,X_{2c})$ and $A = \{\underset{\sim}{w} = (w_2,\ldots,w_c) \mid \sum_{j=2}^{c} w_j - w_b > 0$, $(2 \leqq b \leqq c)$; $\sum_{j=2}^{c} w_j + w_b > 0$, $(2 \leqq b \leqq c)\}$. In the proof that follows we show that (1) $f(\underset{\sim}{\gamma})$ is log concave in $\underset{\sim}{\gamma}$, and (2) $\underset{\sim}{\gamma}(j,\gamma_{11})$ and its permutations form the extreme points of the convex set

$E(\gamma_{11})$.

Our proof that $f(\underset{\sim}{\gamma})$ is log concave is based on a characterization of the log-concavity property for probability measures which are generated by densities and is proved, for example, in Prékopa [5]. The proof that $E(\gamma_{11})$ is the convex hull of $\underset{\sim}{\gamma}(j,\gamma_{11})$ and its permutations is obtained by showing that $\underset{\sim}{\gamma}(j,\gamma_{11})$ majorizes every $\underset{\sim}{\gamma}$ in $E(\gamma_{11})$ in the sense of Hardy, Littlewood and Pólya [3] and the fact that $\underset{\sim}{a}$ majorizes $\underset{\sim}{b}$ if and only if $\underset{\sim}{b}$ is in the convex hull of $\underset{\sim}{a}$ and $\underset{\sim}{a}$'s permutations. For additional details, references, and associated ideas see Marshall and Olkin [4]. The definition of majorization and the lemma characterizing log-concavity in probability measures are stated below for convenient reference.

In n dimensions the vector $\underset{\sim}{b}$ is said to majorize the vector $\underset{\sim}{a}$ (written $\underset{\sim}{a} \prec \underset{\sim}{b}$) if upon reordering components to achieve $a_1 \geqq a_2 \geqq \cdots \geqq a_n$, $b_1 \geqq b_2 \geqq \cdots \geqq b_n$, it follows that $\sum_{i=1}^{k} a_i \leqq \sum_{i=1}^{k} b_i$, $1 \leqq k < n$ and $\sum_{i=1}^{n} a_i = \sum_{i=1}^{n} b_i$. It is known that $\underset{\sim}{a} \prec \underset{\sim}{b}$ if and only if $\underset{\sim}{a}$ is a convex combination of $\underset{\sim}{b}$ and permutations of $\underset{\sim}{b}$.

LEMMA 6.2. <u>Let</u> P <u>be</u> <u>a</u> <u>probability</u> <u>measure</u> <u>on</u> R^n <u>generated</u> <u>by</u> <u>a</u> <u>density</u> $g(\underset{\sim}{x})$, $\underset{\sim}{x} \in R^n$, <u>i.e.</u>, $P\{A\} = \int_A g(\underset{\sim}{x})d\underset{\sim}{x}$ <u>for</u> <u>any</u> <u>Borel</u> <u>set</u> $A \subset R^n$. <u>Then</u> P <u>satisfies</u>

$$(6.4) \qquad P\{\alpha A_0 + (1-\alpha)A_1\} \geqq [P\{A_0\}]^{\alpha}[P\{A_1\}]^{1-\alpha} \text{ for all } 0 \leqq \alpha \leqq 1$$

<u>if</u> <u>and</u> <u>only</u> <u>if</u> $g(\underset{\sim}{x})$ <u>is</u> <u>log-concave</u>. (<u>In</u> (6.4) <u>it</u> <u>is</u> <u>assumed</u> <u>that</u> <u>all</u> <u>events</u> <u>are</u> <u>measurable</u>.)

For a proof of Lemma 6.2 see [5]. In our problem the joint distribution of $\underset{\sim}{X}$ is non-singular (c-1)-variate multivariate normal, and is thus log concave. For any (c-1)-vectors $\underset{\sim}{\gamma}_1, \underset{\sim}{\gamma}_2$ and $0 \leqq \alpha \leqq 1$ let $\underset{\sim}{\gamma} = \alpha\underset{\sim}{\gamma}_1 + (1-\alpha)\underset{\sim}{\gamma}_2$. Then $\{X + \alpha\underset{\sim}{\gamma}_1 + (1-\alpha)\underset{\sim}{\gamma}_2 \in A\} = \{X \in \alpha(A-\underset{\sim}{\gamma}_1) + (1-\alpha)(A-\underset{\sim}{\gamma}_2)\}$ since A is convex. Thus from Lemma 6.2 we have

$$
\begin{aligned}
f(\underset{\sim}{\gamma}) &= P\{\underset{\sim}{X} + \underset{\sim}{\gamma} \in A\} \\
&= P\{X \in \alpha(A-\underset{\sim}{\gamma}_1)+(1-\alpha)(A-\underset{\sim}{\gamma}_2)\} \\
&\geqq [P\{\underset{\sim}{X} \in A-\underset{\sim}{\gamma}_1\}]^{\alpha}[P\{\underset{\sim}{X} \in A-\underset{\sim}{\gamma}_2\}]^{1-\alpha} \\
&\geqq \min\{f(\underset{\sim}{\gamma}_1),\ f(\underset{\sim}{\gamma}_2)\}.
\end{aligned}
$$

Hence the infimum of the PCS occurs at an extreme point of $E(\gamma_{11})$. Since $f(\underset{\sim}{\gamma})$ is constant under permutations it suffices to show that $\underset{\sim}{\gamma}(j,\gamma_{11}) \in E(\gamma_{11})$ and $\underset{\sim}{\gamma}(j,\gamma_{11}) \succ \underset{\sim}{\gamma}$ for all $\underset{\sim}{\gamma} \in E(\gamma_{11})$ to complete the proof of Lemma 6.1.

Now $\underset{\sim}{\gamma}(j,\gamma_{11}) \in E(\gamma_{11})$ since $(c-2-j)(\gamma_{11}-\delta^*) + d_j + j(\delta^*-\gamma_{11}) = \gamma_{11}$ and $\gamma_{11} - \delta^* \geqq d_j \geqq \delta^* - \gamma_{11}$ provided that $\frac{c-2-2j}{c-1-2j}\gamma_{11} > \delta^* \geqq \frac{c-4-2j}{c-3-2j}\gamma_{11}$ which is the defining condition (6.1) for j. Select an arbitrary $\underset{\sim}{\gamma} \in E(\gamma_{11})$ and without loss of generality assume $\gamma_2 \geqq \cdots \geqq \gamma_c$; hence $\delta^* - \gamma_{11} \leqq \gamma_c \leqq \gamma_2 \leqq \gamma_{11}-\delta^*$ and $\sum_{j=2}^{c}\gamma_j = \gamma_{11}$. We shall show that $\underset{\sim}{\gamma} \prec \underset{\sim}{\gamma}(j,\gamma_{11})$.

a) For all m ($2 \leqq m \leqq c-1-j$) we have

$$\sum_{i=2}^{m}\gamma(j,\gamma_{11})_i=(m-1)(\gamma_{11}-\delta^*)\geqq\sum_{i=2}^{m}\gamma_i \text{ since } \gamma_i < \gamma_{11}-\delta^*.$$

b) For all m ($c-j+1 \leqq m \leqq c$) we have

$$\sum_{i=m}^{c}\gamma(j,\gamma_{11})_i=(c-m+1)(\delta^*-\gamma_{11}) \leqq \sum_{i=m}^{c}\gamma_i \text{ since } \gamma_i \geqq \delta^*-\gamma_{11}.$$

Thus $\gamma_{11} - \sum_{i=m}^{c}\gamma(j,\gamma_{11})_i \geqq \gamma_{11} - \sum_{i=m}^{c}\gamma_i$,

i.e., $\sum_{i=2}^{m-1}\gamma(j,\gamma_{11})_i \geqq \sum_{i=2}^{m-1}\gamma_i \quad (c-j \leqq m-1 \leqq c-1)$

and we have completed the proof that $\underset{\sim}{\gamma}(j,\gamma_{11})$ is the LF-configuration over $E(\gamma_{11})$.

REMARK 6.2. Alternatively, Lemma 6.1 could have been proved by applying results in Marshall and Olkin [4] to show that $f(\underset{\sim}{\gamma})$ is a Schur-concave function of $\underset{\sim}{\gamma}$ and hence must attain its minimum at an extreme point of $E(\gamma_{11})$. We have shown the (stronger) property that $f(\underset{\sim}{\gamma})$ is log-concave; some of our later results depend on this stronger property of $f(\underset{\sim}{\gamma})$.

Geometry of the 2×5 case. Consideration of the geometry of the set $E(\gamma_{11})$ over which $f(\underset{\sim}{\gamma})$ is to be minimized w.r.t $\gamma_2, \gamma_3, \ldots, \gamma_c$ gives added insight into the dependence of the LF-configuration $\underset{\sim}{\gamma}(j,\gamma_{11})$ on the relationship between δ^* and γ_{11}. We use the 2×5 case for illustrative purposes since 5 is the smallest c-value when r = 2 for which there is <u>more than one</u> (in this case <u>two</u>) distinct $\underset{\sim}{\gamma}(j,\gamma_{11})$'s.

When c = 5 the set $E(\gamma_{11})$ is isomorphic to the convex set $E'(\gamma_{11})$ in 3-space defined by

$$E'(\gamma_{11}) = \{(\gamma_2,\gamma_3,\gamma_4) \mid \delta^* - \gamma_{11} \leq \gamma_2, \gamma_3, \gamma_4 \leq \gamma_{11} - \delta^*;$$
$$\delta^* \leq \gamma_2 + \gamma_3 + \gamma_4 \leq 2\gamma_{11} - \delta^*\},$$

since $\sum_{i=2}^{5} \gamma_i = \gamma_{11}$ in $E(\gamma_{11})$. Now $E'(\gamma_{11})$ is the intersection of the cube C centered at (0,0,0) with faces parallel to the planes formed by the coordinates axes and with edges of length $2(\gamma_{11}-\delta^*)$, and the slab in 3-space bounded by the parallel planes $\gamma_2+\gamma_3+\gamma_4 = \delta^*$ and $\gamma_2+\gamma_3+\gamma_4 = 2\gamma_{11}-\delta^*$. There are two possible shapes for $E'(\gamma_{11})$ depending on the relationship between δ^* and γ_{11}. The differences between these two shapes as seen in the plane $\gamma_4 = \gamma_{11}-\delta^*$ are displayed as the shaded regions in Figures 6.1 and 6.2 which are valid for $0<\delta^*<\gamma_{11}/2$ and $\gamma_{11}/2<\delta^*<3\gamma_{11}/4$, respectively; when $\delta^* = \gamma_{11}/2$ the two figures are identical and the line $\gamma_2+\gamma_3 = \gamma_{11}$ passes through the point $(\gamma_{11}-\delta^*,\gamma_{11}-\delta^*)$. Note that in the plane $\gamma_4 = \gamma_{11}-\delta^*$ the planes $\gamma_2+\gamma_3+\gamma_4 = \delta^*$ and $\gamma_2+\gamma_3+\gamma_4 = 2\gamma_{11}-\delta^*$ become the lines $\gamma_2+\gamma_3 = 2\delta^*-\gamma_{11}$ and $\gamma_2+\gamma_3 = \gamma_{11}$, respectively. The plane $\gamma_2+\gamma_3+\gamma_4 = \delta^*$ always intersects the cube C, and hence always plays a role in determining $E'(\gamma_{11})$; the plane $\gamma_2+\gamma_3+\gamma_4 = 2\gamma_{11}-\delta^*$ intersects the cube only when $0 < \delta^* < \gamma_{11}/2$, and hence only plays a role in determining $E'(\gamma_{11})$ in that situation.

The key to understanding the role that the shape of $E'(\gamma_{11})$

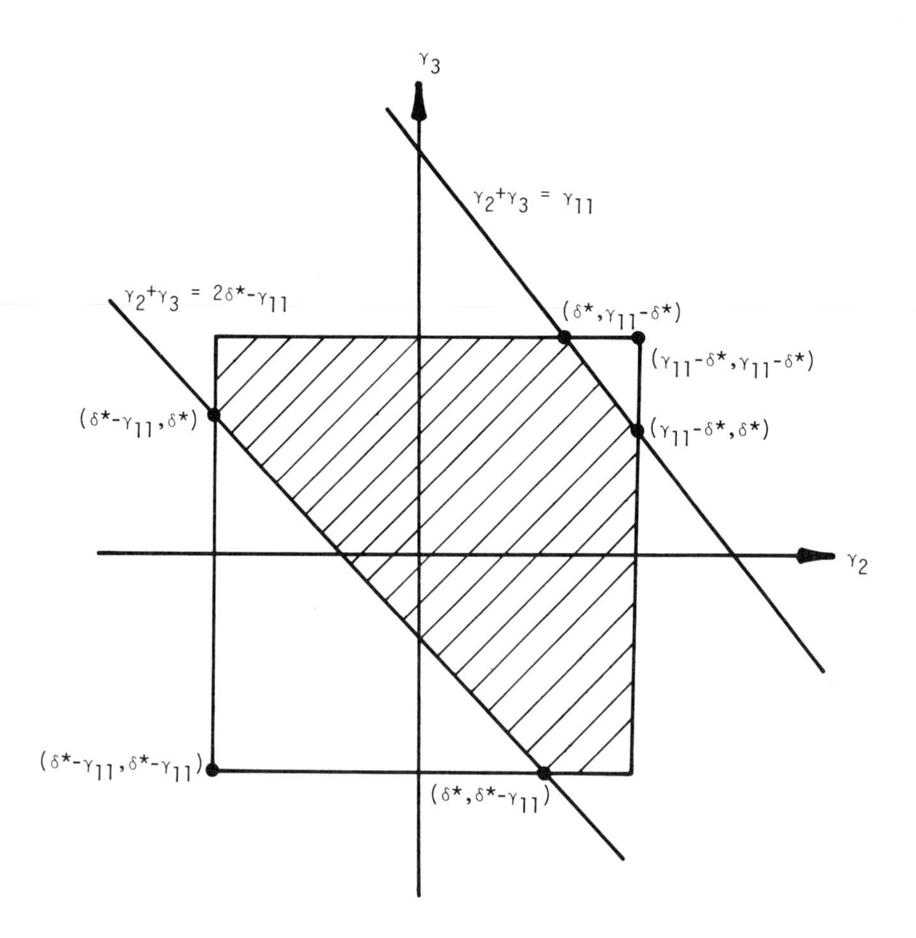

Figure 6.1. The Section of $E'(\gamma_{11})$ in the Plane $\gamma_4 = \gamma_{11}-\delta^*$ when $0 < \delta^* < \gamma_{11}/2$.

plays in the minimization problem is contained in the equation

$$\text{(6.5)} \qquad \inf_{\underset{\sim}{\gamma}\varepsilon E'(\gamma_{11})} f(\underset{\sim}{\gamma}) = \inf_{t\varepsilon I} \inf_{\underset{\sim}{\gamma}\varepsilon S(t)} f(\underset{\sim}{\gamma})$$

where $S(t) = \{(\gamma_2,\gamma_3,\gamma_4)|\gamma_2+\gamma_3+\gamma_4 = t;\ \underset{\sim}{\gamma} \ \varepsilon\ C\}$ and

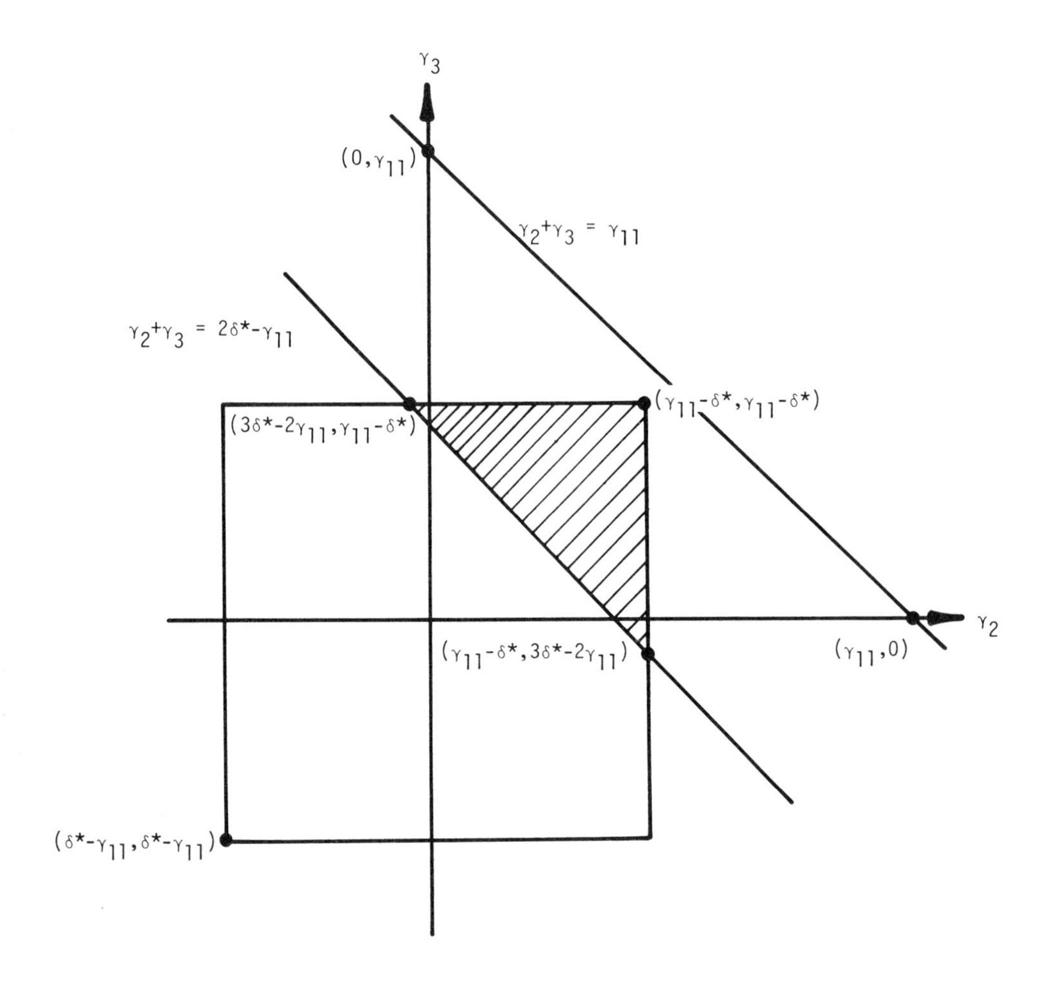

Figure 6.2. The Section of $E'(\gamma_{11})$ in the Plane $\gamma_4 = \gamma_{11}-\delta^*$ when $\gamma_{11}/2 < \delta^* < 3\gamma_{11}/4$.

$$I = \begin{cases} [\delta^*,2\gamma_{11}-\delta^*] \text{ for } 0 < \delta^* < \gamma_{11}/2 \\ [\delta^*,3(\gamma_{11}-\delta^*)] \text{ for } \gamma_{11}/2 < \delta^* < 3\gamma_{11}/4. \end{cases}$$

The majorization argument of Lemma 6.1 shows that on the plane

$\gamma_2+\gamma_3+\gamma_4 = t$, the value of $f(\underset{\sim}{\gamma})$ decreases as $\underset{\sim}{\gamma}$ moves away from the point $\gamma_2 = \gamma_3 = \gamma_4 = t/3$, and its minimum occurs at the three equivalent points where the plane intersects the edges of C, i.e., at

$$(6.6)\qquad \begin{array}{c}(\gamma_{11}-\delta^*,\gamma_{11}-\delta^*,t-2\gamma_{11}+2\delta^*)\\ \text{or}\\ (\gamma_{11}-\delta^*,t-2\gamma_{11}+2\delta^*,\gamma_{11}-\delta^*)\\ \text{or}\\ (t-2\gamma_{11}+2\delta^*,\gamma_{11}-\delta^*,\gamma_{11}-\delta^*).\end{array}$$

Also, $\inf_{\gamma\in S(t)} f(\underset{\sim}{\gamma})$ is a decreasing function of t. Thus the global minimum of $f(\underset{\sim}{\gamma})$ for $\underset{\sim}{\gamma} \in E'(\gamma_{11})$ occurs at the (equivalent) points (6.6) closest to the vertex $V = (\gamma_{11}-\delta^*,\gamma_{11}-\delta^*,\gamma_{11}-\delta^*)$ of C. For $0 < \delta^* < \gamma_{11}/2$ the restriction $\gamma_2+\gamma_3+\gamma_4 \leq 2\gamma_{11}-\delta^*$ causes V to be infeasible, and the minimizing configuration is (6.6) with $t = 2\gamma_{11}-\delta^*$; for $\gamma_{11}/2 < \delta^* < 3\gamma_{11}/4$ the vertex V is feasible, and the minimizing configuration is (6.6) with $t = 3(\gamma_{11}-\delta^*)$, i.e., $(\gamma_{11}-\delta^*,\gamma_{11}-\delta^*,\gamma_{11}-\delta^*)$.

In addition, the above argument shows that $\sup_{\underset{\sim}{\gamma}\in E'(\gamma_{11})} f(\underset{\sim}{\gamma})$ occurs along $\gamma_2 = \gamma_3 = \gamma_4 = t/3$ where $\delta^* \leq t \leq \min\{3(\gamma_{11}-\delta^*), 2\gamma_{11}-\delta^*\}$. A direct argument shows that f is concave along this ray with the maximum occurring at $(\gamma_2,\gamma_3,\gamma_4) = (\gamma_{11}/4,\gamma_{11}/4,\gamma_{11}/4)$ which is termed the most favorable (MF-) configuration.

The Fundamental Theorem. Let $\underset{\sim}{\gamma}_e(t) = \underset{\sim}{\gamma}(t,\delta^*(c-3-2t)/(c-4-2t))$ where $\underset{\sim}{\gamma}(t,\gamma_{11})$ is defined by (6.2). Thus $\underset{\sim}{\gamma}_e(t)$ denotes the extreme point of $E(\gamma_{11})$ when γ_{11} is at the right hand boundary of the t-interval for γ_{11} given in (6.1) when $p \geq 1$ and $t < p$.

THEOREM 6.1. <u>For the</u> $2\times c$ <u>case</u> $(c \geq 3)$ <u>and any specified</u> $\{\Delta^*,\delta^*\}$ <u>with</u> $0 < \Delta^* < \infty$, $0 < \delta^* < (c-2)\Delta^*/(c-1)$ <u>define</u> j <u>by</u> (6.1) <u>with</u> γ_{11} <u>set equal to</u> Δ^*.

A) <u>If</u> $j = p$ (<u>i.e.,</u> $\Delta^* \geq \delta^*(c-1-2p)/(c-2-2p)$) <u>then denoting by</u> $\underset{\sim}{\gamma}^*$ <u>the</u> LF-<u>configuration of the</u> $\underset{\sim}{\gamma}$ <u>over</u> E, <u>we have</u> $\underset{\sim}{\gamma}^* = \underset{\sim}{\gamma}(p,\Delta^*) = (\Delta^*-\delta^*,\ldots,\Delta^*-\delta^*,d_p,\delta^*-\Delta^*,\ldots,\delta^*-\Delta^*)$ <u>where</u>

$d_p = \delta^*(c-2-2p)-\Delta^*(c-3-2p)$, and there are c-p-2, 1 and p elements $\Delta^*-\delta^*, d_p, \delta^*-\Delta^*$, respectively.

B) If $j = t$ where $t < p$ then the LF-configuration, γ^*, over E occurs at one of the following points: $\gamma(t,\Delta^*)$, $\gamma_e(t)$, $\gamma_e(t+1)$, $\ldots, \gamma_e(p-1)$.

REMARK 6.3. Empirical studies indicate that the LF-configuration in (B) can be any of these vectors, the minimizing one(s) depending on c, δ^* and Δ^*.

COROLLARY 6.1.

For c = 3 we have

$$\gamma^* = (\Delta^*-\delta^*, \delta^*) \qquad \text{for } 0 < \delta^* < \frac{1}{2}\Delta^*.$$

For c = 4 we have

$$\gamma^* = (\Delta^*-\delta^*, \Delta^*-\delta^*, 2\delta^*-\Delta^*) \qquad \text{for } 0 < \delta^* < \frac{2}{3}\Delta^*.$$

For c = 5 we have

$$\gamma^* = (\Delta^*-\delta^*, \Delta^*-\delta^*, \delta^*, \delta^*-\Delta^*) \qquad \text{for } 0 < \delta^* \leqq \frac{1}{2}\Delta^*;$$

$$\gamma^* = (\Delta^*-\delta^*, \Delta^*-\delta^*, \Delta^*-\delta^*, 3\delta^*-2\Delta^*)$$

$$\text{or } (\delta^*, \delta^*, \delta^*, -\delta^*) \qquad \text{for } \frac{1}{2}\Delta^* < \delta^* < \frac{3}{4}\Delta^*.$$

For c = 6 we have

$$\gamma^* = (\Delta^*-\delta^*, \Delta^*-\delta^*, \Delta^*-\delta^*, 2\delta^*-\Delta^*, \delta^*-\Delta^*) \quad \text{for } 0 < \delta^* \leqq \frac{2}{3}\Delta^*;$$

$$\gamma^* = (\Delta^*-\delta^*, \Delta^*-\delta^*, \Delta^*-\delta^*, \Delta^*-\delta^*, 4\delta^*-3\Delta^*)$$

$$\text{or } (\delta^*, \delta^*, \delta^*, \delta^*, -2\delta^*) \qquad \text{for } \frac{2}{3}\Delta^* < \delta^* < \frac{4}{5}\Delta^*.$$

PROOF OF THEOREM 6.1. Let $h(\gamma_{11}) = \inf_{E(\gamma_{11})} P\{CS\} = P\{S(j,\gamma_{11})\}$ where $S(j,\gamma_{11})$ is the event of correct selection which takes place when (6.7a) and (6.7b) occur.

$$\text{(6.7a)} \quad \begin{aligned} &\sum_{i=2}^{c} X_{2i}-X_{2b} > -\delta^*, \quad \sum_{i=2}^{c} X_{2i}+X_{2b} > -(2\gamma_{11}-\delta^*), (2 \leqq b \leqq c-1-j); \\ &\sum_{i=2}^{c} X_{2i}-X_{2b} > \delta^*(c-2-2j)-\gamma_{11}(c-2-2j), \qquad (b=c-j); \end{aligned}$$

$$\sum_{i=2}^{c} X_{2i}+X_{2b} > \gamma_{11}(c-4-2j)-\delta^*(c-2-2j), \qquad (b=c-j);$$

(6.7b)

$$\sum_{i=2}^{c} X_{2i}-X_{2b} > -(2\gamma_{11}-\delta^*), \sum_{i=2}^{c} X_{2i}+X_{2b} > -\delta^*, \quad (c-j+1 \leqq b \leqq c)$$

For any $\Delta^* \leqq \alpha < \beta \infty$ it can easily be checked that $S(p,\alpha) \subset S(p,\beta)$ and the result in (A) follows.

Now suppose $\frac{c-1-2t}{c-2-2t}\delta^* < \Delta^* \leqq \frac{c-3-2t}{c-4-2t}\delta^*$; we will study $h(\gamma_{11})$ for $\gamma_{11} \geqq \Delta^*$.

(a) For $\Delta^* \leqq \gamma_{11} \leqq \frac{c-3-2t}{c-4-2t}\delta^*$ there is some α $(0 \leqq \alpha \leqq 1)$ such that $\underset{\sim}{\gamma}(t,\gamma_{11}) = \alpha\underset{\sim}{\gamma}(t,\Delta^*) + (1-\alpha)\underset{\sim}{\gamma}_e(t)$ and hence

$$h(\gamma_{11}) = f(\underset{\sim}{\gamma}(t,\gamma_{11})) \geqq [f(\underset{\sim}{\gamma}(t,\Delta^*))]^{\alpha}[f(\underset{\sim}{\gamma}_e(t))]^{1-\alpha}$$

$$\geqq \min\{h(\Delta^*), h(\delta^*(c-3-2t)/(c-4-2t))\}$$

where the first inequality above follows from the log-concavity of $f(\underset{\sim}{\gamma})$ and the second from the definition of $\underset{\sim}{\gamma}_e(t)$.

(b) For $\frac{c-1-2j}{c-2-2j}\delta^* < \gamma_{11} \leqq \frac{c-3-2j}{c-4-2j}\delta^*$ $(t < j < p)$ an argument similar to (a) shows that $h(\gamma_{11})$ can be no smaller than the minimum of h evaluated at the two end points of the interval containing γ_{11}.

(c) For $\frac{c-1-2p}{c-2-2p}\delta^* < \gamma_{11}$ the argument used in the proof of (A) shows that $h(\cdot)$ is nondecreasing, and hence bounded below by $h(\delta^*(c-1-2p)/(c-2-2p)) = f(\underset{\sim}{\gamma}_e(p-1))$. This completes the proof of Theorem 6.1.

REMARK 6.4. By virtue of the special form of the PCS in the 3×3 case (see (5.4)) an argument similar to the above can be used to show that the LF-configuration for the 3×3 case is $\gamma^*_{22} = \gamma^*_{23} = \gamma^*_{32} = \Delta^* - \delta^*$, $\gamma^*_{33} = 3\delta^* - 2\Delta^*$ for all δ^* $(0 < \delta^* < 3\Delta^*/4)$. For details see the Appendix. We conjecture that the 2×3, 2×4, and 3×3 cases are the only ones of the general r×c case for which there is only <u>one</u> distinct $\underset{\sim}{\gamma}^*$ independent of δ^*/Δ^*.

7. *Directions of future research.* In the 2×c result of

Theorem 6.1 it would be useful to be able to determine analytically for each of $\underset{\sim}{\gamma}(t,\Delta^*),\ldots,\underset{\sim}{\gamma}_e(p-1)$ in (B) the corresponding range of c, Δ^*, δ^* values for which that point is the global infimum of the PCS.

Having determined the LF-configuration of the γ_{ij}, it is necessary to evaluate the exact PCS (5.1) numerically in order to implement P. It is straightforward to do this using existing tables for the cases $(r = 2, c = 3)$, $(r = 2, c = 4)$, and $(r = c = 3)$. However, no appropriate tables have thus far been computed for the $2\times c$ $(c \geqq 5)$ or more general cases. Thus new tables, easily computable approximations, or lower bounds would have to be developed.

Other meaningful goals in addition to Goal I (3.1) can also be considered. For example,

$$\text{GOAL II: "To select the factor-level combination associated with } \max\{|\gamma_{ij}|\ (1 \leqq i \leqq r,\ 1 \leqq j \leqq c)\}. \tag{7.1}$$

Goal I and Goal II can each be modified to deal with the rc+1 decision problem in which the largest positive (for I) or absolute (for II) interaction is to be selected if it is sufficiently large relative to zero (which is here the standard), and <u>no</u> interaction is to be selected if the largest positive (for I) or absolute (for II) interaction is sufficiently close to zero. The approach here is similar to the one adopted in [2]. Two-stage procedures could be used to handle the case of common <u>unknown</u> variance.

8. *Acknowledgment.* We would like to thank Yosef Rinott for some helpful conversations, and to Thomas Wong for noting an error in an earlier version of this paper.

Appendix: LF-configuration of the γ_{ij} for the 3×3 case. We parallel the arguments and notation used in the analysis of the $2\times c$ case. From (5.4) we see that the PCS can be expressed as

$$f(\underset{\sim}{\gamma}) = P\{\sum_{i=2}^{3}\sum_{j=2}^{3}(X_{ij}+\gamma_{ij})-(X_{ab}+\gamma_{ab}) > 0,\ (2\leqq a,b\leqq 3)\} = P\{\underset{\sim}{X}+\underset{\sim}{\gamma}\varepsilon A\}$$

where A is convex and $\underset{\sim}{\gamma} = \begin{pmatrix} \gamma_{22} & \gamma_{23} \\ \gamma_{32} & \gamma_{33} \end{pmatrix}$. For $\gamma_{11} \geqq \Delta^*$ let $E(\gamma_{11})$ denote that section of the preference zone having $\sum_{i=2}^{3} \sum_{j=2}^{3} \gamma_{ij} = \gamma_{11}$ and $E = \bigcup_{\gamma_{11} \geqq \Delta^*} E(\gamma_{11})$. Specifically

$$E(\gamma_{11}) = \{\begin{pmatrix} \gamma_{22} & \gamma_{23} \\ \gamma_{32} & \gamma_{33} \end{pmatrix} \mid \sum_{i=2}^{3} \sum_{j=2}^{3} \gamma_{ij} = \gamma_{11};\ \gamma_{ab} \leqq \gamma_{11} - \delta^*,\ (2 \leqq a,b \leqq 3);$$

$$\gamma_{22} + \gamma_{23} \geqq \delta^* - \gamma_{11}; \gamma_{22} + \gamma_{32} \geqq \delta^* - \gamma_{11};$$

$$\gamma_{32} + \gamma_{33} \geqq \delta^* - \gamma_{11};\ \gamma_{23} + \gamma_{33} \geqq \delta^* - \gamma_{11}\}$$

It is easily checked that $E(\gamma_{11})$ is the convex hull of the row and column permutations of $\underset{\sim}{\gamma}(\gamma_{11}) \equiv \begin{pmatrix} \gamma_{11} - \gamma^* & \gamma_{11} - \delta^* \\ \gamma_{11} - \delta^* & 3\delta^* - 2\gamma_{11} \end{pmatrix}$. From the log-concavity of $f(\underset{\sim}{\gamma})$ and its invariance under row-column permutations of $\underset{\sim}{\gamma}$ we conclude that $\underset{\sim}{\gamma}(\gamma_{11})$ is the LF-configuration of the PCS over the section $E(\gamma_{11})$. If $S(\gamma_{11})$ is the event CS evaluated at $\underset{\sim}{\gamma}(\gamma_{11})$, i.e., $f(\underset{\sim}{\gamma}(\gamma_{11})) \equiv P\{S(\gamma_{11})\}$ then it can be checked that for $\Delta^* \leqq \alpha < \beta$ we have $S(\alpha) \subset S(\beta)$ and hence the PCS decreases in γ_{11}. Thus our final result is that $\underset{\sim}{\gamma}^* = \begin{pmatrix} \Delta^* - \delta^* & \Delta^* - \delta^* \\ \Delta^* - \delta^* & 3\delta^* - 2\Delta^* \end{pmatrix}$ is the global LF-configuration for the problem.

References

[1] Bechhofer, R. E. (1954). A single-sample multiple decision procedure for ranking means of normal populations with known variances. *Ann. Math. Statist.* 25, 16-39.

[2] Bechhofer, R. E. and Turnbull, B. W. (1975). A (k+1)-decision single-stage selection procedure for comparing k normal means with a fixed known standard: the case of common known variance. Tech. Report No. 242, Dept. of Operations Research, Cornell University.

[3] Hardy, G. H., Littlewood, J. E. and Pólya, G. (1952). *Inequalities*, (2nd Ed.). Cambridge University Press, Cambridge.

[4] Marshall, A. W. and Olkin, I. (1974). Majorization in multivariate distributions. *Ann. Statist.* 2, 1189-1200.

[5] Prékopa, A. (1973). On logarithmic concave measures and functions. *Acta. Sci. Math.* 34, 335-343.

IMPROVED MINIMAX ESTIMATORS OF NORMAL MEAN VECTORS FOR CERTAIN TYPES OF COVARIANCE MATRICES

By James O. Berger* and M. E. Bock**
Purdue University

1. *Introduction.* Assume $X = (X_1, \ldots, X_p)^t$ is a p-dimensional random vector ($p \geq 3$) which is normally distributed with mean vector $\theta = (\theta_1, \ldots, \theta_p)^t$ and positive definite covariance matrix Σ. It is desired to estimate θ by an estimator $\delta = (\delta_1, \ldots, \delta_p)^t$ under the quadratic loss

$$L(\delta, \theta, \Sigma) = (\delta - \theta)^t Q(\delta - \theta)/tr(Q\Sigma), \tag{1.1}$$

where Q is a positive definite ($p \times p$) matrix.

The usual minimax and best invariant estimator for θ is $\delta^0(X) = X$. Since Stein [15] first showed that δ^0 could be improved upon for $Q = \Sigma = I_p$, considerable effort by a number of authors has gone into finding significant improvements upon δ^0. For known Σ a wide variety of minimax estimators (better than δ^0) have now been found. (See Bhattacharya [7], Hudson [11], Bock [8], Berger [3], [4].) For unknown Σ, however, the results are very incomplete. Indeed until recently only the situation where Σ is known up to a multiplicative constant had been considered. Berger and Bock [5] made a further step in dealing with unknown Σ, finding minimax estimators of θ under the assumptions that Σ was an unknown diagonal matrix and Q was diagonal. Recently, however, problems have come to our attention which cannot be handled by the results of [5]. These include estimating θ in the above mentioned problem when some but not all of the unknown variances are equal, estimating in the diagonal problem when prior information leads one to think that certain linear restrictions hold, and

---

*The work of this author was supported by the National Science Foundation under Grant #MPS75-07017.

**The work of this author was supported by the National Science Foundation under Grant #MPS75-08554.

estimation when combining regression problems with possibly different unknown variances. This paper extends the results of [5] to cover the above situations, and also introduces a larger class of minimax estimators than was found in that paper. The remainder of this section will introduce the needed notation. Section 2 will present the major theorems and Section 3 will deal with applications.

If $Z = (Z_1,\ldots,Z_k)^t$, let $D(Z)$ be the $(p\mathrm{x}p)$ diagonal matrix with diagonal elements

$$D_i(Z) = \sum_{j=1}^{k} d_j^i Z_j = Z^t d^i$$

where the $d^i = (d_1^i,\ldots,d_k^i)^t$, $i = 1,\ldots,p$, are vectors of known nonnegative constants. In this paper it will be assumed for each j, that d_j^i is nonzero for at least one i. Define

$$\Sigma(Z) = AD(Z)A^t,$$

where A is a known $(p\mathrm{x}p)$ nonsingular matrix.

It will be assumed that the covariance matrix, Σ, is of the form $\Sigma = \Sigma(\underline{\sigma}^2)$, where $\underline{\sigma}^2 = (\sigma_1^2,\ldots,\sigma_k^2)^t$ is an unknown vector of positive numbers. Also assume that estimates V_i for σ_i^2 are available, where the V_i are independently distributed (of each other and of X) and (V_i/σ_i^2) is $\chi_{n_i}^2$ (i.e. a chi-square random variable with n_i degrees of freedom). Assume $n_i \geq 3$ for $i = 1,\ldots,k$. For notational convenience, let $W_i = V_i/(n_i-2)$ and $W = (W_1,\ldots,W_k)^t$.

The estimators considered will be of the form

$$\delta(X,W) = \left(I_p - \frac{r(X,W)Q^{-1}B\Sigma^{-1}(W)}{X^t\Sigma^{-1}(W)C\Sigma^{-1}(W)X}\right)X, \tag{1.2}$$

where r is a measurable real function, and B and C are fixed positive definite $(p\mathrm{x}p)$ matrices. Conditions on r, B, and C will be developed under which the above estimator is minimax.

The notation $E(Z)$ will be used for the expectation of Z.

Subscripts on E will refer to parameter values, while superscripts on E will refer to the random variables with respect to which the expectation is to be taken. When obvious, subscripts and superscripts will be omitted.

For an estimator, δ, define the risk function

$$R(\delta,\theta,\underline{\sigma}^2) = E^{X,W}_{\theta,\underline{\sigma}^2} L(\delta(X,W),\theta,\underline{\sigma}^2).$$

Noting that the risk of the usual estimator, $\delta^0(X,W) = X$, is $R(\delta^0,\theta,\underline{\sigma}^2) = 1$, it is clear that an estimator δ is minimax if $R(\delta,\theta,\underline{\sigma}^2) \leq 1$ for all θ and $\underline{\sigma}^2$.

Finally, if S is a (pxp) matrix, let $ch_{max}(S)$ denote the maximum characteristic root of S and tr(S) denote the trace of S.

2. *Basic results.* In this section it will be assumed that A = I and that B and C in (1.2) are diagonal, with diagonal elements $\{b_1,\ldots,b_p\}$ and $\{c_1,\ldots,c_p\}$ respectively.

For use in the following theorem let $I(\alpha)$ denote the usual indicator function on $(0,\infty)$ (i.e. $I(\alpha) = 1$ if $\alpha > 0$, and $I(\alpha) = 0$ if $\alpha \leq 0$), and define

$$r_i = \sum_{j=1}^{p} I(d_j^i), \qquad H_i = \max_{1\leq j\leq k} \{\frac{I(d_j^i)}{\chi^2_{n_j}}\},$$

$$(2.1)\quad \alpha_i = \sum_{j=1}^{p} b_j I(d_i^j), \qquad \beta_i = \max_{1\leq \ell\leq p} \{[\frac{2\alpha_i}{(n_i-2)} + b_\ell]I(d_i^\ell)\},$$

$$T = \max_{1\leq i\leq k} \{\beta_i(n_i-2)/\chi^2_{n_i}\}, \text{ and } \tau = E(T + 4\sum_{\{j:r_j>1\}} b_j H_j).$$

THEOREM 1. <u>Assume that</u> A = I, <u>that</u> δ <u>is of the form</u> (1.2) <u>with</u> B <u>and</u> C <u>diagonal, and that</u>

(i) $0 \leq r(X,W) \leq 2[tr(B) - 2\tau]/ch_{max}[C^{-1}B^tQB]$,

(ii) $r(X,W)$ <u>is nondecreasing in</u> $|X_i|$, $i = 1,\ldots,p$,

(iii) $r(X,W)$ <u>is nonincreasing in</u> W_i, $i = 1,\ldots,k$,

(iv) $r(X,W)/(X^tD^{-1}(W)CD^{-1}(W)X)$ _is nondecreasing in_ W_i _for_ $i = 1,\ldots,k$.

Then δ _is a minimax estimator of_ θ.

PROOF. Throughout the proof it will be assumed that all first order partial derivatives of r exist. The generalization to r merely nondecreasing or nonincreasing in the various coordinates can be done analogously by treating all integrals as Riemann integrals.

For notational convenience define

$$||X||_W^2 = X^tD^{-1}(W)CD^{-1}(W)X = \sum_{i=1}^{p} [c_iX_i^2/(W^td^i)^2],$$

and

$$\Delta(\theta) = [R(\delta,\theta,\underline{\sigma}^2) - R(\delta^0,\theta,\underline{\sigma}^2)]\mathrm{tr}(QD(\underline{\sigma}^2)).$$

To show that δ is minimax, it is clearly only necessary to show that $\Delta(\theta) \leq 0$ for all θ. Now

$$\Delta(\theta) = E_{\theta,\underline{\sigma}^2}\left\{\left[(X-\theta) - \frac{r(X,W)Q^{-1}BD^{-1}(W)X}{||X||_W^2}\right]^tQ\left[(X-\theta) - \right.\right.$$

$$(2.2)\qquad \left.\left.\frac{r(X,W)Q^{-1}BD^{-1}(W)X}{||X||_W^2}\right] - (X-\theta)^tQ(X-\theta)\right\}$$

$$= E\left\{\frac{-2r(X,W)(X-\theta)^tBD^{-1}(W)X}{||X||_W^2} + \frac{r(X,W)^2X^tD^{-1}(W)BQ^{-1}BD^{-1}(W)X}{||X||_W^4}\right\}.$$

As in Berger [4], integration by parts with respect to $X_i (1 \leq i \leq p)$, together with Condition (ii), gives that

$$E\left\{\frac{-2r(X,W)(X-\theta)^tBD^{-1}(W)X}{||X||_W^2}\right\} \leq E\left\{\frac{-2r(X,W)\mathrm{tr}[BD^{-1}(W)D(\underline{\sigma}^2)]}{||X||_W^2}\right\}$$

$$(2.3)\qquad + E\left\{\frac{4r(X,W)X^tD^{-1}(W)CD^{-1}(W)D(\underline{\sigma}^2)BD^{-1}(W)X}{||X||_W^4}\right\}.$$

An integration by parts (first noticed by Efron and Morris [10] verifies that if h is an absolutely continuous function from

$(0,\infty)$ into R^1 and its derivative h' exists, then

(2.4) $$E[(n_i-2)W_i h(W_i)/\sigma_i^2] = E[n_i h(W_i)] + 2E[W_i h'(W_i)],$$

provided the expectations exist. Define h_j by

$$h_j(W_i) = r(X,W)/[||X||_W^2 \, W^t d^j],$$

considering the other variables fixed. Clearly this is absolutely continuous and differentiable in W_i, except possibly when $W^t d^j = d_i^j W_i$ and $X_j = 0$ (an event of measure zero). Defining $T_i(W)$ as the (pxp) diagonal matrix with diagonal elements $T_i^\ell(W) = d_i^\ell / W^t d^\ell$, calculation gives

$$h_j'(W_i) = \frac{-d_i^j r(X,W)}{||X||_W^2 (W^t d^j)^2} + \frac{2r(X,W)[X^t D^{-1}(W) C T_i(W) D^{-1}(W) X]}{||X||_W^4 \, W^t d^j}$$
$$+ \frac{1}{||X||_W^2 \, W^t d^j} \{\frac{\partial}{\partial W_i} r(X,W)\}.$$

Using this in (2.4) gives

$$E\left\{\frac{(n_i-2)W_i r(X,W)}{\sigma_i^2 ||X||_W^2 \, W^t d^j}\right\} = E\left\{\frac{n_i r(X,W)}{||X||_W^2 \, W^t d^j}\right\} - 2E\left\{\frac{d_i^j W_i r(X,W)}{||X||_W^2 (W^t d^j)^2}\right\}$$
$$+ 4E\left\{\frac{r(X,W)[X^t D^{-1}(W) C T_i(W) D^{-1}(W) X] W_i}{||X||_W^4 \, W^t d^j}\right\} + 2E\left\{\frac{W_i [\frac{\partial}{\partial W_i} r(X,W)]}{||X||_W^2 \, W^t d^j}\right\}.$$

Multiplying through by $d_i^j \sigma_i^2/(n_i-2)$ and rearranging gives

$$E\{\frac{r(X,W) d_i^j \sigma_i^2}{||X||_W^2 W^t d^j}\} = E\{\frac{r(X,W) W_i d_i^j}{||X||_W^2 W^t d^j}\} - 2E\{\frac{d_i^j \sigma_i^2 W_i [\frac{\partial}{\partial W_i} r(X,W)]}{(n_i-2)||X||_W^2 W^t d^j}\}$$

(2.5) $$- 4E\{\frac{r(X,W)[X^t D^{-1}(W) C T_i(W) D^{-1}(W) X] W_i d_i^j \sigma_i^2}{(n_i-2)||X||_W^4 W^t d^j}$$

$$- E\{\frac{2r(X,W)d_i^j\sigma_i^2}{(n_i-2)||X||_W^2 W^t d^j}[1 - \frac{W_i d_i^j}{W^t d^j}]\}.$$

Define

$$\gamma_i^j(W) = \frac{W_i d_i^j \sigma_i^2}{(n_i-2)W^t d^j}, S_j(W) = \sum_{i=1}^k \gamma_i^j(W)T_i(W) \text{ and } L_j(W) = \sum_{i=1}^k \gamma_i^j W_i^{-1}[1 - \frac{W_i d_i^j}{W^t d^j}].$$

Summing in (2.5) over i ($1 \leq i \leq k$) and using Condition (iii) then gives

$$E\{\frac{r(X,W)(\underline{\sigma}^2)^t d^j}{||X||_W^2 W^t d^j}\} \geq E\{\frac{r(X,W)}{||X||_W^2}\} - 4E\{\frac{r(X,W)[X^t D^{-1}(W)CS_j(W)D^{-1}(W)X]}{||X||_W^4}\}$$

$$- 2E\{\frac{r(X,W)L_j(W)}{||X||_W^2}\}.$$

Combining this with (2.2) and (2.3) and noting that

$tr[BD^{-1}(W)D(\underline{\sigma}^2)] = \sum_{j=1}^p [b_j(\underline{\sigma}^2)^t d^j / W^t d^j]$ gives

$$\Delta(\underline{\theta}) \leq E\{\frac{-2r(X,W)tr(B)}{||X||_W^2}\} + E\{\frac{8r(X,W)}{||X||_W^4}[\sum_{j=1}^p (b_j X^t D^{-1}(W)CS_j(W)D^{-1}(W)X]\}$$

$$(2.6) \qquad + E\{\frac{4r(X,W)}{||X||_W^4}[X^t D^{-1}(W)CD^{-1}(W)D(\underline{\sigma}^2)BD^{-1}(W)X]\}$$

$$+ E\{\frac{4r(X,W)}{||X||_W^2}[\sum_{j=1}^p b_j L_j(W)]\} + E\{\frac{r(X,W)^2}{||X||_W^4}[X^t D^{-1}(W)BQ^{-1}BD^{-1}(W)X]\}.$$

Defining $S = [2\sum_{j=1}^p b_j S_j(W)] + D^{-1}(W)D(\underline{\sigma}^2)B$, (2.6) becomes

$$(2.7) \qquad \Delta(\theta) \leq E\{\frac{r(X,W)}{||X||_W^2}[-2tr(B) + \frac{4}{||X||_W^2}(X^t D^{-1}(W)CSD^{-1}(W)X) + \frac{r(X,W)}{||X||_W^2}(X^t D^{-1}(W)BQ^{-1}BD^{-1}(W)X) + 4\sum_{j=1}^p b_j L_j(W)]\}.$$

Using the fact that for any symmetric matrix A_1,

$$\frac{[D^{-1}(W)X]^t A_1 [D^{-1}(W)X]}{||X||_W^2} \leq ch_{max}[C^{-1}A_1],$$

it is clear that (2.7) implies

$$(2.8)\quad \Delta(\theta) \leq E\{\frac{r(X,W)}{||X||_W^2}[-2tr(B)+4\,ch_{max}(S)+r(X,W)ch_{max}(C^{-1}BQ^{-1}B) + 4\sum_{j=1}^{p} b_j L_j(W)]\}.$$

By construction, S is a diagonal matrix with ℓ*th* diagonal element

$$(2\sum_{j=1}^{p}\sum_{i=1}^{k} b_j\gamma_i^j \frac{d_i^\ell}{W^t d^\ell}) + \frac{b_\ell(\underline{\sigma}^2)^t d^\ell}{W^t d^\ell}.$$

Clearly

$$(2.9)\qquad \gamma_i^j(W) = \frac{\sigma_i^2 W_i d_i^j}{(n_i-2)W^t d^j} \leq \frac{\sigma_i^2 I(d_i^j)}{(n_i-2)}.$$

Recalling the definitions in (2.1), it follows that the ℓ*th* diagonal element of S is no larger than

$$\sum_{i=1}^{k}\{[\frac{2\alpha_i}{(n_i-2)} + b_\ell]\frac{d_i^\ell\sigma_i^2}{W^t d^\ell} \leq \max_{1\leq i\leq k}\{[\frac{2\alpha_i}{(n_i-2)} + b_\ell]\frac{I(d_i^\ell)\sigma_i^2}{W_i}\}.$$

Another use of the definitions in (2.1) finally gives

$$ch_{max}(S) \leq \max_{1\leq i\leq k}\{\beta_i\sigma_i^2/W_i\}.$$

Defining $H_j(W) = \max_{1\leq i\leq k}\{\sigma_i^2 I(d_i^j)/[(n_i-2)W_i]\}$, a similar application of (2.9) shows that $\sum_{j=1}^{p} b_j L_j(W) \leq \sum_{\{j:r_j>1\}} b_j H_j(W)$. Then defining

$q(W) = \max_{1\leq i\leq k}\{\beta_i\sigma_i^2/W_i\} + \sum_{\{j:r_j>1\}} b_j H_j(W)$, denoting $ch_{max}(C^{-1}BQ^{-1}B)$ by ρ, and using (2.8) gives

$$(2.10)\qquad \Delta(\theta) \leq E\{\frac{r(X,W)}{||X||_W^2}[-2tr(B) + 4q(W) + r(X,W)\rho]\}.$$

Now q(W) is nonincreasing in W_1, and by Condition (iii), r(X,W) is nonincreasing in W_1. Hence

$$[2\ tr(B) - 4q(W) - r(X,W)\rho]$$

is nondecreasing in W_1. By Condition (iv),

$$[r(X,W)/||X||_W^2]$$

is also nondecreasing in W_1. In general, if f and g are measurable real valued nondecreasing functions whose expectations exist, then

$$E^Y[f(Y)g(Y)] \geq E^Y[f(Y)]E^Y[g(Y)].$$

Hence

$$E^{W_1}_{\sigma_1^2}\{\frac{r(X,W)}{||X||_W^2}[2tr(B) - 4q(W) - r(X,W)\rho]\}$$

$$\geq (E^{W_1}_{\sigma_1^2}\{r(X,W)/||X||_W^2\})(E^{W_1}_{\sigma_1^2}\{2tr(B) - 4q(W) - r(X,W)\rho\}). \tag{2.11}$$

Using the independence of the W_i and proceeding as above with the two terms on the right hand side of (2.11) (as functions of W_2) gives

$$E^{W_2}_{\sigma_2^2}[(E^{W_1}_{\sigma_1^2}r(X,W)/||X||_W^2\})(E^{W_1}_{\sigma_1^2}\{2tr(B)-4q(W)-r(X,W)\rho\})]$$

$$\geq (E^{W_1,W_2}_{\sigma_1^2,\sigma_2^2}\{r(X,W)/||X||_W^2\})(E^{W_1,W_2}_{\sigma_1^2,\sigma_2^2}\{2tr(B)-4q(W)-r(X,W)\rho\}).$$

Continuing in the obvious manner verifies that

$$E^X_{\theta,\underline{\sigma}^2}[E^W_{\underline{\sigma}^2}\{r(X,W)||X||_W^{-2}(2tr(B)-4Q(W)-r(X,W)\rho)\}]$$

$$\geq E^X_{\theta,\underline{\sigma}^2}[(E^W_{\underline{\sigma}^2}\{r(X,W)/||X||_W^2\})(E^W_{\underline{\sigma}^2}\{2tr(B)-4q(W)-r(X,W)\rho\})]$$

(2.12)

$$= E^X_{\theta,\underline{\sigma}^2}[(E^W_{\underline{\sigma}^2}\{r(X,W)/||X||^2_W\})(2tr(B) - 4\tau - E^W_{\underline{\sigma}^2}[r(X,W)]\rho)].$$

(The last step follows from (2.1) and the definition of $q(W)$, noting that $(\sigma_i^2/W_i) = (n_i-2)/(V_i/\sigma_i^2) = (n_i-2)/X^2_{n_i}$.) Condition (i) ensures that $r(X,W)/||X||^2_W \geq 0$ and that $\{2tr(B) - 4\tau - E^W_{\underline{\sigma}^2}[r(X,W)]ch_{max}(C^{-1}BQ^{-1}B)\} \geq 0$. Combining this with (2.10) and (2.12) verifies that $\Delta(\theta) \leq 0$ and hence that δ is minimax. ||

It should be remarked that the upper bound on $r(X,W)$ in Condition (i) of Theorem 1 is undoubtedly smaller than is really needed to ensure minimaxity of the estimator. Many of the inequalities used in the proof were fairly crude, resulting in a τ which is too big. Unfortunately we were unable to obtain sharper bounds.

To actually apply Theorem 1 it is clearly necessary to calculate τ. It can be shown as in [5], that if $\min\{n_i\} \to \infty$, then $\tau \to \max\{b_i\}$. For large n_i this would be a reasonable approximation, especially in light of the preceeding paragraph. An exact formula for $E(T)$ (and hence $E(H_i)$ and τ) can be given when the n_i are even. Indeed a straightforward calculation then verifies that

$$(2.13)\quad E(T) = \sum_{\ell=1}^{k} \lambda_\ell \sum\{[\prod_{i\neq\ell} \frac{(\lambda_i m_i)^{(m_i-j(i))}}{(m_i-j(i))!}] \frac{(m-J(\ell)-2)!\,m_\ell^{m_\ell}}{(m_\ell-1)!\,M^{(m-J(\ell)-1)}}\},$$

where $\lambda_i = \beta_i(n_i-2)/n_i$, $m_i = n_i/2$, $m = \sum_{i=1}^{k} m_i$, $J(\ell) = \sum_{i\neq\ell} j(i)$, $M = \sum_{i=1}^{k} (m_i\lambda_i)$, and the inner summation is over all combinations $(j(1), j(2),\ldots,j(\ell-1), j(\ell+1),\ldots,j(k))$ where the $j(i)$ are integers between 1 and m_i inclusive. Table 1 gives values of $E(T)$ (to two decimal places) for the special case $n_i = n$ and $\beta_i = n/(n-2)$ for all i. Note that if $\beta_i = \rho n/(n-2)$, then $E(T)$ is the value given in the table multiplied by ρ.

Before proceeding to the applications, one final result will be presented which is helpful in dealing with transformed versions of the problem. The result will hold for arbitrary Q and Σ.

Consider the transformation $Y = RX - r^0$, where R is an $(m\times p)$ matrix $(m \leq p)$ of rank m, and r^0 is an $(m\times 1)$ vector. Define $\eta = R\theta - r^0$, $\Sigma^* = R\Sigma R^t$, and $Q^* = (RQ^{-1}R^t)^{-1}$. Clearly Y is normally distributed with mean η and covariance matrix Σ^*. The problem of estimating η by an estimator $\delta^*(Y) = Y+\gamma(Y)$ (γ: $R^m \to R^m$) under the quadratic loss

$$L^*(\delta^*,\eta,\Sigma^*) = (\delta^*-\eta)^t Q^*(\delta^*-\eta)/\mathrm{tr}(Q^*\Sigma^*),$$

will now be shown to be equivalent (in terms of minimaxity) to the original problem of estimating θ using the estimator

$$\delta(X) = X + Q^{-1}R^t(RQ^{-1}R^t)^{-1}\gamma(RX-r^0). \tag{2.14}$$

THEOREM 2. _δ^* is a minimax estimator of η under loss L^* if and only if δ is a minimax estimator of θ under the original loss L._

PROOF. For convenience, let $T = Q^{-1}R^t(RQ^{-1}R^t)^{-1} = Q^{-1}R^tQ^*$. Also define

$$\Delta^*(\eta) = E_{\eta,\Sigma^*}\{L^*(\delta^*(Y),\eta,\Sigma^*) - L^*(Y,\eta,\Sigma^*)\}\mathrm{tr}(Q^*\Sigma^*),$$

and

$$\Delta(\theta) = E_{\theta,\Sigma}\{L(\delta(X),\theta,\Sigma) - L(X,\theta,\Sigma)\}\mathrm{tr}(Q\Sigma).$$

Clearly

$$\begin{aligned}\Delta(\theta) &= E_{\theta,\Sigma}\{[X+T\gamma(RX-r^0)-\theta]^tQ[X+T\gamma(RX-r^0)-\theta]-[X-\theta]^tQ[X-\theta]\}\\ &= E_{\theta,\Sigma}\{-2(X-\theta)^tQT\gamma(RX-r^0) + \gamma(RX-r^0)^tT^tQT\gamma(RX-r^0)\}\\ &= E_{\theta,\Sigma}\{-2(X-\theta)^tR^tQ^*\gamma + \gamma^tQ^*RQ^{-1}QQ^{-1}R^tQ^*\gamma\}\\ &= E_{\theta,\Sigma}\{-2([RX-r^0]-[R\theta-r^0])^tQ^*\gamma(RX-r^0)+\gamma(RX-r^0)^t\\ &\qquad Q^*\gamma(RX-r^0)\}\\ &= E_{\eta,\Sigma^*}\{-2(Y-\eta)^tQ^*\gamma(Y) + \gamma(Y)^tQ^*\gamma(Y)\} = \Delta^*(\eta).\end{aligned}$$

The conclusion follows immediately. ||

Table 1

Values of E(T)

n	k = 3	4	5	6	7	8	9	10	11	12
4	3.66	4.22	4.79	5.26	5.15	6.15	6.63	7.06	7.50	7.99
5	2.88	3.29	3.65	3.97	4.30	4.57	4.85	5.09	5.31	5.53
6	2.45	2.78	3.01	3.27	3.51	3.69	3.89	4.05	4.22	4.35
7	2.21	2.46	2.66	2.85	3.03	3.17	3.31	3.44	3.57	3.67
8	2.04	2.26	2.42	2.59	2.74	2.85	2.97	3.08	3.18	3.26
9	1.94	2.14	2.24	2.39	2.51	2.61	2.71	2.79	2.89	2.96
10	1.84	2.01	2.11	2.24	2.35	2.43	2.52	2.58	2.66	2.73
11	1.76	1.89	2.02	2.13	2.23	2.30	2.38	2.44	2.51	2.57
12	1.69	1.83	1.93	2.04	2.12	2.19	2.26	2.32	2.38	2.43
13	1.64	1.77	1.86	1.96	2.04	2.10	2.17	2.22	2.28	2.32
14	1.60	1.72	1.81	1.90	1.97	2.03	2.09	2.14	2.19	2.23
15	1.57	1.68	1.77	1.85	1.91	1.97	2.03	2.07	2.12	2.15
16	1.54	1.64	1.72	1.80	1.86	1.91	1.96	2.01	2.05	2.08
17	1.51	1.61	1.69	1.76	1.82	1.87	1.91	1.96	2.00	2.03
18	1.49	1.59	1.66	1.73	1.78	1.83	1.87	1.91	1.95	1.98
19	1.47	1.56	1.63	1.69	1.75	1.79	1.83	1.87	1.91	1.93
20	1.45	1.53	1.60	1.66	1.71	1.76	1.80	1.84	1.87	1.90
21	1.43	1.51	1.58	1.64	1.69	1.73	1.77	1.80	1.84	1.86
22	1.41	1.50	1.56	1.62	1.66	1.70	1.74	1.77	1.81	1.83
23	1.40	1.48	1.55	1.60	1.64	1.68	1.72	1.75	1.78	1.80
24	1.39	1.45	1.53	1.58	1.62	1.65	1.69	1.72	1.75	1.77
25	1.37	1.45	1.51	1.56	1.60	1.64	1.67	1.70	1.73	1.75
26	1.36	1.44	1.49	1.54	1.58	1.62	1.65	1.68	1.71	1.73
27	1.35	1.42	1.48	1.53	1.57	1.63	1.66	1.66	1.68	1.71
28	1.35	1.41	1.47	1.52	1.55	1.58	1.61	1.64	1.66	1.68
29	1.34	1.41	1.46	1.51	1.54	1.57	1.60	1.63	1.65	1.67
30	1.33	1.40	1.45	1.49	1.53	1.56	1.59	1.61	1.63	1.65
35	1.30	1.36	1.40	1.44	1.48	1.50	1.53	1.55	1.57	1.59
40	1.27	1.32	1.37	1.40	1.43	1.45	1.48	1.50	1.52	1.53
45	1.25	1.30	1.34	1.37	1.40	1.42	1.44	1.46	1.48	1.49
50	1.23	1.28	1.31	1.35	1.37	1.39	1.41	1.43	1.44	1.45
55	1.22	1.26	1.29	1.33	1.35	1.37	1.38	1.40	1.41	1.43
60	1.21	1.25	1.28	1.31	1.33	1.35	1.36	1.38	1.39	1.40
65	1.20	1.24	1.27	1.29	1.31	1.33	1.34	1.36	1.37	1.38
70	1.19	1.23	1.25	1.28	1.30	1.31	1.33	1.34	1.35	1.36
75	1.18	1.22	1.24	1.27	1.28	1.30	1.31	1.33	1.34	1.34

3. *Applications.* Application 1. Consider the problem of combining p independent estimation problems with unknown variances. Thus assume

$$\ddagger = \ddagger(\underline{\sigma}^2) = \begin{pmatrix} \sigma_1^2 I_{p_1} & & & 0 \\ & \sigma_2^2 I_{p_2} & & \\ & & \ddots & \\ 0 & & & \sigma_k^2 I_{p_k} \end{pmatrix}$$

where $\sum_{i=1}^{k} p_i = p$ and the σ_i^2 are unknown. (If $p_i = 1$ for $i = 1,\ldots,k$, this is the problem considered in [5].) For simplicity we only consider estimators (1.2) in which $B = I_p$. An easy calculation using (2.1) then gives that

$$\beta_i = 1 + 2p_i/(n_i-2) \quad \text{and } r_i = 1, \tag{3.1}$$

and hence $\tau = E(T)$, which can be calculated from (2.13) for even n_i. Note that if the $n_i = n$ and $p_i = p/k$ for all i, then

$$\beta_i = [1 + \frac{2p}{(n-2)k}] = [\frac{(n-2)}{n} + \frac{2p}{nk}][\frac{n}{(n-2)}],$$

so E(T) could be found by using the value in Table 1 multiplied by $\rho = [(n-2)/n + 2p/(nk)]$.

If Q also happened to be diagonal, Theorem 1 gives that the simple estimator

$$\delta(X,W) = (I_p - \frac{r(X,W)Q^{-1}\ddagger^{-1}(W)}{X^t\ddagger^{-1}(W)Q^{-1}\ddagger^{-1}(W)X})X \tag{3.2}$$

is minimax, providing $0 \leq r(X,W) \leq 2(p-2\tau)$ and $r(X,W)$ satisfies Conditions (ii), (iii), and (iv) of the theorem. For a diagonal estimator such as (3.2), Berger and Bock [6] suggest choosing $r(X,W) \equiv c$ $(0 \leq c \leq 2(p-2\tau))$ and then using the "positive part" version, δ^+, of the estimator. The ith coordinate of δ^+ is given by

$$\delta_i^+(X,W) = (1-c/[\{X^t\ddagger^{-1}(W)Q^{-1}\ddagger^{-1}(W)X\}q_iW_{k(i)}])^+X_i, \tag{3.3}$$

where $k(i)$ is the index of the variance of X_i and "+" stands for the usual positive part. The results of [6] can be used to show that $\delta+$ has smaller risk than δ, and hence is also minimax. As an example of the improvement in risk that can be obtained using (3.3), the risk function was numerically calculated for $p = 4$, $p_i = 1$ $(i = 1,\ldots,4)$, $n_i = 16$ $(i = 1,\ldots,4)$, $\sigma_i^2 = q_i = 1$ $(i = 1, \ldots,4)$, and $c = 2(p - 2\tau)$. Figure 1 depicts this risk along the coordinate axes. (The constant line $c = 1$ is the risk of the usual estimator $\delta^0(X) = X$.) Clearly significant savings are achievable.

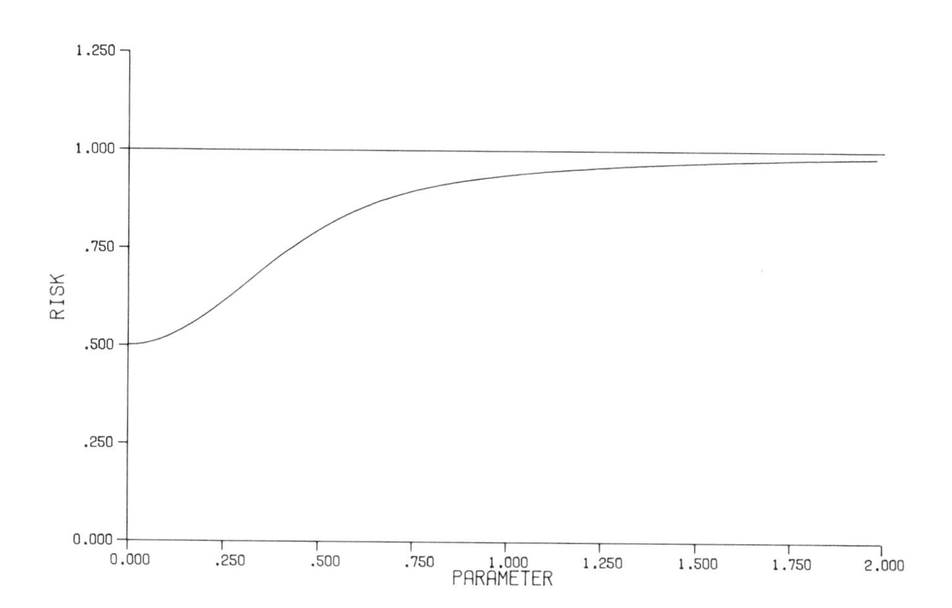

Figure 1

An alternative method of improving on the choice $r(X,W) \equiv c$ is to use the estimator.

$$(3.4) \qquad \delta^*(X,W) = \left(I_p - \frac{\min\{X^t\ddagger^{-1}(W)X, c\}Q^{-1}\ddagger^{-1}(W)}{X^t\ddagger^{-1}(W)Q^{-1}\ddagger^{-1}(W)X}\right)X.$$

The method of [6] can again be used to show that this results in an improvement over the original minimax estimator. The risks of

(3.3) and (3.4) have been very similar in numerical studies, so the choice of which estimator to use is largely a matter of convenience.

Both estimators (3.3) and (3.4) have risk functions which are smallest near $\theta = 0$. If prior information is available in the nature of a linear restriction, it pays to use versions of these estimators which will have smallest risks for θ which nearly satisfy the linear restriction. Indeed assume the prior linear restriction is of the form

$$R\theta = r^0,$$

where R is an (mxp) matrix of rank m with at most one nonzero entry in each column, and r^0 is an (mx1) vector. The matrix R would typically be determined by constructing a set of linear restraints of the form $\sum_{j=1}^{p} r_{ij}\theta_j = r_i^0$, and the condition on R essentially says that each θ_i can occur in at most one such restraint. This situation should be quite common.

To obtain the desired estimator it is necessary to look at the transformed problem of estimating $\eta = R\theta - r^0$ by an estimator $\delta^*(Y)$, where $Y = RX - r^0$ and is normally distributed with mean η and covariance matrix $\ddagger^*(\underline{\sigma}^2) = R\ddagger(\underline{\sigma}^2)R^t$. It is easy to check that for R as above, the matrices $\ddagger^*(\underline{\sigma}^2)$ and $Q^* = (RQ^{-1}R^t)^{-1}$ are diagonal. (RDR^t will be diagonal for any diagonal matrix D.) Hence Theorem 1 and [6] can again be used to show that for estimating η under loss

$$L^*(\delta^*,\eta,\underline{\sigma}^2) = (\delta^*-\eta)^t Q^*(\delta^*-\eta)/tr[Q^*\ddagger^*(\underline{\sigma}^2)],$$

the estimator

$$\delta^*(Y,W) = \left(I_m - \frac{\min\{Y^t\ddagger^{*-1}(W)Y,c\}Q^{*-1}\ddagger^{*-1}(W)}{Y^t\ddagger^{*-1}(W)Q^{*-1}\ddagger^{*-1}(W)Y}\right)Y$$

is minimax providing $0 \leq c \leq 2(p-2\tau)$. Finally, Theorem 2 shows that the estimator

$$(3.5) \quad \delta(X,W) = X - \frac{\min\{(RX-r^0)^t \Sigma^{*-1}(W)(RX-r^0), c\}}{(RX-r^0)^t \Sigma^{*-1}(W) Q^{*-1} \Sigma^{*-1}(W)(RX-r^0)} Q^{-1} R^t \Sigma^{*-1}(W)(RX-r^0)$$

is a minimax estimator of θ. It can easily be seen that this estimator has minimum risk at $R\theta - r^0 = 0$, which was the desired behavior.

The above ideas can also be applied to the situation where R is not of the required form, but can be transformed to satisfy the conditions. Indeed if there exists a nonsingular (mxm) matrix T such that $R' = TR$ has at most one nonzero entry in each column, then (3.5), with $RX - r^0$ replaced by $(R'X - Tr^0)$, will be a minimax estimator which again has minimum risk at $R\theta - r^0 = 0$.

Application 2. Assume that $\Sigma(\underline{\sigma}^2)$ is block diagonal, i.e. that

$$\Sigma(\underline{\sigma}^2) = \begin{pmatrix} \sigma_1^2 \Sigma_1 & & 0 \\ & \ddots & \\ 0 & & \sigma_k^2 \Sigma_k \end{pmatrix},$$

where the Σ_i are known $(p_i \times p_i)$ matrices. In the general notational setup of Section 1, this corresponds to choosing

$$A = \begin{pmatrix} \Sigma_1^{1/2} & & 0 \\ & \ddots & \\ 0 & & \Sigma_k^{1/2} \end{pmatrix} \text{ and } D(Z) = \begin{pmatrix} Z_1 I_{p_1} & & 0 \\ & \ddots & \\ 0 & & Z_k I_{p_k} \end{pmatrix}.$$

Consider the estimator

$$(3.6) \qquad \delta(X,W) = \left(I_p - \frac{cQ^{-1}A^{-1}BA^{-1}D^{-1}(W)}{X^t D^{-1}(W) A^{-1} C A^{-1} D^{-1}(W) X}\right)X.$$

THEOREM 3. The estimator δ given by (3.6) is minimax under the following two conditions:

(i) B and C are block diagonal and commute (i.e., there exists a block diagonal orthogonal matrix Γ such that $\Gamma B \Gamma^t = B^*$ and $\Gamma C \Gamma^t = C^*$ are both diagonal);

(ii) $0 \leq c \leq 2[\mathrm{tr}(B^*) - 2\tau]/ch_{max}[C^{-1}B^t A^{-1} Q^{-1} A^{-1} B]$, where τ is *defined by* (2.1) *with the* b_j *replaced by the* b_j^*.

PROOF. Considering the transformed problem resulting from the transformation $Y = \Gamma A^{-1}X$, it is straightforward to apply Theorems 1 and 2 to give the desired conclusion. ||

In typical block diagonal problems, Q is also appropriately block diagonal. If so, then by choosing B = I and $C = A^{-1}Q^{-1}A^{-1}$, it is clear from Theorem 3 that

$$(3.7) \qquad \delta(X,W) = \left(I_p - \frac{cQ^{-1}\Sigma^{-1}(W)}{X^t\Sigma^{-1}(W)Q^{-1}\Sigma^{-1}(W)}\right)X$$

is minimax providing $0 \le c \le 2(p-2\tau)$. It is easy to check that τ is again given by (2.1), (2.13), and (3.1). The results of [6] can also be applied to (3.7) to get the improved minimax estimator

$$(3.8) \qquad \delta^*(X,W) = \left(I_p - \frac{\min\{X^t\Sigma^{-1}(W)X, c\}Q^{-1}\Sigma^{-1}(W)}{X^t\Sigma^{-1}(W)Q^{-1}\Sigma^{-1}(W)X}\right)X.$$

If the linear restriction $R\theta - r^0 = 0$ is thought to hold, using an estimator like (3.5) is again preferrable. If R is itself block diagonal (with the same block structure as the original problem), it follows as before from Theorems 2 and 3 and [6] that (3.5) defines a minimax estimator. Matrices R which are not block diagonal will sometimes work also. Basically, the needed condition is that $R\Sigma(\underline{\sigma}^2)R^t$ must have some block diagonal structure with the blocks known up to unknown constant multipliers.

The most obvious application of the block diagonal situation is in combining independent regression problems. Thus suppose $Y^i = A^i\beta^i + \varepsilon^i$ is the *ith* regression equation ($i = 1,\dots,k$) where Y^i is an $(m_i \times 1)$ vector of observations, A^i is an $(m_i \times p_i)$ design matrix ($m_i - p_i \ge 3$), β^i is a $(p_i \times 1)$ vector of unknown parameters, and ε^i is an $(m_i \times 1)$ random vector which is normally distributed with mean vector zero and covariance matrix $\sigma_i^2 I_{p_i}$, σ_i^2 being unknown. The sufficient statistic for (β^i, σ_i^2) is $(\hat{\beta}^i, S_i^2)$, where

$$\hat{\beta}^i = [(A^i)^t A^i]^{-1}(A^i)^t Y^i$$

and

$$S_i^2 = (Y^i - A^i\hat{\beta}^i)^t(Y^i - A^i\hat{\beta}^i)/(m_i - p_i).$$

Now $\hat{\beta}^i$ is normally distributed with mean β^i and covariance matrix

$$\sigma_i^2 \ddagger_i = \sigma_i^2[(A^i)^t A^i]^{-1},$$

while S_i^2 is independent of $\hat{\beta}^i$ and $(m_i - p_i)S_i^2/\sigma_i^2$ is $\chi^2_{(m_i-p_i)}$.
Assume that the loss in the ith problem of estimating β^i by δ^i is

$$L(\delta^i,\beta^i) = (\delta^i-\beta^i)^t Q_i(\delta^i-\beta^i).$$

The least squares estimate of $\theta=(\beta^1,\ldots,\beta^k)$ is $X=(\hat{\beta}^1,\ldots,\hat{\beta}^k)$. Clearly X has the block diagonal covariance matrix $\ddagger(\underline{\sigma}^2)$. Adding the losses in the component problems and multiplying by the factor $[tr(Q\ddagger(\underline{\sigma}^2))]^{-1}$ yields a loss in estimating θ by δ of the form $(\delta-\theta)^t Q(\delta-\theta)/[tr(Q\ddagger(\underline{\sigma}^2))]$ where Q is the block diagonal matrix with blocks Q_i. Finally, letting $n_i = m_i-p_i$ and defining $W_i = n_iS_i^2/(n_i-2)$, the problem fits exactly the block diagonal formulation. It can be concluded that (3.8) gives an estimator at least as good as the least squares estimator, provided $0 \leq c \leq 2(p-2\tau)$.

References

[1] Alam, Khursheed (1973). A family of admissible minimax estimators of the mean of a multivariate normal distribution. *Ann. Statist.* 1, 517-525.

[2] Baranchik, A. J. (1970). A family of minimax estimators of the mean of a multivariate normal distribution. *Ann. Math. Statist.* 41, 642-645.

[3] Berger, J. (1976a). Admissible minimax estimation of a multivariate normal mean with arbitrary quadratic loss. *Ann. Statist.* 4, 223-226.

[4] Berger, J. (1976b). Minimax estimation of a multivariate normal mean under arbitrary quadratic loss. *J. Multivariate Anal.* 6, 256-264.

[5] Berger, J. and Bock, M. E. (1976a). Combining independent normal mean estimation problems with unknown variances. *Ann. Statist.* 4, 642-648.

[6] Berger, J. and Bock, M. E. (1976b). Eliminating singularities of Stein-type estimators of location vectors. *J. Roy. Statist. Soc. Ser. B*, 38, 166-170.

[7] Bhattacharya, P. K. (1966). Estimating the mean of a multivariate normal population with general quadratic loss function. *Ann. Math. Statist.* 37, 1819-1827.

[8] Bock, M. E. (1975). Minimax estimators of the mean of a multivariate normal distribution. *Ann. Statist.* 3, 209-218.

[9] Efron, B. and Morris, C. (1973). Stein's estimation rule and its competitors - an empirical Bayes approach. *J. Amer. Statist. Assoc.* 68, 117-130.

[10] Efron, B. and Morris, C. (1976). Families of minimax estimators of the mean of a multivariate normal distribution. *Ann. Statist.* 4, 11-21.

[11] Hudson, M. (1974). Empirical Bayes estimation. Technical Report #58, Stanford University.

[12] James, W. and Stein, C. (1960). Estimation with quadratic loss. *Proc. Fourth Berkeley Symp. Math. Statist. Prob.* 1, 361-379.

[13] Lin, Pi-Erh, and Tsai, Hui-Liang (1973). Generalized Bayes minimax estimators of the multivariate normal mean with unknown covariance matrix. *Ann. Statist.* 1, 142-145.

[14] Stein, C. (1955). Inadmissibility of the usual estimator for the mean of a multivariate normal distribution. *Proc. Third Berkeley Symp. Math. Statist. Prob.* 1, 197-206.

[15] Strawderman, W. E. (1973). Proper Bayes minimax estimators of the multivariate normal mean vector for the case of common unknown variances. *Ann. Statist.* 1, 1189-1194.

ON SELECTING A SET OF GOOD POPULATIONS

By P. J. Bickel* and J. A. Yahav**
University of California, Berkeley

1. *Introduction.* In many experimental situations the experimenter is faced with a problem of selecting one or more out of k possible treatments. The treatments are usually characterized by a numerical parameter and the experimenter is interested in choosing the treatments with the larger parameter value. Conventionally such problems are dealt with by tests of homogeneity and multiple comparison methods. In the statistical literature one finds many research reports dealing directly with the problem of selecting good populations. However one finds little evidence of their use in practice. The existing literature on this subject follows two lines of attack. The first, proposed by Bechhofer [1] is the so called "indifference-zone approach". The second proposed by Gupta [5], [6] is the so called "subset selection approach". Both approaches have a common target, the so called "probability of correct selection".

Our line of attack is somewhat similar to the "subset selection approach". We are looking for an optimal invariant subset for a given loss function. The loss function that we use is a sum of two components. The first is the difference between the best population performance and the average performance of the subset that is selected. The second is a penalty for not including the best population. The solutions we arrive at also solve the problem of maximizing the average performance of the subset subject to a given probability of correct selection. (i.e.

*Supported by NSF MPS-73-08698 and ONR N00014-75-C-0444 at University of California, Berkeley.

**Supported by ONR, N00014-75-C-0560 at the Department of Mathematical Statistics, Columbia Univeristy and ONR, N00014-75-C-0444 at University of California, Berkeley.

including the best population in the subset.)

We assume first that the parameters of the k populations are known but we do not know which population is associated with which parameter. For simplicity we deal with normal populations with known equal variances where we assume the parameters of interest are the means. (However it can be seen that the methods can be generalized to monotone likelihood ratio families.) We investigate the optimal solution when k (the number of populations) goes to infinity under the assumption that the "empirical distribution" of the means $\mu_i^{(k)}$, i=1,...,k, converges in a suitable sense to a smooth limiting probability distribution. The asymptotic solution turns out to be "Select the populations that generated the first 100 λ_0 percent of the order statistics", where λ_0 depends on the limiting distribution of the $\mu_i^{(k)}$ and on the penalty associated with a "wrong" selection. In Section 5 we discuss the case where the limiting distribution of the $\mu_i^{(k)}$ is not known.

2. *Notation and statement of the problem.* Let $\Pi_1,\ldots,\Pi_k$ denote populations (probability distributions). For simplicity we shall suppose that Π_i is normal with mean μ_i and variance 1 for $i = 1,2,\ldots,k$. We assume that $\underset{\sim}{\mu} = (\mu_1,\mu_2,\ldots,\mu_k)$ is an unknown permutation of a known vector $\mu^* = (\mu_{(1)},\mu_{(2)},\ldots,\mu_{(k)})$ where $\mu_{(1)} \geq \mu_{(2)} \geq \ldots \geq \mu_{(k)}$. We wish to select a subset of these populations "which have large means". We adopt a decision theoretic framework in which our possible decisions are the 2^k-1 nonempty subsets of $\{\Pi_1,\Pi_2,\ldots,\Pi_k\}$. If $S = \{\Pi_{i_1},\Pi_{i_2},\ldots,\Pi_{i_r}\}$ is such a subset we adopt as a measure of loss

$$(2.1) \qquad \ell(\underset{\sim}{\mu},S) = \mu_{(1)} - \frac{\sum_{j=1}^{r}\mu_{i_j}}{r} + K\, I_{\{\Pi_{(1)} \notin S\}}$$

where $\Pi_{(j)}$ is the population with mean $\mu_{(j)}$, K is a positive constant, and I_A is the indicator function of A. This loss function represents the loss of performance due to using the populations in the subset instead of using the best population plus a loss incurred by a "wrong selection" i.e. by not including the best population. Let $\underset{\sim}{X} = (X_1, X_2, \ldots, X_k)$ be the observations, one observation from each population, and let $\underset{\sim}{Z} = (Z_1, \ldots, Z_k)$ be an ordered permutation of $\underset{\sim}{X}$ with $Z_1 \geq Z_2 \geq \ldots \geq Z_k$. Let $D_1, D_2, \ldots, D_k$ denote the antiranks of $X_1, X_2, \ldots, X_k$ defined by $X_{D_i} = Z_i$ $i = 1, 2, \ldots, k$. For future use we also define the variables $\tilde{\pi}_i$, $\tilde{\mu}_i$ by

$$\tilde{\Pi}_i = \Pi_{D_i} \tag{2.2}$$

$$\tilde{\mu}_i = \mu_{D_i} \tag{2.3}$$

If $\mathcal{B}$ denotes the set of all possible decisions, a pure decision function d is any measurable map from k dimensional Euclidean space to $\mathcal{B}$. Since our problem is invariant under permutations of the observations it seems natural to consider invariant d. What this means is that if $i_1, i_2, \ldots, i_k$ is any permutation of $(1, 2, \ldots, k)$ and $(j_1, j_2, \ldots, j_k)$ is its inverse permutation then,

$$\Pi_i \in d(X_1, X_2, \ldots, X_k) \Leftrightarrow \Pi_{j_i} \in d(X_{i_1}, X_{i_2}, \ldots, X_{i_k}).$$

Let,

$$R(\underset{\sim}{\mu}, d) = E[\ell(\underset{\sim}{\mu}, d)] \tag{2.4}$$

be the risk of a rule. Recall that since the group of permutations operating on our parameter space is transitive R does not depend on $\underset{\sim}{\mu}$. Since both the decision and parameter space are finite there exists an optimal (uniformly best) invariant procedure d*, and d* is minimax among all procedures. (See Ferguson [3]). We show how to construct d* in the next section. d* also has the following important property. All permutation invariant procedures have constant probability of correctly selecting the population with the largest mean, $P_{\underset{\sim}{\mu}}[\Pi_{(1)} \in d(\underset{\sim}{X})]$. Clearly d* maximizes the expected value of the average selected mean

$$E_{\underset{\sim}{\mu}}\left[\frac{1}{r(d)} \sum_{j=1}^{r(d)} \{\mu_{i_j} : i_j \in d(\underset{\sim}{X})\}\right], \tag{2.5}$$

where r(d) is the number of elements in $d(\underset{\sim}{X})$, among all invariant procedures d which have:

$$(2.6) \qquad P_{\underset{\sim}{\mu}}[\Pi_{(1)} \in d(\underset{\sim}{X})] \geq P_{\underset{\sim}{\mu}}[\Pi_{(1)} \in d^*(\underset{\sim}{X})].$$

Again by a standard argument, d* maximizes the minimum expected value of the average selected mean among all d which satisfy (2.6) for all $\underset{\sim}{\mu}$.

3. *The optimal invariant procedure.* It is easy to see that invariant procedures are those having the structure,

$$(3.1) \qquad d(\underset{\sim}{X}) = \{\tilde{\Pi}_j : j \in S(\underset{\sim}{Z})\}$$

where S is a measurable set function of $\underset{\sim}{Z}$ taking as values the 2^k-1 nonempty subsets of $\{1,2,\ldots,k\}$. That is, given the order statistics $\underset{\sim}{Z}$, we select the populations which led to the order statistics with indeces in S. Since invariant procedures have constant risk we can carry out all computations under $\underset{\sim}{\mu} = \mu^*$. To simplify notation we write μ_i for $\mu_{(i)}$ and suppress the subscript μ^* on expectations and probabilities. Since,

$$R(\underset{\sim}{\mu},d) = E\{E[\ell(\underset{\sim}{\mu},d)/\underset{\sim}{Z}]\}$$

and given $\underset{\sim}{Z}$ an invariant procedure is specified by a subset of $\{1,2,\ldots,k\}$ it is clear that

$$(3.2) \qquad d^*(\underset{\sim}{X}) = \{\tilde{\Pi}_j : j \in S^*(\underset{\sim}{Z})\}$$

where $S^*(\underset{\sim}{Z})$ minimizes the "posterior" risk

$$(3.3) \quad \mu_1 - \frac{1}{r} E[\Sigma_{j=1}^{r} \tilde{\mu}_{i_j}/\underset{\sim}{Z}]+K\, P[\mu_1 \notin \{\tilde{\mu}_{i_1},\tilde{\mu}_{i_2},\ldots,\tilde{\mu}_{i_r}\}/\underset{\sim}{Z}]$$

over all sets $S(\underset{\sim}{Z}) = \{i_1,i_2,\ldots,i_r\}$

Let,

$$(3.4) \qquad \alpha(i,j) = P(D_j = i/\underset{\sim}{Z})$$

Then (3.3) can be rewritten,

$$(3.5) \qquad \mu_1 - \frac{1}{r}\sum_{j=1}^{r}\sum_{i=1}^{k} \alpha(i,i_j)\mu_i + K(1-\sum_{j=1}^{r} \alpha(1,i_j)).$$

We now can state our,

THEOREM 3.1. <u>The optimal invariant procedure</u> d* <u>is given by</u>,

$$(3.6) \qquad d^*(\underset{\sim}{X}) = \{\tilde{\Pi}_1,\tilde{\Pi}_2,\ldots,\tilde{\Pi}_{r^*}\}$$

<u>where</u> r* <u>maximizes</u>

(3.7) $$V(r) = \frac{1}{r} \sum_{j=1}^{r} \sum_{i=1}^{k} \alpha(i,j)\mu_i + K \sum_{j=1}^{r} \alpha(1,j)$$

The proof of this theorem requires

LEMMA 3.1. <u>For</u> $\ell < m$,

(3.8) $$\sum_{i=1}^{c} \alpha(i,\ell) \geq \sum_{i=1}^{c} \alpha(i,m)$$

<u>That is, the distribution of</u> D_ℓ <u>given</u> $\underset{\sim}{Z}$, <u>is stochastically smaller than that of</u> D_m <u>given</u> $\underset{\sim}{Z}$.

PROOF.

(3.9) $$\begin{aligned} &P(D_\ell \leq c/\underset{\sim}{Z}) - P(D_m \leq c/\underset{\sim}{Z}) \\ &= \Sigma_{a \leq c < b}[P(D_\ell = a,\ D_m = b/\underset{\sim}{Z}) - P(D_m = a,\ D_\ell = b/\underset{\sim}{Z})] \\ &= \Sigma\ P[D_r = i_r,\ r = 1,2,\ldots,k,\ i_\ell = a,\ i_m = b/\underset{\sim}{Z}] \\ &- P[D_r = i_r,\ r = 1,2,\ldots,k,\ i_\ell = b,\ i_m = a/\underset{\sim}{Z}] \end{aligned}$$

where the sum runs over all permutations of $(1,2,\ldots,k)$ in which a,b appear in the ℓ, m places. But the summands on the right hand of (3.9) are proportional to,

(3.10) $$\Pi_{r \neq \ell,m}\varphi(Z_r - \mu_{i_r})[\varphi(Z_\ell - \mu_a)\varphi(Z_m - \mu_b) - \varphi(Z_m - \mu_a)\varphi(Z_\ell - \mu_b)]$$

which is nonnegative since $\mu_a > \mu_b$, $Z_\ell > Z_m$ and the normal location family has a monotone likelihood ratio. The lemma follows.

PROOF OF THE THEOREM. Let

(3.11) $$W(i_1,i_2,\ldots,i_r) = \frac{1}{r} \sum_{j=1}^{r} \sum_{i=1}^{k} \alpha(i,i_j)\mu_i + K \sum_{j=1}^{r} \alpha(1,i_j)$$

Since minimization of (3.5) over all r is equivalent to maximization of W and $W(1,\ldots,r) = V(r)$ the theorem follows from,

(3.12) $$W(i_1,i_2,\ldots,i_r) \leq W(1,2,\ldots,r) \text{ for } r = 1,2,\ldots,k.$$

But,

(3.13) $$\begin{aligned} W(i_1,i_2,\ldots,i_r) = \sum_{j=1}^{r} \{&\frac{1}{r}\,[\sum_{i=1}^{k-1}(\mu_i - \mu_{i+1})P(D_{i_j} \leq i|\underset{\sim}{Z}) + \mu_k] \\ &+ K\ P(D_{i_j} \leq 1|Z)\} \end{aligned}$$

and in view of Lemma 3.1 is clearly maximized by $(i_1, i_2, \ldots, i_r) = (1, 2, \ldots, r)$.

NOTE. 1) A related loss structure was considered by Goel and Rubin [4] in a report which we saw only after this paper was partially prepared. They derive a theorem for their loss function similar to our Theorem 3.1 using some results of Eaton [2]. This technique of proof applies to our theorem as well as the generalizations we cite below. We have chosen to retain our direct argument for the sake of simplicity.

2) By considering standardized sample means we can expand Theorem 3.1 to the situation where we take n observations from normal populations $\Pi_1, \Pi_2, \ldots, \Pi_k$ with equal known variance σ^2 and means $\mu_1, \mu_2, \ldots, \mu_k$ known up to a permutation.

3) More generally the theorem goes through verbatim if $\Pi_1, \Pi_2, \ldots, \Pi_k$ have densities $f(x, \mu_i)$ $i = 1, 2, \ldots, k$ where $(\mu_1, \mu_2, \ldots, \mu_k)$ is known up to a permutation and $f(x, \mu)$ is a monotone likelihood ratio family in μ.

4) Finally we note that the structure of the optimal procedures as given in (3.6) is valid under the monotone likelihood ratio assumptions of 3) provided that our loss function ℓ is invariant and obeys (3.5) of [2]. That is, we interpret g as a permutation of $\mu_1, \ldots, \mu_k$ when acting on the parameter space and as a permutation of $\Pi_1, \ldots, \Pi_k$ when acting on the decision space. We require,

$$\ell(g\underset{\sim}{\mu}, g^{-1}S) = \ell(\underset{\sim}{\mu}, S)$$

$$\ell(\underset{\sim}{\mu}, S) \geq \ell(\underset{\sim}{\mu}, S')$$

whenever $\mu_i \geq \mu_j$, S and S' have the same number of elements $i \in S'$, $j \in S$, $S - \{j\} = S' - \{i\}$.

4. *The number of populations is large.* In the previous section we gave an optimal solution for the unrealistic situation that μ^* is known. The solution is comparatively easy to calculate for moderately small k. This leaves two important questions: (1) Can we find approximate solutions for k large? (2) What can be

done if μ^* is unknown? We deal with the first question in this section.

Let,

(4.1) $$F_k(x) = k^{-1}\{\text{number of coordinates of } \tilde{\mu} \leq x\}.$$

$F_k(x)$ is the d.f. of the measure placing mass $\frac{1}{k}$ at each μ_i, the "empirical" distribution of $\tilde{\mu}$. We need the following assumptions.

CONDITION A1. Suppose that

(4.A.1) $$F_k \overset{\mathcal{L}}{\to} F \quad \text{as } k \to \infty.$$

Further we suppose F is the d.f. of a proper probability measure which is,

(a) Bounded above ($\exists\, x_0$ s.t. $F(x_0) = 1$).

(b) Continuous

(c) Strictly increasing on $\{x: 0 < F(x) < 1\}$.

Without loss of generality we shall take $\inf\{x: F(x) = 1\} = 0$.

CONDITION A2. Suppose that

(4.A.2) $$\lim_{n\to-\infty} \int_{-\infty}^{n} x dF_k(x) = 0$$

uniformly in k.

The last assumption requires more notation. For $1 \leq h \leq k$, $0 \leq i \leq [\frac{k}{h}]-1$, let

$$\mu(i,h) = h^{-1}\Sigma_{j=ih+1}^{(i+1)h} \mu_j$$

$$S^2(i,h) = \Sigma_{j=ih+1}^{(i+1)h}(\mu_j-\mu(i,h))^2$$

$$M(\epsilon,h) = \max\{S^2(i,h): 0 \leq i \leq \{(1-\epsilon)\tfrac{k}{h}]-1\}.$$

CONDITION A3. Suppose that for every $\epsilon > 0$ we can find a sequence $\{h(k)\}$ such that for every $\delta > 0$,

(4.A.3) $$ke^{-\delta h} \to 0$$

and

(4.A.4) $$M(\epsilon,h) \to 0.$$

We note here that (4.A.3) and (4.A.4) are readily satisfied if for some $M(\epsilon)$ independent of k, $1 \le j \le (1-\epsilon)k$, $(\mu_{(j)}-\mu_{(j+1)}) \le \frac{M(\epsilon)}{k}$. Then $M(\epsilon,h) = O(h^3k^{-2})$ and $h(k) \sim k^{\alpha}$, where $0 < \alpha < 2/3$, is a sequence which satisfies (4.A.3) and (4.A.4). In particular if the $\mu_{(j)}$ are the order statistics of a sample of size k from F, where F has a density bounded away from 0 on compacts contained in the set $0 < F < 1$ and F has finite expectation, the conditions (4.A.1) - (4.A.4) are satisfied almost surely.

Define

$$H(x) = \int_{-\infty}^{\infty} \Phi(x-u)dF(u). \tag{4.2}$$

Then $H(\cdot)$ is continuous and strictly increasing. For $0 \le \lambda \le 1$ define a family of procedures $d_k(\cdot,\lambda)$ by,

$$\begin{aligned} d_k(\underset{\sim}{X},\lambda) &= \{\tilde{\Pi}_1,\ldots,\tilde{\Pi}_{[\lambda k]}\} \quad \text{if } [\lambda k] \ge 1 \\ &= \{\tilde{\Pi}_1\} \quad\quad\quad\quad\; \text{if } [\lambda k] < 1. \end{aligned} \tag{4.3}$$

Define functions,

$$\begin{aligned} \tilde{V}_1(\lambda) &= \lambda^{-1}\int_{-\infty}^{0} x[1-\Phi(H^{-1}(1-\lambda)-x)]dF(x), \quad 0 < \lambda \le 1, \\ \tilde{V}_1(0) &= 0. \end{aligned} \tag{4.4}$$

$$\tilde{V}_2(\lambda) = 1-\Phi(H^{-1}(1-\lambda)). \tag{4.5}$$

$$\tilde{V}(\lambda,K) = \tilde{V}_1(\lambda) + K\,\tilde{V}_2(\lambda). \tag{4.6}$$

LEMMA 4.1. *The function $\tilde{V}_1$ is continuous and strictly decreasing in λ on* [0,1].

PROOF. Verify that,

$$\tilde{V}_1(\lambda) = \lambda^{-1}\int_{1-\lambda}^{1} [H^{-1}]'(t)\int_{-\infty}^{\infty} z\,\varphi(H^{-1}(t)-z)dF(z)dt. \tag{4.7}$$

Note that the family of probability measures $\{P_t\}$ defined by,

$$\frac{dP_t}{dF} = [H^{-1}]'(t) \quad (z-H^{-1}(t)) \tag{4.8}$$

is a monotone likelihood ratio family in t. Hence $\int_{\infty}^{\infty} z dP_t(z)$ is increasing in t and

$$(4.9) \qquad \tilde{V}_1(\lambda) = \frac{1}{\lambda}\int_{1-\lambda}^{1}\int_{-\infty}^{\infty} z\, dP_t(z)dt$$

is decreasing in λ. For $\lambda \to 0$ we have $\tilde{V}_1(\lambda)$ converging to

$$(4.10) \qquad \lim_{t\uparrow 1} \int_{-\infty}^{\infty} zdP_t(z) = (F) \text{ ess. sup.} z = 0.$$

As a consequence of this lemma we see that $\tilde{V}(\lambda,K)$ attains its supremum on $[0,1]$. Let $\lambda_0(K)$ be the smallest maximum point.

THEOREM 4.1. _Let_ $r_k^*(Z)$ _be as in Theorem_ 3.1 _and suppose conditions_ A1-A3 _hold. Then,_

$$(4.11) \qquad \frac{r_k^*(\underset{\sim}{Z})}{k} \overset{P}{\to} \lambda_0$$

$$(4.12) \qquad \lim_k P(\Pi_{(1)} \in d_k^*(\underset{\sim}{X})) = \lim_k P(\Pi_{(1)} \in d_k(\underset{\sim}{X},\lambda_0)) = V_2(\lambda_0).$$

$$(4.13) \qquad \lim_k E[\frac{1}{r_k^*(\underset{\sim}{Z})} \Sigma_{j=1}^{r_k^*} \mu_{(j)}]=\lambda_0^{-1}\lim_k \frac{1}{k} E[\Sigma_{j=1}^{[\lambda_0 k]} \mu_{(j)}]=V_1(\lambda_0).$$

Thus the rule $d_k(X,\lambda_0)$ is asymptotically optimal in a pointwise sense (4.11) and globally,

$$(4.14) \qquad \lim_k R(\tilde{\mu},d_k(\underset{\sim}{X},\lambda_0))=\lim_k R(\tilde{\mu},d_k^*(\underset{\sim}{X}))=\tilde{V}(\lambda_0(K),K).$$

Let $0 < \gamma < 1$ be a specified desired lower bound to the probability of selection of $\Pi_{(1)}$ in the subset (A lower bound for the probability of "correct selection"). Let,

$$(4.15) \qquad \lambda(\gamma) = 1 - H(\Phi^{-1}(1-\gamma)).$$

THEOREM 4.2. _Under conditions_ A1-A3 _the procedure_ $d_k(\underset{\sim}{X},\lambda(\gamma))$ _has the properties_

$$(4.16) \qquad \lim_k P_{\underset{\sim}{\mu}}[\Pi_{(1)} \in d_k(\underset{\sim}{X},\lambda(\gamma))] = \gamma.$$

_If_ $d_k(\underset{\sim}{X})$ _is any sequence of invariant procedures such that_

$$(4.17) \qquad \liminf_k P_{\underset{\sim}{\mu}}[\Pi_{(1)} \in d_k(\underset{\sim}{X})] \geq \gamma$$

then

$$(4.18) \qquad \limsup_k E_{\underset{\sim}{\mu}}[(\#d_k)^{-1}\Sigma\{\mu_i\colon \Pi_i \in d_k(\underset{\sim}{X})\}]$$

$$\leq \lim_k E_{\underset{\sim}{\mu}}(\# d_k(\underset{\sim}{X},\lambda(\gamma)))^{-1}\Sigma\{\mu_i: \Pi_i \in d_k(\underset{\sim}{X},\lambda(\gamma))\}$$

where $\# d_k$ is the number of elements in d_k.

Theorem 4.2 is a consequence of Theorem 4.1 and Lemma 4.2.

LEMMA 4.2. *The function* $\lambda_0(K)$

(i) *is increasing in* K

(ii) *satisfies* $\lambda_0(0) = 0$, $\lambda_0(\infty) = 1$.

PROOF. Part (i) follows from Lemma 4.1 by recalling that $\tilde{V}_1$ is decreasing in λ and $\tilde{V}_2$ is increasing in λ. Part (ii) follows from the definition of $\tilde{V}_1$ and $\tilde{V}_2$ and $\lambda_0(K)$ and Lemma 4.1.

PROOF OF THEOREM 4.2. By Theorem 4.1 it suffices to exhibit a K such that,

$$\lambda(\gamma) = \lambda_0(K) \tag{4.19}$$

or at least

$$\tilde{V}(\lambda(\gamma),K) = \tilde{V}(\lambda_0(K),K). \tag{4.20}$$

From Lemma 4.2 either (4.19) holds or there exists K such that $\lambda_0(K-) \leq \lambda(\gamma) \leq \lambda_0(K^+)$. This implies $\tilde{V}(\lambda_0(K-),K) = \tilde{V}(\lambda_0(K+),K) = \tilde{V}(\lambda_0(K),K)$ and then (4.20) is satisfied.

We now proceed with the proof of Theorem 4.1. Define the stochastic process,

$$A(\lambda_1,\lambda_2) = \Sigma_{j=1}^{[\lambda_2 k]} \alpha(< \lambda_1 k >, j) \tag{4.21}$$

where $<\lambda_1 k> = [\lambda_1 k] + 1$.

LEMMA 4.3. *Under conditions* A1 *and* A3

$$\sup\{|A(\lambda_1,\lambda_2) - (1-\Phi(H^{-1}(1-\lambda_2)-F^{-1}(1-\lambda_1)))|: \\ 0 < \epsilon_2 \leq \lambda_2 \leq 1 - \epsilon_2, 0 \leq \lambda_1 \leq 1 - \epsilon_1\} \overset{P}{\to} 0. \tag{4.22}$$

PROOF.

$$A(\lambda_1,\lambda_2) = P(X_{<\lambda_1 k>} \geq Z_{[\lambda_2 k]}/\underset{\sim}{Z}) \tag{4.23}$$

By a theorem of Shorack [8]

(4.24) $\sup\{|Z_{[\lambda_2 k]}-H^{-1}(1-\lambda_2)|: \epsilon_2 \leq \lambda_2 \leq 1 - \epsilon_2\} \overset{P}{\to} 0.$

The conditions of the theorem are satisfied since $\frac{1}{k}\Sigma_{i=1}^{k} \Phi(x-\mu_i) \to H(x)$ a.s. uniformly in x as $k \to \infty$ and H is continuous and strictly increasing. By theorem A1 of the appendix

(4.25) $$\sup\{|P(X_{<\lambda_1 k>} \geq t|\underset{\sim}{Z})-P(X_{<\lambda_1 k>} \geq t)|: -\infty < t < \infty,$$

$$0 \leq \lambda_1 \leq 1 - \epsilon_1\} \overset{P}{\to} 0.$$

The lemma then follows from (4.23), (4.24) and (4.25).

LEMMA 3.4. Under conditions A1-A3,

(4.26) $$\sup\{|\frac{1}{\lambda_2 k} \Sigma_{j=1}^{[\lambda_2 k]} \Sigma_{t=1}^{k}\alpha(t,j)\mu_t - \frac{1}{\lambda_2} \int_{-\infty}^{0} x(1-\Phi(H^{-1}(1-\lambda_2)-x))\, dF(x)|$$

$$: \epsilon \leq \lambda_2 \leq 1 - \epsilon\} \overset{P}{\to} 0.$$

PROOF.

(3.27) $$\frac{1}{\lambda_2 k} \Sigma_{j=1}^{[\lambda_2 k]} \Sigma_{t=1}^{k} \alpha(t,j)\mu_t = \frac{1}{\lambda_2} [\frac{1}{k} \Sigma_{t=1}^{k} \mu_t A(\frac{t}{k},\lambda_2)]$$

$$= \frac{1}{\lambda_2} \int_{-\infty}^{0} x\, A(1-F_k(x),\lambda_2)dF_k(x).$$

By Lemma 4.3,

(4.28) $$\sup\{|\int_{\{x:F_k(x)\geq\epsilon_1^k\}} x[A(1-F\ (x),\lambda_2)-\Phi(H^{-1}(1-\lambda_2)-F^{-1}(F_k(x)))]dF_k(x)| : \epsilon_2 \leq \lambda_2$$

$$\leq 1 - \epsilon_2\} \overset{P}{\to} 0.$$

Now apply a) of condition A1 and condition A2 to show that $\{x:F_k(x) \geq \epsilon_1\}$ can be replaced by $(-\infty,0)$.

Finally using (4.A.1) and (4.A.2)

(4.29) $$\int_{-\infty}^{0} x\Phi(H^{-1}(1-\lambda_2) - F^{-1}(F_k(x)))dF_k(x)$$

$$\to \int_{-\infty}^{0} x\Phi(H^{-1}(1-\lambda_2)-x)dF(x)$$

which completes the proof.

Theorem 4.1 follows from Lemmas 4.3 and 4.4.

NOTE. In general if σ and "$\mu_{(1)}$" are known the optimal procedure is applied to the equivalent observations $\frac{X_i-\mu_{(1)}}{\sigma}$. Equivalently, λ_0 and $\lambda(\gamma)$ are calculated from $\tilde{V}_1,\tilde{V}_2$ based on the limiting distribution of $\frac{\mu_i-\mu_{(1)}}{\sigma}$ which is given by $F(\mu_{(1)}+\sigma x)$.

5. *Unknown μ^*: Bayes, empirical Bayes and compound theory.* If the $\mu_{(i)}$ are not known the procedure of Section 3 cannot be implemented. We can consider a hierarchy of possible models where the knowledge of the $\mu_{(i)}$ varies but $\underline{\sigma \text{ is known}}$.

1. THE BAYES MODEL. Here we suppose the μ_i are a sample from a known distribution F. This corresponds to specifying an exchangeable product Bayes prior on $(\mu_1,\ldots,\mu_k)$.

2. THE EMPIRICAL BAYES MODEL. We make the same assumptions on the generation of the μ_i as in 1 but suppose F is wholly or partly unknown.

3. THE COMPOUND MODEL. We assume the μ_i are arbitrary wholly or partly unknown constants.

In each of these models we can seek different goals. If the loss structure leading to Theorem 4.1 is adopted it seems reasonable in the first two models to average over the μ_i using F. If the formulation leading to Theorem 4.2 is adopted it seems reasonable to continue to measure risks conditional on $\mu_1,\ldots,\mu_k$.

Model 1 is easy to deal with. If the sequence of empirical distributions of $\mu_1,\ldots,\mu_k$ satisfy conditions A1-A3 with probability 1 then clearly $d_k(\underset{\sim}{X},\lambda_0)$ asymptotically minimizes the Bayes risk. The regularity conditions A1-A3 are satisfied if ess $\sup_F x < \infty$, $\int x\, dF(x) > -\infty$ and F possesses a density bounded away from 0 on $\{x: \epsilon \leq F(x) < 1\}$ for every $\epsilon > 0$. Models 2 and 3 are about equally hard. In either case we suppose that the μ_i (with probability 1 in model 2) satisfy conditions A1-A3. Then what we can do depends on an estimation and prior knowledge of F. An important special case is worth singling out.

2',3': ess $\sup_F x$ is known.

Then we can get an optimal rule for the situation of Theorem 4.2. Assume without loss of generality that ess $\sup_F x = 0$. We need only to estimate $\lambda(\gamma)$ and this is naturally done by,

$$\hat{\lambda}(\gamma) = 1-\hat{H}(\Phi^{-1}(1-\gamma)) \tag{5.1}$$

where $\hat{H}$ is the empirical distribution of $X_1, X_2, \ldots, X_k$.

THEOREM 5.1. <u>Suppose models 2' or 3' are valid and the</u> μ_i <u>satisfy conditions</u> A1-A3. <u>Then the rule</u> $d_k(\underset{\sim}{X}, \hat{\lambda})$ <u>has the same properties as the rule</u> $d_k(\underset{\sim}{X}, \lambda(\gamma))$.

PROOF. The argument is essentially the same as that for Theorem 4.2. □

The rule $d_k(\underset{\sim}{X}, \hat{\lambda}(\gamma))$ can be put in a very simple form: "Select those populations for which $X_i \geq \Phi^{-1}(1-\gamma)$. If there is no such population select the population corresponding to $X_{(1)}$".

If we only know an upper bound $\bar{m}$ on ess $\sup_F x$ and/or σ is not known it is still possible to obtain a conservative approximation to $d_k(\cdot, \lambda(\gamma))$. The rule is, "Select those populations with $(X_i-\bar{m})/s \geq \Phi^{-1}(1-\gamma)$," where

$$s^2 = (n-1)^{-1}\Sigma(X_i-\bar{X})^2.$$

This approximation has probability of selection at least $(1-\gamma)$ asymptotically since s^2 tends to a limit strictly larger than σ^2 and replacement of ess $\sup_F x$ by $\bar{m}$ can only increase the probability of correct selection.

If we do not know ess $\sup_F x$, or if we want to implement the aims of Theorem 4.1 we need an estimate of F rather than just of H. It is known that for this problem F is identifiable and an estimate of F can be obtained in various ways, see e.g. Maritz [7]. Asymptotically optimal procedures can thus be constructed in principle. Their potential usefulness should be studied.

Appendix. Suppose that (4.A.1)-(4.A.4) hold. To avoid confusion we shall add the superscript k to the X_i, Z_i and μ_i. Thus $\{X_1^{(k)},\ldots,X_k^{(k)}\}$ $k \geq 1$ is a double array of independent variables, $X_i^{(k)} \sim N(\mu_i^{(k)}, 1)$ $i = 1,2,\ldots,k$. Without loss of generality let $0 \geq \mu_1^{(k)} \geq \mu_2^{(k)} \geq \ldots \geq \mu_k^{(k)}$. We let $\underset{\sim}{Z}^{(k)} = (Z_1^{(k)}, Z_2^{(k)},\ldots, Z_k^{(k)})$ be the ordered $X_i^{(k)}$ with $Z_1^{(k)} \geq Z_2^{(k)} \geq \ldots \geq Z_k^{(k)}$.

THEOREM A1. Under conditions A1 and A3

$$\text{(A.1)} \quad \sup\{|P(X_j^{(k)} \leq t/\underset{\sim}{Z}^{(k)}) - \Phi(t-\mu_j^{(k)})|: 1 \leq j \leq (1-\epsilon)k, -\infty < t < \infty\} \overset{P}{\to} 0$$

The proof proceeds by a series of lemmas and an auxiliary Theorem A.2. We need some further notation. Let $V_k(i,h)$ be the ordered set of variables $X_{(ih+1)}^{(k)},\ldots,X_{(i+1)h}^{(k)}$ $i=1,\ldots,[\frac{k}{h}]-1$ and $\nu_i^{(k)}(x)$ be the empirical distribution of $V_k(i,h)$ (i.e. assigning mass $\frac{1}{h}$ to each member of $V_k(i,h)$).

LEMMA A.1. Suppose $ih+1 \leq j \leq (i+1)h$. Then

$$\text{(A.2)} \quad P[X_j^{(k)} \in A/\underset{\sim}{Z}^{(k)}] = E(P(X_j \in A/V_k(i,h))/\underset{\sim}{Z}^{(k)}).$$

PROOF. Given $V_k(i,h)$, $\underset{\sim}{Z}^{(k)}$ and $X_j^{(k)}$ are independent.

THEOREM A.2. Under conditions A1 and A3,

$$\text{(A.3)} \quad \sup\{|P(X_j^{(k)} \leq t/\underset{\sim}{Z}^{(k)}) - E(\nu_i^{(k)}(t)/\underset{\sim}{Z}^{(k)})|: ih+1 \leq j \leq (i+1)h,$$
$$0 \leq i \leq [(1-\epsilon)\frac{k}{h}], -\infty < t < \infty\} \overset{P}{\to} 0.$$

The proof of this theorem requires Lemma A.1 and two further lemmas. We proceed to study these conditional probabilities. Let $\mu_1 \geq \mu_2 \geq \ldots \geq \mu_k$ and $z_1 \geq z_2 \geq \ldots \geq z_h$ be given numbers. Define,

$$\text{(A.4)} \quad S(z,\mu) = \frac{1}{h!}\Sigma\{\exp[\Sigma_{j=1}^h z_j\mu_{i_j}]: \text{all permutations } (i_1,i_2,\ldots,i_h) \text{ of } (1,2,\ldots,h)\}.$$

$$\text{(A.5)} \quad S_{i\ell}(z,\mu) = \frac{1}{(h-1)!}\Sigma\{\exp[\Sigma_{j=1}^h z_j\mu_{i_j}]: \text{all permutations}$$

$(i_1,\ldots,i_h)$ of $(1,\ldots,h)$ with $i_\ell = i\}$.

(A.6) $$P_{i\ell}(z,\mu) = h^{-1}\, S_{i\ell}/S$$

LEMMA A.2. Let $z. = \frac{1}{h}\,\Sigma_{i=1}^{h}\, z_i, \mu. = \frac{1}{h}\,\Sigma_{i=1}^{h}\,\mu_i.$ Then

(A.7) $$|\lg S(z,\mu)-h\, z.\cdot\mu.| \leq [\Sigma(z_i-z.)^2\cdot\Sigma(\mu_i-\mu.)^2]^{1/2}$$

(A.8) $$|\lg S_{i\ell}(z,\mu)-h\, z.\mu.| \leq \frac{h}{h-1}\,[\Sigma(z_i-z.)^2\Sigma(\mu_i-\mu.)^2]^{1/2}$$

PROOF.

(A.9) $$S(z,\mu)e^{-hz.\mu.} = \frac{1}{h!}\,\Sigma\{\exp[\Sigma_{j=1}^{h}(z_j-z.)(\mu_{i_j}-\mu.)]: \text{all perm.}\}$$

Apply the Schwartz inequality to $\Sigma_{j=1}^{h}(z_j-z.)(\mu_{i_j}-\mu.)$ and evaluate the resulting upper and lower bounds in (A.9) to get (A.7). For (A.8) note that,

(A.10) $$S_{i\ell}\, e^{-hz.\mu.} = \exp[\frac{h}{h-1}\,(z_i-z.)(\mu_\ell-\mu.)]x$$
$$\cdot\ \Sigma\{\frac{1}{(h-1)}!\exp[\sum_{\substack{j=1\\ j\neq\ell}}^{h}(z_j-z.(\ell))(\mu_{j_i}-\mu(i))]$$
$$: \text{all perm. } (i,\ldots,i_h) \text{ of } (1,\ldots,h) \text{ where } i_\ell = i\},$$

where $z.(\ell) = (h-1)^{-1}\Sigma_{\substack{j=1\\ j\neq\ell}}^{h} z_j$ and $\mu.(i) = (h-1)^{-1}\Sigma_{\substack{j=1\\ j\neq i}}^{h}\mu_j$.

Arguing as before we get

(A.11) $$|\lg S_{i\ell}-hz.\mu.| \leq [\sum_{\substack{j=1\\ j\neq\ell}}^{h}(z_j-z.(\ell))^2 + \frac{h}{h-1}\,(z_i-z.)^2]^{1/2}x$$
$$\cdot\ [\sum_{\substack{j=1\\ j\neq i}}^{h}(\mu_j-\mu._{(i)})^2 + \frac{h}{h-1}\,(\mu_j-\mu.)^2]^{1/2}$$
$$\leq \frac{h}{h-1}\,[\Sigma_{j=1}^{h}(\mu_j-\mu.)^2]^{1/2}[\Sigma_{j=1}^{h}(z_i-z.)^2]^{1/2}.$$

LEMMA A.3. *For* $h \geq 2$,

(A.12) $$\max_{1 \leq \ell < h} \Sigma_{i=1}^{h} |P_{i\ell} - \frac{1}{h}| \leq |\exp 3[\Sigma_{j=1}^{h}(\mu_j - \mu.)^2 \Sigma_{j=1}^{h}(z_j - z.)^2]^{1/2} - 1|$$

PROOF. By the definition of (A.6) and Lemma A.2 for $h \geq 2$

$$|P_{i\ell} - \frac{1}{h}| = \frac{1}{h}| \frac{S_{i\ell}}{S} - 1| \leq \frac{1}{h} \{|\exp[\frac{2h-1}{h-1} [\Sigma_{j=1}^{h}(z_j - z.)^2 \cdot$$

$$\cdot \Sigma_{j=1}^{h}(\mu_j - \mu.)^2]^{1/2}] - 1|\}$$

and the lemma follows.

LEMMA A.4. *Let* U_i $i = 1,2,\ldots,m$ *be* i.i.d. *variables distributed as chi-square with* ℓ *degrees of freedom.* (*Here* ℓ *may depend on* m.) *Let* $\{a_{im}\}$ *be a double array of nonnegative constants. Let*

(A.11) $$a_m = \max_i a_{im}$$

Suppose that as $m \to \infty$

(A.12) $$a_m \to 0$$

and that for every $\epsilon > 0$

(A.13) $$me^{-\epsilon\ell/a_m} \to 0.$$

Then

(A.14) $$E[\max_{1 \leq i \leq m} \exp[a_{im} U_i]^{1/2}] = 1 + 0(1).$$

PROOF. We have

(A.15) $$E[\max_{1 \leq i \leq m} \exp[a_{im} U_i]^{1/2} - 1] = \int_0^\infty P[\max_{1 \leq i \leq m} U_i \geq a_m^{-1} 1g^2 (1+t)]dt$$

$$\leq \delta + m \int_\delta^\infty P[U_1 \geq a_m^{-1} 1g^2(1+t)]dt.$$

By Chernoff's inequality,

$$P[U_1 \geq u] \leq u^\ell \exp[-\frac{\ell u}{2} + \ell] \leq \exp[-\frac{\ell u}{4}] \text{ for } u \geq u_0$$

where $u_0 = 4(1g\ u_0 + 1)$. By (A.12) for each fixed $\delta > 0$ we can find $m(\delta)$ such that for $m \geq m(\delta)$, $a_m^{-1} 1g^2(1+\delta) \geq u_0$, $a_m^{-1} \geq 8$.

Then for $m \geq m(\delta)$, $t \geq \delta$, $\epsilon = 1g^2(1+\delta)/8$,

(A.16) $P[U_1 \geq a_m^{-1} \lg^2(1+t)] \leq \exp\{[-\frac{\epsilon \ell}{a_m}] - \lg^2(1+t)\}$.

The lemma now follows from (A.15) and (A.16).

The following lemma is due to Singh [9].

LEMMA A.5. Let W_i, $i = 1,2,\ldots,n$ be independent random variables with W_i having d.f. F_i. Let,

(A.17) $$\nu(t) = n^{-1}\Sigma_{i=1}^{n} I[W_i \leq t].$$

Then there exist constants $c < \infty$ and $\epsilon > 0$ independent of n and the F_i such that

(A.18) $$P\{\sup_t |\nu(t) - E(\nu(t)]| \geq x\} \leq cn^{1/2}e^{-\epsilon nx^2}.$$

PROOF OF THEOREM A.2. Fix h. Suppose $V_k(0,h) = (z_1,z_2,\ldots,z_h)$ and let $\mu_1,\ldots,\mu_h$ in the definition of $P_{i\ell}$ be $\mu_1^{(k)},\ldots,\mu_h^k$. Then if $i \leq j \leq h$, $P(X_j^{(k)} = z_\ell/V_k(0,1)] = P_{j\ell}$. Under this correspondence, by Lemma A.3,

(A.19) $$\sup\{|P[X_j^{(k)} \leq t/V(0,h)]-V_0^{(k)}(t)| = 1 \leq j \leq h, \infty<t<\infty\} \leq$$

$$\max_{1\leq j\leq h} \Sigma_{\ell=1}^{h} |P_{j\ell} - \frac{1}{h}|$$

$$\leq \exp\{3[\Sigma_{\ell=1}^{h}(X_\ell^{(k)}-X^{(k)}(0,h))^2$$

$$\cdot \Sigma_{\ell=1}^{h}(\mu_\ell^{(k)}-\mu_.^{(k)}(0,h))^2]^{1/2}\}$$

where $X_.^{(k)}(i,h) = h^{-1}\Sigma_{\ell=ih+1}^{(i+1)h} X^{(k)}$ and $\mu_.^{(k)}(i,h) = h^{-1}\Sigma_{\ell=ih+1}^{(i+1)h}\mu_\ell^{(k)}$.

The first sum term in the product on the R.H.S. of (A.19) has a noncentral χ_{h-1}^2 distribution with the noncentrality parameter being the second sum term. After some manipulations we find that the R.H.S. of (A.19) is bounded by,

$$\exp 3\sqrt{2}\ \{\Sigma_{\ell=1}^{h}(\mu_\ell^{(k)}-\mu_.^{(k)}(0,h))^2+U_1^{1/2}[\Sigma_{\ell=1}^{h}(\mu_\ell^{(k)}-\mu_.^{k}(0,h))^2]^{1/2}\}$$

where U_1 has a chi-square distribution with h-1 d.f. and depends on $V_k(0,h)$ only.

We can apply the same argument to the variables in each

$V_k(r,h)$ and obtain,

$$\text{(A.20)}\quad \sup\{|P[X_j^{(k)} \le t/V_k(i,h)]-V_i^{(k)}(t)|: ih+1 \le j \le (i+1)h,$$
$$0 \le i \le [(1-\epsilon)\tfrac{k}{h}],\ -\infty < t < \infty\} \le$$
$$\le \max\{\exp 3\sqrt{2}\,[\Sigma_{j=ih+1}^{(i+1)h}(\mu_j^{(k)}-\mu^{(k)}(i,h))^2]$$
$$+\ U_i^{1/2}[\Sigma_{j=ih+1}^{(i+1)h}(\mu_j^{(k)}-\ ^{(k)}(i,h))^2]^{1/2}:$$
$$0 \le i \le [(1-\epsilon)\tfrac{k}{h}]\}.$$

where the U_i are $\chi^2_{(h-1)}$ variables and U_i depends on $V_k(i,h)$ only. We now bound the expectation (unconditional!) of the L.H.S. of (A.20) by that of the R.H.S. Now suppose that $h(k)$ is a sequence having properties (4.A.3) and (4.A.4). We apply Lemma A.4 to conclude that the expectation of the R.H.S. of (A.20) converges to 0 and hence so does that of the L.H.S. But this expectation is bigger than the expectation of the L.H.S. of (A.3). Theorem A.2 follows.

PROOF OF THEOREM A.1. We apply Theorem A.2 with the sequence $\{h(k)\}$ satisfying (4.A.3) and (4.A.4). We clearly need only check that

$$\text{(A.21)}\quad E[\sup\{|\nu_i^{(k)}(t)-h^{-1}\Sigma_{j=ih+1}^{(i+1)h}\Phi(t-\mu_j^{(k)})|: -\infty<t<\infty,\ 0\le i\le[\tfrac{(1-\epsilon)k}{h}]\}$$
$$\to 0$$

and

$$\text{(A.22)}\quad \sup\{|h^{-1}\ \Sigma_{\rho=ih+1}^{(i+1)h}[\Phi(t-\mu_\rho^{(k)})-\Phi(t-\mu_j^{(k)})]|:$$
$$-\infty<t<\infty,\ ih+1 \le j \le (i+1)h,\ 0\le i\le[(1-\epsilon)\tfrac{k}{h}]\} \to 0.$$

Since $|\Phi(t-\mu_\ell^{(k)})-\Phi(t-\mu_j^{(k)})| \le \varphi(0)|\mu_\ell^{(k)}-\mu_j^{(k)}|$ we can check that the expression in (A.20) is bounded by $2\ \varphi(0)M^{1/2}(\epsilon,h)$ and (A.20) follows from (4.A.4). Finally by Lemma A.5 and 4.A.3 the expression on the left in (A.21) is bounded by

(A.23) $\delta + cn^{1/2}\left(\frac{(1-\epsilon)k}{n} + 1\right)\int_{\delta}^{\infty} e^{-\epsilon x^2 h} dx = \delta + O^k e^{-\epsilon \frac{\delta^2}{2} h}) = \delta + o(1).$

Therefore (A.21) holds and the theorem is proved.

References

[1] Bechhofer, R. E. (1954). A single-sample multiple decision procedure for ranking means of normal populations with known variances. *Ann. Math. Statist.* 25, 16-39.

[2] Eaton, M. L. (1967). Some optimum properties of ranking procedures. *Ann. Math. Statist.* 38, 124-137.

[3] Ferguson, T. S. (1966). *Mathematical Statistics: A decision theoretic approach.* Academic Press.

[4] Goel, P. and Rubin, H. (1975). On selecting a subset containing the best population. Purdue University, Department of Statistics Mimeo. Series # 432.

[5] Gupta, S. S. (1956). On a decision rule for a problem in ranking means. Ins. Stat. Mimeo Ser. No. 150, University of North Carolina, Chapel Hill, N.C.

[6] Gupta, S. S. (1965). On some multiple decision (selection and ranking) rules. *Technometrics* 7, 225-45.

[7] Maritz, J. S. (1966). *Empirical Bayes Methods.* Methhuen's monographs on applied probability and statistics.

[8] Shorack, G. (1973). Convergence of reduced empirical and quantile processes. *Ann. Statist.* 1, 146-152.

[9] Singh, R. S. (1975). On the Glivenko Cantelli theorem for weighted empiricals based on independent random variables. *Ann. Prob.* 3, 371-374.

CLOSURE THEOREMS FOR SEQUENTIAL-DESIGN PROCESSES*

By Lawrence D. Brown
Rutgers University

1. *Introduction.* Stochastic control and decision processes occur in a variety of theoretical and applied contexts, for example: "statistical decision problems" (sequential and nonsequential), "stochastic dynamic programming problems", "gambling processes", "optimal stopping problems", "stochastic adaptive control processes", etc. It has long been recognized that these are all mathematically closely related. Since this is the case, all of these decision processes can be viewed as variations on a single theoretical formulation. Such a formulation is the first goal of this paper.

The second goal of the paper is to set forth general conditions under which optimal policies are guaranteed to exist.

The given theoretical formulation is flexible enough to include most variants of the types of processes named above. At the same time we have tried to be parsimonious in our construction and economical in our notation. We hope the reader will find our formulation convenient for many purposes, and particularly for proving theoretical results like the optimal policy theorems mentioned above.

Characteristics of the formulation: In our formulation a countably additive stochastic process is to be observed. At discrete times, or stages, of the process various types of decisions or actions may be taken. (The underlying stochastic process may be either a discrete-time or a continuous-time process.)

At each stage of the process the "statistician" or "controller" chooses from among a set of available actions. This choice may depend on the past history of the process and on the past actions taken. The decision procedure for choosing this action

*Work supported in part by NSF Grant MPS-72-05075-A02.

may be randomized, if desired.

The actions available at any given stage may depend on what actions have been taken in the past. There is no explicit provision in the formulation which allows the set of available actions also to depend on the past observations in the stochastic process, however, such a situation may be included within the formulation by the simple trick described in Section 6. An application to a gambling problem of Dubins and Savage [6] is given in Section 7.

The actions taken at each stage may be of several types. They may involve a decision to stop the process - or to stop and make some sort of terminal decision, as in the classical sequential testing problems. If the action involves a continuation of the observed process it may also involve the stochastic law for the future of the observed process. (This law also may depend on the value of the "parameter" as discussed below.) The word "design" appears in the title of our paper for this reason since a statistician might describe this possibility by saying that the current action includes the experimental design for future observations. As another possibility, the action may concern the time(s) at which future action(s) may be taken.

Finally, the value of any given sequence of actions is measured by either a loss function or a gain function. This loss, or gain, may depend on the actions taken and on the observations in the stochastic process which is observed. Naturally, in problems of a statistical nature the loss, or gain, may depend also on the true state of nature. We have assumed that the loss function is bounded below by a constant. (Equivalently, a gain function must be bounded above.) However this assumption can be weakened as explained in Remark 2.14.

There are certain technical measure theoretic assumptions in addition to the assumptions mentioned above. These assumptions are all of the standard variety, except for Assumption 2.21.

In statistical problems the distribution of the observed variables depends on the true (but unknown) value of the "parameter".

For this purpose our formulation also includes a parameter space. The parameter space has no topological or other structure here; it is merely a set indexing the possible distributions. Hence the formulation is not restricted to those problems known in the statistical literature as parametric problems.

In non-statistical contexts the distribution does not depend on an unknown parameter. All such problems may be included in our formulation by the device of choosing the parameter space to consist of only one point, corresponding to the given distribution.

Existence of optimal decision procedures: Sections 3 and 4 of this paper culminate in Theorems 4.4 and 4.7 which establish the existence of "optimal policies". These theorems involve several extra assumptions. In the usual applications these assumptions require that:

(a) The loss function depends on the actions in a lower semi-continuous way. (Equivalently, the gain is an upper semi-continuous function of the actions taken.)

(b) For all actions whose loss is anywhere finite the possible distributions of the observed process through any finite stage form a dominated family whose densities have a suitable continuity property as a function of the actions taken.

(c) Either the loss function is finitary (Assumption 3.17 (ii)) or sampling without stopping results in an infinite loss.

These consequences are more fully described in Section 4; see especially Discussion 4.6.

The existence of Bayes procedures in the design-decision context is proved in Section 5. The dominatedness assumption of Sections 3 and 4 (referred to in (b), above) is not required.

Our optimality proof is quite different from the usual proofs of such results in "stochastic dynamic programming" settings, including "optimal stopping" problems such as those treated in Chow, Robbins, and Siegmund [5]. (Schäl [13] is an exception.) The "usual proof" proceeds by writing down the basic dynamic programming equation (backward induction equation) for the process, and then demonstrating that this equation has a measurable solution satisfying appropriate boundary conditions. This demonstration proceeds from an examination of the structure of the basic equation. (Generally, one must actually write down a sequence of basic equations for truncated versions of the process, show that each of these truncated processes possesses an optimal policy, and that this sequence of policies converges to an appropriate solution for the original process.) These basic equations do not appear at all in our approach.

The basic equation referred to above is the "pointwise version", in that at each stage of the process and for each possible past sample point one takes an optimal action. There also exists a "global version" of the equations in which at each stage of the process one chooses an optimal procedure - i.e. a procedure which maximizes the expected future gain.

Our optimal policies must, a fortiori, satisfy the "global" version of the basic equations, but not the usual "pointwise" version. However, whenever the pointwise version has a measurable solution then the two versions have the same solution; and the optimal policies whose existence we have established here can be computed by solving the usual basic equations.

[We conjecture that under the technical conditions of our paper it is always the case that the pointwise version really does have a solution. The conjecture appears especially plausible under the conditions of Theorems 3.18 and 4.4 as applied to the dynamic programming setting. See Discussion 6.2.]

In this context our results are closely related to those of Schäl [13], where optimal policies of the type we describe are called weakly optimal policies.

As we have noted, the main theorems in Sections 3 and 4 usually require that the finite-stage distributions of the observed process form a dominated family. Such an assumption is particularly undesirable for certain general applications of the theory. One such application is to the theory of gambling processes and related dynamic programming processes. Section 7 closes with an optimal policy theorem which does not contain any dominatedness assumptions.

This theorem is proved by using a trick which transfers the problem to become a special case of Theorem 3.18. Presumably, this trick could be used to generate other optimality results which do not contain the dominatedness assumption.

2. *Description of the process.*

2.1. Decision stages: Decisions may be made at discrete stages of the process. The set of stages is labelled T, and T is either $1,\ldots,n$; or $1,2,\ldots$; or $1,2,\ldots,\infty$.

2.2. Action space: The space of available actions is denoted by A. It is assumed that $A \subset K = \underset{t\varepsilon T}{\times} K_t$.

Let π_t and $\pi_{(t)}$ denote projection maps with $\pi_t : A \to K_t$, $\pi_{(t)} : A \to \underset{\tau \leq t}{\times} K_\tau$. Then $\pi_t(a) = a_t$ describes the action taken (under a) at stage $t\varepsilon T$, and $\pi_{(t)}(a) = a_1,\ldots,a_t$ are the actions taken (under a) at stages $1,\ldots,t$.

2.3. Discussion: Later elements of the formulation are arranged so as to allow various types of actions, such as to continue sampling or to stop sampling and make a certain terminal decision. As will be seen, other types of actions are also possible. (In previous literature on the subject, various classes of actions have usually been separated out and given separate notations; and the action space has been divided into explicit subsets - e.g. a "stopping time" and a "terminal decision".)

Clearly, not all actions are available at all stages of the process. In fact, if actions $a_1,\ldots,a_{t-1}$ have been taken at stages $1,\ldots,t-1$ then the only actions which can be taken at stage t are those in $\pi_t((a_1,\ldots,a_{t-1},K_t,K_{t+1},\ldots) \cap A)$. It may be that this set of available actions at stage t consists only of a single value - say α_t. In this case the set of actions now available at stage t is trivial. The statistician is not required to make an actual decision as to which action to follow at this stage of the process. (In the classical sequential testing problem it may be that the particular action a_{t-1} corresponds to the decision to stop and make a certain terminal decision at stage t-1. In that case the set of available actions at stages t, t+1,... will all be trivial - each a_τ, $\tau \geq t$, corresponds to the situation "the process has already stopped" - so that no further actual decisions can be made after stage t-1.)

The general formulation allows an action to be taken at stage 1. This will be before any random variables have been observed. However, in many specific problems the set of actions available at this stage will be trivial. In other problems only two actions may be available, e.g. "Begin!" or "Do not begin!"

2.4. Assumptions: Assume henceforth that the spaces K_t are compact and second countable. Let $\mathcal{A}_t$ denote the Borel field on K_t, and $\mathcal{A}$ the Borel field on K. It is convenient to use the symbol $\mathcal{A}_t$ also to denote the σ-field on A induced by the projection map π_t onto K_t, $\mathcal{A}_t$. In the very few instances where this notation could be ambiguous it will be clarified by writing, for example, $S \in \mathcal{A}_t$, $S \subset A$.

Assume $A \in \mathcal{A}$. Let $\bar{A}$ denote the closure of A in K. The symbol $\mathcal{A}$ will also be used to denote the σ-field $\mathcal{A}$ on K as restricted to A. Again, there should be no confusion. Similarly, $\mathcal{A}_{(t)}$ denotes the σ-field induced on $\mathcal{A}$ (or on K, or on $K_{(t)} = \times_{\tau \leq t} K_\tau$, etc.) by the map $\pi_{(t)}$.

2.5. Sample space: The sample space is $X = \times_{t \in T} X_t$. Each co-ordinate space X_t is endowed with a σ-field, $\mathcal{B}_t$, and $\mathcal{B}$ denotes the product σ-field on X. $\mathcal{B}_{(t)}$ is the product σ-field $\times_{\tau \leq t} \mathcal{B}_\tau$. For any particular point $x \in X$ the value $x_t = \pi_t(x)$ is to be thought of as the value of the random variable observed at stage t.

2.6. Observable events: The events which can be observed at stage t are controlled by the actions taken at stages 1,...,t. Thus, for each $a \in A$ there is a σ-field $\mathcal{B}_t(a) \subset \mathcal{B}_t$. $\mathcal{B}_t(a)$ depends only on $\pi_{(t)}(a)$ - i.e. if $\pi_{(t)}(\alpha) = \pi_{(t)}(\beta)$ then $\mathcal{B}_t(\alpha) = \mathcal{B}_t(\beta)$. $\mathcal{B}_t(a)$ describes the events observable at stage t of the process given that actions $a_1,\ldots,a_t$ have been taken at stages $1,\ldots,t$. (It may be that $\mathcal{B}_t(a)$ is the trivial σ-field, meaning that when actions $a_1,\ldots,a_t$ have been taken there is nothing interesting to be observed at stage t.) For convenience, let $\mathcal{B}_0(a)$ denote the trivial σ-field.

Again, the symbol $\mathcal{B}_t(a)$ will also be used to denote the σ-field on X induced by the projection map π_t onto X_t, $\mathcal{B}_t(a)$.

$\mathcal{B}_{(t)}(a)$ denotes the σ-field on X generated by $\{\mathcal{B}_\tau(a):\tau=1,\ldots, t\}$. The same symbol will also be used to denote the projection of this σ-field on $\times_{\tau\leq t} X_\tau$. Let $\mathcal{B}(a) = \mathcal{B}_{(\infty)}(a)$ denote the σ-field on X generated by $\{\mathcal{B}_\tau(a):\tau\varepsilon T\}$.

Finally, if $C \subset A$, $C \,\varepsilon\, \mathcal{A}_{(t)}$, let $\mathcal{B}_{(t)}(C)$ denote the σ-field generated by the collection $\{\mathcal{B}_{(t)}(a):a\varepsilon C\}$.

2.7. Parameter space: The parameter space is denoted by Θ. For now, and throughout much of the following development this "space" is really just an index set, having no further structure.

2.8. Possible distributions: The possible distributions on X are denoted by $F_\theta(\cdot|a)$, $\theta\varepsilon\Theta$, $a\varepsilon A$. The following standard assumptions are made:

(i) $F_\theta(\cdot|a)$ is a probability distribution on $\mathcal{B}$ for each $\theta\varepsilon\Theta$.

(ii) $F_\theta(B|\cdot)$ is $\mathcal{A}$ measurable for each $B\varepsilon\mathcal{B}$.

In addition it is required that the distribution through stage t should depend (in a measurable way) only on actions taken through stage t. Thus assume

(iii) $F_\theta(B|\cdot)$ is $\mathcal{A}_{(t)}$ measurable for each $B\varepsilon\mathcal{B}_{(t)}$, $t\varepsilon T$.

2.9. Discussion: The preceding assumptions require that the probabilities of certain non-observable events be defined. Thus, $F_\theta(B|a)$ must be defined for $B\varepsilon\mathcal{B}_{(t)}$ even if $B\notin\mathcal{B}_{(t)}(a)$. This requirement may seem counterintuitive from some points of view. Yet, it can be motivated by thinking of $B\varepsilon\mathcal{B}_{(t)}$ as an event that some other observer (possibly, "Nature") might be looking at -- this observer would be observing the same realization of the stochastic process $X_1, X_2, \ldots$ but would be using a different action for which B was observable.

2.10. Conditional distributions: We have found it necessary to also assume the existence of appropriately measurable versions of the conditional distribution on $\mathcal{B}_t$ given $\mathcal{A}$, $\mathcal{B}_{(t-1)}$. Formally,

(i) $F_\theta^t(\cdot|a,x)$ is a version of the conditional probability distribution on $\mathcal{B}_t$ given $\mathcal{A} \times \mathcal{B}_{(t-1)}$.

(ii) $F_\theta^t(B|\cdot,\cdot)$ is $\mathcal{A}_{(t)} \times \mathcal{B}_{(t-1)}$ measurable for all $B \epsilon \mathcal{B}_t$.

[It appears plausible that 2.10 (i) and (ii) should be implied by 2.8 (i) and (iii), at least when X_t, $t \epsilon T$ are Polish spaces. However, we do not know a general theorem to this effect.]

2.11. Sequential decision procedures: A stagewise conditional decision procedure[1] at stage $t \epsilon T$ is a conditional probability measure δ_t satisfying:

(i) $\delta_t(\cdot|a,x)$ is a probability measure on $\mathcal{A}_t$.

(ii) $\delta_t(C|\cdot,\cdot)$ is $\mathcal{A}_{(t-1)} \times \mathcal{B}_{(t-1)}$ measurable for each $C \epsilon \mathcal{A}_t$.

(iii) $\delta_t(C|a,\cdot)$ is $\mathcal{B}_{(t-1)}(a)$ measurable for each $C \epsilon \mathcal{A}_t$, $a \epsilon A$.

(Note that $\delta_t(C|a,\cdot)$ really depends only on $\pi_{(t-1)}(a)$ because of (ii). It is therefore justifiable to abuse the notation slightly by writing $\delta_t(C|a_1,\ldots,a_{t-1};x_1,\ldots,x_{t-1})$ when this is convenient, and to use this notation for $C \subset K_t$, $C \epsilon \mathcal{A}_t$ as well as for $C \subset A$, $C \epsilon \mathcal{A}_t$.)

A sequence over $t \epsilon T$ of such rules determines a sequential decision procedure. We use the notation δ or $\{\delta_t\}$ for a sequential decision procedure. Let $\mathcal{D} = \{\delta\}$.

2.12. The observed process: The preceding suffices to guarantee the existence of a stochastic process of observed actions and sample points at each stage, when $\theta \epsilon \Theta$ is true and the sequential procedure δ has been used. This process is defined on

1. See the note on terminology at the end of this paper.

$\times_{t\in T}(K_t,X_t)$ with the corresponding product σ-field. The probability of a cylinder set $(C_1,B_1)\times\ldots\times(C_n,B_n)\times(K_{n+1},X_{n+1})\times\ldots$ is given by

$$(1)\quad \Delta_{\theta,\delta}\Big(\mathop{\times}_{t=1}^{n}(C_t,B_t)\mathop{\times}_{t>n}(K_t,X_t)\Big)$$

$$=\int_{C_1}\int_{B_1}\cdots\cdots\int_{C_n}\int_{B_n} F_\theta^n(dx_n|a_1,\ldots,a_n;x_1,\ldots,x_{n-1})$$

$$\delta_n(da_n|a_1,\ldots,a_{n-1};x_1,\ldots,x_{n-1})\cdot\ \cdot\ \cdot F_\theta(dx_1|a_1)\delta_1(da_1).$$

It is straightforward to check that the probability measure, $\Delta_{\theta,\delta}$, is well defined by (1), defines a stochastic process on $K_1\times X_1\times K_2\times X_2\times\ldots.$ with the corresponding σ-fields, and that the decision component of this process is concentrated on $A\subset\times_{t\in T}K_t$.

(If $\infty\in T$ then, of course, n may be ∞ in (1) - and similar expressions to follow - but only a finite number of C_t, B_t should be different from K_t, X_t.)

2.13. Loss function: The loss function is a map $L:\Theta\times A\times X\to[0,\infty]$ such that $L(\theta,\cdot,\cdot)$ is $\mathcal{A}\times\mathcal{B}$ measurable.

2.14. Remark: In traditional statistical settings the loss is independent of the observation $x\in X$. However the general theory developed in this paper applies to many other types of problems. In many of these settings - e.g. stochastic adaptive control processes - the loss naturally depends partly or entirely on $x\in X$. Note also that the loss may depend on unobserved (e.g. future) values of x, as well as observed values.

The fact that L is bounded below by the value zero is inessential to the theory, however it is important that the loss be bounded below for each $\theta\in\Theta$ and the value zero is chosen merely for convenience. In settings in which a "gain function" appears one should, of course, define L = -G, and it is then important that G be bounded above for each $\theta\in\Theta$.

We mention an important relaxation of this assumption. The condition $L\geq 0$ can be replaced throughout this paper by the following condition:

$$E_{\Delta_{\theta,\delta}} (\inf_{a \in \mathcal{A}} L^-(\theta,a,x)) > -\infty \quad \text{for all } \delta \in \mathcal{D}$$

The reader may easily check the validity of this assertion. In the appropriate "optimal stopping" context this condition is precisely the condition, "$E(\sup x_n^+) < \infty$", of Chow, Robbins, Siegmund [5, especially Theorems 4.5 and 4.5'].

2.15. Risk: The risk function $R: \Theta \times \mathcal{D} \to [0,\infty]$ is defined to be the expectation of the loss - $E(L(\theta,a,x))$ - computed under the measure $\Delta_{\theta,\delta}$.

As usual the risk functions serve to define a partial ordering, " $\succ$ ", on $\mathcal{D}$ by $\delta_2 \succ \delta_1$ if $R(\theta,\delta_2) \leq R(\theta,\delta_1)$ for all $\theta \in \Theta$. Maximal elements in this partial ordering are called admissible. When Θ contains just one point the maximal elements are called optimal policies.

2.16. The set of risk functions: Let $\Gamma = \{R(\cdot,\delta): \delta \in \mathcal{D}\}$. Γ may be thought of as a subset of the compact product space $T = \times_{\theta \in \Theta} [0,\infty]$. Let

$$\hat{\Gamma} = \{r \in T: \exists \delta \in \mathcal{D}\ R(\theta,\delta) \leq r(\theta)\ \forall\ \theta \in \Theta\}.$$

2.17. Discussion: The primary goal of the remainder of the theoretical part of this paper is to provide conditions under which $\hat{\Gamma}$ is compact. For when $\hat{\Gamma}$ is compact various conclusions follow - The following are a few of the most important such conclusions:

(1) There exists a minimal complete class (in abstract terms - every $r \in \Gamma$ is dominated by a maximal element under $\succ$). Thus, if Θ contains just one point, then there exist optimal policies.

(2) There exists a minimax procedure, and a least favorable sequence of prior distributions.

(3) If Θ is finite, then the Bayes procedures form a complete class. (This conclusion holds, more generally, if Θ is compact and all risk functions are continuous and real-valued.)

(4) The Stein-Le Cam necessary and sufficient condition for admissibility is valid. See Stein [14] and Farrell [7].

Proofs of the above statements are of a topological character, and are well known. For (1)-(3) see for example, Wald [15] and LeCam [8].

2.18. Formula: A sequential decision procedure, $\delta=\{\delta_t : t\varepsilon T\}$, defines a conditional probability measure on $\mathcal{A}$ given $x\varepsilon X$ through the formula (1), below, which gives the measure of cylinder sets. To avoid overburdening the notation we also use the symbol δ as a general notation for this measure.

(1)
$$\delta(A \cap (C_1 \times \ldots \times {}_t \times K_{t+1} \times \ldots) | x)$$
$$= \int_{a_1 \varepsilon C_1} \int_{a_{t-1} \varepsilon C_{t-1}} \delta_t(C_t | a_1, \ldots, a_{t-1}; x)$$
$$\delta_{t-1}(da_{t-1} | a_1, \ldots, a_{t-2}; x) \ldots \delta(da_1),$$
$$C_i \; \varepsilon \; \mathcal{A}_i, \quad i = 1, \ldots, t\varepsilon T.$$

THEOREM 2.19. Let $\{\delta_t\}$ be a sequential decision procedure. Then the conditional measure δ defined by 2.18 (1) satisfies:

(1) $\delta(\cdot | x)$ is a probability measure on A, $\mathcal{A}$, for each $x\varepsilon X$; and

(2) For any $C\varepsilon\mathcal{A}_{(t-1)}$, $D\varepsilon\mathcal{A}_t$ the function $\delta(C \cap D | \cdot)$ is measurable with respect to $\mathcal{B}_{(t-1)}(C)$.

PROOF. Condition (1) is obvious from 2.18 (1). For (2) note that $\mathcal{B}_{(n)}(\pi_{(n)}(C)) \subset \mathcal{B}_{(t-1)}(C)$ for $n \leq t-1$. Consider $\delta_n(E_n | a_{(n-1)}; \cdot)$ for $n \leq t$, $E_n \varepsilon \mathcal{A}_n$, and $a_{(n-1)} \varepsilon \pi_{(n-1)}(C)$. By definition it is $\mathcal{B}_{(n-1)}(\pi_{(n-1)}(C))$ measurable, and thus it is $\mathcal{B}_{(t-1)}(C)$ measurable by the preceding remark. The assertion (2) then follows from standard Fubini-type theorems on iterated integration; as given in Meyer [10] or Neveu [11].

2.20. Decision procedures: A measurable conditional probability distribution satisfying conditions 2.19 (1) and (2) is called a decision procedure (as contrasted to a sequential decision procedure as defined in 2.11).

Theorem 2.19 establishes that every sequential decision

procedure determines a corresponding decision procedure. In Sections 3 and 4 we find it convenient to work with decision procedures, rather than with sequential decision procedures. Before we can be justified in doing this, it is necessary that we establish conditions under which every decision procedure corresponds to a sequential decision procedure (up to $\{F_\theta\}$ equivalence). This is done in the remainder of this section.

Actually, the following theorem has two important limitations. One of these is Assumption 2.21, which appears to be essentially necessary in any case. The second is that $\{F_\theta(\cdot|a):\theta\varepsilon\Theta,\ a\varepsilon A\}$ should form a dominated family - dominated by the measure ν. In this way, "$\{F_\theta\}$-equivalence" becomes merely ν-equivalence for the fixed measure ν, and standard conditional probability results can be invoked in the proof of Theorem 2.22. We do not know how far Theorem 2.22 can be extended without this dominatedness assumption; however the dominatedness assumption is in any event a natural one to make in the important Proposition 3.2, and in the remaining theory of Sections 3 and 4. See Assumptions 3.1 or 3.17.

2.21. Assumption: Throughout the remainder of the paper assume that

$$\mathcal{B}_{(t)}(a) = \cap\,\{\mathcal{B}_t(C):\pi_{(t)}(a)\varepsilon C, C\subset \mathcal{Q}_{(t)}, C \text{ is open}\},\ \text{for each } a\varepsilon A,\ t\varepsilon T.$$

[This assumption is a kind of semi-continuity assumption on the σ-fields $\mathcal{B}_{(t)}(a)$. It can be interpreted as saying that if an event is observable at stage t on any neighborhood of $a_{(t)}$ then it should be observable at $a_{(t)}$ itself.]

THEOREM 2.22. *Let ν be a given probability measure on $X,\mathcal{B}$, and let δ be a given decision procedure. Then there exists a sequential decision procedure $\{\delta_t:t\varepsilon T\}$ such that*

$$(1)\quad \Delta((C\cap D)\times B)=\int_{x\varepsilon B}\delta(C\cap D|x)\nu(dx)=\int_{a_{(t-1)}\varepsilon\pi_{(t-1)}(C),x\varepsilon B}\cdots\int\delta_t(D|a_1,\ldots,a_{t-1};x)\delta_{t-1}(da_{t-1}|$$

$a_1,\ldots,a_{t-2};x)\ldots.\delta_1(da_1)\nu(dx),$

$C\varepsilon\mathcal{A}_{(t-1)},\ D\varepsilon\mathcal{A}_t,\ B\varepsilon\mathcal{B}.$

In other words, δ_t is a determination of the conditional distribution of Δ given $\mathcal{A}_{(t-1)}\times\mathcal{B}$ with $\Delta = \nu \circ \delta$ defined on $\mathcal{A}\times\mathcal{B}$ by (1). To see this equivalence more clearly, rewrite (1), by induction, as

(2) $$\Delta((C\cap D)\times B) = \int_{a_{(t-1)}\varepsilon\pi_{(t-1)}(C),x\varepsilon B} \delta_t(D|a_{(t-1)};x)\Delta(da,dx),\ C,\ D,\ B \text{ as in (1).}$$

PROOF. Fix $t\varepsilon T$. We will construct a σ-field $\mathcal{B}' \subset \mathcal{A}_{(t-1)}\times\mathcal{B}$ such that:

(3) $g:A\times X\to R$ is $\mathcal{B}'$ measurable only if $g(a,\cdot)$ is $\mathcal{B}_{(t-1)}(a)$ measurable for all $a\varepsilon A$; and

(4) For each $D\varepsilon\mathcal{B}_t$ there is a determination of $E_\Delta(\chi_D|\mathcal{A}_{(t-1)}\times\mathcal{B})$ which is $\mathcal{B}'$ measurable.

Once $\mathcal{B}'$ has been constructed, define δ_t as the measurable conditional probability measure of Δ on $\mathcal{A}_t$ given $\mathcal{B}'$. This conditional probability measure exists since A, $\mathcal{A}_t$ is standard Borel (see, e.g., Neveu [11]), and it satisfies (1) and the defining properties of a sequential decision procedure (2.11) because of (3) and (4). It remains to construct $\mathcal{B}'$ and check (3) and (4).

Let ρ be a metric on K. If $B\subset K$ let $\rho(B)$ denote the diameter of B. Let $\mathcal{B}'_\varepsilon$ be the least σ-field on $\mathcal{A}\times X$ containing all sets of the form $C\times B$ where $C\varepsilon\mathcal{A}_{(t-1)}$, $\rho(C)\le\varepsilon$, and $B\varepsilon\mathcal{B}_{(t-1)}(C)$. Let $\mathcal{B}' = \bigcap_{\varepsilon>0} B'_\varepsilon$. Then, $\mathcal{B}'$ satisfies (3) because of Assumption 2.21.

Fix $D\varepsilon\mathcal{B}_t$. Let $\{C_i:i=1,\ldots\}$ be a basis for the topology on $A_{(t-1)}$; and let $\mathcal{A}^*_k$ denote the σ-field generated by the sets $\{\pi^{-1}_{(t-1)}(C_i):i=1,\ldots,k\}$. $\mathcal{A}^*_k$ is actually an algebra of sets generated by a finite number of atoms, say $\{C^*_{k,j}:j=1,\ldots,J(k)\}$. Furthermore

(5) $$\lim_{k\to\infty}\sup_j \rho(C^*_{k,j}) = 0.$$

One may write

$$(6) \qquad E_\Delta(\chi_D | \mathcal{A}^*_k \times \mathcal{B}) = \sum_{j=1}^{J(k)} \delta(D \cap C^*_{k,j} | x) \chi_{C^*_{k,j}}(a).$$
$$= g_k(a,x), \text{ say.}$$

Define $g = \limsup_{k\to\infty} g_k$. For each $a \varepsilon \mathcal{A}$, $g_k(a,\cdot)$ is $\mathcal{B}'_\varepsilon$ measurable where $\varepsilon = \sup_j \rho(C^*_{k,j})$. It follows from (5) that g is $\mathcal{B}'$ measurable.

Finally, it can be checked that $\{g_k\}$ is a Martingale relative to Δ. By the Martingale convergence theorem g is a version of $E_\Delta(\chi_D | \mathcal{A}_{(t-1)} \times \mathcal{B})$ since $\mathcal{A}_{(t-1)} \times \mathcal{B}$ is the least σ-field containing $\{\mathcal{A}^*_k \times \mathcal{B}: k=1,\ldots\}$. Thus $\mathcal{B}'$ has the desired properties and the theorem is proved.

3. *Basic theorems for a compact action space.* It is proved in this section that under certain assumptions $\hat{\Gamma}$ is compact. The basic assumptions in addition to the compactness of A are that $\{F\}$ or $\{F^{(t)}\}$ be a dominated family of distributions whose densities have a suitable continuity property, and that the loss, $L(\theta,\cdot,x)$, be lower semi-continuous on A. The following assumption is somewhat relaxed in Assumption 3.17.

3.1. Assumption: The family $\{F_\theta(\cdot|a): \theta\varepsilon\Theta, a\varepsilon A\}$ is a dominated family of distributions. In this case it is always possible to choose the dominating σ-finite measure to be a probability measure which is a linear combination of measures in the family. (See Lehmann [9].) Assume this has been done. Denote the dominating probability measure by ν, and let

$$f_\theta(\cdot|a) = \frac{dF_\theta(\cdot|a)}{d\nu} \quad \theta\varepsilon\Theta,\ a\varepsilon A.$$

Assume that $f_\theta(\cdot|\cdot)$ is $\mathcal{B} \times \mathcal{A}$ measurable. (This measurability condition follows from a Martingale argument in Meyer [10] if X is a Polish space and $\mathcal{B}$ is its Borel field.)

Since ν is a linear combination from $\{F_\theta(\cdot|a)\}$, the conditional distributions of ν given $\mathcal{B}_{(t)}$ exist.

PROPOSITION 3.2. Let Assumption 3.1 be satisfied. Let δ be

<u>any decision procedure</u>. <u>According to Theorem</u> 1.22, δ <u>corresponds to an essentially unique</u> (a.e.Δ) <u>sequential decision procedure</u>, δ^*, <u>say</u>. <u>Then</u>, $R(\theta,\delta^*)$ (<u>as defined in</u> (1.15)) <u>may be computed as</u>

(1) $$R(\theta,\delta^*) = \iint L(\theta,a,x,)f_\theta(x|a)\delta(da|x)\nu(dx).$$

PROOF. This proposition follows directly from Theorem 2.22 and the properties of conditional probability after one writes out explicitly the expressions for $E_{\Delta_{\theta,\delta^*}}(L)$ and $E_\Delta(L)$. These explicit expressions are rather lengthy since they involve repeated expressions like $F_\theta^t(dx_t|a_{(t)},x_{(t-1)}) = c(a_{(t)},x_{(t-1)})f_\theta(x|a)\nu(dx_t|a_{(t)},x_{(t-1)})$. We omit further details.

3.3. <u>Convention</u>: Theorems 2.19, and 2.22 and Proposition 3.2 show that there is a one-one correspondence between equivalence classes of decision procedures and of sequential decision procedures, and this equivalence preserves risk functions.

Hereafter in the context of Assumption 3.1 we will use the symbol $\mathcal{D}$ interchangeably to represent the set of all sequential decision procedures or the set of all decision procedures; and the symbol $R(\theta,\delta)$ will be used when δ is either type of decision procedure.

3.4. <u>Definition</u>: A <u>generalized decision procedure</u> taking decisions on a subset $D \subset K$ is a measurable conditional probability distribution on D given the σ-field $\mathcal{B}$ on X. Such decision procedures may "depend on the future".

Define a topology on the set of such procedures according to the convergence definition: $\delta_\alpha \to \delta$ if

(1) $$\int\delta_\alpha(da|x)f(x)c(a)\nu(dx) \to \int\delta(da|x)f(x)c(a)\nu(dx) \text{ for all}$$
$$f\varepsilon L_1(X,\mathcal{B},\nu) = L_1, c\varepsilon C(K)$$

where $C(K)$ denotes the continuous functions on K. This topology can equivalently be viewed as the weak topology generated by the maps $\int\delta(da|x)f(x)c(a)\nu(dx)$, $f\varepsilon L_1$, $c\varepsilon C(K)$.

The preceding topology is not Hausdorff. Hence it is convenient to introduce the equivalence relationship: $\delta_1 \sim \delta_2$ if

$\int \delta_1(da|x)f(x)c(a)\nu(dx) = \int \delta_2(da|x)f(x)c(a)\nu(dx)$ for all $f \in L_1$, $c \in C(K)$. Let $\mathcal{D}(K)$ denote the set of equivalence classes of generalized decision procedures on K with the above topology. Let $\mathcal{D}(D)$ denote the subset of $\mathcal{D}(K)$ consisting of the equivalence classes of generalized procedures on D.

Let $\mathcal{D}_p(D)$ be the subset of $\mathcal{D}(D)$ consisting of those equivalence classes of $\mathcal{D}(D)$ such that some member of the class is a decision procedure on D in the sense of Definition 2.20. Thus $\mathcal{D}_p(A)$ is a topological space whose elements are equivalence classes of members of $\mathcal{D}$. When there is no confusion we write the symbol, δ, for either an element of $\mathcal{D}$ or an equivalence class in $\mathcal{D}_p(A)$ (or in $\mathcal{D}(K)$, etc.). The following basic facts can be verified by standard probabilistic arguments.

PROPOSITION 3.5. (i) *$\delta_1 \sim \delta_2$ if and only if $\delta_1(C|\cdot) = \delta_2(C|\cdot)$ almost everywhere (ν) for all $C \in \mathcal{A}$, $C \subset K$; and equivalently if and only if*

$$\int \delta_1(da|\cdot)c(a) = \int \delta_2(da|\cdot)c(a) \quad \text{a.e.}(\nu) \tag{1}$$

for all $c \in C(K)$.

(ii) *If $\delta_1(C|\cdot) \in \underline{\mathcal{B}}' \subset \mathcal{B}$ for all $C \in \mathcal{A}$, $C \subset K$, where $\underline{\mathcal{B}}'$ denotes the intersection of $\mathcal{B}$ with the ν-completion of some given $\mathcal{B}' \subset \mathcal{B}$, then there is a $\delta_2 \sim \delta_1$ such that $\delta_2(C|\cdot) \in \mathcal{B}'$ for all $C \in \mathcal{A}$, $C \subset K$.*

(iii) *If D is a closed subset of K then $\delta \in \mathcal{D}(D)$ if and only if $\int \delta(da|\cdot)c(a) = 0$ a.e. (ν) for all $c \in C(K)$ with $\Sigma_c \cap D = \phi$ where*

$$\Sigma_c = \{a : c(a) > 0\} \tag{2}$$

(iv) *Under Assumption 3.1 (or Assumption 3.17) if $\delta_1, \delta_2 \in \mathcal{D}$ and $\delta_1 \sim \delta_2$ then $R(\theta,\delta_1) = R(\theta,\delta_2)$ for all $\theta \in \Theta$.*

(v) *If $\delta \in \mathcal{D}(K)$ then $\delta \in \mathcal{D}_p(K)$ if and only if*

$$\int \delta(da|\cdot)c(a)d(a) \in L_1(X, \underline{\mathcal{B}}_{(t-1)}(\Sigma_c), \nu) \tag{3}$$

for all $c, d \in C(K)$ with c being $\mathcal{A}_{(t-1)}$ measurable and d being $\mathcal{A}_t$ measurable.

Proposition 3.5 enables the proof of the following basic

theorem.

THEOREM 3.6. <u>If</u> D <u>is closed in</u> K <u>then</u> $\mathcal{D}_p(D)$ <u>is compact</u>.

PROOF. $\mathcal{D}(K)$ is compact by Le Cam [11, Theorem 2]. By Proposition 3.5 (iii) $\delta\varepsilon\mathcal{D}(D)$ if and only if $\int\delta(da|x)c(a)f(x)\nu(dx) = 0$ for all $f\varepsilon L_1(X,\mathcal{B},\nu)$. Hence $\mathcal{D}(D)$ is closed in $\mathcal{D}(K)$.

If $\delta'\varepsilon\mathcal{D}(D) - \mathcal{D}_p(D)$ then there exists $c,d\varepsilon C(K)$ as in 3.5(3) such that $\int\delta'(da|\cdot)c(a)d(a)\notin L_1(X,\underline{\mathcal{B}}_{(t-1)}(\Sigma_c),\nu)$. There is then an $f'\varepsilon L_1(X,\mathcal{B},\nu)$ such that $\int\delta'(da|x)c(a)d(a)f'(x)\nu(dx) = 1 > 0 = \int\delta(da|x)c(a)d(a)f'(x)\nu(dx)$ for all $\delta\varepsilon\mathcal{D}_p(D)$. Hence there is a neighborhood of δ' which is disjoint from $\mathcal{D}_p(D)$. It follows that $\mathcal{D}_p(D)$ is closed in $\mathcal{D}(D)$; hence compact.

It remains only to discuss some technical methodology which is required for the proof of the first basic statistical result in Theorem 3.14.

3.7. <u>Definition</u>: Let $Q_k = \{q:A\times X\to R:q(a,x) = \sum_{i=1}^{n} c_i(a)h_i(x)$ with $c_i\varepsilon C(K)$, $h_i\varepsilon L_k(X,\mathcal{B},\nu)$, $c_i,h_i \geq 0$, $n < \infty\}$. The two cases of interest are $k=1(L_1)$ and $k = \infty(L_\infty)$. The respective norms in these spaces will be denoted by $||\cdot||_1$ and $||\cdot||_\infty$.

Here is the second major assumption which is required.

3.8. <u>Assumption</u>: For each $\theta\varepsilon\Theta$ the map of $A\to L_1(X,\mathcal{B},\nu)$ defined by $a\to f_\theta(\cdot|a)$ is continuous. (Thus $||f_\theta(\cdot|a_i) - f_\theta(\cdot|a)||_1 \to 0$ as $a_i\to a$.)

The importance of this assumption is that it implies the conclusion of the following fundamental lemma.

LEMMA 3.9. <u>Suppose Assumption</u> 3.8 <u>is satisfied and</u> A <u>is compact. Then for each</u> $\theta\varepsilon\Theta$ <u>there is a sequence</u> $<q_i> \varepsilon Q_1$ <u>such that</u> $q_i(a,\cdot)\to f_\theta(\cdot|a)$ <u>in</u> $L_1(X,\mathcal{B},\nu)$, <u>uniformly in</u> $a\varepsilon A$. (That is, for every $\varepsilon > 0$ there is a $q\varepsilon Q_1$ such that $||q_i(a,\cdot) - f_\theta(\cdot|a)||_1 < \varepsilon$ for all $a\varepsilon A$.)

PROOF. Since A is compact, Assumption 3.8 implies that the set $\{f_\theta(\cdot|a):a\varepsilon A\}$ is compact. Furthermore, for an $h\varepsilon L_1$ $\{a:||f_\theta(\cdot|a) - h(\cdot)||_1 < \varepsilon\}$ is open in A. It follows that for any $\varepsilon > 0$ there is a finite collection $h_1,\ldots,h_n$ and a "partition of unity" $c_1,\ldots,c_n$ of continuous functions in $C(K)$ with $\Sigma_{c_i} \subset \{a: ||f_\theta(\cdot|a)-h_i(\cdot)||_1 < \varepsilon\}$ such that $||f_\theta(\cdot,a) - \Sigma c_i(a)h_i(\cdot)||_1 < \varepsilon$ for all $a\varepsilon A$.

The following important result provides the second basic "product representation".

THEOREM 3.10. <u>Suppose</u> D <u>is closed in</u> K <u>and</u> $\ell: D \times X \to [0,\infty]$ <u>is</u> $\mathcal{A} \times \mathcal{B}$ <u>measurable with</u> $\ell(\cdot,x)$ <u>lower semi continuous on</u> D <u>for each</u> $x\varepsilon X$. <u>Then there is a non-decreasing sequence</u> $\langle q_i\rangle \ \varepsilon\ Q_\infty$ <u>such that</u> $q_i(\cdot,x) \uparrow \ell(\cdot,x)$ <u>for</u> a.e.(ν) $x\varepsilon X$.

PROOF. Let $Q_\infty^* = \{q\varepsilon Q_\infty : q(a,x) = \sum_{i=1}^{n} c_i(a)\chi_{S_i}(x)$, and $S_i \cap S_j = \phi$ for $i \neq j\}$. Clearly $Q_\infty^* \subset Q_\infty$. Let $q^{(1)}, q^{(2)} \ \varepsilon\ Q_\infty^*$. Then

$$\max(q^{(1)},q^{(2)}) = \sum_{i=1}^{n^{(1)}} \sum_{j=1}^{n^{(2)}} \max(c_i^{(1)}(a),c_j^{(2)}(a))\chi_{S_i\cap S_j}(x) \ \varepsilon Q_\infty^*$$

Hence Q_∞^* is an (increasing) lattice.

Let $\{d_i\}$ be a countable dense subset of D, and let ρ denote a metric on D. Let $M = \{m_i\}$ denote the non-negative rational numbers. Let $c_{ij}(a) = m_i - j\rho(a,d_i)$. Consider the set

$$S'_{ij} = \{x : c_{ij}(a) \leq \ell(a,x) \quad \text{for all } a\varepsilon D\}$$

By standard projection theorems, this set is universally measurable. (See, e.g., Meyer [10].) In particular there exists a set $S_{ij} \subset S'_{ij}$ such that $S_{ij} \ \varepsilon\ \mathcal{B}$ and $\bar{\nu}(S'_{ij}-S_{ij}) = 0$ where $\bar{\nu}$ denotes outer ν measure.

It is well known that $\ell(a,x) = \sup c_{ij}(a)\chi_{S'_{ij}}(x)$ since $\ell(\cdot,x)$ is lower semi-continuous. It follows that $\ell(\cdot,x) = \sup c_{ij}(\cdot)\chi_{S_{ij}}(x)$ for almost all (ν) $x\varepsilon X$. The Theorem follows since $\{c_{ij}(a)\chi_{S_{ij}}(x)\} \subset Q_\infty^* \subset Q_\infty$ and Q_∞^* is a lattice.

[If $X,\mathcal{B}$ is a Polish space with its Borel field then Brown and Purves [4] yields that the S'_{ij} are $\mathcal{B}$ measurable. Hence, in this case, the qualification, "for almost all $(\nu)x\varepsilon X$" may be dropped from the conclusion of the theorem.]

When $\mathcal{D}_p(A)$ is compact then compactness of $\hat{\Gamma}$ is implied by coordinatewise lower semi-continuity of the maps $\rho_\theta = R(\theta,\cdot):\mathcal{D}_p(A) \to [0,\infty]$. This is the content of the following result.

LEMMA 3.11. *Suppose $\mathcal{D}_p(A)$ is compact. Suppose each of the maps $\rho_\theta:\mathcal{D}_p(A) \to [0,\infty]$, $\theta\varepsilon\Theta$, defined by $\rho_\theta(\delta) = R(\theta,\delta)$ is lower semi-continuous. Then $\hat{\Gamma}$ is compact.*

PROOF. Let $s_\alpha(\cdot)$ be any net in $\hat{\Gamma}$ such that $s_\alpha \to s \;\varepsilon \times_{\theta\varepsilon\Theta} [0,\infty]$. By the definition of $\hat{\Gamma}$, there is a $\delta_\alpha \varepsilon\; \mathcal{D}_p(A)$ such that $R(\theta,\delta_\alpha) \leq s_\alpha(\theta)$. Let $\{\delta_{\alpha'}\}$ denote a convergent subnet, and $\delta_{\alpha'} \to \delta$. Then $R(\theta,\delta) \leq \liminf R(\theta,\delta_{\alpha'}) \leq \liminf s_{\alpha'}(\theta) = s(\theta)$. Hence $s\varepsilon\hat{\Gamma}$.

The following corollary is required later in this paper.

COROLLARY 3.12. *Suppose V is a closed subset of $\mathcal{D}_p(A)$, $\mathcal{D}_p(A)$ is compact and each ρ_θ is lower semi-continuous. Let*

$$\hat{\Gamma}(V) = \{r(\cdot):r(\cdot)\varepsilon \times_{\theta\varepsilon\Theta} [0,\infty],\ \exists\delta\varepsilon V \ni R(\theta,\delta) \leq r(\theta)\,\forall\theta\varepsilon\Theta\}.$$

Then $\hat{\Gamma}(V)$ is compact.

The remaining assumption required is

3.13. Assumption: $L(\theta,\cdot,x)$ is lower semi-continuous (as a map to $[0,\infty]$) for each $\theta\varepsilon\Theta$, $x\varepsilon X$.

Here is the main result in this case. See Discussion 2.17 for some further consequences.

THEOREM 3.14. *Let Assumptions 3.1, 3.8, and 3.13 be satisfied. If A is compact in K then the maps ρ_θ, $\theta\varepsilon\Theta$ are lower semi-continuous, and $\hat{\Gamma}$ is compact.*

PROOF. Fix $\theta\varepsilon\Theta$. By Theorem 3.10 there is a sequence $\langle q_i\rangle$ in Q_∞ such that $q_i(\cdot,\cdot) \uparrow L(\theta,\cdot,\cdot)$. Then,

$$\rho_\theta(\delta) = R(\theta,\delta) = \sup_{i\to\infty} \int q_i(a,x)f_\theta(x|a)\delta(da|x)\nu(dx). \tag{1}$$

By Lemma 3.9 there is a sequence $\langle q_j'\rangle \;\varepsilon\; Q_1$ such that $q_j'(a,\cdot)\to f(\cdot|a)$ in L_1 uniformly in a.

Each expression of the form

$$\int q_i(a,x)q_j'(a,x)\delta(da|x)\nu(dx)$$

describes a continuous map: $\mathcal{D}_p(A) \to [0,\infty)$. Furthermore

$$|\int c_k(a)h_k(x)(f_\theta(x|a) - q_j'(a,x))\delta(da|x)\nu(dx)|$$
$$\leq \sup_{a\in A}|c_k(a)| \cdot ||h_k||_\infty \cdot \sup_{a\in A}||f_\theta(\cdot|a)$$
$$- q_j'(a,\cdot)||_1 \to 0 \quad \text{as } j \to \infty.$$

Hence each expression in the supremum on the right of (1) is a continuous function: $\mathcal{D}_p(A) \to [0,\infty)$.

It follows that $\rho_\theta(\cdot)$ is lower semi-continuous since it is written in (1) as the supremum of continuous functions. The second assertion of the theorem follows from Lemma 3.11.

3.15. Remark: The fact that $\delta\varepsilon\mathcal{D}_p(A)$ rather than merely $\delta\varepsilon\mathcal{D}(A)$ is not used in the semi-continuity part of the proof of the theorem. It is used to connect this assertion to the risk $R(\theta,\cdot)$. Therefore, it is also true that the function defined on $\delta\varepsilon\mathcal{D}(A)$ by $\int L(\theta,a,x)f_\theta(x|a)\delta(da|x)\nu(dx)$ is lower semi-continuous under the other assumptions of the theorem. This function has no natural statistical meaning; but this fact is needed in the proof of Theorem 4.7.

In the more rare circumstance where every loss function is continuous and bounded then a stronger conclusion is possible:

COROLLARY 3.16. *Suppose the assumptions of Theorem* 3.12 *are satisfied and, in addition,* $L(\theta,\cdot,\cdot)$ *is bounded for each* $\theta\varepsilon\Theta$ *and* $L(\theta,\cdot x)$ *is continuous for each* $\theta\varepsilon\Theta$, $x\varepsilon X$. *Then* ρ_θ *is continuous for each* $\theta\varepsilon\Theta$ *and* Γ, *itself, is compact.*

PROOF. Apply the proof of the theorem to L and to $-L(\theta,\cdot,\cdot)+\sup L(\theta,\cdot,\cdot)$ to see that ρ_θ is a continuous map of $\mathcal{D}_p(A) \to [0,\infty)$. The Corollary follows from this.

[This Corollary can also be proved another way. Under the given continuity assumptions it is possible to prove a version of Theorem 3.10 in which the sequence $\{q_i\}$ is uniformly convergent. (This alternate version is actually much easier to establish than is Theorem 3.10, which we needed for the proof of Theorem 3.14.) Substituting this version of Theorem 3.10 in the proof of Theorem

3.14 then proves Corollary 3.16. This proof appears in Portnoy [12].]

If the loss function satisfies another mild structural condition then the above assumption that $\{F_\theta(\cdot|a):\theta\varepsilon\Theta,a\varepsilon\mathfrak{A}\}$ be a dominated family (Assumption 3.1) can be significantly relaxed in one respect. The application of the following theorem is further clarified by the Discussions 4.1 and 4.6.

3.17. Assumptions:

(i) Let $F_\theta^{(t)}(\cdot|a)$ denote the marginal distribution of F_θ on $X_{(t)}$. Assume the family $\{F_\theta^{(t)}(\cdot|a):\theta\varepsilon\Theta,a\varepsilon A\}$ satisfies Assumptions 3.1 and 3.8 for each $t\varepsilon T$.

(ii) Suppose there exists a non-decreasing sequence of functions $L^{(t)}:\Theta\times A_{(t)}\times X_{(t)}\to[0,\beta]$ such that $L^{(t)}(\theta,b,y)\leq \inf\{L(\theta,a,x):\pi_{(t)}(a)=b,\ \pi_{(t)}(x)=y\}$, and $L^{(t)}(\theta,\cdot,y)$ is lower semi-continuous for each θ,y and t, and

$$\lim_{t\to\infty} L^{(t)} = L. \tag{1}$$

[As an example, suppose $L(\theta,a,x) = \sum_{t=1}^{\infty} \ell_t(\theta,a_{(t)},x_{(t)})$ where each $\ell_t(\theta,\cdot,x_t)\geq 0$ is lower semi-continuous. Then (ii) is satisfied with $L^{(t)} = \sum_{\tau=1}^{t} \ell_\tau$. Note that (ii) implies the lower semi-continuity assertion of Assumption 3.13.]

THEOREM 3.18. <u>Let Assumption 3.17 be satisfied. Then the maps</u> ρ_θ, $\theta\varepsilon\Theta$, <u>are lower semi-continuous, and</u> $\hat{\Gamma}$ <u>is compact.</u>

PROOF. Let $\wp^{(t)}$ denote the original problem truncated to the index set $T^{(t)} = 1,\ldots,t$, and with loss function $L^{(t)}$. Let $\rho_\theta^{(t)}$ denote the corresponding maps in this problem. Thanks to Assumption 3.17 these maps are lower semi-continuous by Theorem 3.14. Now, $\rho_\theta^{(t)}\uparrow\rho_\theta$ since $L^{(t)}\uparrow L$. Hence ρ_θ is lower semi-continuous. As before, $\hat{\Gamma}$ is then compact by Lemma 3.11.

The following is analagous to Corollary 3.16.

COROLLARY 3.19. <u>Suppose the assumptions of Theorem</u> 3.18 <u>are</u>

<u>satisfied and, in addition,</u> $L(\theta,\cdot,\cdot)$ <u>is bounded, the functions</u> $L^{(t)}(\theta,\cdot,x)$ <u>of Assumption</u> 3.16 <u>are continuous for each</u> $\theta\varepsilon\Theta$, $x\varepsilon X$, <u>and the convergence in</u> 3.16 (1) <u>is uniform in</u> a <u>for each</u> $\theta\varepsilon\Theta$, $x\varepsilon X$. <u>Then</u> ρ_θ <u>is continuous for each</u> $\theta\varepsilon\Theta$ <u>and</u> Γ <u>is compact</u>.

PROOF. The assumptions guarantee that L and -L+b satisfy the assumptions of Theorem 3.18, with $b=b(\theta) = \sup L(\theta,\cdot,\cdot)$. (Note, however that $(-L+b)^{(t)} \neq -L^{(t)}+b$.) The corollary then follows from the theorem.

4. *Basic theorems for general action spaces.*

4.1. <u>Discussion</u>: Because of the combined effect of Assumptions 3.1, 3.8, and the assumption that A be compact, Theorem 3.14 is unsuitable for virtually all unbounded-stage sequential applications. In many problems of all general types A is not compact and so neither Theorem 3.14 or 3.18 can be directly applied.

In sequential problems the action space often contains a sequence $\{\alpha_n:n=1,2,\ldots\}$ corresponding to the decisions "stop at stage n". The only natural limit for such a sequence would be an action (say α_∞) corresponding to the decision "do not stop". But if Assumption 3.1 is to be satisfied this latter action often cannot possibly be in A because there is no suitable choice for $F_\theta(\cdot|\alpha_\infty)$ which would be dominated by ν. This possibility alone need not prevent the application of Theorem 3.18, however.

Even when the preceding difficulty does not occur, it may be that A is not compact simply because certain "natural" actions are prohibited. For example in many fixed sample size estimation problems $A = (-\infty,\infty)$, and $L(\theta,a,x) = (\theta-a)^2$. In this case the actions $\pm\infty$ are "natural", though prohibited, along with the interpretation $L(\theta,\pm\infty,x) = \infty$. (It is possible, and equally "natural", to use the one-point compactification $(-\infty,\infty) \cup \{i\}$ of A, again with the interpretation $L(\theta,i,x) = \infty$.)

The following additional assumptions are required to adapt Theorem 3.18 to the case where A is not compact. Further conditions are needed to adapt Theorem 3.14, and these are given later.

4.2. Assumptions: There is a well formulated statistical problem with action space $\bar{A}$ which "extends" the original problem as described in (i)-(iv), below. Call this problem $\bar{\mathcal{P}}$, and denote its elements by $\bar{\mathcal{B}}$, $\bar{F}$, etc.

There is a map $m:\bar{A} \to A$ with $m(a) = a$ if $a \varepsilon A$ such that

(i) m is measurable as map: $(\bar{A},\bar{\mathcal{A}}_t) \to (A,\mathcal{A}_t)$, $\forall\, t\varepsilon T$.

(ii) $\mathcal{B}_t(m(a)) = \bar{\mathcal{B}}_t(a)$ $\forall\, a\varepsilon\bar{A}$, $t\varepsilon T$.

(iii) $F_\theta^t(\cdot|m(a),\cdot) = \bar{F}_\theta^t(\cdot|a,\cdot)$ $\forall\, a\varepsilon\bar{A}$, $t\varepsilon T$.

(iv) $L(\theta,m(a),x) \leq \bar{L}(\theta,a,x)$ $\forall\, \theta\varepsilon\Theta$, $a\varepsilon\bar{A}$, $x\varepsilon X$.

Assume also that $\bar{L}$ satisfies the lower semi-continuity condition - Assumption 3.13 - on $\bar{A}$. Let $\bar{\rho}_\theta$ denote the map ρ_θ in problem $\bar{\mathcal{P}}$.

4.3. Remark: It is required above that $\bar{L}(\theta,\cdot,x)$ be lower semi-continuous. It can be shown that the formula

(1) $\bar{L}(\theta,b,x) = \sup\{\inf\{L(\theta,a,x):a\varepsilon G \cap A\}:\ G \subset K,\ G\varepsilon\mathcal{G},\ b\varepsilon G\}$

defines the largest lower semi-continuous extension of $L(\theta,\cdot,x)$ to $\bar{A}$, where $\mathcal{G}$ denotes a countable basis for the topology on A. Furthermore, if X is standard Borel then this extension is $\mathcal{A} \times \mathcal{B}$ measurable. (Use the projection theorem in Brown and Purves [4].) In this case there would be no loss of generality in assuming that the function $\bar{L}$ in Assumption 4.2 were defined by (1).

In contrast, the existence of the function, m, in Assumption 4.2 is not guaranteed. m must be constructed in each example by examining $\bar{L}$ and L. Its existence is not otherwise guaranteed, and may depend also upon the choice for the compactification of A. (The compactification is in turn determined by the choice for the space K in which A is imbedded.)

The following theorem extends Theorem 3.18. The extension of Theorem 3.14 is given in Theorem 4.7.

THEOREM 4.4. *Suppose that Assumption 4.2 is satisfied and that the hypotheses of Theorem 3.18 are satisfied in problem $\bar{\mathcal{P}}$. Then the maps ρ_θ and $\bar{\rho}_\theta$, $\theta\varepsilon\Theta$, are lower semi-continuous. Furthermore $\hat{\Gamma} = \hat{\bar{\Gamma}}$ is compact.*

[Concerning Assumption 3.17 (ii) for problem $\bar{\mathcal{P}}$ note that if the original loss function L satisfies this condition then $\bar{L}$, defined by 4.3 (1) will also if $\bar{A} = \underset{t\varepsilon T}{\times} \bar{A}_t$. (Define $\overline{L^{(t)}}$ from $L^{(t)}$ by 4.3 (1).)]

PROOF. $\bar{\rho}_\theta$ is lower semi-continuous by Theorem 3.18. Hence $\hat{\bar{\Gamma}}$ is compact.

Now, let $\bar{\delta}\varepsilon\bar{\mathcal{D}}$. Define δ by $\delta(B|x) = \bar{\delta}(m^{-1}(B)|x)$, $B\varepsilon\mathcal{a}$. It is straightforward to check from 4.2 (i)-(ii) that $\delta\varepsilon\mathcal{D} \subset \bar{\mathcal{D}}$ and from 4.2 (iii)-(iv) that $R(\theta,\delta) \leq \bar{R}(\theta,\bar{\delta})$. Hence $\hat{\Gamma} = \hat{\bar{\Gamma}}$, and so $\hat{\Gamma}$ is also compact. The additional assumption needed for the extension of Theorem 3.14 is

4.5. Assumption:

(i) For each $\theta\varepsilon\Theta$ there is locally compact subset $A' = A'(\theta) \subset \bar{A}$ such that Assumptions 3.1 and 3.8 are satisfied relative to the problem with action space $A'(\theta)$ and parameter space $\{\theta\}$.

(ii) For each $\theta\varepsilon\Theta$ and every $c < \infty$ there is a compact subset $D \subset A'(\theta)$ such that

(1) $\bar{L}(\theta,a,x) > c$ for $a\varepsilon\bar{A}-D$, $x\varepsilon X$.

In particular, (1) and the semi-continuity of $\bar{L}$ imply that

(2) $\bar{L}(\theta,a,x) = \infty$ for $a\varepsilon\bar{A}-A'(\theta)$.

[Such a condition is not required in Theorem 4.4. On the other hand, Theorem 4.7 does not require the special Assumption 3.17 on the structure of L. Assumption 3.17 is not satisfied in the most common sequential statistical problems, and so only Theorem 4.7 can be applied in such problems.]

4.6. Discussion: Many common statistical decision problems involve i.i.d. observations from some distribution indexed by $\theta\varepsilon\Theta$ (and each θ indexes a distinct distribution), and the decision $a\varepsilon\bar{A}$ concerns when to stop sampling, and, possibly, other things such as a terminal decision. One or more points in $\bar{A}$ correspond to the situation where sampling never stops. Let α_∞ denote such a point. Then all the distributions $F_\theta(\cdot|\alpha_\infty)$ are mutually singular. It follows that one may have $\alpha_\infty\varepsilon A'(\theta)$ for at most a countable number of points $\theta\varepsilon\Theta$, since $\{\bar{F}_\theta(\cdot|\alpha_\infty):\alpha_\infty\varepsilon A'(\theta)\}$ must be a dominated family of distributions. If $\alpha_\infty \notin A'(\theta)$ then $\bar{L}(\theta,\alpha_\infty,x) = \infty$ by 4.5 (2). (In fact, 4.5 (1) says that $\bar{L}(\theta,a,x) \to \infty$ uniformly in x as $a \to \alpha_\infty$.) Suppose Θ is uncountable. It then follows from Assumption 4.2 (ii)-(iii) that if $\bar{A}$ contains a point under which sampling never stops, then A must also contain at least one such point. To summarize: in such situations some sort of decision under which sampling never stops must be "available" (i.e. in A). Furthermore, the cost of such a decision must be infinite (uniformly in x) for all but at most a countable number of parameter values.

There are many examples in which the conclusion of Theorem 4.7 fails because A fails to contain an action under which sampling never stops (even though the cost of such an action is infinite). However, we do not know whether the second part of the above summary statement is logically necessary (in the special i.i.d. case) for the conclusion of Theorem 4.7.

Assumption 4.5 is phrased in sufficiently general terms to allow other sorts of applications. For example, Assumption 4.5 would be satisfied in the following simple two stage, design and estimation, problem: $A_1 = (0,k]$, $A_2 = (-\infty,\infty) = \Theta$, $F^1_\theta(\cdot|a_1)$ is the normal distribution with mean θ and variance a_1. The process stops after choosing a_1 observing x_1 and estimating by a_2. (Formally, all other co-ordinates of A and X are trivial.) To be specific set $L(\theta,a,x) = (\theta-a_2)^2 + a_1^{-1}$. Then this problem satisfies Assumption 4.5 with $A'(\theta) \equiv (0,k] \times [-\infty,\infty] \subset \bar{A} = [0,k] \times [-\infty,\infty]$.

[Note that $k < \infty$ is required in the preceding example. If $A_1 = (0,\infty)$ then Assumption 4.5 cannot be satisfied and, in fact, the conclusion of Theorem 4.7 fails: No procedure has $R(0,\delta)=0= \lim R(0,\delta_i)$ where δ_i takes action (i,0) with probability one. (In this problem it is an open question whether a minimal complete class exists even though $\hat{\Gamma}$ is not compact.)]

THEOREM 4.7. <u>Let Assumptions</u> 4.2 <u>and</u> 4.5 <u>be satisfied</u>. <u>Then</u>

<u>for each</u> $\theta\varepsilon\Theta$ <u>the map</u> ρ_θ <u>is lower semi-continuous</u>. <u>Furthermore,</u> $\hat{\Gamma}$ <u>is compact</u>.

PROOF. Fix $\theta\varepsilon\Theta$. Let $G \subset \bar{G} \subset A'$ where G is open in $\bar{A}$ and $G \neq A'$. Let $V_G(k) = \{\delta: \int_G \bar{L}(\theta,a,x)\bar{f}_\theta(x|a)\delta(da|x)\nu(dx) \leq k\}$. Pick $a_0 \varepsilon A'-G$. Let $G_0 = \bar{G} \cup \{a_0\}$. For any $\delta\varepsilon\mathcal{D}(\bar{A})$ define δ^* by $\delta^*(x,C) = \delta(x,C \cap \bar{G}) + (1-\delta(x,C \cap \bar{G}))\chi_C(a_0)$. Then $\delta^*\varepsilon\mathcal{D}(G_0)$. (It need not be true that $\delta^*\varepsilon\mathcal{D}_p(G_0)$.) $\mathcal{D}(G_0)$ is compact by Theorem 3.6. Furthermore

$$\int c(a)h(x)\delta(da|x)\nu(dx) = \int c(a)h(x)\delta^*(da|x)\nu(dx) \tag{1}$$

for all $h\varepsilon L_1$, $c\varepsilon C(K)$ with $\Sigma_c \subset G$.

Let $<\delta_\alpha>$ be a convergent net in $V_G(k)$ with $\delta_\alpha \to \delta$. Then $<\delta^*_\alpha>$ is a net in $\mathcal{D}(G_0)$ and so has an accumulation point, say δ^*_0.

It follows from (1) that

$$\int_G \bar{L}(\theta,a,x)\bar{f}_\theta(x|a)\delta_\alpha(da|x)\nu(dx) = \int_G \bar{L}(\theta,a,x)\bar{f}_\theta(x|a)\delta^*_\alpha(da|x)\nu(dx) \tag{2}$$

Hence,

$$\int \bar{L}(\theta,a,x)\bar{f}_\theta(x|a)\delta^*_0(da|x)\nu(dx) \leq k$$

by Remark 3.15 since $\bar{L}(\theta,\cdot,x)\chi_G(\cdot)$ is lower semi-continuous on G_0. It also follows from (1) and the definitions of convergence on $\mathcal{D}(\bar{A})$ and $\mathcal{D}(G_0)$ that (2) holds with δ in place of δ_α and δ^*_0 in place of δ^*_α. (Note: It does not follow that $\delta^* = \delta^*_0$, and this need not be true.) Hence $\delta\varepsilon V_G(k)$. This proves that $V_G(k)$ is closed.

If $H \subset A'$, H closed, define $V'_H(k) = \{\delta: \int_H \bar{f}_\theta(x|a)\delta(da|x)\nu(dx) \geq k\}$. Choose G, H so that $H \subset G \subset \bar{G} \subset A'$, G open. Let $<\delta_\alpha> \subset V'_H(k)$. It follows from (1) that

$$\int_H \bar{f}_\theta(x|a)\delta_\alpha(da|x)\nu(dx) = \int_H \bar{f}_\theta(x|a)\delta^*_\alpha(da|x)\nu(dx) \tag{3}$$

with equality also for $\delta = \lim \delta_\alpha$ and δ^*_0, respectively.

By Remark 3.15, $\int_H \bar{f}_\theta(x|a)\delta^*_\alpha(da|x)\nu(dx)$ is upper semi-continuous on $\mathcal{D}(G_0)$. It follows that $V'_H(k)$ is closed.

Let $G_i \subset \bar{G}_i \subset A'$, $\bigcup_{i=1}^{\infty} G_i = A'$, G_i open $G_i \neq A'$. Then, for $k < \infty$

$$\{\delta : \bar{\rho}_\theta(\delta) \le k\}$$
$$= \left(\bigcap_{i=1}^{\infty} V_{G_i}(k)\right) \cap \left(\bigcap_{i=1}^{\infty} V'_{\bar{G}_i}(1-\varepsilon_i)\right)$$

where $2k\varepsilon_i^{-1} = \inf\{\bar{L}(\theta,a,x) : a \varepsilon \bar{A} - \bar{G}_i\}$. (To see this, note that the choice of ε_i guarantees that $\bar{\rho}_\theta(\delta) \ge 2k$ for all $\delta \notin V'_{\bar{G}_i}(1-\varepsilon_i)$. Also, 4.5 (1) implies that $\varepsilon_i \to 0$. Thus $\delta \varepsilon \bigcap_{i=1}^{\infty} V'_{\bar{G}_i}(1-\varepsilon_i)$ only if $\int_{A'} \bar{f}_\theta(x|a)\delta(da|x)\nu(dx) \ge 1$. [Actually, values > 1 are impossible.])

The sets appearing on the right of (4) are closed. Hence $\{\delta : \bar{\rho}_\theta(\delta) \le k\}$ is closed in $\mathcal{D}(\bar{A})$, which proves the lower semi-continuity of $\bar{\rho}_\theta$ on $\mathcal{D}(\bar{A})$.

As noted, $\bar{R}(\theta,\delta) = \bar{\rho}_\theta(\delta)$. It follows that $\hat{\bar{\Gamma}}$, the dominated-risk set in problem $\bar{\mathcal{P}}$, is compact. Compactness of $\hat{\Gamma}$ follows exactly as in Theorem 4.4.

5. *Bayes procedures.* It is well known that the existence of Bayes procedures may be deduced from dynamic programming results like those described in Discussion 7.2. However we will take an easier path, proceeding directly from Section 4.

5.1. Definitions and Assumptions: Assume Θ is a measure space with σ-field $\mathcal{C}$, say, and L is $\mathcal{C} \times \mathcal{G} \times \mathcal{B}$ measurable.

Let P denote a probability measure on Θ, $\mathcal{C}$. δ_P is a Bayes procedure if

$$\int R(\theta,\delta_P)P(d\theta) = \inf_{\delta \varepsilon \mathcal{D}} \int R(\theta,\delta)P(d\theta).$$

In addition to the technical Assumptions 2.8 (ii) and 2.10 (ii) assume that $F_{\cdot}(B|\cdot)$ is $\mathcal{C} \times \mathcal{G}$ measurable for each $B \varepsilon \mathcal{B}$ and that each $F^t_{\cdot}(B|\cdot,\cdot)$ is $\mathcal{C} \times \mathcal{G}_{(t)} \times \mathcal{B}_{(t-1)}$ measurable for each $B \varepsilon \mathcal{B}_t$.

Let $G = P \circ F_\theta$ be the measure which is defined on rectangles by

$$G(D \times B|a) = \int_{\theta \varepsilon D} F_\theta(B|a)P(d\theta) \qquad D\varepsilon C,\ B\varepsilon \mathcal{B}.$$

It can be checked that $G(\cdot|a)$ satisfies Assumption 2.10. Let G^* be the marginal measure:

$$G^*(B|a) = G(\Theta \times B|a).$$

Clearly, it may be the case that $\{G^*(\cdot|a):a\varepsilon A\}$ is a dominated family even though $\{F_\theta(\cdot|a):\theta\varepsilon\Theta, a\varepsilon A\}$ is not. It can be shown that if $\{G^*(\cdot|a):a\varepsilon A\}$ is a dominated family then it satisfies Assumption 3.8 if X is a Borel subset of a compact metric space $\bar{X}$ and if $F_\theta(\cdot|\cdot)$ defines a continuous function of A to $P(\bar{X})$ the Borel probability measures on $\bar{X}$ with the weak-$\star$ topology.

THEOREM 5.2. _Let_ P _be given. Suppose that Assumption_ 5.1 _is satisfied. Suppose further that the remaining Assumptions of Theorems_ 4.4 _or_ 4.7 _are satisfied for the problem with the family of distributions_ $\{G^*(\cdot|a):a\varepsilon A\}$ _for all_ $\theta\varepsilon\Theta$ _in place of the family_ $\{F_\theta(\cdot|a)\}$. _Then there is a Bayes procedure_, δ_P.

PROOF. Consider a new statistical problem, $\mathcal{P}'$, with a trivial parameter space - $\{\theta'\}$; with $X' = \Theta \times X$, and $\mathcal{B}_t'(a) = \mathcal{J} \times \mathcal{B}_t(a)$ where $\mathcal{J}$ denotes the trivial σ-field on Θ; with $F_{\theta'} = G$ and with $L'(\theta',a,(\theta,x)) = L(\theta,a,x)$. [In other words, make Θ an unobservable part of the sample space, and use the distributions, $\{G(\cdot|a):a\varepsilon A\}$.]

$\hat{\Gamma}'$ is closed in $\mathcal{P}'$ by Theorem 4.4 or 4.7. Thus, an optimal policy exists in problem $\mathcal{P}'$. It must not depend on $\theta\varepsilon\Theta$, and is thus seen to correspond to a Bayes procedure in the original problem.

5.3. _Remark_: Note that Theorem 5.2 does not guarantee that a Bayes procedure minimizes the posterior risk (a.e.). Nevertheless in fixed-sample, non-design settings where such a concept is appropriate, it can be shown that this indeed the case. See, for example, Brown and Purves [4] (whose technical conditions can be generalized slightly to include Theorem 5.2 in the fixed-sample non-design case).

6. *Available actions dependent on past observations.* It was pointed out in the introduction that the formulation in Section 2 makes no explicit provision allowing the set of available actions at a given stage to depend on past observations. (The available actions at a given stage may, however, depend on past actions since A need not equal $\underset{t\varepsilon T}{\times} A_t$.) This section describes a simple trick which allows for such a possibility and also gives an optimal policy theorem in this setting.

6.1. Definitions: Let $\Phi(x)$, $x\varepsilon X$, be a non-empty subset of A such that $\pi_{(t)}(\Phi(x))$ depends only on $\pi_{(t-1)}(x)$. The set $\Phi(x)$ is to be interpreted as the set of available actions when x is observed. Thus the K_t-cross-section of $\pi_{(t)}(\Phi(x))$ at $a_1,\ldots,a_{t-1}$ is the set of actions available to the experimenter at stage t given that $\pi_{(t-1)}(x)$ has been observed and actions $a_1,\ldots,a_{t-1}$ have been taken through stage t-1. This set will be denoted by $\Phi_t(x;a)$; and is, of course, a function only of $\pi_{(t-1)}(x)$, $\pi_{(t-1)}(a)$.

Let $\mathcal{D}^*(\Phi)$ denote the set of procedures $\delta\varepsilon\mathcal{D}$ for which $\delta(\Phi(x)|x) = 1$ for almost all $x\varepsilon X$ $(\Delta_{\theta,\delta})$. In such a problem the experimenter must limit himself to procedures $\delta\varepsilon\mathcal{D}^*(\Phi)$.

For convenience, let Φ denote the set $\Phi = \{x,a:a\varepsilon\Phi(x)\} \subset X\times A$. The introduction of $\mathcal{D}^*(\Phi)$ is the promised "trick". The reader who is disappointed by this trick may look to the proof of Theorem 6.4 for a better one.

6.2. Remarks: Pathologies can occur. Even though the set $\Phi(x)$ is non-empty for all x it may be that $\mathcal{D}(\Phi)$ is empty for measurability reasons. See Blackwell and Dubins [2]. In order to omit consideration of such situations here we assume directly that $\mathcal{D}^*(\Phi) \neq \phi$ whenever such a condition is desirable.

The following theorem extends the optimality result of Section 4.

THEOREM 6.4. Let the assumptions of Theorem 4.4 or 4.7 be

satisfied. Suppose Φ is $\mathcal{B} \times \mathcal{A}$ measurable and $\Phi(x)$ is relatively closed in A for each $x \in X$. Then $\hat{\Gamma}(\mathcal{D}^*(\Phi))$ is compact. (Hence there exist optimal policies in $\mathcal{D}^*(\Phi)$ (so long as $\mathcal{D}^*(\Phi) \neq \phi$), etc.)

PROOF. Let $\bar{\Phi}(x)$ denote the closure of $\Phi(x)$ in $\bar{A}$. $\Phi(x) = \bar{\Phi}(x) \cap A$. $\bar{\Phi} = \{(a,x): a \in \bar{\Phi}(x)\}$ is measurable with respect to $\mathcal{A} \times \hat{\mathcal{B}}$ where $\hat{\mathcal{B}}$ denotes the universal completion of $\mathcal{B}$. (This can be deduced from 4.3 (1).) As a consequence of Corollary 3.11 (applied to $\mathcal{D}(\bar{A})$) and the proof of Theorem 4.5, it is sufficient to show that $\mathcal{D}^*(\Phi)$ is closed in $\mathcal{D}(\bar{A})$ in the topology of Section 3.

Introduce an (artificial) statistical problem with decision space $\bar{A}$; and with X and ν as in the original problem; with $\Theta = \{\theta_0\}$; $f_{\theta_0} \equiv 1$; and where the loss function satisfies $L(\theta_0, a, x) = 0$ if $a \in \bar{\Phi}(x)$, $= \infty$ otherwise for almost every $x(\nu)$. By Theorem 2.14 the map ρ_{θ_0} is lower semi-continuous in this problem. Hence $\{\delta: R(\theta_0, \delta) = 0\} = \mathcal{D}^*(\Phi)$ is closed in $\mathcal{D}(\bar{A})$, as desired. ||

7. *Dynamic programming.*

7.1. Definition (Dynamic programming): Following Blackwell [3], and others, a problem of the type defined in Section 1 is a (stochastic) dynamic programming problem (a la Bellman [1]) when:

(i) Θ is trivial (contains one point).

(ii) X_t, $\mathcal{B}_t(a) \approx X_1$, $\mathcal{B}_1$ for all t, a; and X_1, $\mathcal{B}_1$ is standard Borel (i.e. X_1 is imbeddable as a Borel subset of a Polish space and $\mathcal{B}_1$ is the Borel σ-field).

(iii) A_1 is trivial (contains one point only).

(iv) $F^t(\cdot|x,a) \equiv q(\cdot|x_{t-1}, a_t)$, $t \geq 2$, where q is specified and is $\mathcal{B}_{t-1} \times \mathcal{A}_t$ measurable.

[Note that (ii) implies inter alia that the problem has no "design" aspect.]

According to Blackwell [3] any problem of the above form can be modified in an unimportant way to yield an equivalent gambling problem. Here a gambling problem is a problem satisfying (i)-(iv) and the following additional conditions:

(v) $A_t \subset P(X_t)$, the probability distributions on X_t, $\mathcal{B}_t$; and $q(\cdot|x_{t-1},a_t) = a_t$.

(vi) $\delta \varepsilon \mathcal{D}^*(\Phi)$ (see Definition 6.1) where (consonant with (i)-(v)) $\pi_t(\Phi(x))$ depends only on x_{t-1} and Φ is $\mathcal{B} \times \mathcal{Q}$ measurable.

By (ii) and (vi), $\Phi(x)$ can be rewritten as $\Phi(x) = \{a:(x_{t-1}, a_t)\varepsilon\hat{\Phi}(x)$ for $t \geq 2\}$ where $\hat{\Phi}$ is a given $\mathcal{B} \times \mathcal{Q}$ measurable set. (As usual, $\hat{\Phi} = \{(x_1,a_2):a_2\varepsilon\hat{\Phi}(x_1)\}$.)

(vii) In all such problems the objective is to maximize the expectation of some given gain function $G = G(a,x)$.

7.2. Discussion: Theorems 3.14, 3.18, 4.4 and 4.7 all yield the existence of optimal policies in dynamic programming problems. Theorem 3.18 appears to be identical in this important limited context to Schäl [13, Theorem 6.5], and the proofs are very similar. (Our statement of the requisite regularity conditions is more explicit - thus better - in that Schäl's paper contains no analog of Theorem 3.10 - one of his regularity conditions is the statement that $L(\theta,\cdot,\cdot)$ satisfy the conclusion of our Theorem 3.10. We should also point out that our dominatedness assumption-Assumption 3.17 (i) - is not explicit in Schäl's paper, but it is a consequence of his other assumptions.)

7.3. Relaxing the dominatedness assumption: Note that the optimality results of Section 3 require that $\{F^{(t)}(\cdot|a):a\varepsilon A\}$ be a dominated family for each $t\varepsilon T$. For certain types of applications this assumption is both too strong, and unnecessary. For example, in gambling problems $q(\cdot|x_{t-1},a_t)$ represents the distribution of the payoffs at stage t if the fortune at stage t-1 is x_{t-1} and the bet a_t is made for stage t. In many examples such distributions do not form a dominated family.

In order to fill this gap in our results we present the following theorem. A nearly equivalent result was previously given in Schäl [13, Theorem 5.5]. Schäl gives an elegant, independent proof for his theorems; instead we use a trick which reduces it to a corollary of Theorem 3.18.

We give only the corollary result to Theorem 3.18 in the

gambling problem setting, because it is easier to state. However, it seems reasonable that other results can be proved from Theorems 4.4 and 4.7 by means of this trick, including results outside the dynamic programming context of this section.

We adopt the following point of view. Since X is a Borel subset of $\bar{X}_t$, the distributions $P(X_t)$ have a natural imbedding in $P(\bar{X}_t)$. Hence we consider A_t as a subset of $P(\bar{X}_t)$ for topological purposes.

7.4. THEOREM: Consider a problem of the form 7.1 (i)-(vii). Let $A_t \subset P(\bar{X}_t)$, $t \geq 2$, have the weak*-topology. Suppose $\hat{\Phi}$ is closed in $X_1 \times A_2$. Suppose $G(\cdot,\cdot)$ is bounded and is a jointly semi-continuous function on $A \times X$. Then there is an optimal policy.

[The boundedness assumption on G may be relaxed by invoking the (gain-function version of the) condition in Remark 2.14.]

PROOF. Since $\bar{X}_1$ is compact and metrizable it is homeomorphic to a closed subset of [0,1]. Therefore, there is no loss of generality in assuming that $X_1 = [0,1]$, and we shall do so.

Consider a (new) statistical problem, $\wp'$, in which the sample space is $X' = \underset{t\varepsilon T}{\times} [0,1]_t$ and the observations $(x_1', x_2', \ldots.)$ are all independent with x_1' having the same distribution as did x_1 in the original problem and with $x_2', x_3', \ldots$ each having the uniform distribution on [0,1]; irrespective of the actions taken. The action space in $\wp'$ is $a_1 \times (\underset{t\geq 2}{\times} P(X_t))$ (as in the original gampling problem.

For each $a \varepsilon P(X_t)$, $x \varepsilon X_t$, define the map

$$C_a(x) = [a([0,x)),\ a([0,x])].$$

(C_a may be called the cumulative distribution map of a.) In problem $\wp'$ define the gain function, G, by

$$G'(a,x') = \sup\{G(a,x) : x_t' \varepsilon C_{a_t}(x_t),\ t \geq 2\}; \tag{1}$$

and define the procedure-restriction via the set function

$$\Phi'(x') = \{a : \exists x \ni x_t' \varepsilon C_{a_t}(x_t),\ t \geq 2, \text{ and } a \varepsilon \Phi(x)\}. \tag{2}$$

From (1) and the joint upper semi-continuity of G follows that $G'(\cdot,x')$ is upper semi-continuous for each $x' \varepsilon X'$.

Consider the set $S_1 = \{(a_1,a_2,\ldots,x_1,x_2,\ldots,x_1',x_2',\ldots): x_t' \varepsilon C_{a_t}(x_t)\}$. This is a closed subset of $A \times X \times X'$. Furthermore, $S_2 = S_1 \cap (\Phi \times X')$ is also closed. Now $\Phi' = \{(x',a): a \varepsilon \Phi'(x')\} = \{(x',a): \exists x \ni (a,x,x') \varepsilon S_2\}$. Thus (except for permutation of co-ordinates) Φ' is the projection of S_2 on the A, X' axes. It follows that Φ' is $\mathcal{B}' \times \mathcal{A}$ measurable and that $\Phi'(x')$ is closed for each $x \varepsilon X$. In fact, Φ' is closed in $X' \times A$; and it follows from this that it contains a Borel graph so that $\mathcal{D}'^*(\Phi') \neq \phi$.

Theorem 6.4 yields the existence of optimal policies in problem $\mathcal{P}'$. Since the original problem, $\mathcal{P}$, and $\mathcal{P}'$ are essentially equivalent via the correspondence $x_t \to C_{a_t}^{-1}(x_t)$ it follows that optimal policies also exist for $\mathcal{P}$.

[The hypotheses of the above theorem are virtually the same as those in an optimal policy theorem of Dubins and Savage [6, Theorem 16.1] (with their improved "upper semi-continuity" condition in place of their condition (e)).]

Note on terminology. A reader has commented that our terminology here may be confusing.

It should be emphasized that the procedure $\delta_t(\cdot|\cdot,\cdot)$ describes the distribution on $\mathcal{A}_t$ "given the observable past". However, it is not obvious in making this definition that $\delta(\cdot|\cdot,\cdot)$ is really the conditional distribution on $\mathcal{A}_t$ of some distribution with respect to some σ-field. It is a consequence of Theorem 1.22 and its proof that if $\{F_\theta(\cdot|a)\} \ll \nu$ then this is in fact the case. To generalize slightly, it follows from Theorem 1.22 that if for given $\theta \varepsilon \Theta$ there is a measure ν_θ such that $\{F_\theta(\cdot|a): a \varepsilon \mathcal{A}\} \ll \nu_\theta$ then $\delta(\cdot|\cdot,\cdot)$ is a version of the conditional distribution on $\mathcal{A}_t$ of $\Delta_{\theta,\delta}$ given $\mathcal{B}'$; where $\mathcal{B}' = \mathcal{B}'(t)$ is defined as in the proof of Theorem 1.22.

On the other hand, the measures $\delta(\cdot|\cdot)$ (defined in 1.18) are not necessarily equal to the conditional distribution on $\mathcal{A}$ of $\Delta_{\theta,\delta}$

given $\mathcal{B}$. For this reason the symbolism, "$\delta(\cdot|\cdot)$" which we have adopted may be misleading.

However, note that for any $C \in \mathcal{A}_{(t-1)}$, $D \in \mathcal{A}_t$, $\delta(C \cap D|x)$ is the conditional probability on $C \cap D$ given $\mathcal{B}_{(t-1)}(C)$. In this sense $\delta(\cdot|\cdot)$ may properly be thought of as describing the conditional distribution of the present given "the observable past" (but not the future or unobservable past). To take a more precise point of view, it can be shown that if $F_\theta(\cdot|a)$ is independent of $a \in \mathcal{A}$ then $\delta(\cdot|\cdot)$ is indeed the conditional distribution on $\mathcal{A}$ given $\mathcal{B}$ of $\Delta_{\theta,\delta}$. However, when $F_\theta(\cdot|a)$ depends on a then $\delta(\cdot|\cdot)$ may not equal this conditional distribution of $\Delta_{\theta,\delta}$.

References

[1] Bellman, R. (1957). *Dynamic Programming*, Princeton University Press, Princeton, New Jersey.

[2] Blackwell, D. and Dubins, L. E. (1975). On existence and non-existence of proper, regular, conditional distributions. *Ann. Prob.* 3, 741-752.

[3] Blackwell, D. (1976). The stochastic processes of Borel gambling and dynamic programming. *Ann. Statist.* 4, 370-374.

[4] Brown, L. and Purves, R. (1973). Measurable selections of extrema. *Ann. Statist.* 1, 902-912.

[5] Chow, Y. S., Robbins, H., and Siegmund, D. (1971). *Great Expectations: The Theory of Optimal Stopping*. Houghton Mifflin Co., Boston.

[6] Dubins, L. E., and Savage, L. J. (1965). *How to Gamble if You Must.* McGraw-Hill Co., New York.

[7] Farrell, R. (1968), Towards a theory of generalized Bayes tests. *Ann. Math. Statist.* 39, 1-22.

[8] Le Cam, L. (1955). An extension of Wald's theory of statistical decision functions. *Ann. Math. Statist.* 26, 69-81.

[9] Lehmann, E. (1959). *Testing Statistical Hypotheses*. J. Wiley and Sons, Inc., New York.

[10] Meyer, P. A. (1966). *Probabilities and Potentials* Blaisdell Publishing Co., Waltham, Massachusetts.

[11] Neveu, J. (1965). *Mathematical Foundations of the Calculus of Probability*. Holden-Day, San Francisco.

[12] Portnoy, S. (1973). Notes on Decision Theory. Statistics Department, Harvard University.

[13] Schäl, M. (1975). On dynamic programming: Compactness of the space of policies. *Stochastic Processes and Their Applications* 3, 345-364.

[14] Stein, C. (1955). A necessary and sufficient condition for admissibility. *Ann. Math. Statist.* 26, 518-522.

[15] Wald, A. (1950). *Statistical Decision Functions*. J. Wiley and Sons, Inc., New York.

A SUBSET SELECTION PROBLEM EMPLOYING A NEW CRITERION

By Herman Chernoff and Joseph Yahav
Massachusetts Institute of Technology
Tel Aviv University and Columbia University

1. *Introduction and summary.* The extensive literature on Ranking and Selection procedures depends heavily on the use of two alternative criteria which are philosophically out of tune with the decision theoretic framework in which these papers are written. We propose a new criterion for selecting a subset of k populations and evaluate two subset selection procedures introduced respectively by Gupta [5] and by Desu and Sobel [2] in terms of this criterion. These results based on a Monte Carlo studies show that the Gupta procedure is highly efficient and are followed in Section 7 by a polemic criticizing the use of mathematically convenient criteria which implicitly imply peculiar and discontinuous loss functions. The criterion used in this paper is advanced in terms of a natural loss function for only one potential application and is not intended to be universally relevant. Rather, one would expect that specific applications would dictate related relevant criteria or loss functions.

Calculations are carried out in an invariant Bayesian framework using normal distributions and normal priors. Comparisons are based on Monte Carlo simulations for the case of $k = 8$ populations.

2. *Loss function.* Suppose that a farmer has a set of k varieties of treatments available for use on his land. He carries out an experiment on each treatment. The resulting data are used to select a subset of the available treatments. His land will be equally divided into portions to be treated by the elements of this subset and the ensuing crop determines the best one among the

*Prepared under Contract N00014-75-C-0555 (NR 042-331) for the Office of Naval Research.

treatments of this subset. This best treatment is applied to his land for the next several years.

Suppose $\mu_1, \mu_2, \ldots, \mu_k$ are the unknown means corresponding to the treatments. Let $K = \{1,2,\ldots,k\}$ and let J be a subset of K. The value associated with the subset J is

$$V(\mu,J) = \bar{\mu}(J) + r\tilde{\mu}(J) \tag{2.1}$$

where

$$\bar{\mu}(J) = \sum_{i\varepsilon J} \mu_i / \sum_{i\varepsilon J} 1 \tag{2.2}$$

is the average of the means associated with the subset J and represents the expected yield for the first crop and

$$\tilde{\mu}(J) = \max_{i\varepsilon J} \mu_i \tag{2.3}$$

is the maximum of those means and r is a positive constant. Here $\tilde{\mu}(J)$ could represent the expected yield for each of the next r years. Finally let the loss be given by

$$L(\mu,J) = (1 + r)\tilde{\mu}(K) - V(\mu,J) \tag{2.4}$$

which compares the value $V(\mu,J)$ with the best achieved by the farmer who correctly guesses the best treatment. Note that $L(\mu,J)$ is not affected if each μ_i is decreased by the same amount, i.e. $L(\mu,J)$ depends on $\mu = (\mu_1,\ldots,\mu_k)$ only through the differences $\mu_i - \mu_j$.

3. *Normal model.* We phrase the problem in terms of k normal populations with unknown means and common known variance. Each population is sampled equally often. The unknown means are assumed to have a joint normal prior distribution.

We incorporate in our prior distribution information about the order of magnitude of the differences among the means but we do not use information indicating any prior preference for one population over another.

Let the vector of population means be given by

$$\mu = \mu^{(1)} + \nu e$$

where $e' = (1,1,\ldots,1)$, ν is a scalar independent of $\mu^{(1)}$,

$$\mathcal{L}(\mu^{(1)}) = N(0,\sigma_0^2 I),$$

and

$$\mathcal{L}(\nu) = N(\nu_0,\sigma_\nu^2).$$

Then

$$\mathcal{L}(\mu) = N(\nu_0 e,\sigma_0^2 I+\sigma_\nu^2 U)$$

where

$$U = ee'.$$

The vector of sample means is given by

(3.1) $$X = \mu + \varepsilon$$

where

(3.2) $$\mathcal{L}(\varepsilon) = N(0,\sigma_\varepsilon^2 I).$$

Then the covariance matrix of X is given by

$$\sum_{XX} = (\sigma_0^2 + \sigma_\varepsilon^2)I + \sigma_\nu^2 U$$

which is easy to invert since

$$(I + bU)^{-1} = I - \frac{b}{1+kb} U.$$

The posterior distribution of μ is of the form

(3.3) $$\mathcal{L}(\mu|X) = N(\hat{\mu},\sum{}^*)$$

where $\hat{\mu}$, the Bayes estimate of μ is given by

(3.4) $$\hat{\mu} = \sum_{\mu X}\sum_{XX}^{-1} (X - \nu_0 e) + \nu_0 e$$

and

$$\sum{}^* = \sum_{\mu\mu} - \sum_{\mu X}\sum_{XX}^{-1}\sum_{X\mu}$$

can be expressed as $(\sigma_0^{-2} + \sigma_\varepsilon^{-2})^{-1} I$ plus a multiple of U. The stochastic behavior of $L(\mu,J)$ depends on the distribution of the differences of the components of μ and is not affected if a possible random multiple of e is subtracted from μ and a multiple of U is added to $\sum^*$. Moreover if the subset selection procedure $J(\hat{\mu})$ is invariant in the sense $J(\hat{\mu} + ae) = J(\hat{\mu})$, and the Bayes procedure is so invariant, then the conditional and unconditional stochastic behavior of $L(\mu,J(\hat{\mu}))$ is not affected if a possibly random multiple of e is also subtracted from $\hat{\mu}$. A detailed calculation of $\hat{\mu}$, $\sum_{\mu X}\sum_{XX}^{-1}\sum_{X\mu}$ and $\sum^*$ shows that the behavior of the losses for invariant subset selection procedures are the same as if

$$\mathcal{L}(\mu|X) = N(\hat{\mu}, \tau^2 I) \tag{3.5}$$

and

$$\mathcal{L}(\hat{\mu}) = N(0, \sigma^2\tau^2 I) \tag{3.6}$$

where

$$\tau^2 = (\sigma_0^{-2} + \sigma_\varepsilon^{-2})^{-1} \tag{3.7}$$

and

$$\sigma^2 = \sigma_0^2/\sigma_\varepsilon^2. \tag{3.8}$$

Note that ν_0 and σ_ν^2 do not appear in the above expressions.

The parametrization in terms of τ and σ were introduced partly because they were convenient for the Monte Carlo computations to be described subsequently. In those computations τ was normalized to be 1. For later reference we observe

$$\sigma_0^2 = \tau^2(1 + \sigma^2) \tag{3.9}$$

$$\sigma_\varepsilon^2 = \tau^2(1 + \sigma^{-2}) \tag{3.10}$$

and according to the above "assumptions" on the distributions of μ and $\hat{\mu}$,

$$\mathcal{L}(\mu) = N(0, \sigma_0^2 I). \tag{3.11}$$

It is sometimes convenient to recall that σ^2 represents the ratio of the amount of information in the sample yielding X to the information in the prior distribution.

4. *Computations.* To compare alternative approaches we calculate the conditional expected value and loss given $\hat{\mu}$. Suppose that $\hat{\mu}_1 \geq \hat{\mu}_2 \geq \ldots \geq \hat{\mu}_k$, and that $J = J_j(\hat{\mu}) = \{1,2,\ldots,j\}$. Then $\bar{\mu}(J) = \bar{\mu}_j = j^{-1}(\mu_1 + \ldots + \mu_j)$ and

(4.1) $$E(\bar{\mu}_j|X) = j^{-1}(\hat{\mu}_1+\ldots+\hat{\mu}_j).$$

The conditional distribution of $\tilde{\mu}_j = \tilde{\mu}(J)$ is

(4.2) $$F_j(y) = \prod_{i=1}^{j} \Phi\left(\frac{y-\hat{\mu}_i}{\tau}\right) = \prod_{i=1}^{j} \Phi_i(y)$$

(where Φ is the standard normal c.d.f., and $\Phi_i(y) = \Phi[(y-\hat{\mu}_i)/\tau]$ with expectation

$$E(\tilde{\mu}_j|X) = \int_{-\infty}^{\infty} y\, dF_j(y) = \int_0^{\infty} [1-F_j(y)]dy - \int_{-\infty}^{0} F_j(y)dy.$$

The relations

(4.3) $$E(\tilde{\mu}_{j+1}|X) = E(\tilde{\mu}_j|X) + \int_{-\infty}^{\infty} [F_j(y)-F_{j+1}(y)]dy$$
$$E(\tilde{\mu}_1|X) = \hat{\mu}_1$$

permit us to compute $E(\tilde{\mu}_j|X)$ as a function of $\hat{\mu}$ recursively. This computation uses the successive numerical integrations of $F_j(y) - F_{j+1}(y) = F_j(y)[1 - \Phi_{j+1}(y)]$. Combining (4.1) and (4.3) we obtain the conditional expectations of $V(\mu,J)$ and $L(\mu,J)$.

(4.4) $$E\{V(\mu,J)|X\} = E\{\bar{\mu}_j|X\} + r\, E\{\tilde{\mu}_j|X\}$$

and

(4.5) $$E\{L(\mu,J)|X\} = (1+r)E\{\tilde{\mu}_k|X\} - E\{\bar{\mu}_j|X\} - r\, E\{\tilde{\mu}_j|X\}.$$

It is also relatively easy to calculate the conditional probability that J contains the largest of the k populations. This is given by

(4.6) $$P_j(X) = P\{\tilde{\mu}_j = \tilde{\mu}_k|X\} = \sum_{i=1}^{j} \int_{-\infty}^{\infty} F_k(y)d[\log \Phi_i(y)]$$

5. *Comparison of strategies.* The Bayes strategy consists of selecting that $J = J_B$ which minimizes the conditional expectation of $L(\mu,J)$, or equivalently, which maximizes the conditional expectation of $V(\mu,J)$. The fixed subset size method of Desu and Sobel [2] consists of selecting as elements of $J = J_F(f)$ the

subscripts of the f largest sample means where f is a specified integer. In our normal Bayesian formulation this is equivalent to using the f largest $\hat{\mu}_i$. The Gupta fixed difference method [5] selects as elements of $J = J_D(d)$ the subscripts of all sample means within some specified distance of the largest sample mean. In our formulation this is the same as using all the $\hat{\mu}_i$ within some specified distance of the largest $\hat{\mu}_i$.

These three methods for subset selection produce conditional losses and values whose expectations can be computed by Monte Carlo methods and then compared. These comparisons depend on the parameters σ_0^2, σ_ε^2, r and k. Given these parameters for a specific problem one has a precise basis for comparison (assuming our model). This basis is somewhat unsatisfying if one desires a general overview, for the number of parameters is uncomfortably large. A second difficulty in seeking a general overview divorced from specific application with specified parameter values is the difficulty in selecting an appropriate measure for the relative efficiency of alternative procedures. There are several candidate measures but little pressing reason to prefer one over the others.

To be more specific, we introduce some notation. Let the subscripts B, F, D, and ∞ stand for the methods which use the Bayes solution, the fixed subset size method, the fixed difference procedure, and a hypothetical wizard whose guesses are infallible. Thus $V_F(f) = E[V(\mu,J)|\hat{\mu}]$ where $J = J_F(f)$ is the set of subscripts corresponding to the f largest $\hat{\mu}_i$, $V_B = \max_{1 \le f \le k} V_F(f)$, $V_D(d) = E[V(\mu,J)|\hat{\mu}]$ where $J = J_D(d)$ is the set of subscripts j for which $\max_i \hat{\mu}_i - \hat{\mu}_j \le d$, and $V_\infty = (1+r)\tilde{\mu}(K)$. Furthermore, $L_B = V_\infty - V_B$, $R_F(f) = V_B - V_F(f)$, and $R_D(d) = V_B - V_D(d)$. Finally, lower case letters represent expectations. For example $v_F(f) = EV_F(f)$, $\ell_B = v_\infty - v_b$, $r_D(d) = v_b - v_D(d)$, and $v_\infty = (1+r)E[\tilde{\mu}(K)] = (1+r)\sigma_0 e_k$ where e_k is the expectation of the maximum of k i.i.d N(0,1) random variables. The values v_B, $v_F(f)$, $v_D(d)$, v_∞ describe the alternate methods and their relations. To be favorable to the F and

D approaches we may compare $v_F = \max_{1 \leq f \leq k} v_F(f)$ and $v_D = \max_{d \geq 0} v_D(d)$ with v_B and v_∞.

Typically, a concept of relative efficiency designed to compare alternate approaches uses a natural "bench mark" as a standard to aim at. In this problem there are several possibilities. For example if we use v_∞ as a standard, we might compare F with B by the measure $(v_\infty - v_B)/(v_\infty - v_F) = \ell_B/(\ell_B + r_F)$. If we use the optimum v_B, as the standard the above measure would degenerate, but we could compare F with D using $(v_B - v_D)/(v_B - v_F) = r_D/r_F$ as an alternate to $(v_\infty - v_D)/(v_\infty - v_F) = (\ell_B + r_D)/(\ell_B + r_F)$. A third standard could be zero, in which case we might use v_F/v_D. However, zero seems to have little to recommend it as a goal in our context where the random choice of any single population would achieve that goal.

We propose to use a fourth measure of efficiency which has its roots in a conventional approach in many statistical problems where it is appropriate to compare methods of inference in terms of the ratio of the sample sizes required to obtain equally good inferences. The precision of the sample mean is measured by σ_ε^{-2}. If the sample size is increased by a factor, the precision σ_ε^{-2} will be increased by the same factor. Thus we inquire as to the extent to which an increase in σ_ε^{-2} offsets the lack of optimality of v_D. That is, by what factor must σ_ε^{-2} be increased from σ_ε^{*-2} so that $v_D = v_B^*$ where v_D corresponds to the new σ_ε^{-2} and v_B^* to σ_ε^{*-2}, the old value. The reciprocal of this factor represents the <u>relative efficiency</u> e_{DB} of the fixed difference to the Bayes procedure. A similar definition yields e_{FB}.

In principle these efficiencies depend on the parameters r, k, σ_0^2 and σ_ε^{-2}. However it is easy to see that for fixed r and k we may write v as a function of τ and σ and

(5.1) $$v_B = v_B(\tau,\sigma) = \tau v_B(1,\sigma) = \sigma_0 v_B(\sigma)$$

where $v_B(\sigma) = (1+\sigma^2)^{-1/2}\, v_B(1,\sigma)$. A similar notation may be applied to v, ℓ, and r and with subscripts D, F, ∞. In particular

$$v_\infty(\sigma) = (1+r)e_k. \tag{5.2}$$

Moreover σ^2 increases by the same factor as σ_ε^{-2} and n do when the sample size increases. Thus $e_{DB} = \sigma^{*2}/\sigma^2$ where $v_B(\sigma^*) = v_D(\sigma)$. For fixed r and k, e_{DB} depends on σ_0^2 and σ_ε^2 or on τ^2 and σ^2 only as a function of $\sigma^2 = \sigma_0^2/\sigma_\varepsilon^2$. The same applies to e_{FB}.

In applying the Monte Carlo calculations to estimate these efficiencies, an additional difficulty appears. The fixed difference method is highly efficient and the appropriate values of σ and σ^* would be close to each other and the graphs used to estimate these values would not be sufficiently refined to do the task directly. The following analysis makes graphical analysis more feasible.

$$\begin{aligned} v_B(\sigma) - v_D(\sigma) &= v_B(\sigma) - v_B(\sigma^*) \\ - r_D(\sigma) &= \ell_B(\sigma) - \ell_B(\sigma^*) \\ - r_D(\sigma) &\approx \frac{d\ell_B(\sigma)}{d(\log \sigma^2)} [\log \sigma^2 - \log(\sigma^*)^2] \end{aligned}$$

Thus the ratio of the sample sizes required for $v_D(\sigma) = v_B(\sigma^*)$ is given by

$$\frac{n^*}{n} = \frac{\sigma^{*2}}{\sigma^2} \approx e^{-\rho(\sigma) r_D(\sigma)}$$

where

$$\rho(\sigma) = -\left[\frac{d\ell_B(\sigma)}{d(\log \sigma^2)}\right]^{-1} \tag{5.3}$$

In summary we approximate the relative efficiencies of the D and F methods to the Bayes method by

$$e_{DB} = e^{-\rho(\sigma) r_D(\sigma)} \tag{5.4}$$

and

(5.5) $$e_{FB} = e^{-\rho(\sigma) r_F(\sigma)}$$

In the case of the F method, these relative efficiencies are often rather low and σ and σ^* are far apart. Then the primitive method based on drawing $v_F(\sigma)$ and $V_B(\sigma^*)$ (or equivalently $\ell_F(\sigma)$ and $\ell_B(\sigma^*)$) may be preferable to the above method which depends on a graphical estimate of $\rho(\sigma)$ and the assumption that σ is close to σ^*.

6. *Estimates based on Monte Carlo Calculations.* The basic parameters on which the operating characteristics of the various procedures mentioned depend are k, r, σ, and σ_ε or alternatively k, r, σ, τ. After normalization such as (5.1) we may reduce the effective parameters to k, r, and σ. Table 1 was compiled using Monte Carlo calculations and smoothing. This lists $v_\infty(\sigma) = (1+r)e_8 = 1.4236(1+r)$, and estimates of $\ell_B(\sigma) = v_\infty(\sigma) - v_B(\sigma)$, $r_F(\sigma) = v_B(\sigma) - v_F(\sigma) = \ell_F(\sigma) - \ell_B(\sigma)$, $r_D(\sigma) = v_B(\sigma) - v_D(\sigma) = \ell_D(\sigma) - \ell_B(\sigma)$, the probabilities p_B, p_F, p_D that the respective methods yield subsets containing the largest population, the optimal fixed subset size f_0, the optimal fixed difference d_0, π_{BD}, the probability that the Bayes and optimal fixed difference procedures yield the same subset, $\rho(\sigma)$, and the efficiencies e_{FB} and e_{DB}.

Note that the dependence on k and r has been suppressed in our notation and that all computations were carried out for k = 8. Also, d_0 is subject to the same notational equations as v_B, ℓ_B, r_F, etc., i.e. $d_0(\tau,\sigma) = d_0(1,\sigma) = \sigma_0 d_0(\sigma)$, while f_0, p_B, p_F, p_D, π_{BD}, $\rho(\sigma)$, e_{FB}, and e_{DB} are independent of τ. We present graphs of $\ell_B(\sigma)$, $r_F(\sigma)$, $r_D(\sigma)$, p_B, ρ, e_{FB}, e_{DB} and π_{BD} in Figures 1 to 8.

As a byproduct of the Monte Carlo Computation we present a decomposition of the (σ,r) space into regions in which f_0 assumes specified values (Figure 9).

The computations indicate that $d_0(1,\sigma)$ is extremely insensitive to σ and can be closely approximated (over the range considered) by a linear function of log r. Moreover for small values of r, $d_0(1,\sigma)$ seems to behave like a multiple of log (1+r). (See Figures 10a, 10b).

Given a problem which can be formulated as ours, the Bayes procedure is optimal and other procedures are at a disadvantage. The criticism we shall pose of the D and F methods may be countered by criticism of the model we proposed. Even assuming a Bayesian and normal framework, as well as the special loss structure we formulated, the Bayes procedure should be checked for robustness before it is seriously considered. A limited number of calculations were carried out to evaluate how sensitive the Bayes procedure is to incorrect specification of the parameters r, σ_0 and σ_ε. In the first case let r be the correct value and r* be the incorrectly specified value. The efficiencies of using the Bayes procedure in these cases is tabulated in Table 2 for a few values of r, σ and r*. Since $\mathcal{L}(\mu|\hat{\mu}) = N(\hat{\mu},\tau^2)$, incorrectly specifying σ for fixed τ has no effect on the procedure. On the other hand incorrectly specifying τ^* in place of τ has an effect depending on σ and τ^*/τ which is partly tabulated in Table 3.*

The estimates in this section were calculated with the intention of presenting a rough picture of how well the various methods perform and how they compare. Details concerning the Monte Carlo calculations appear in the appendix. The individual estimates presented in this section should not be considered highly reliable and were not intended to be used as precisely calculated operating characteristics. Rather the object was to serve as a basis for criticizing the philosophy in current vogue in Ranking and Selection.

7. *Polemics*. In the late 1940's it was understood that the hypothesis testing method of Analysis of Variance was not strictly appropriate in a subset of problems where one tested for the

---

*An Empirical Bayes approach could use the observed variance of the components of X to estimate $\sigma_0^2 + \sigma_\varepsilon^2$ and to modify the specification of τ based on prior belief.

equality of k means and had apriori reason to expect the means to be unequal. Paulson [7] posed the problem as the multiple decision problem of classifying the k unknown means into a superior and inferior group.

Bechhofer [1] considered, together with some generalizations, the problem of selecting the t largest of k means. In the case where it is desired to obtain the largest mean and where an equal number of observations is taken from each of k normal populations with common (known) variance, the appropriate (invariant) decision procedure is simply to select the population corresponding to the largest sample mean. The "problem" gains substance when one asks how large a sample size is appropriate. Bechhofer [1] proposed the indifference zone criterion. Given $\delta > 0$ and P^*, $0 < P^* < 1$, select the sample size so that the probability of correctly selecting the population with the largest mean exceeds P^* when the greatest mean exceeds all others by at least δ. This criterion has the mathematical beauty that the "least favorable" situation occurs when all population means except the largest are exactly δ units below the largest and it suffices to consider the sample size sufficient to treat that case.

From a decision-theoretic point of view this criterion implicitly assumes a discontinuous and rather unsatisfying cost function which assigns zero loss to the selection of a population whose mean is within δ of the largest and some constant loss r if the mean of the selected population is more than δ units away from the largest mean. The rather meager objective, leading to a prescription which selects one number, the sample size, leaves little scope for a criticism designed to show that the criterion is inappropriate. Any number obtained by an alternative approach could be equally well obtained by application of the indifference zone criterion with some suitably selected P^* and δ.

Greater scope for criticism is permitted when the problem is enlarged to select a subset of the populations. Here, two types of formulation are regularly considered. One, an indifference zone formulation, is the following: Given P*, δ and a subset size f, find the sample size for which the probability that the subset of populations corresponding to the f largest sample means contains the population with the largest (population) mean with probability at least P* when the largest mean exceeds all others by at least δ. (See Desu and Sobel [2]). A second formulation is to produce a decision rule for selecting a non-empty subset of the population of minimal expected sample size subject to the requirement that it contains the best population with probability at least P* no matter what are the values of the k means. (See Gupta [5]). In connection with this criterion, Gupta finds it relatively easy to evaluate rules of the form: Select the i-th population if $\bar{X}_i$ is within some constant d (depending on the sample size and variance σ^2) of the largest sample mean. The Desu-Sobel and Gupta formulations relate in obvious ways to the methods we labeled $J_F(f)$ and $J_D(d)$.

Both of these approaches implicitly assume discontinuous loss functions. In particular the Gupta version assumes that the cost of missing the best population does not depend on how superior it is to the ones obtained. It also assumes a cost proportional to the number of populations in the subgroup whereas it is not easy to visualize a problem with this cost characteristic.

We have an almost paradoxical situation. The rationale of Decision Theory, introduced as a generalization of current theories of inference lies in a pragmatic cost conscious view of statistical problems. In the literature on ranking and selection the approach to this basically practical problem is commonly regarded as a decision theoretic approach and yet it seems to have been so contaminated by the need to find a mathematically tractable or elegant solution that the cost structures tends to be left implicit. When made explicit, they are peculiar and discontinuous and are seldom if ever closely related to the real costs of any problem.*

This lack of reality in the cost structure casts great doubt on the relevance of the solutions in the related literature. To be fair, to cast doubts does not prove that the procedures are bad. While the authors ought to carry the burden of proof they generally decline this by evaluating how well the procedures satisfy the ad hoc criteria they formulated rather than by attempting to show how well they accomplish what is really desired in real problems. In fact, Desu and Sobel [2] claim that it is not possible to compare the F and D procedures but that claim ignores the possibility that both may be candidates for a given problem in which case both can be compared on how well they perform on that problem.

The burden of proof shifts somewhat to those who undertake to criticize. This shifted burden is a heavy one for the indifference zone approach leaves little scope for criticism since it offers so little in inferential content. However our computations indicate rather clearly that the D procedure with the proper choice of d can be remarkably efficient for the Bayesian problem we posed. On the other hand the F procedure is generally quite inefficient. That the D procedure would do better than F is no surprise for it at least does the data the courtesy of using them in a meaningful way. The F procedure seems to have little justification unless circumstances outside the data determine f, in which case the Desu-Sobel paper serves only as a table to describe

*After completion of the manuscript, the authors' attention was directed to the work of Somerville [8] who formulated a similar loss function with a similar rationale for the problem of selecting the best of k populations. Somerville obtained the sample size which minimizes the maximum risk and also considered two stage sampling schemes for the case of three populations. A Bayesian version of the one stage problem of selecting the largest of k means as well as minimax and indifference approaches of this problem are considered by Dunnett [3]. Some recent papers (for example see Goel and Rubin [4]) have begun to attack ranking and selection problems in terms of more natural decision theoretic problems.

limited properties of a paralyzed procedure.

To return to the D procedure, its relatively high efficiency for our problem leads one to conjecture that it would be robust against a wide range of loss functions where the cost of increasing the size of the subset does not increase too rapidly. Moreover it can be readily adapted with slight modification to different priors so that invariance with respect to permutations of the population is not required in the prior. Heuristic considerations lead us to believe that for low values of k the efficiency of the D method is not altered much if the prior is not normal.

Although the D method seems to be quite good the literature fails to point this out effectively. Admittedly the present paper does little to add deeply to an understanding of why it should be so good and when it may fail.

The literature on ranking and selection is extensive. An overwhelming part of it consists of relatively minor generalizations of one sample and two sample theory, contributing little in theoretical insight and avoiding difficult questions. For example, the important question of how to use incoming data to sample the alternative populations selectively so as to concentrate on the more likely candidates receives little emphasis. Surprisingly little of the published work on this field directed toward a practical solution has shown up in applications. Instead there has been a flourishing use of Analysis of Variance, supplemented by the theory of simultaneous confidence intervals which bears a close resemblance to the D method.

One is tempted to explain the apparent lack of applications by the inappropriateness of the formulation. In view of the efficiency of the D method it may be even more relevant that the D method and the tables derived provide little help on how to proceed in a practical problem to select a good value of d for that problem. To be widely used a statistical recipe should be easy to interpret and to apply. We submit that a choice of P^* or d based on an intuitive feeling of what is appropriate is likely to be highly inefficient. Our computations show that substantial changes in d lead to substantial inefficiencies. Nor is it likely that the user will appreciate the meaning of P* without understanding the consequences of missing the best population.

Admittedly the Bayes procedure developed for a special cost structure in this paper, and requiring numerical integrations, is hardly likely to become a widely applied statistical recipe. On the other hand we may ask how the relatively efficient D method may be made available to potential users. A potential difficulty is the large number of relevant parameters involved. These are k, the sample size n, σ_0^2, σ_ε^2 and r (in our formulation), d and δ (in the Gupta formulation). We suggest a careful tabulation of

$$E\,\bar{\mu}_{J_D},\ E\,\tilde{\mu}_{J_D},\ EP_{J_D},\ E\,\bar{\mu}^2_{J_D},\ E\,\tilde{\mu}^2_{J_D}$$

as functions of k, σ^2, and d would form useful building blocks on which to construct reasonable invariant procedures.

In summary, our polemic has fizzled somewhat. The proper use of the D method can be quite efficient. Nevertheless, a reasonable use of it or any sensible procedure requires some knowledge of the cost structure of the problem and would be facilitated by tabulating appropriate operating characteristics. The F method has little to recommend it outside of the possibility of dire necessity. The variability of p_F in our tables suggests that setting P* by intuition is likely to lead to inefficiencies in any reasonable problem.

APPENDIX

Details of Monte Carlo Calculations. Closed form calculations seem to be neither readily attainable nor likely to be very instructive. Thus Monte Carlo calculations were applied to furnish the tables and graphs of Section 6.

The basic parameters on which the operating characteristics of the procedures depend are k, r, σ_0 and σ_ε, or alternatively k, r, σ, τ. One hundred simulations were carried out for each of

100 parameter sets where $k = 8$, $\tau = 1$, $\sigma = (1.5)^{i-4}$ with $i = 1$ to 10 and $r = 2^{j-2}$ with $j = 1$ to 10. Each simulation involved the generation of a random vector $\hat{\mu}$ with the $N(0,\sigma^2 I)$ distribution and the ensuing computations by numerical integration (using the integrals described in Section 4) of the conditional expectations $E(\bar{\mu}_j|X)$, $E(\tilde{\mu}_j|X)$ and $P_j(X)$, $j = 1,2,\ldots,8$. Introducing the value of r, one may compute the conditional values of $V_F(f)$, $f=1,2,\ldots,8$, V_B, V_∞, and $V_D(d)$ for various values of d. The averages of 100 such simulations provide estimates of $E(\bar{\mu}_j)$, $E(\tilde{\mu}_j)$, the probability of correct selection using $J_F(f)$, $f = 1,2,\ldots,8$; $v_F(\sigma)$, $v_B(\sigma)$, $v_\infty(\sigma)$ (which is known to be $(1+r)e_8$), $v_D(\sigma)$, p_F, p_B and p_D as well as d_0 and f_0, the optimal d and f. In addition we estimate π_{BD} by the proportion of times that $J_B = J_D(d_0)$, which means that the Bayes and best fixed difference procedure coincide.

The above estimates of v_∞, v_B, v_F and v_D are used to estimate $\ell_B = v_\infty - v_B$, $r_F = v_B - v_F$ and $r_D = v_B - v_D$.

A few points require some elaboration here.

1. The optimal value f_0 of f may be computed by maximizing the estimate of $EV_F(f)$ with respect to f. Alternatively it is obvious that f_0 is nondecreasing in r, and $EV_F(f) > EV_F(f+1)$ if $E\{\bar{\mu}_f + r\tilde{\mu}_f\} > E(\bar{\mu}_{f+1} + r\tilde{\mu}_{f+1})$. Thus f is preferred to $f + 1$ if and only if $r < E(\bar{\mu}_f - \bar{\mu}_{f+1})/E(\tilde{\mu}_{f+1} - \tilde{\mu}_f)$ which is directly estimated in terms of our averages. Thus we estimate for each Σ those values of r for which each possible value of f_0 is appropriate (see figure 9).

2. The random values of $\hat{\mu}$ generated for a given value of σ and r are reused in the calculations of the V and v values for each of the other values of r. Thus there is an element of dependence among the various quantities estimated for a given σ with different values of r.

3. The quantity ℓ_B is generally small compared to v_B and v_∞. Two alternatives for estimating ℓ_B are to take the difference between the two estimates of v_B and v_∞ or to subtract the estimate

of v_B from the known value of v_∞. It was judged that the version of the first, which constitutes an average of the small quantities $v_\infty - v_B$, each the difference of two highly correlated variables, was preferable to the second. The latter involves the difference of two large terms, one of which is highly variable and the other is known. Then the final estimate of v_B used was the known value of v_∞ minus the estimate of ℓ_B. The same principle gives estimates of r_F and r_D as average values of $V_B - V_F(f_0)$ and $V_B - V_D(d_0)$.

4. The optimal value d_0 of d was computed by maximizing $v_D(d)$ using a somewhat unreliable iterative procedure. Fortunately, the results of preliminary calculations had indicated how d_0 depends on r and these results were used to obtain good initial approximations to d_0. As a result there were very few cases where the iterative procedure, which carried some diagnostic warnings, failed.

5. The value of $\rho(\sigma)$ involves the derivative $d\ell_B(\sigma)/d(\log \sigma^2)$. It was proposed to calculate these derivatives by drawing ℓ_B as a function of $\log \sigma$ and graphically estimating $d\ell_B(\sigma)/d(\log \sigma)$. However, for a major portion of the (r,σ) values considered, the appropriate procedures often involve subsets of size 1 or 2. In those limiting cases where this is to be expected, namely r small or σ large, we find that the Bayes expected loss is approximated by

$$\ell_B(\tau,\sigma) \approx (1+r)e_k(\sigma_0 - \tau\sigma) = (1+r)e_k\ \tau[\sqrt{1+\sigma^2} - \sigma]$$

(Incidentally, for large r, $\ell_B(\tau,\sigma) \approx e_k\sigma_0$, and $\ell_B(\sigma) = e_k$.) Thus, in the interest of the refined comparison of the graphs it was felt expedient to graph $\log \ell_{B1}(\sigma)$ versus $\log \sigma$ where

$$\ell_{B1}(\sigma) = \frac{\ell_B(1,\sigma)}{\sqrt{1+\sigma^2} - \sigma} = \ell_B(\sigma)\,\frac{\sqrt{1+\sigma^2}}{\sqrt{1+\sigma^2} - \sigma}$$

(Note that for large r, $\ell_{B1}(\sigma) \approx e_k\sqrt{1+\sigma^2}/[\sqrt{1+\sigma^2} - \sigma]$). Then

$$\frac{d\ell_B(\sigma)}{d \log \sigma} = \ell_B(\sigma) \frac{d(\log \ell_B)}{d(\log \sigma)} = \ell_{B1}(\sigma) \frac{[\sqrt{1+\sigma^2} - \sigma]}{\sqrt{1+\sigma^2}} \left[D_1 - \frac{\sigma(\sqrt{1+\sigma^2}+\sigma)}{1+\sigma^2}\right]$$

where

$$D_1 = \frac{d(\log \ell_{B1})}{d(\log \sigma)}$$

is estimated from the graphs. By this device we avoid the error in the graphical estimation of the major term adjacent to D_1 in the bracketed expression above.

6. As the authors carried it out, the graphical estimation of r_F, r_D, ℓ_B, etc., depended on a combination of several policies. First, log ℓ_{B1} was graphed in place of ℓ_B as described above so that some advantage could be taken of the known limiting behavior which constituted a major part of the quantities being estimated. The authors felt free to make smooth freehand approximations taking advantage of known monotonicities.

Several additional computer runs were used. One used 100 simulations for each of the parameter values of $k = 8$, $\tau = 1$, $\sigma = 1.5^{i-3.5}$, $i = 1,\ldots,10$, $r = 2^{j-1.5}$, $j = 1,2,\ldots,10$. The values of σ and r used in this and in our main computation are equally spaced in logarithmic scales. The four estimated values of ℓ_{B1} corresponding to those (σ,r) values in this supplementary run next to that of a (σ,r) value in the original run were used to provide additional information on the value of ℓ_{B1} at the (σ,r) value in the original run. This policy was carried out less systematically for r_F and r_D. Because the variances were larger for larger σ, some additional supplementary runs were made for larger σ. These were rarely used because by accident the r values chosen were inconvenient. Finally, the main computer run was preceded by many preliminary computations which provided considerable (if sometimes spotty) information. These were used rarely except for the estimates of robustness under the specification bias which were included in Section 6. They also included evaluations of the alternative procedures when the true means were in arithmetic progression. The results of those calculations have not been systematically collected and analyzed.

The study of r_D could be improved by the technique to be described. As it stands now r_D is clearly much smaller than r_F. If more precision were desired one could use larger sample sizes. Alternatively we note that $V_B = V_D$ with probability π_{BD} which is usually very high. Our current estimates of r_D are essentially based on the very few cases (typically less than 10) where $V_B \neq V_D$. If we observe that

$$r_D = (1 - \pi_{BD})E[V_B - V_D | V_B \neq V_D],$$

we see that the two factors in the above expression could be estimated separately. Thus $1 - \pi_{BD}$ could be estimated using the frequency of $V_B > V_D$ at the (r,σ) point being studied and at nearby values using smoothing techniques. One would expect that such a smoothed estimate of $1 - \pi_{BD}$ and a corresponding smoothed estimate of $E(V_B - V_D | V_B \neq V_D)$ would lead to an improvement of the estimate of r_D. It is also possible that the method of importance sampling in Monte Carlo simulations (see [6]) could help get more precise estimates.

All averages carried out in the computer simulations were accompanied by sample standard deviations. In the case of r_D these were typically based on a sample of 100 observations, most of which were zero, frequently yielding unreliable estimates.

References

[1] Bechhofer, R. E. (1954). A single sample multiple decision procedure for ranking means of normal populations with known variances. *Ann. Math. Statist.* 25, 16-39.

[2] Desu, M. M., and Sobel, M. (1968). A fixed subset-size approach to a selection problem. *Biometrika*, 55, 401-410.

[3] Dunnett, C. W. (1960). On selecting the largest of k normal population means (with discussion) *Journ. Royal Statist. Soc.* B, 22, 1-40.

[4] Goel, P. K. and Rubin, H. (1976). On selecting a subset containing the best population - a Bayesian approach. Purdue Department of Statistics Mimeo. Series, No. 432.

[5] Gupta, S. S. (1965). On some multiple decision (selection and ranking) rules. *Technometrics*, 7, 225-245.

[6] Hammersely, J. M. and Handscomb, D. C. (1964). *Monte Carlo Methods*. John Wiley and Sons, N. Y.

[7] Paulson, E. (1949). A multiple decision procedures for certain problems in the analysis of variance. *Ann. Math. Statist.* 20, 95-98.

[8] Somerville, P. N. (1954). Some problems of optimum sampling. *Biometrika* 41, 420-429.

Table 1: Operating Characteristics of the Bayes, the Optimal Fixed Subset Size and the Optimal Fixed Difference Methods.

The entries v_∞, ℓ_B, r_F and r_D are normalized according to the rule illustrated for r_D; $r_D = \sigma_0 r_D(\sigma)$. The entry $d_0(1,\sigma) = d_0/\tau$ where $\tau^2 = (\sigma_0^{-2} + \sigma_\varepsilon^{-2})^{-1}$ and $\sigma = \sigma_0/\sigma_\varepsilon$.

r	σ	$v_\infty(\sigma)$	$\ell_B(\sigma)$	$r_F(\sigma)$	$r_D(\sigma)$	p_B	p_F	p_D	f_o	$d_o(1,\sigma)$	π	ρ	e_{FB}	e_{DB}
.5	.30	2.14	1.2833	.0413	.0038	.68	.67	.67	4	.41	.73	5.9	.78	.98
	.44		1.0990	.0751	.0018	.60	.47	.59	2	.39	.85	3.6	.76	.99
	.67		.8666	.0672	.0011	.59	.57	.60	2	.41	.89	3.0	.80	1.00
	1.00		.5841	.0531	.0001	.62	.45	.62	1	.39	.98	3.1	.83	1.00
	1.50		.3519	.0364	.0002	.70	.56	.71	1	.40	.97	4.0	.75	1.00
	2.25		.1628	.0132	.0001	.79	.70	.79	1	.39	.99	6.6	.90	1.00
	3.38		.0856	.0077	.0000	.84	.77	.84	1	.39	1.00	12.9	.91	1.00
	5.06		.0439	.0033	.0000	.87	.83	.87	1	.37	.99	26.9	.92	1.00
	7.59		.0182	.0025	.0000	.94	.89	.94	1	.41	1.00	58.0	.89	1.00
	11.39	2.14	.0069	.0002	.0000	.95	.94	.95	1	.42	1.00	127.0	.88	1.00
1.0	.30	2.85	1.3586	.0267	.0083	.86	.87	.88	6	.67	.56	8.1	.80	.94
	.44		1.2146	.0672	.0070	.75	.74	.74	4	.61	.65	3.8	.80	.97
	.67		.9802	.0856	.0044	.72	.72	.72	3	.64	.77	2.7	.76	.99
	1.00		.6876	.1152	.0040	.70	.67	.68	2	.61	.83	2.7	.73	.99
	1.50		.4095	.1082	.0013	.79	.56	.79	1	.63	.89	3.5	.70	1.00
	2.25		.1931	.0415	.0004	.83	.70	.83	1	.63	.97	5.8	.77	1.00
	3.38		.1029	.0215	.0001	.86	.77	.86	1	.62	.99	11.1	.79	1.00
	5.06		.0530	.0099	.0001	.89	.83	.89	1	.61	.99	23.1	.81	1.00
	7.59		.0213	.0064	.0000	.95	.89	.95	1	.61	.99	49.3	.74	1.00
	11.39	2.85	.0084	.0010	.0000	.96	.94	.96	1	.60	1.00	107.3	.73	1.00
2.0	.30	4.27	1.3962	.0252	.0062	.96	.94	.96	7	.87	.72	25.3	.53	.85
	.44		1.3159	.0408	.0161	.90	.91	.92	6	1.01	.50	5.1	.81	.93
	.67		1.0972	.0915	.0147	.84	.82	.80	4	.80	.65	2.9	.77	.96
	1.00		.8074	.1243	.0065	.83	.81	.82	3	.95	.79	2.5	.71	.98
	1.50		.4752	.1523	.0029	.86	.79	.85	2	.91	.87	3.0	.58	.99
	2.25		.2356	.1163	.0004	.88	.70	.89	1	.97	.96	4.8	.55	1.00
	3.38		.1279	.0588	.0003	.91	.77	.91	1	.91	.98	9.1	.59	1.00
	5.06		.0665	.0279	.0002	.93	.83	.93	1	.89	.98	18.7	.59	1.00
	7.59		.0259	.0157	.0000	.96	.89	.96	1	.85	1.00	39.9	.51	1.00
	11.39	4.27	.0109	.0032	.0000	.97	.94	.97	1	.96	.99	86.7	.55	1.00
4.0	.30	7.12	1.4160	.0109	.0000	.99	.94	1.00	8	1.46	.75	72.4	.42	.69
	.44		1.3677	.0344	.0098	.97	.96	.97	7	1.27	.62	7.3	.78	.90
	.67		1.2008	.0889	.0162	.92	.95	.93	6	1.23	.52	3.3	.75	.95
	1.00		.9288	.1328	.0159	.90	.89	.88	4	1.24	.69	2.4	.71	.97
	1.50		.5580	.2161	.0146	.91	.90	.90	3	1.16	.87	2.7	.52	.99
	2.25		.2908	.2455	.0034	.92	.88	.91	2	1.22	.94	4.1	.38	.99
	3.38		.1582	.1529	.0023	.93	.77	.93	1	1.24	.95	7.5	.32	.99
	5.06		.0811	.0762	.0005	.96	.83	.96	1	1.21	.99	15.2	.31	1.00
	7.59		.0318	.0374	.0000	.97	.89	.97	1	1.19	.99	32.1	.31	1.00
	11.39	7.12	.0140	.0096	.0002	.98	.94	.98	1	1.21	.99	69.1	.54	1.00
8.0	.30	12.81	1.4251	.0018	.0026	1.00	1.00	1.00	8	1.46	.94	195.5	.46	.60
	.44		1.4005	.0283	.0112	.99	1.00	1.00	8	1.68	.71	12.4	.72	.88
	.67		1.2872	.0735	.0139	.97	.98	.98	7	1.56	.65	4.0	.74	.94
	1.00		1.0423	.1224	.0153	.95	.95	.95	5	1.58	.65	2.4	.72	.97
	1.50		.6519	.2452	.0101	.94	.90	.90	3	1.48	.79	2.4	.52	.98
	2.25		.3557	.3132	.0068	.95	.88	.88	2	1.71	.84	3.5	.35	.98
	3.38		.1946	.2147	.0016	.96	.94	.94	2	1.58	.94	6.3	.24	.99
	5.06		.0957	.1874	.0004	.97	.83	.83	1	1.54	.96	12.3	.15	1.00
	7.59		.0392	.0854	.0004	.98	.89	.89	1	1.53	.96	25.6	.14	1.00
	11.39	12.81	.0179	.0244	.0003	.99	.94	.94	1	1.41	.96	54.7	.30	1.00

Table 1 (continued)

r	σ	$v_\infty(\sigma)$	$\ell_B(\sigma)$	$r_F(\sigma)$	$r_D(\sigma)$	p_B	p_F	p_D	f_o	$d_o(1,\sigma)$	π	ρ	e_{FB}	e_{DB}
16.0	.30	24.20	1.4266	.0003	.0003	1.00	1.00	1.00	8	1.80	.99	297.2	1.00	.92
	.44		1.4130	.0158	.0046	1.00	1.00	1.00	8	1.90	.90	34.7	.57	.88
	.67		1.3388	.0734	.0158	.99	.98	.99	7	1.89	.71	5.2	.72	.93
	1.00		1.1220	.1288	.0241	.98	.98	.95	6	1.59	.59	2.9	.69	.93
	1.50		.7464	.2605	.0223	.97	.96	.97	4	1.90	.74	2.3	.55	.96
	2.25		.4258	.3372	.0088	.97	.95	.95	3	1.57	.75	3.1	.36	.97
	3.38		.2303	.2644	.0018	.98	.94	.97	2	1.82	.93	5.3	.19	.99
	5.06		.1107	.2564	.0005	.99	.96	.99	2	1.86	.96	10.1	.08	.99
	7.59		.0492	.1862	.0002	.99	.89	.98	1	1.81	.98	20.6	.02	1.00
	11.39	24.20	.0219	.0581	.0000	.99	.94	.99	1	1.89	1.00	44.1	.04	1.00
32.0	.30	46.98	1.4269	.0000	.0000	.99	1.00	1.00	8	2.11	1.00	623.3	1.00	1.00
	.44		1.4199	.0088	.0029	1.00	1.00	1.00	8	2.11	.94	49.4	.64	.88
	.67		1.3729	.0491	.0063	1.00	1.00	1.00	8	2.09	.85	7.4	.69	.95
	1.00		1.1857	.1369	.0233	1.00	.99	.99	7	2.17	.62	3.1	.69	.93
	1.50		.8355	.2574	.0242	.99	.98	.98	5	2.21	.70	2.2	.56	.95
	2.25		.4943	.3516	.0112	.99	.99	.98	4	2.23	.85	3.1	.36	.97
	3.38		.2689	.3410	.0094	.99	.98	.98	3	2.01	.82	4.6	.21	.98
	5.06		.1272	.3230	.0014	.99	.96	.99	2	2.16	.96	8.5	.06	.99
	7.59		.0625	.3092	.0006	.99	.98	.99	2	2.23	.98	17.5	.01	.99
	11.39	46.98	.0263	.1289	.0000	.99	.94	1.00	1	2.23	.99	37.7	.00	1.00
64.0	.30	92.53	1.4269	.0000	.0000	1.00	1.00	1.00	8	2.41	1.00	623.3	1.00	1.00
	.44		1.4247	.0041	.0006	1.00	1.00	1.00	8	2.33	.98	365.9	.23	.86
	.67		1.3940	.0280	.0094	1.00	1.00	1.00	8	2.43	.89	11.7	.70	.93
	1.00		1.2403	.1506	.0227	1.00	.99	1.00	7	2.46	.70	3.4	.68	.93
	1.50		.9146	.2545	.0219	.99	1.00	.99	6	2.49	.76	2.2	.58	.93
	2.25		.5573	.3688	.0188	.99	.99	.99	4	2.54	.76	2.9	.36	.95
	3.38		.3057	.3766	.0099	.99	.98	.99	3	2.49	.84	4.0	.22	.96
	5.06		.1460	.4185	.0041	1.00	.98	.99	3	2.38	.91	7.5	.05	.97
	7.59		.0750	.3650	.0044	1.00	.98	.99	2	2.41	.91	15.8	.00	.96
	11.39	92.53	.0311	.2542	.0025	1.00	1.00	1.00	2	2.27	.95	32.9	.00	.99
128.0	.30	183.64	1.4269	.0000	.0000	1.00	1.00	1.00	8	2.71	1.00	623.3	1.00	1.00
	.44		1.4265	.0022	.0022	1.00	1.00	1.00	8	2.78	.98	1022.1	.36	1.00
	.67		1.4054	.0167	.0042	1.00	1.00	1.00	8	2.51	.91	23.4	.66	.91
	1.00		1.2845	.1091	.0191	1.00	1.00	1.00	8	2.80	.74	4.2	.65	.92
	1.50		.9822	.2521	.0387	1.00	1.00	1.00	6	2.79	.73	2.2	.58	.92
	2.25		.6254	.3993	.0254	1.00	1.00	.99	5	2.76	.77	2.8	.35	.94
	3.38		.3480	.4591	.0049	1.00	1.00	1.00	4	2.79	.89	3.6	.20	.96
	5.06		.1679	.4738	.0039	1.00	.99	1.00	3	2.74	.90	6.6	.05	.96
	7.59		.0849	.4708	.0038	1.00	1.00	1.00	3	2.60	.94	13.7	.00	.95
	11.39	183.64	.0365	.2618	.0000	1.00	1.00	1.00	2	2.76	1.00	29.2	.00	.98
256.0	.30	365.86	1.4269	.0000	.0003	1.00	1.00	1.00	8	3.01	1.00	623.3	1.00	1.00
	.44		1.4277	.0010	.0000	1.00	1.00	1.00	8	3.01	.99	1022.1	1.00	1.00
	.67		1.4123	.0097	.0010	1.00	1.00	1.00	8	3.13	.94	59.4	.66	.79
	1.00		1.3128	.0809	.0044	1.00	1.00	1.00	8	2.95	.81	5.6	.61	.92
	1.50		1.0559	.2370	.0156	1.00	1.00	1.00	7	3.01	.68	2.3	.58	.91
	2.25		.6902	.4228	.0411	1.00	1.00	1.00	5	3.06	.71	2.8	.34	.90
	3.38		.3861	.5078	.0381	1.00	.99	1.00	4	3.00	.87	3.3	.23	.95
	5.06		.1893	.5482	.0158	1.00	1.00	1.00	4	3.02	.93	6.0	.06	.96
	7.59		.0951	.5106	.0040	1.00	.99	1.00	3	3.13	.99	12.2	.01	.95
	11.39	365.86	.0434	.2810	.0007	1.00	1.00	1.00	2	3.06	.95	25.3	.00	.97

Table 2: Sensitivity of Bayes Procedure to Incorrect Specification of r

Entries are efficiencies when r* is incorrectly specified in place of r.

$\sigma=1$

r \ r*	1.6	2.0	2.6
1.6		.98	
2.0	.98		.98
2.6		.98	

$\sigma=1.5$

r \ r*	3	4	6
3		.97	
4	.97		.97
6		.96	

$\sigma=2$

r \ r*	1.6	2.0	2.6
1.6		.99	
2.0	.99		.97
2.6		.96	

$\sigma=1$

r \ r*	3	4	6
3		.98	
4	.96		.97
6		.95	

$\sigma=1$

r \ r*	8	10	13
8		.99	
10	.99		.99
13		.99	

$\sigma=1.5$

r \ r*	7.5	10	15
7.5		.98	
10	.96		.97
15		.96	

$\sigma=2$

r \ r*	8	10	13
8		.98	
10	.98		.98
13		.97	

$\sigma=1$

r \ r*	7.5	10	15
7.5		.98	
10	.98		.97
15		.97	

Table 3: Sensitivity of Bayes Procedure to Incorrect Specification of τ

Entries are efficiencies when τ^* is incorrectly specified in place of τ.

$\tau = (\sigma_0^{-2} + \sigma_\varepsilon^{-2})^{-1}$ $\qquad$ $\sigma = \sigma_0/\sigma_\varepsilon$.

	τ^*/τ	efficiency		τ^*/τ	efficiency
r=2	.80	.91	r=10	.67	.60
σ=1	1.25	.91	σ=1	.76	.75
				.08	.89
r=2	.80	.92		1.25	.92
σ=2	1.25	.93		1.30	.85
				1.50	.66
r=4	.67	.69			
σ=1.5	.75	.78	r=10	.67	.57
	1.33	.80	σ=1.5	.75	.76
	1.50	.66		1.33	.77
				1.50	.58
r=4	.67	.81			
σ=3	.75	.84	r=10	.80	.80
	1.33	.86	σ=2	1.25	.82
	1.50	.76			

Figure 1: Expected Loss for Bayes Procedure

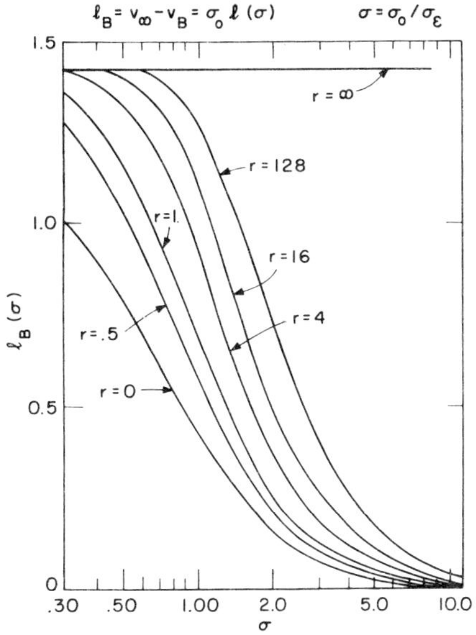

$\ell_B = v_\infty - v_B = \sigma_0 \ell(\sigma) \qquad \sigma = \sigma_0/\sigma_\varepsilon$

Figure 2: Probability of Correct Selection for Bayes Procedure

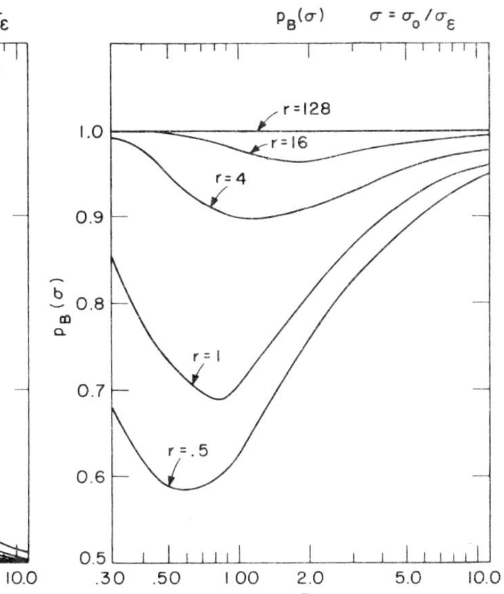

Figure 3: Risk for Optimal Fixed Subset Size Procedure

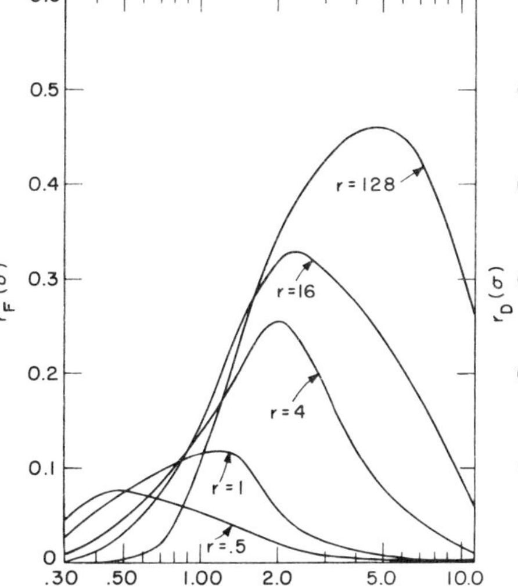

Figure 4: Risk for Optimal Fixed Difference Procedure

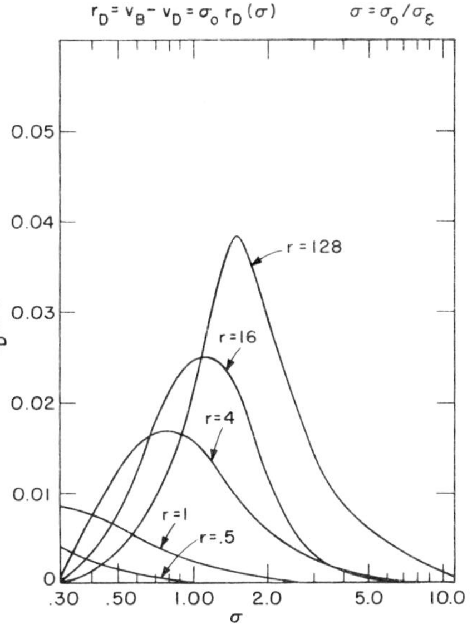

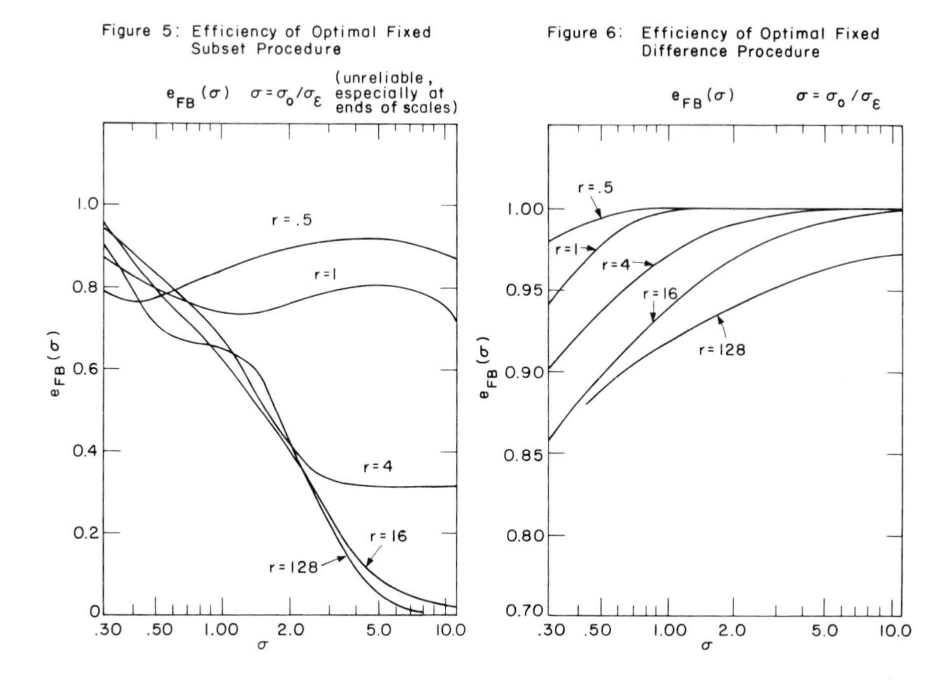
Figure 5: Efficiency of Optimal Fixed Subset Procedure
$e_{FB}(\sigma)$ $\sigma = \sigma_0/\sigma_\varepsilon$ (unreliable, especially at ends of scales)
r = .5
r = 1
r = 4
r = 16
r = 128
$e_{FB}(\sigma)$
σ
Figure 6: Efficiency of Optimal Fixed Difference Procedure
$e_{FB}(\sigma)$ $\sigma = \sigma_0/\sigma_\varepsilon$
r = .5
r = 1
r = 4
r = 16
r = 128
$e_{FB}(\sigma)$
σ

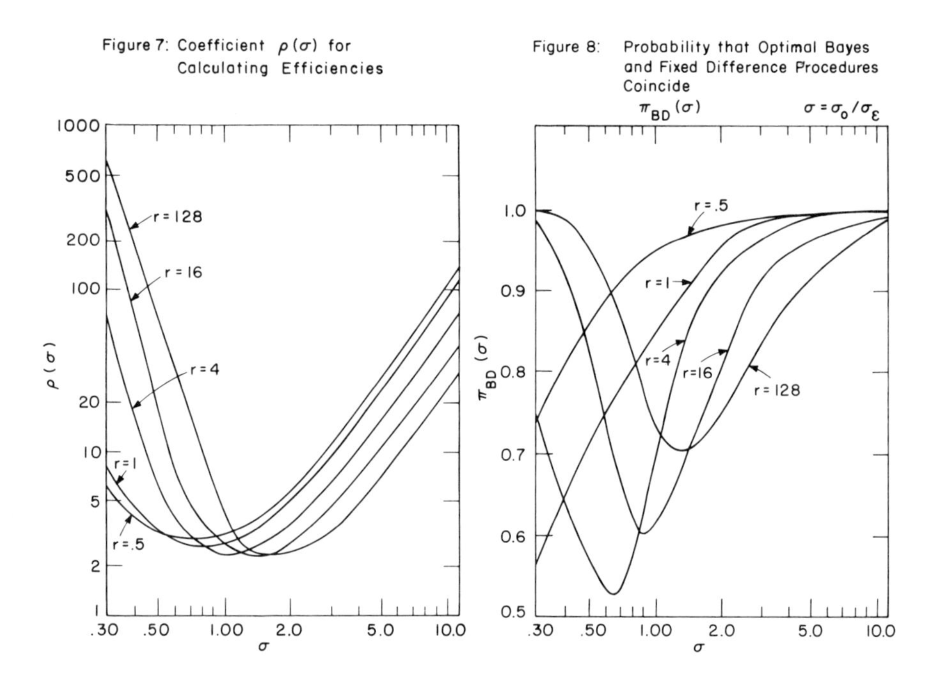
Figure 7: Coefficient $\rho(\sigma)$ for Calculating Efficiencies
r = 128
r = 16
r = 4
r = 1
r = .5
$\rho(\sigma)$
σ
Figure 8: Probability that Optimal Bayes and Fixed Difference Procedures Coincide
$\pi_{BD}(\sigma)$ $\sigma = \sigma_0/\sigma_\varepsilon$
r = .5
r = 1
r = 4
r = 16
r = 128
$\pi_{BD}(\sigma)$
σ

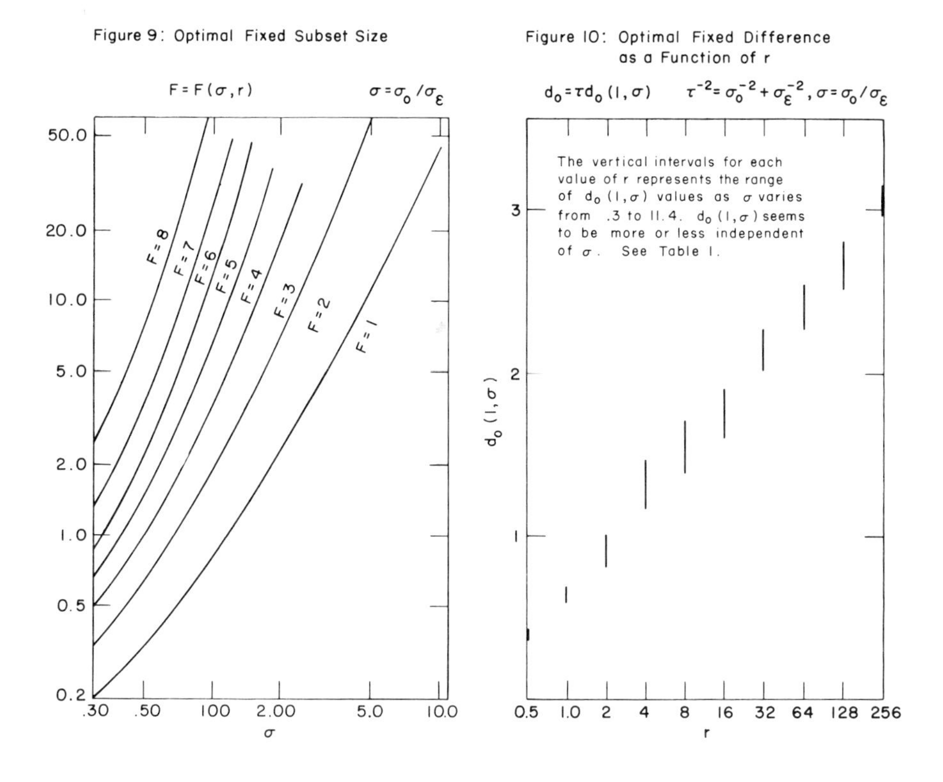
Figure 9: Optimal Fixed Subset Size
F = F(σ, r)
σ = σ₀/σ_ε
50.0
20.0
10.0
5.0
2.0
1.0
0.5
0.2
.30 .50 100 2.00 5.0 10.0
σ
F = 8
F = 7
F = 6
F = 5
F = 4
F = 3
F = 2
F = 1
Figure 10: Optimal Fixed Difference as a Function of r
d₀ = τd₀(1, σ)   τ⁻² = σ₀⁻² + σ_ε⁻², σ = σ₀/σ_ε
The vertical intervals for each value of r represents the range of d₀(1, σ) values as σ varies from .3 to 11.4. d₀(1, σ) seems to be more or less independent of σ. See Table 1.
d₀(1, σ)
3
2
1
0.5 1.0 2 4 8 16 32 64 128 256
r

EXAMPLES RELEVANT TO THE ROBUSTNESS OF APPLIED INFERENCES*

By A. P. Dempster
Harvard University

1. *Introduction.* The purpose of this paper is to give elementary examples of a kind of theoretical study which I believe merits extensive development. The basic idea is to simulate processes of inference as applied directly to data, where the data set (or perhaps class of data sets) is the unit of study, and one examines the variation in inferences implied by differing methods and models. Another approach to a similar objective is attempted in [4]. The present paper extends a viewpoint described in [3].

To set the stage, I describe three basic types of activity which applied statisticians engage in while interacting with data. These modes exclude major parts of statistical practice, for example, issues related to the quality of data or to the design of studies, but they are intended to cover the main processes of statistical inference in everyday use. The three activities are ordered in the sense that (ii) depends on (i), and (iii) depends on (i) and (ii), but in practice there is generally much backing and filling. The activities are:

(i) *Formulating models.* The elements here are (a) study of the background and context of a data set, aimed at uncovering the current state of scientific opinion on the natural and social systems underlying the data, especially as they relate to needs or goals to be served by the data analysis, (b) exploratory data analysis engineered to turn up signs or clues as to which of many possible statistical models are likely to be adequately descriptive, and (c) the detailed specification of a class of models. The activities of model formulation proceed informally and are relatively uncodified.

*Research supported in part by National Science Foundation Grant MCS75-01493.

(ii) *Testing models.* Having a model, or more typically a class of possible models, it is prudent to ask: do the data fit the model? Informal tests of fit are carried out by seeking meaningful differences between empirical quantities and their theoretical counterparts. When differences are noticeable but their size is of borderline significance, the need for formal testing tools becomes apparent. Difficult questions arise concerning choices of statistics designated to measure lack of fit and concerning roles for probability in the assessment of sizes of deviations. Ordinary tail area tests of significance are the most common type of test in practice. This paper will illustrate the use of likelihood as a measure of relative fit of hypotheses to data, in the spirit of Fisher [5, p. 68].

(iii) *Using models.* Once accepted, a model combines with data to produce assessments of the values of quantities which cannot be directly observed, such as parameters of physical or social systems, or risks and expectations associated with future events. These activities can be loosely described by the generic term estimation. Estimation may be carried out either for reasons of intrinsic scientific interest, or because choices among actions depend on the results. Bayesian inference and Bayesian decision analysis are fundamental tools here.

Probabilities are used quite differently in the three different types of activity. First, they appear as essential elements of mathematical models formulated to describe random phenomena, or more broadly as descriptors of a priori uncertainty in both observable phenomena and systems hypothesized to underlie phenomena. In the case of testing fit of models, it is necessary to judge whether the observed data imply the occurrence of a too improbable outcome, as defined by a small tail area, or a small relative likelihood, under a hypothesized a priori model. For purposes of estimation and decision-making, the canonical tools are the computation of posterior distributions and posterior expected risks, whether one follows Wald's theory to a choice among admissible procedures, or follows the direct subjectivist approach of interpreting posterior distributions and posterior expected costs and benefits as reflections of the decision maker's uncertainties and utilities.

It has become traditional in mathematical statistics to adopt the model as a basic unit for research, in the sense that papers are typically concerned with the analysis of a class of data sets associated with a given model. It is surprising that this tradition has dominated even the subjectivist Bayesian literature. I am advocating here a shift toward theoretical studies which treat the observed data as fixed, as in a given application, and analyze the effects of passing many models over the fixed data set. One consequence of the advocated approach is a painful awareness of dependence on prior information, not only on what is traditionally called a prior distribution in Bayesian inference, but on all kinds of prior understanding of what is more or less likely to be true or important about the phenomena under study, and about the mechanisms which selected and recorded the data. In particular, as in the examples below, one is led to study the dependence of particular Bayesian inferences on soft a priori choices of sampling models.

2. *Bayesian robustness*. The concept of Bayesian robustness adopted here elaborates and extends the concept of inference robustness introduced by Box and Tiao [1]. The idea is that after going through the process described above of formulating, testing, and using a model to reach tentative Bayesian conclusions or choices of actions, one is then obliged to assess the softness of these end results implied by the softness of the model. Robustness is a property of a given analysis applied to a given real world situation comprised of background knowledge, a set of data, and the intended uses of the data. The practical aim is to obtain a sense of a degree of robustness rather than a simple black or white judgment of robust or nonrobust.

Checking robustness requires:

(A) Identifying an adequate range of models appropriate to the given situation,

(B) Carrying out reanalyses of the given data across the range of alternative models, and

(C) Judging the importance of variation in the results of the analyses (B).

Brief comments will be made on these points, taken in the order (C), (A), (B).

Bayesian decision analysis suggests an obvious criterion for judging robustness. The robustness of the analysis based on a model P against an alternative model P_a is measured by the difference between the posterior expected loss of the P_a-optimum decision and the posterior expected loss of the P-optimum decision, computed under the P_a model.[1] After computing the difference, it is necessary to judge its importance, since nonrobustness can be declared only if the difference is large enough to cause concern. Thus, the formal losses used in the analysis must have an assessable relationship with real world costs or utilities.

How to define an adequate range of appropriate models? The objective is to be broad enough to escape criticism from the proposer of yet another a priori plausible model. Fortunately, there are two mechanisms which serve to narrow the range. First, it will be widely agreed that models which fail sensible tests of

[1] I am implicitly assuming that robustness means inference robustness in the sense of robustness against reasonable changes in the probability model, assuming a fixed loss structure. Such robustness depends on the loss structure in the sense that, if the loss structure changes, but is still common across P and P_a, the judgment of the robustness of P relative to P_a may change. It would also be of interest to consider a concept of utility robustness where the loss structure is varied from L to L_a while P remains unchanged. Most generally, the robustness of (P,L) against the alternative (P_a,L_a) could be studied.

conformity with the data need not be considered. Here I part company with many Bayesians who pretend not to see the universally surrounding medium of goodness-of-fit testing, presumably because of its awkwardly unBayesian character. Second, models can be omitted if they are trivially different from models already included, in the sense that the consequences of using an omitted model do not differ importantly from the consequences of using a corresponding included model. Both restriction principles lead to the possibility of errors of the first or second kind. In the case of the second, or substitution, principle, an error of the first kind would mean the inclusion of a model which makes no essential difference in the robustness analysis. An example would be the study of models which discretize continuous models to allow for rounding errors in the data, except in circumstances where such concern for rounding error importantly changes the analysis. The more dangerous errors of the second kind occur when plausible models which do make a difference are omitted, whether through oversight or fatigue.

A valid Bayesian objection to significance tests is that they lead many statisticians to fail to take seriously alternative models which are not significantly different from an included model, thus laying themselves open to errors of the second kind on the substitution principle, especially when the data set is small so that important differences cannot be detected by significance testing. I agree that it is dangerous to use acceptance of a null hypothesis as grounds for omitting or rejecting alternative hypotheses. But I do believe it is valid to use tests, as suggested above, to reject models directly via tests of fit to data, even when there is serious nonrobustness of a nominated analysis against the models thus rejected.

A major inhibitor of robustness analyses is that they are difficult and expensive to perform, even after the difficult exercise of judgment required to formulate models, tests, and decision problems. Step (B) above implies the performance of many Bayesian analyses, any one of which is likely to be challenging. There is a large scope for the development of improved numerical and analytical techniques.

The ultimate goal of research on Bayesian robustness should be to classify applied situations so that a plausible prepackaged robustness analysis appropriate within each class will be available. I believe that only the faintest beginnings have been made on this task.

3. *Examples*. Two very small data sets are discussed, the size being dictated by the decision to use a hand calculator (HP-65). Unfortunately, the examples are divorced from real world connections, so that a priori plausibility of the models considered is largely conjectural. Furthermore, it is simply assumed that maximum likelihood estimates are optimal choices in an undescribed decision problem, and that judgments about the importance of the differences among estimates obtained under various models can be made directly by inspecting the differences. Despite these limitations, interesting insights can be obtained by passing a range of models through a prescribed data set.

Both examples refer to a small sample of units, which may be individuals or matched pairs, such that each unit has a pair of observations resulting from Treatment A and Treatment B. In the case of Example 1, the observations are continuous measurements, while in Example 2 they are 0 or 1 responses.

EXAMPLE 1. The data here, originally from Cushny and Peebles [2], were used by Student [6] in his famous paper on the t-distribution. The data consist of the ten numbers

0.0, 0.8, 1.0, 1.2, 1.3, 1.3, 1.4, 1.8, 2.4, 4.6

which represent extra hours of sleep obtained by ten subjects under one medication as opposed to another. The main concern is taken here to be the robustness of the sample mean against various failures of the normal sampling model. The concern is prompted by the fact that data exploration quickly points to the observation 4.6 as a possible "outlier."

Four families of parametric models are analyzed. Prior distributions for the parameters are not formally introduced, since the absence of background information renders the intelligent choice of priors meaningless, and since the objective is primarily to study the effects of varying traditional specifications of sampling models. The families are:

(i) *Scale outlier model.* 9 observations are drawn at random from a $N(\theta,\sigma^2)$ population, while 1 is drawn from a $N(\theta,k\sigma^2)$ population for some $k > 1$, where the prior probability is .1 that each of the 10 units is the outlier.

(ii) *Shift outlier model.* Same as (i), except that $N(\theta,k\sigma^2)$ is replaced by $N(\theta+a,\sigma^2)$ for $a \neq 0$.

(iii) *Symmetric long-tailed model.* A random sample of 10 units from a t-distributed population with location θ, scale σ, and degrees of freedom r.

(iv) *Nonsymmetric contamination model.* A random sample of 10 units from a mixture with proportions p and (1-p) of $N(\theta,\sigma^2)$ and $N(\theta+a,\sigma^2)$ populations.

Figure 1 shows maximum likelihood estimates for θ, based on model (i), which run from 1.58 which is the mean of 10 data points to 1.24 which is the mean of the smallest 9 data points. L(k) is the likelihood function maximized over θ and σ^2.[2] Interpreting L(k) as a measure of relative fit to the data of models with different values of k, one sees that the data do not speak strongly against any value of $k \geq 1$, not even the null value $k = 1$ which corresponds to no outlier. It is evident that the choice of a final weighted estimate $\hat{\theta}$ is quite sensitive to the choice of a

---

[2]Bayesians would strictly prefer to integrate out θ and σ^2, thus making use of the width of the full likelihood in the θ and σ^2 dimensions, but also introducing dependence of the prior density on θ and σ^2 given k. Since the conclusions are unlikely to change much, I have consistently eliminated nuisance parameters by maximization.

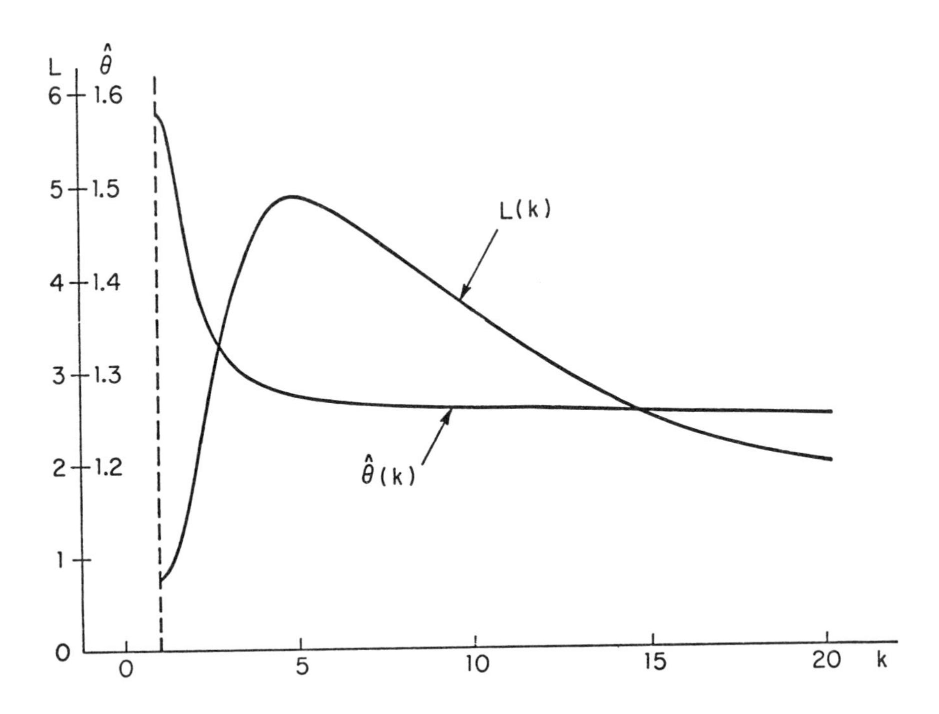

Figure 1. Estimated θ, *and likelihood, as a function of* k, *for Example 1, Model (i).*

prior distribution for k. If locating θ within the range (1.24, 1.56) is important, then the user has no option but to supply the requisite prior distribution for k.

Figure 2 is the analog of Figure 1 based on Model (ii) instead of Model (i). The estimated θ is no longer confined to the interval (1.24, 1.58) and draws attention to the obvious fact that users whose prior beliefs about a were far out, would adopt more extreme $\hat{\theta}$. For example, if my prior distribution for a were uniform on (.4, .5), then $\hat{\theta}$ = 1.14 is a reasonable estimate. The likelihood L(a) is much more strongly against the null value, about 40:1 for a = .34 vs. a = 0, than was the case in Figure 1. But this judgment reflects the assumption that the outlier has the

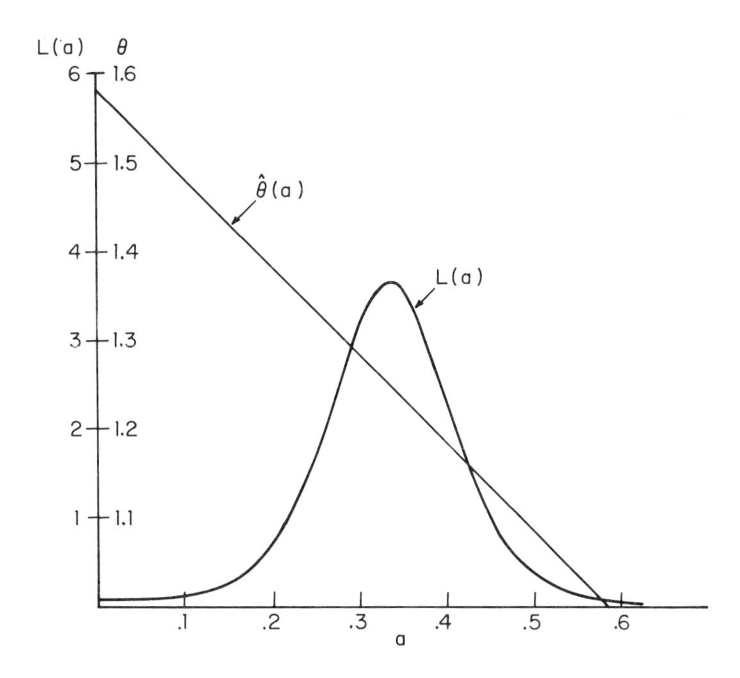

Figure 2. Estimated θ*, and likelihood, as a function of* a*, for Example 1, Model (ii).*

same variance σ^2 as the main population, and unless this assumption has a firm basis in fact, then L(a) is questionable as evidence that $a \neq 0$.

Figure 3 shows the result of fitting t-distributions by maximum likelihood for all values of the degrees of freedom r. Somewhere near r = 1, the results start to become pathological, and become increasingly so as $r \downarrow 0$. The standard t-distribution for r close to 0 has a very large spike near 0, and $\hat{\theta}(r)$ approaches 1.3 as $r \downarrow 0$ only because 1.3 is a repeated observation in the sample. If values of r close to zero have any credibility, then the joint prior distribution for r and σ is obviously heavily implicated in any sensible estimate for θ, and likewise, it becomes necessary to replace the maximized L(r) by a marginal density of

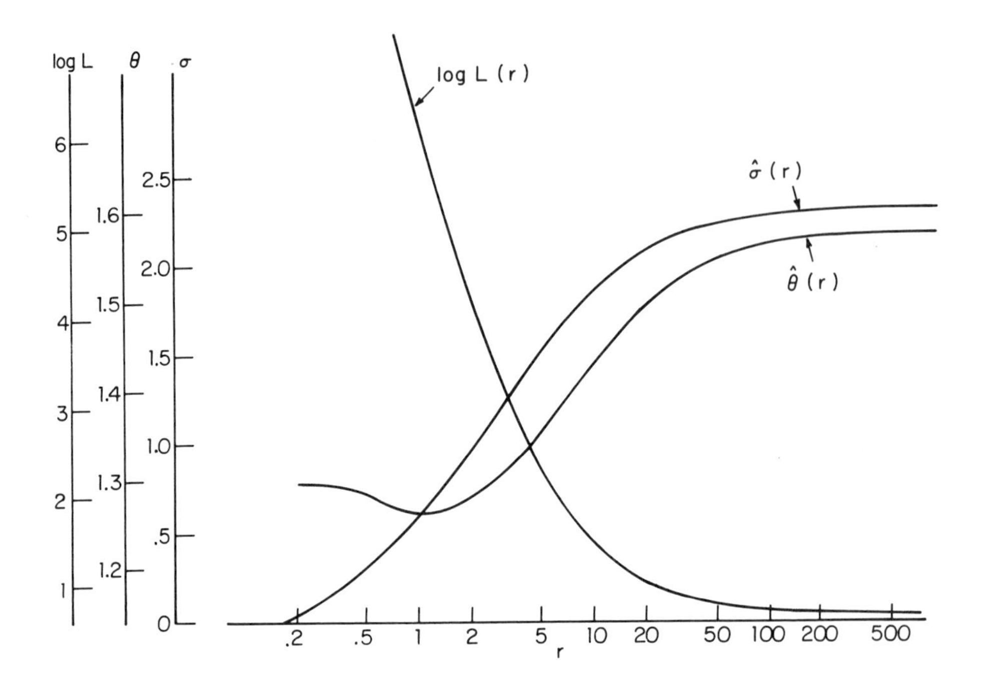

Figure 3. Estimated θ and σ, and log likelihood, as a function of r, for Example 1, Model (iii).

r if likelihood is to be interpreted as measuring concordance between the data and various values of r. Otherwise, for more reasonable values of r, the conclusions are similar to those based on Model (i), in the sense that the final choice of an estimate $\hat{\theta}$ on the range (1.26, 1.58) requires a genuine choice of prior distribution for r.

Model (iv) is a modification of Model (ii) where, instead of assuming that there is one outlier, the number of outliers (all from the same shifted normal) is binomially distributed with parameter p. Figure 4a shows the estimated mean θ of the uncontaminated part of the population for p = .001, .01, .1, .2, .4 and

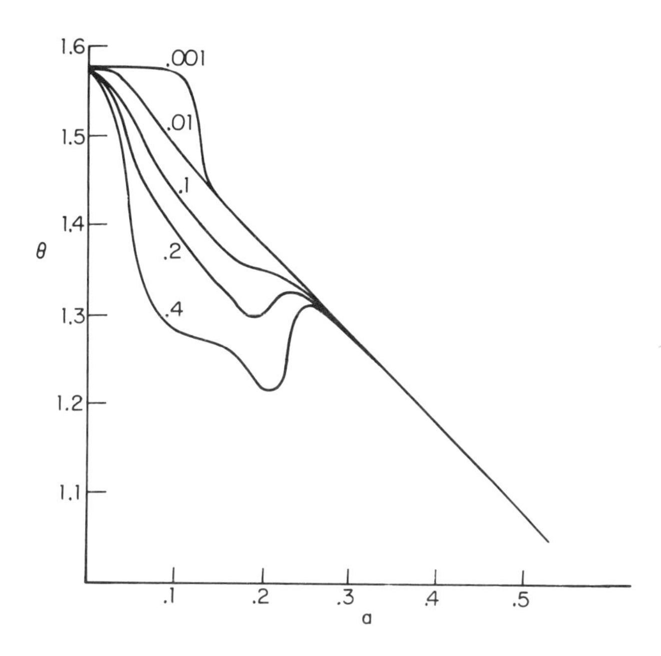

Figure 4a. Estimated θ, *as a function of* p *and* a, *for Example 1, Model (iv).*

for a range of a. The likelihoods exhibited in Figure 4b are broader and shallower than that in Figure 2, but still suggest that (.2, .5) is a plausible range for a, and remains so over a considerable range of p values.

If the objective is to estimate the population mean of the mixture, then the relevant estimates are those shown in Figure 4c rather than in Figure 4a. When $p = .1$, the sample mean continues to be very close to the maximum likelihood estimate, except possibly for a close to .1. However, if $p > .1$, then the sample mean should be increased, while if $p < .1$, the sample mean should be decreased. Since the given data are consistent with a range of values of p centered at .1, it appears that whether to raise or lower the sample mean is dependent essentially on prior judgment

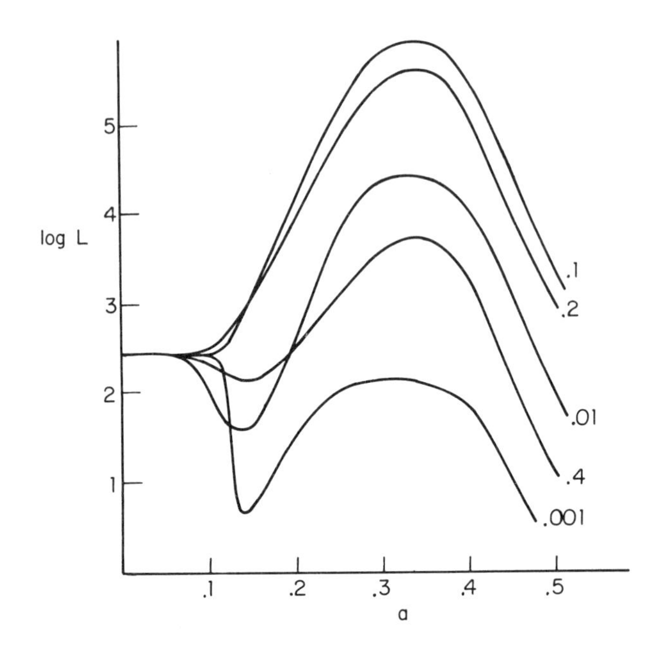

Figure 4b. Log likelihood, as a function of p *and* a, *for Example 1, Model (iv).*

of whether p is greater or less than .1. It is unlikely that Cushny and Peebles could have possessed prior knowledge on such an issue, had they chosen to regard Model (iv) as plausible, and hence one could conclude that the sample mean is unstable even as to direction of shift against Model (iv).

EXAMPLE 2. The data here are artificial counts representing 40 matched pairs, where each pair responds 0 or 1 (e.g., live or die) on each of treatments A and B:

	0 on B	1 on B
0 on A	11	13
1 on A	1	15

The result is a common 2×2 table of counts, with the extra feature that the diagonal counts 11 and 15 are sensibly called the

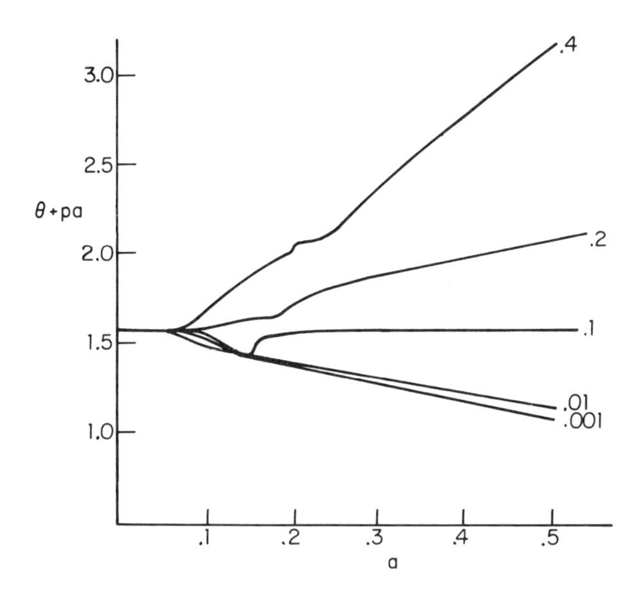

Figure 4c. Estimated θ+pa, *as a function of* p *and* a, *for Example 1, Model (iv).*

consistent pairs (both 0 or both 1) while the off-diagonal counts 1 and 13 are called inconsistent pairs. There is little exploration to be done with such simple data, beyond computing marginal totals (24 and 16 for A, and 12 and 28 for B) and noting that under independence the consistent pairs have an excess of 3.8 counts over expected while the inconsistent pairs have corresponding deficiencies of 3.8. The Yates corrected χ^2 for testing independence is 5.40, which indicates significance at approximately the .02 level.

To construct a model for the data, suppose that the ith pair is characterized by an unknown pair of "physical" probabilities (p_{iA}, p_{iB}), with the interpretation that, were those values known, then before the experiment the pair member to receive treatment A has probability p_{iA} of a result 1 and $1-p_{iA}$ of result 0, while the pair member to receive treatment B has independently the probabilities p_{iB} and $(1-p_{iB})$ of results 1 and 0. Suppose further that the 40 experimental pairs are randomly sampled from a large population of pairs, so that the unknown (p_{iA},p_{iB}) for i=1,2,..,40 are randomly drawn from a distribution over the unit square. It follows that the observed counts were sampled from a quadrinomial distribution with 2×2 cell probabilities

$$p_{00} = E\,(1-p_{iA})(1-p_{iB})$$

$$p_{01} = E\,(1-p_{iA})p_{iB}$$

$$p_{10} = E\,p_{iA}(1-p_{iB})$$

$$p_{11} = E\,p_{iA}p_{iB},$$

where p_{ij} denotes the population fraction of pairs scoring i on A and j on B, and E denotes expectation under the bivariate distribution of (p_{iA},p_{iB}). Similarly, it is easily shown[3] that

$$p_{00}p_{11} - p_{01}p_{10} = \mathrm{COV}(p_{iA},p_{iB}),$$

where COV is the covariance operator defined by E.

The experimental situation is arranged so that a positive correlation is expected between p_{iA} and p_{iB}, since this was the purpose of matching. Thus the expectation is that the product of counts for consistent pairs exceeds that for inconsistent pairs. A specific mathematical device often used to achieve the desired positive correlation is the logistic shift model which assumes that

$$\log \frac{p_{iB}}{1-p_{iB}} - \log \frac{p_{iA}}{1-p_{iA}} = \lambda,$$

where λ is the same for all pairs i.

To carry out a Bayesian analysis, it is necessary to go a further step and introduce completely specified distributions for (p_{iA},p_{iB}). My suggestion for a reasonable choice consistent with the logistic shift model is to make $\log \frac{p_{iA}}{1-p_{iA}}$ and $\log \frac{p_{iB}}{1-p_{iB}}$ linearly depend on a common logistic variable with different

[3]Personal communication from J. Cornfield.

locations and a common scale. More precisely, let μ_i be uniformly distributed on (0,1), and let

$$\log \frac{p_{iA}}{1-p_{iA}} = \alpha_A + \sigma \log \frac{\mu_i}{1-\mu_i}, \quad \text{and}$$

$$\log \frac{p_{iB}}{1-p_{iB}} = \alpha_B + \sigma \log \frac{\mu_i}{1-\mu_i}.$$

Note that $\alpha_B-\alpha_A = \lambda$, as defined above. The numerical results reported below adopted a 10 point uniform distribution over .05 (.1).95, for ease of calculation.

The likelihood of the data was computed for a grid of (α_A,α_B) points for each of $\sigma = 0, 2, 4, 6$. Curves of constant likelihood were then interpolated and plotted in Figure 5. For each value of σ, the (α_A,α_B) point of maximum likelihood was found, and two contours were plotted, one where the log likelihood is .5 less than the maximum, and one where the log likelihood is 2.0 less than the maximum. For large samples, these contours should be concentric ellipses with the second having radius double the first. Figure 5 shows that the asymptotic prediction is visually confirmed by the appearance of 4 pairs of roughly elliptical, similar and concentric curves, whose scale increases as σ moves through 0, 2, 4, 6. The values of the maximized log likelihood corresponding to $\sigma = 0, 2, 4, 6$ are -51.36, -47.28, -48.96, -50.88, also shown on Figure 5. In addition, the maximum likelihood estimates $\hat{\alpha}_A = -.832$, $\hat{\alpha}_B = 1.731$, $\hat{\sigma} = 1.684$, where the associated log likelihood is -47.21, were computed and the point located in Figure 5 on the curve of maximum likelihood estimates of α_A and α_B for given σ.

It is interesting that the maximized likelihood as a function of σ changes fairly slowly, so that even $\sigma = 6$ is not beyond all credibility. The estimated $\hat{\alpha}_A$, $\hat{\alpha}_B$ as well as their confidence contours grow roughly in proportion to σ. Figure 5 illustrates a little understood general property of logistic models, namely that as the number of terms in the multiple logistic grows, in our

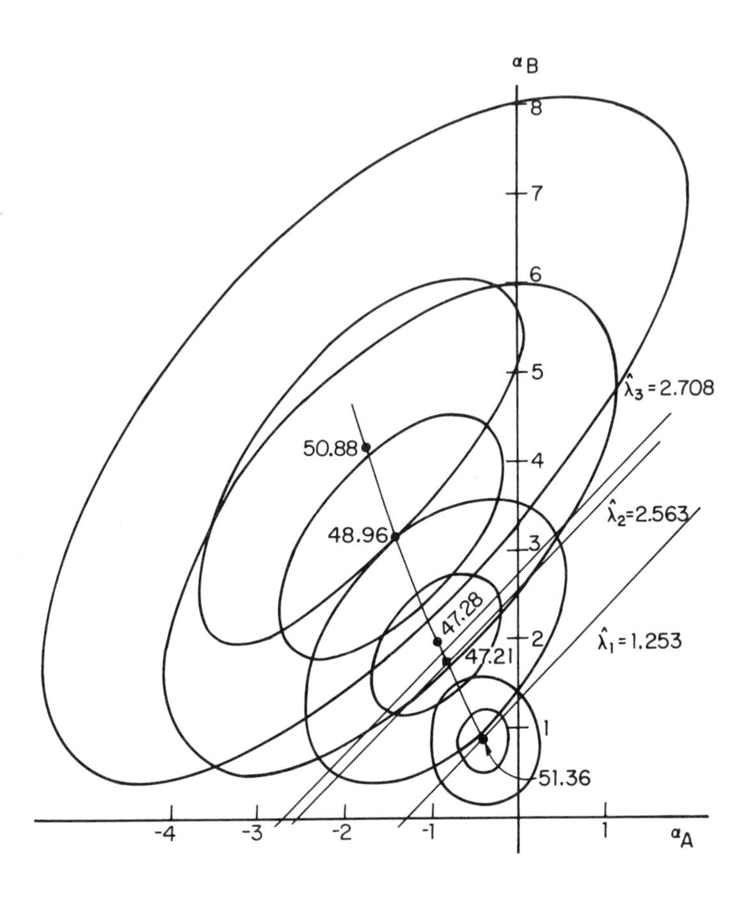

Figure 5. Likelihood contours and estimates of λ, for Example 2.

case by introducing random effects terms, the estimated logistic parameters also tend to grow. Thus, if logistic parameters are estimated for many competing risks, the estimated odds ratio multiplier for any one chosen risk factor may turn out to be quite large. There is a strong need for Bayesian robustness studies of the sensitivity of these estimated odds ratio multipliers to plausible failures of the logistic model.

Assuming that the logistic shift model holds, great interest attaches to estimation of the parameter λ. If it can be assumed that independence holds, i.e., $\sigma = 0$, then the maximum likelihood estimate is

$$\hat{\lambda}_1 = \log \frac{\hat{p}_{iB}}{1-\hat{p}_{iB}} - \log \frac{\hat{p}_{iA}}{1-\hat{p}_{iA}} = \log \frac{7}{3} - \log \frac{6}{4}$$

$$= 1.253.$$

When all 3 parameters are estimated by maximum likelihood, we find

$$\hat{\lambda}_2 = \hat{\alpha}_B - \hat{\alpha}_A = 1.731 - (-.832)$$

$$= 2.563.$$

Another widely used technique is to use the restricted likelihood based on the inconsistent matched pairs. Given that a pair responds either 0, 1 or 1, 0, the relative probabilities of these outcomes are $(1-p_{iA})p_{iB} : p_{iA}(1-p_{iB})$, or $e^{\lambda} : 1$. Given 14 such pairs, distributed in the ratio 13:1 the conditional likelihood $[e^{\lambda}/(1=e^{\lambda})]^{13}[1/(1+e^{\lambda})]$ can be maximized to yield

$$\hat{\lambda}_3 = \log 14 = 2.708.$$

Clearly, $\hat{\lambda}_1$ corresponds to a model which fits badly, while $\hat{\lambda}_2$ and $\hat{\lambda}_3$ differ little. Again, as in Example 1, the final choice of $\hat{\lambda}$ clearly depends more on the choice of prior distributions for α_A, α_B and σ and on the choice between maximum likelihood and conditional maximum likelihood.

The analyses just performed apply equally well to 2×2 tables which are a multiple of the given table, for example, 10 times or 100 times as large. The same λ estimates and the same likelihood contours would apply, the main change being that differences among log likelihood would be multiplied by the log of the multiplier. Accepting the model, the likelihoods for $\hat{\lambda}_2$ and $\hat{\lambda}_3$ would differ by less than .1 even if there were 4×10^5 matched pairs instead of 40. Robustness concerns in larger samples would focus much more on the validity of the sampling model, than on the conventional prior distribution.

4. *Concluding remark.* It will be relatively easy to carry out analyses like the foregoing for all 2×2 tables, since such

tables are characterized by only 3 characteristics, and important qualitative changes will come relatively slowly as the characteristics change. Most real data sets require many more characteristics for an adequate summary, so that an important problem will be to identify those characteristics which are most sensitive as indicators of nonrobustness. One principle might be to focus on the sensitivity to changes in data points on the outer edges of a sample. For example, examples like Example 1 should be repeated with "outliers" differing in number and extremity.

References

[1] Box, G. E. P. and G. C. Tiao (1964). A note on criterion robustness and inference robustness. *Biometrika* 51, 169-173.

[2] Cushny, A. R. and A. R. Peebles (1905). The action of optical isomers. II. Hyocines. *J. Physiology* 32, 501-510.

[3] Dempster, A. P. (1975). A subjectivist look at robustness. Technical Report S-33, Department of Statistics, Harvard University. (Invited paper, 40th Session, International Statistical Instiute, Warsaw, September 1975.)

[4] Dempster, A. P., Martin Schatzoff, and Nanny Wermuth (1977). A simulation study of alternatives to ordinary least squares. *J. Amer. Statist. Assoc.*, to appear.

[5] Fisher, R. A. (1958). *Statistical Methods and Scientific Inferences*. Hafner, New York.

[6] "Student" (1908). The probable error of a mean. *Biometrika* 6, 1-25.

ON SOME Γ-MINIMAX SELECTION AND MULTIPLE COMPARISON PROCEDURES*

By Shanti S. Gupta and Deng-Yuan Huang
Purdue University and Academia Sinica, Taipei, Taiwan

1. *Introduction.* Except in rare situations, information concerning the a priori distribution of a parameter is likely to be incomplete. Hence the use of a Bayes rule on some systematically produced choice for an a priori distribution, as advocated by the Bayesian school, is difficult to justify. This appears to be the case sometimes even if the a priori distribution is known fairly accurately (see Blum and Rosenblatt [1]). Robbins [14] has suggested that attention be paid to the case in which it is known only that the distribution of the parameter is a member of some given family of distributions. The use of partial or incomplete prior information in statistical inference has led to the development of the Γ-minimax criterion, a term initially employed by Blum and Rosenblatt [1]. Although it is often very difficult for the statistician to have the a priori information needed to completely determine the prior distribution of the parameter of interest, it is likely that he may have enough information to specify that the prior is a member of a subset Γ of the class of all priors. The Γ-minimax criterion requires one to select a decision rule that minimizes the maximum expected risk over Γ. Note that if Γ consists of a single prior, then the Γ-minimax decision rule is the Bayes rule for that prior. At the other extreme, if Γ consists of all priors then the Γ-minimax rule is the usual minimax rule. Some contributions have been made to develop decision rules which are hybrids between the Bayes and minimax criteria by Hodges and Lehmann [6], Menges [11], Blum and Rosenblatt [1], Jackson, O'Donover, Zimmer and Deely [8], Randles

*This research was supported by Office of Naval Research Contracts N00014-67-A-0226-0014 and N00014-75-C-0455 at Purdue University.

and Hollander [13], Solomon [16], [17], Watson [20], DeRouen and Mitchell [3], Stäel von Holstein [19], Gupta and Huang [4] and Huang [7].

In this paper, we are concerned with the problems of selection procedures for the "best" population and of multiple comparison procedures which are optimal in the sense of the Γ-minimax criterion. It should be pointed out that Deely and Gupta [2] discussed some Bayesian subset selection procedures using a different approach. Some minimax subset selection procedures have also been considered by Schaafsma [15].

2. *Selection procedures for the "best" populations.* Let $\Pi_1, \Pi_2,\ldots,\Pi_k$ be k populations with probability densities $f_{\theta_1}(\cdot), f_{\theta_2}(\cdot),\ldots,f_{\theta_k}(\cdot)$, respectively. We are given one observation from each population π_i, $1 \leq i \leq k$. The vector of observations $\underline{x} = (x_1,\ldots,x_k)$ is assumed to have density $f_{\underline{\theta}}(\underline{x})$ (with respect to some σ-finite measure μ). Let Ω be the whole parameter space. We use τ_{ij} as a measure of distance between Π_i and Π_j. Define $\tau_i = \max_{\substack{1\leq j\leq k \\ j\neq i}} \tau_{ij}$. Let $\Omega_i = \{\underline{\theta} | \tau_i \leq \tau_0\}$, $1 \leq i \leq k$, where τ_0 is a given constant. The parameter space Ω is partitioned into k+1 mutually exclusive subsets $\Omega = \Omega_0 \cup \Omega_1 \cup\ldots\cup \Omega_k$, where Ω_0 is an indifference zone ($\Omega_0 = \phi$ is permitted).

Let the function $\delta_1, \delta_2,\ldots,\delta_k$ be referred to as the individual selection probabilities. The population Π_i associated with $\tau_i = \min_{1\leq j\leq k} \tau_j$ is called the "best" and a selection of the best population is called a correct selection. If $\underline{\theta} \in \Omega_0$, the selection of any one population is correct.

2.1. *Γ-minimax selection procedures.* We define the loss function for $\delta = (\delta_1,\ldots,\delta_k)$ as follows:

$$L(\underline{\theta},\delta) = \sum_{i=1}^{k} \sum_{j=1}^{k} L^{(i)}(\underline{\theta},\delta_j)$$

where, for any i, $1 \leq i \leq k$, and $\underline{\theta} \in \Omega_i$, $L^{(r)}(\underline{\theta},\delta_j) = 0$ for all $r \neq i$,

$$L^{(i)}(\underline{\theta},\delta_j) = \begin{cases} w_{ij}\delta_j & \text{for } j \neq i, \\ 0 & \text{for } j = i, \end{cases}$$

and $w_{ij}(j \neq i)$ are given positive numbers and $w_{ii} = 0$. For $\underline{\theta} \in \Omega_0$, the loss is zero.

If $\rho(\underline{\theta})$ is a distribution over Ω, then the risk of a procedure is

$$r(\rho,\delta) = \int_{\Omega} \int_{E^k} L(\underline{\theta},\delta) f_{\underline{\theta}}(\underline{x}) d\mu(\underline{x}) d\rho(\underline{\theta})$$

$$= \sum_{i,j=1}^{k} \int_{\Omega_i} \int_{E^k} L^{(i)}(\underline{\theta},\delta_j) f_{\underline{\theta}}(\underline{x}) d\mu(\underline{x}) d\rho(\underline{\theta})$$

$$= \sum_{i=1}^{k} \int_{\Omega_i} \int_{E^k} \sum_{j=1}^{k} w_{ij}\delta_j f_{\underline{\theta}}(\underline{x}) d\mu(\underline{x}) d\rho(\underline{\theta}),$$

where μ is the Lebesgue measure defined on the k-dim. Euclidean sapce E^k. Assume that partial prior information is available in the selection problem, so that we are able to specify $\pi_i = P[\underline{\theta} \in \Omega_i]$, $\sum_{i=1}^{k} \pi_i \leq 1$. Define

$$\Gamma = \{\rho(\underline{\theta}) | \int_{\Omega_i} d\rho(\underline{\theta}) = \pi_i, \ i=1,\dots,k\}.$$

A decision rule δ^0 is called Γ-minimax if

$$\sup_{\rho \in \Gamma} r(\rho,\delta^0) = \inf_{\delta} \sup_{\rho \in \Gamma} r(\rho,\delta).$$

THEOREM 2.1. _For any_ i, $1 \leq i \leq k$, _let_ $\underline{\theta}_i^* \in \Omega_i$ _be such that_

$$(2.1)\qquad \sup_{\underset{-}{\theta}\in\Omega_i} \int_{E^k} \sum_{j=1}^{k} w_{ij}\delta_j^0 f_{\underset{-}{\theta}}(\underset{-}{x})d\mu(\underset{-}{x}) = \int_{E^k} \sum_{j=1}^{k} w_{ij}\delta_j^0 f_{\underset{-}{\theta}_i^*}(\underset{-}{x})d\mu(\underset{-}{x}),$$

where

$$\delta_i^0(\underset{-}{x}) = \begin{cases} 1 & \text{if } L_i(\underset{-}{x}) < \min\limits_{\substack{1\le j\le k\\ j\ne i}} L_j(\underset{-}{x}), \\ a_i & \qquad\qquad = \qquad\qquad , \\ 0 & \qquad\qquad > \qquad\qquad , \end{cases}$$

$L_i(\underset{-}{x}) = \sum_{j=1}^{k} w_{ji}\pi_j f_{\underset{-}{\theta}_j^*}(\underset{-}{x})$, $1 \le i \le k$, $\sum_{i=1}^{k} a_i = 1$. Then $\delta^0 = (\delta_1^0,\ldots,\delta_k^0)$ is a Γ-minimax procedure.

PROOF. For $\rho_0 \in \Gamma_0$, $\Gamma_0 = \{\rho(\underset{-}{\theta})|P[\underset{-}{\theta} = \underset{-}{\theta}_i^*] = \pi_i,\ 1 \le i \le k\}$ and for any decision rule δ,

$$\sup_{\rho\in\Gamma} r(\rho,\delta) \ge r(\rho_0,\delta)$$

$$= \sum_{i=1}^{k} \int \sum_{j=1}^{k} w_{ij}\pi_i\delta_j f_{\underset{-}{\theta}_i^*}(\underset{-}{x})d\mu(\underset{-}{x})$$

$$= \sum_{j=1}^{k} \int \delta_j \sum_{i=1}^{k} w_{ij}\pi_i f_{\underset{-}{\theta}_i^*}(\underset{-}{x})d\mu(\underset{-}{x})$$

$$= \sum_{j=1}^{k} \int \delta_j(\underset{-}{x})L_j(\underset{-}{x})d\mu(\underset{-}{x})$$

$$\ge \sum_{j=1}^{k} \int \delta_j^0(\underset{-}{x})L_j(\underset{-}{x})d\mu(\underset{-}{x}) \ge \sup_{\rho\in\Gamma} r(\rho,\delta^0).$$

This proof is complete.

Existence of $\underset{-}{\theta}_i^*$ *for the condition* (2.1). Let $\Pi_1, \Pi_2,\ldots,\Pi_k$ be k independent populations with densities $f_{\theta_1}(\cdot)$, $f_{\theta_2}(\cdot),\ldots,$ $f_{\theta_k}(\cdot)$, respectively. Let $f_{\theta_i}(x) = f(x-\theta_i)$ have monotone likelihood ratio in x, $1 \le i \le k$. Let $\tau_{ij} = \theta_j - \theta_i$ and $\tau_0 = -\Delta, \Delta > 0$.

Then $\Omega_i = \{\underline{\theta} | \theta_i \geq \max_{\substack{1 \leq j \leq k \\ j \neq i}} \theta_j + \Delta\}$, $1 \leq i \leq k$.

For any i, $1 \leq i \leq k$, we assume that $w_{ij} = w_i$, $1 \leq j \leq k$, $j \neq i$. Since $\sum_{j=1}^{k} \delta_j^0 = 1$ and $w_{ii} = 0$, hence $\sum_{j=1}^{k} w_{ij}\delta_j^0 f_{\underline{\theta}}(\underline{x}) = w_i (1 - \delta_i^0) f_{\underline{\theta}}(\underline{x})$. Thus the condition (2.1) is equivalent to

$$\inf_{\underline{\theta} \in \Omega_i} E_{\underline{\theta}} \delta_i^0 = E_{\underline{\theta}_i^*} \delta_i^0, \tag{2.2}$$

where $\underline{\theta}_i^* = (\theta_0^*, \ldots, \theta_0^*, \theta_0^* + \Delta, \theta_0^*, \ldots, \theta_0^*)$, θ_0^* fixed.

In order to prove (2.2), we proceed as follows:

$$L_i(\underline{x}) < \min_{j \neq i} L_j(\underline{x})$$

is equivalent to

$$w_j \pi_j \frac{f_{\theta_0^*+\Delta}(x_j)}{f_{\theta_0^*}(x_j)} < w_i \pi_i \frac{f_{\theta_0^*+\Delta}(x_i)}{f_{\theta_0^*}(x_i)}, \text{ for all } j \neq i. \tag{2.3}$$

Hence $\delta_i^0(\underline{x})$ is non-decreasing in x_i and non-increasing in x_j, for all $j \neq i$. Therefore

$$\inf_{\underline{\theta} \in \Omega_i} E_{\underline{\theta}} \delta_i^0 = \inf_{\substack{\theta_j = \theta^*, j \neq i \\ \theta_i = \theta^* + \Delta}} E_{\underline{\theta}} \delta_i^0.$$

To determine θ^*, let

$$f_{\underline{\theta}}(\underline{x}) = \prod_{j=1}^{k} \frac{\sqrt{n}}{\sqrt{2\pi}} e^{-\frac{n}{2}(\bar{x}_i - \theta_i)^2}.$$

For any i, $1 \leq i \leq k$, the condition (2.3) is equivalent to

$$\max_{j \neq i} \bar{x}_j + \frac{1}{n\Delta} \log \frac{w_j \pi_j}{w_i \pi_i} < \bar{x}_i.$$

Since for $\underline{\theta} \in \{\underline{\theta} | \theta_j = \theta^*, j \neq i, \theta_i = \theta^* + \Delta\}$,

$$E_{\underline{\theta}} \delta_i^0 = \int_{-\infty}^{\infty} \prod_{j \neq i} \Phi\left(x + \Delta - \frac{1}{n\Delta} \log \frac{w_j \pi_j}{w_i \pi_i}\right) d\Phi(x)$$

is independent of $\underline{\theta}$, hence we can choose any fixed θ^*. In the more general case of (2.3), let

$$S(c_{ij}) = \{\underline{x} \mid \frac{f_{\theta_0^*+\Delta}(x_j)}{f_{\theta_0^*}(x_j)} \Big/ \frac{f_{\theta_0^*+\Delta}(x_i)}{f_{\theta_0^*}(x_i)} < c_{ij}\}$$

for all $j \neq i$, and $S(c_i) = \bigcap_{j \neq i} S(c_{ij})$. Since $S(1) = \{\underline{x} \mid x_j < x_i\}$ and $S(c_{ij}) \supset S(c'_{ij})$ for $c_{ij} > c'_{ij}$, $\{\underline{x} \mid x_j < x_i + d_{ij}\} \supset \{\underline{x} \mid x_j < x_i + d'_{ij}\}$ for $d_{ij} > d'_{ij}$. Hence for any c_{ij} there is a unique d_{ij} such that

$$S(c_{ij}) = \{\underline{x} \mid x_j < x_i + d_{ij}\}.$$

Thus for any i,

$$\begin{aligned} S(c_i) &= \bigcap_{j \neq i} \{\underline{x} \mid x_j < x_i + d_{ij}\} \\ &= \{\underline{x} \mid x_j < x_i + d_{ij} \text{ for all } j \neq i\}. \end{aligned}$$

It follows that

$$\inf_{\substack{\theta_j = \theta^*, j \neq i \\ \theta_i = \theta^* + \Delta}} E_{\underline{\theta}} \delta_i^0 \text{ is independent of } \underline{\theta}.$$

2.2. *An application of Theorem* 2.1. We now apply Theorem 2.1 to the selection of "superior" populations.

As before, we use τ_{ij} as a measure of distance between Π_i and Π_j and define $\xi_i = \max_{1 \leq j \leq k} \tau_{ij}$. A population Π_i is said to be

superior (or good) if $\xi_i < \Delta$,

inferior (or bad) if $\xi_i \geq \Delta$,

where Δ is a given positive constant.

Our purpose is to construct an optimal procedure to seek out the true superiors while rejecting all inferior ones.

Let $\Omega(r_1, r_2) = \{\underline{\theta} \mid \tau_{ij} \geq \Delta$, for all $i \in r_1$ and some $j \in r_2\}$, where r_1 and r_2 are a partition of $\{1,2,\ldots,k\}$. Then

$$\Omega = \cup\{\Omega(r_1, r_2) \mid r_1 \cup r_2 = \{1,2,\ldots,k\}, \text{ and } r_1 \cap r_2 = \phi\}.$$

Let

$$L(\underline{\theta},\delta) = \sum_{r_1,r_2} \sum_{s_1,s_2} L^{(r_1,r_2)}(\underline{\theta},\delta_{(s_1,s_2)})$$

where

$$L^{(r_1,r_2)}(\underline{\theta},\delta_{(s_1,s_2)}) = \begin{cases} w_{(r_1,r_2),(s_1,s_2)}\delta_{(s_1,s_2)}, & \text{if } \underline{\theta} \in \Omega(r_1,r_2) \\ 0 & \text{otherwise,} \end{cases}$$

and $L^{(t_1,t_2)}(\underline{\theta},\delta_{(s_1,s_2)}) = 0 \;\; \forall\, \underline{\theta} \in \Omega(r_1,r_2)$ with $(t_1,t_2) \neq (r_1,r_2)$, and $w_{(r_1,r_2),(s_1,s_2)}$ $((r_1,r) \neq (s_1,s_2))$ are positive constants and $w_{(r_1,r_2),(r_1,r_2)} = 0$.

Let

$\Gamma = \{\rho(\underline{\theta}) \,|\, \int_{\Omega(r_1,r_2)} d\rho(\underline{\theta}) = \pi_{(r_1,r_2)}$, for all partition (r_1,r_2) of $\{1,2,\ldots,k\}\}$.

For any partition (r_1,r_2), let $\underline{\theta}^*_{(r_1,r_2)} \in \Omega(r_1,r_2)$ be such that

$$\sup_{\underline{\theta}\in\Omega(r_1,r_2)} \int_{E^k} \sum_{s_1,s_2} w_{(r_1,r_2),(s_1,s_2)}\delta^0_{(s_1,s_2)} f_{\underline{\theta}}(\underline{x})d\mu(\underline{x})$$

$$= \int_{E^k} \sum_{s_1,s_2} w_{(r_1,r_2),(s_1,s_2)}\delta^0_{(s_1,s_2)} f_{\underline{\theta}^*_{(r_1,r_2)}}(\underline{x})d\mu(\underline{x}),$$

where

$$\delta^0_{(r_1,r_2)}(\underline{x}) = \begin{cases} 1 & \text{if} \quad L_{(r_1,r_2)} < \min_{(s_1,s_2)\neq(r_1,r_2)} L_{(s_1,s_2)}, \\ b_{(r_1,r_2)} & \qquad\qquad = \qquad\qquad , \\ 0 & \qquad\qquad > \qquad\qquad , \end{cases}$$

and $L_{(s_1,s_2)}(\underline{x}) = \sum_{t_1,t_2} w_{(t_1,t_2),(s_1,s_2)}\pi_{(t_1,t_2)} f_{\underline{\theta}^*_{(t_1,t_2)}}(\underline{x})$, and

$\sum_{r_1,r_2} b_{(r_1,r_2)} = 1$. Then

$$\delta^0 = (\delta^0_{(r_1,r_2)} | r_1 \cup r_2 = \{1,2,\ldots,k\},\ r_1 \cap r_2 = \phi\})$$

is a Γ-minimax procedure. The following example illustrates the partitioning $\Omega(r_1,r_2)$. We consider the case of location parameter so that $f_{\theta_i}(x) = f(x-\theta_i)$, $1 \leq i \leq k$. Let $\tau_{ij} = \theta_j - \theta_i$. Since for any partition r_1 and r_2,

$$\{\tau_{ij} \geq \Delta \text{ for all } i \in r_1 \text{ and for some } j \in r_2\}$$
$$= \{\max_{j \in r_2} \theta_j \geq \max_{i \in r_1} \theta_i + \Delta\}$$
$$= \{\max_{1 \leq j \leq k} \theta_j \geq \max_{i \in r_1} \theta_i + \Delta\}$$

which means that all of θ_i, $i \in r_1$, are inferior populations.

2.3. *Subset selection procedures - Bayesian approach.* We are interested in constructing an optimal selection procedure in the Bayesian sense to select a subset containing the best population. Let $\Omega = \Omega_0 \cup \Omega_1 \cup \ldots \cup \Omega_k$, where $\Omega_i = \{\underline{\theta} | \theta_i > \max_{j \neq i} \theta_j\}$,, $1 \leq i \leq k$, and $\Omega_0 = \Omega - \bigcup_{j=1}^{k} \Omega_j$. The loss function associated with the selection rule $\delta = (\delta_1,\ldots,\delta_k)$ is as follows:

$$L(\underline{\theta},\delta) = \sum_{i=1}^{k} L^{(i)}(\underline{\theta},\delta_i)$$

where

$$L^{(i)}(\underline{\theta},\delta_i) = \begin{cases} L_1(1-\delta_i) & \text{if } \underline{\theta} \in \Omega_i, \\ L_2\delta_i & \text{if } \underline{\theta} \in \bar{\Omega}_i, \\ 0 & \text{otherwise,} \end{cases}$$

where $\bar{\Omega}_i = \bigcup_{\substack{j=1 \\ j \neq i}}^{k} \Omega_j$.

Assume Γ consists of one element $G(\underline{\theta})$ where G is completely known and has joint density $g(\underline{\theta})$. The risk function is

$$R(\underline{\theta},\delta) = \sum_{i=1}^{k} R^{(i)}(\underline{\theta},\delta_i),$$

where

$$R^{(i)}(\underline{\theta},\delta_i) = \int_{E^k} L^{(i)}(\underline{\theta},\delta_i) f_{\underline{\theta}}(\underline{x})d\mu(\underline{x}).$$

Thus $R(\underline{\theta},\delta) = L_1M_1+L_2M_2$, where M_1 (M_2) is the expected number of the best (non-best) population rejected (selected).

Let $f(\underline{\theta}|\underline{x})$ be the posterior distribution of $f_\theta(\underline{x})$, let $f(\underline{x}) = \int_\Omega f_{\underline{\theta}}(\underline{x})g(\underline{\theta})d\underline{\theta}$, and for any i, let $q_i(\underline{x}) = \int_{\Omega_i} f(\underline{\theta}|\underline{x})d\underline{\theta}$, $\bar{q}_i(\underline{x}) = \int_{\bar{\Omega}_i} f(\underline{\theta}|\underline{x})d\underline{\theta}$. Define

$$\delta_i^*(\underline{x}) = \begin{cases} 1 & \text{if } L_2\bar{q}_i(\underline{x}) < L_1q_i(\underline{x}), \\ d_i & \qquad = \qquad , \\ 0 & \qquad > \qquad . \end{cases}$$

We know that the risk is given by

$$r(G,\delta) = \int_\Omega \int_{E^k} L(\underline{\theta},\delta)f_{\underline{\theta}}(\underline{x})d\mu(\underline{x})dG(\underline{\theta}).$$

Hence

$$\begin{aligned} r(G,\delta) = & \sum_{i=1}^{k} \{\int_{E^k} L_1(1-\delta_i)q_i(\underline{x})f(\underline{x})d\mu(\underline{x}) \\ & + \int_{E^k} L_2\delta_i\bar{q}_i(\underline{x})f(\underline{x})d\mu(\underline{x})\} \\ \geq & \sum_{i=1}^{k} \{\int_{E^k} L_1(1-\delta_i^*)q_i(\underline{x})f(\underline{x})d\mu(\underline{x}) \\ & + \int_{E^k} L_2\delta_i^*\bar{q}_i(\underline{x})f(\underline{x})d\mu(\underline{x})\} = r(G,\delta^*). \end{aligned}$$

Thus $\delta^* = (\delta_1^*,\ldots,\delta_k^*)$ is a Bayes rule.

2.4. *Subset selection procedures for slippage problems.* Let $\Omega = \{\underline{\theta}|\theta_j=\theta,\ \forall\, j \neq i,\ \theta_i \geq \theta + \Delta$ for some $i,\ 1 \leq i \leq k\}$, $\Delta > 0$ and let $\Omega_1(i) = \{\underline{\theta}|\theta_i \geq \theta + \Delta,\ \theta_j = \theta\ \forall\, j \neq i\}$ and

$\Omega_2(i) = \bigcup_{\substack{j=1 \\ j \neq i}}^{k} \Omega_1(j)$. We consider the k-sample slippage problem and wish to select a subset containing the population which has "slipped" to the right of the rest by an amount at least Δ. The loss function is

$$L(\underline{\theta},\delta) = \sum_{i=1}^{k} L^{(i)}(\underline{\theta},\delta_i)$$

where

$$L^{(i)}(\underline{\theta},\delta_i) = \begin{cases} L_1(1-\delta_i) & \text{if } \underline{\theta} \in \Omega_1(i), \\ L_2\delta_i & \text{if } \underline{\theta} \in \Omega_2(i), \\ 0 & \text{otherwise} \end{cases} .$$

The risk function is

$$R(\underline{\theta},\delta) = \sum_{i=1}^{k} R^{(i)}(\underline{\theta},\delta_i),$$

where

$$R^{(i)}(\underline{\theta},\delta_i) = \int_{E^k} L^{(i)}(\underline{\theta},\delta_i(\underline{x}))f_{\underline{\theta}}(\underline{x})d\mu(\underline{x}).$$

If $\rho(\underline{\theta})$ is a distribution over Ω then the expected risk of a procedure is

$$r(\rho,\delta) = \sum_{i=1}^{k} r^{(i)}(\rho,\delta_i),$$

where

$$r^{(i)}(\rho,\delta_i) = \int_{\Omega} R^{(i)}(\underline{\theta},\delta_i)d\rho(\underline{\theta}).$$

Assume that some partial prior information is available in the selection problem, so that we are able to specify $p_i = P[\underline{\theta} \in \Omega_1(i)]$, $1 \leq i \leq k$ and $\sum_{i=1}^{k} p_i \leq 1$. Define

$$\Gamma = \{\rho(\underline{\theta}) \mid \int_{\Omega_1(i)} d\rho(\underline{\theta}) = p_i,\ 1 \leq i \leq k\}.$$

THEOREM 2.2. *For any* i, *let* $\underline{\theta}^*(i) \in \Omega_1(i)$ *and* $\underline{\theta}_*^i(j) \in \Omega_1(j)$ *be such that*

$$\inf_{\underline{\theta} \in \Omega_1(i)} \int \delta_i^0(\underline{x}) f_{\underline{\theta}}(\underline{x}) d\mu(\underline{x}) = \int \delta_i^0(\underline{x}) f_{\underline{\theta}^*(i)}(\underline{x}) d\mu(\underline{x})$$

<u>and</u>

$$\sup_{\underline{\theta} \in \Omega_1(j)} \int \delta_i^0(\underline{x}) f_{\underline{\theta}}(\underline{x}) d\mu(\underline{x}) = \sum_{j \neq i} \int \delta_i^0(\underline{x}) f_{\underline{\theta}_*^i(j)}(\underline{x}) d\mu(\underline{x}),$$

<u>where</u>

$$\delta_i^0(\underline{x}) = \begin{cases} 1 & \text{if } L_2 \sum_{j \neq i} p_j f_{\underline{\theta}_*^i(j)}(\underline{x}) < L_1 p_i f_{\underline{\theta}^*(i)}(\underline{x}), \\ \ell_i & \qquad\qquad = \qquad\qquad , \\ 0 & \qquad\qquad > \qquad\qquad . \end{cases}$$

<u>Then</u> $\delta^0 = (\delta_1^0, \ldots, \delta_k^0)$ <u>is a</u> Γ-<u>minimax procedure</u>.

PROOF. For $\rho_0 \in \Gamma$, where $\Gamma_0 = \{\rho(\underline{\theta}) | P[\underline{\theta}=\underline{\theta}^*(i)] = p_i$, $P[\underline{\theta}=\underline{\theta}_*^i(j)] = p_j,\ \forall\, j \neq i,\ 1 \leq i \leq k\}$. For any decision rule δ,

$$\sup_{\rho \in \Gamma} r(\rho,\delta) \geq r(\rho_0,\delta) = \sum_{i=1}^{k} r^{(i)}(\rho_0,\delta_i)$$

$$= \sum_{i=1}^{k} \{L_1 p_i \int [1-\delta_i(\underline{x})] f_{\underline{\theta}^*(i)}(\underline{x}) d\mu(\underline{x})$$

$$+ L_2 \sum_{\substack{j=1 \\ j \neq i}}^{k} p_j \int \delta_i(\underline{x}) f_{\underline{\theta}_*^i(j)}(\underline{x}) d\mu(\underline{x})\}$$

$$\geq \sum_{i=1}^{k} \{L_1 p_i \int [1-\delta_i^0(\underline{x})] f_{\underline{\theta}^*(i)}(\underline{x}) d\mu(\underline{x})$$

$$+ L_2 \sum_{\substack{j=1 \\ j \neq i}}^{k} p_j \int \delta_i^0(\underline{x}) f_{\underline{\theta}_*^i(j)}(\underline{x}) d\mu(\underline{x})$$

$$= \sum_{i=1}^{k} r^{(i)}(\rho_0,\delta_i^0) \geq \sup_{\rho \in \Gamma} r(\rho,\delta^0).$$

REMARK. Let $\Pi_1, \Pi_2, \ldots, \Pi_k$ be independent and let $f_{\underline{\theta}}(\underline{x}) = \prod_{j=1}^{k} f_{\theta_j}(x_j)$. Assume that $f_\theta(x)$ has a monotone likelihood ratio. Let $\underline{\theta}^*(i) = (\theta_0, \ldots, \theta_0, \theta_0+\Delta, \theta_0, \ldots, \theta_0)$ and

$\underline{\theta}_*^i(j) = (\theta_0,\ldots,\theta_0,\ \theta_0+\Delta,\ \theta_0,\ldots,\theta_0)$, θ_0 fixed.

$$L_2 \sum_{\substack{j=1 \\ j\neq i}}^{k} p_j f_{\underline{\theta}_*^i(j)}(\underline{x}) < L_1 p_i f_{\underline{\theta}^*(i)}(\underline{x}) \tag{2.4}$$

is equivalent to

$$\sum_{\substack{j=1 \\ j\neq i}}^{k} \frac{L_2 p_j}{L_1 p_i} \frac{f_{\theta_0}(x_i) f_{\theta_0+\Delta}(x_j)}{f_{\theta_0+\Delta}(x_i) f_{\theta_0}(x_j)} < 1. \tag{2.5}$$

Hence $\delta_i^0(\underline{x})$ is non-decreasing in x_i and non-increasing in $x_j (j\neq i)$, and for any i,

$$\inf_{\underline{\theta}\in\Omega_1(i)} \int \delta_i^0(\underline{x}) f_{\underline{\theta}}(\underline{x}) d\mu(\underline{x}) = \inf_{\substack{\theta_j=\theta^*, j\neq i \\ \theta_i=\theta^*+\Delta}} \int \delta_i^0(\underline{x}) f_{\underline{\theta}}(\underline{x}) d\mu(\underline{x}), \tag{2.6}$$

and

$$\sup_{\underline{\theta}\in\Omega_1(j)} \int \delta_i^0(\underline{x}) f_{\underline{\theta}}(\underline{x}) d\mu(\underline{x}) = \sup_{\substack{\theta_r=\theta^*, r\neq j \\ \theta_j=\theta^*+\Delta}} \int \delta_i^0(\underline{x}) f_{\underline{\theta}}(\underline{x}) d\mu(\underline{x}). \tag{2.7}$$

Note that the above discussion of the existence of (2.6) and (2.7) can be completed in the same way as at the end of Section 2.1.

3. *Multiple Comparison Procedures*. Although the literature on multiple hypothesis testing and multiple comparison methods is vast (see e.g. Miller [12]), the literature on the optimality of the methods is rather scarce. An important contribution was made by Lehmann [9], [10]. Lehmann finds optimal rules among the class of unbiased rules, where optimality means minimizing the expected loss, and where the loss is the sum of the losses from the individual decisions. It has been a common complaint that the powers of separate tests are small when using multiple tests. Spjøtvol [18] directed attention to maximizing the power of the individual test. Instead of using the constraint that the probability of at least one false rejection is smaller than a certain number α, an upper bound $\dot{r}$ on the expected number of false

rejections is used. In this section, we are interested in discussing multiple comparison rules which are optimal in the sense of Γ-minimax criterion.

Let X be a random observation with probability distribution depending upon a parameter θ, $\theta \in \Omega$, Ω being a subset of the real line. Consider a family of hypothesis testing problems

$$(3.1) \qquad H_t: \theta \in \Omega_{0t} \text{ vs. } K_t: \theta \in \Omega_{1t}, \ t \in T,$$

where $\Omega_{it} \subset \Omega$, $i = 0,1$ and T is a finite index set with k elements. A test of the hypotheses (3.1) will be defined to be a vector $\delta(x) = (\delta_1(x),\ldots,\delta_k(x))$, where the elements of the vector are ordinary test functions. When x is observed we reject H_t with probability $\delta_t(x)$, $t \in T$. The loss function is

$$L(\theta,\delta) = \sum_{t=1}^{k} L^{(t)}(\theta,\delta_t)$$

where

$$L^{(t)}(\theta,\delta_t) = \begin{cases} L_1(1-\delta_t) & \text{if } \theta \in \omega_{1t}, \\ L_2\delta_t & \text{if } \theta \in \omega_{0t}, \\ 0 & \text{otherwise,} \end{cases}$$

ω_{it} is a closed subset of Ω_{it}, $i = 0,1$.

The risk function is

$$R(\theta,\delta) = \sum_{t=1}^{k} R^{(t)}(\theta,\delta_t)$$

where

$$R^{(t)}(\theta,\delta_t) = \int L^{(t)}(\theta,\delta_t(x))dF_\theta(x),$$

$F_\theta(x)$ is the cdf corresponding to the density $f_\theta(x)$ of X. Thus $R(\theta,\delta) = L_1N_1 + L_2N_2$, where $N_1(N_2)$ is the expected value of type I (II) error. If $\rho(\theta)$ is a distribution over Ω, then the expected risk of a procedure is

$$r(\rho,\delta) = \sum_{t=1}^{k} r^{(t)}(\rho,\delta_t),$$

where

$$r^{(t)}(\rho,\delta_t) = \int_\Omega R^{(t)}(\theta,\delta_t)d\rho(\theta).$$

Assume that partial prior information is available in the selection problem, so that we are able to specify $\pi_t' = P[\theta \in \omega_{0t}]$ and $\pi_t = P[\theta \in \omega_{1t}]$ so that $\pi_t + \pi_t' \leq 1$ for $t = 1,2,\ldots,k$.
Define

$$\Gamma = \{\rho(\theta) | \int_{\omega_{0t}} d\rho(\theta) = \pi_t' \text{ and } \int_{\omega_{1t}} d\rho(\theta) = \pi_t, \quad \text{for } t = 1,2,\ldots,k\}.$$

Let μ be a σ-finite measure on a measurable space.

THEOREM 3.1. *Let* $\theta_{0t} \in \omega_{0t}$ *and* $\theta_{1t} \in \omega_{1t}$ *be such that*

$$\int \delta_t^0(x) f_{\theta_{1t}}(x)d\mu(x) = \inf_{\theta\in\omega_{1t}} \int \delta_t^0(x) f_\theta(x)d\mu(x),$$

and

$$\int \delta_t^0(x) f_{\theta_{0t}}(x)d\mu(x) = \sup_{\theta\in\omega_{0t}} \int \delta_t^0(x) f_\theta(x)d\mu(x),$$

where

$$\delta_t^0(x) = \begin{cases} 1 & \text{if } L_2\pi_t' f_{\theta_{0t}}(x) < L_1\pi_t f_{\theta_{1t}}(x), \\ m_t & \qquad\qquad = \qquad\qquad , \\ 0 & \qquad\qquad > \qquad\qquad . \end{cases}$$

Then $\delta^0 = (\delta_1^0,\ldots,\delta_k^0)$ *is a* Γ-*minimax procedure*.

PROOF. This proof is similar to the proof of Theorem 2.1 and hence omitted.

Applications to comparison of means of normal random variables with common known variance. Let X_{ij} be $N(\mu_i,1)$, $j = 1,2,\ldots,n_i$, $i = 1,\ldots,r$, and independent. Hence we assume, without loss of generality, that the known variance is 1. Consider the following hypothesis about linear functions in the $\{\mu_i\}$,

$$H_t\colon \sum_{i=1}^r a_{ti}\mu_i = b_t \text{ vs. } K_t\colon \sum_{i=1}^r a_{ti}\mu_i > b_t,$$

for $t \in T$, where $\{a_{ti}\}$ and $\{b_t\}$ are given constants.

By the same method as in Spjøtvol [18], let $\bar{X}_i = \frac{1}{n_i} \sum_{j=1}^{n_i} X_{ij}$, $i = 1,\ldots,r$. For fixed t, we transform $\bar{X}_1,\ldots,\bar{X}_r$ to $Y_1,\ldots,Y_r$ by a nonsingular transformation, $Y_1 = \sum_{i=1}^{r} a_{ti}\bar{X}_i$, $Y_j = \sum_{i=1}^{r} b_{ji}\bar{X}_i$, $j = 2,\ldots,r$, where $\{b_{ji}\}$ are chosen so that $\mathrm{Cov}(Y_1,Y_j) = \sum_{i=1}^{r} a_{ti}b_{ji}/n_i = 0$, $j = 2,\ldots,r$. We have that Y_1 is $N(\sum_{i=1}^{r} a_{ti}\mu_i, \sum_{i=1}^{r} a_{ti}^2/n_i)$ and Y_j, $j = 2,\ldots,r$ is $N(\sum_{i=1}^{r} b_{ji}\mu_i, \sum_{i=1}^{r} b_{ji}^2/n_i)$. Now look at a special case.

Difference between means. Here the hypotheses are

$$H_{ij}: \mu_i = \mu_j \text{ vs. } K_{ij}: \mu_i > \mu_j, \; i \neq j.$$

The pair (i,j) corresponds to the index t, and $k = r(r-1)$. Note that H_{ij} vs. K_{ij} and H_{ji} vs. K_{ji} are two different problems; H_{ij} is the same as H_{ji}, but the alternatives are different. If we had used alternatives $\mu_i \neq \mu_j$, T would have $\frac{1}{2}r(r-1)$ elements. For all $(i,j) \in T$, reject H_{ij} when $(\frac{1}{n_i} + \frac{1}{n_j})^{-\frac{1}{2}}(\bar{X}_i - \bar{X}_j) > Z_{ij}$, where

$$Z_{ij} = \frac{\Delta}{2(\frac{1}{n_i} + \frac{1}{n_j})} + \frac{1}{\Delta} \log \frac{L_2\pi'_{ij}}{L_1\pi_{ij}}.$$

Acknowledgement. The authors wish to thank Professor S. Panchapakesan for valuable comments and conversations and also thank Mr. R. Berger for a critical reading of this paper.

References

[1] Blum, J. R. and Rosenblatt, J. (1967). On partial a priori information in statistical inference. *Ann. Math. Statist.* 38, 1671-1678.

[2] Deely, J. J. and Gupta, S. S. (1968). On the properties of subset selection procedures. *Sankhyā* 30, 37-50.

[3] DeRouen, T. A. and Mitchell, T. J. (1974). A G_1-minimax estimator for a linear combination of binomial probabilities. *J. Amer. Statist. Assoc.* 69, 231-233.

[4] Gupta, S. S. and Huang, D. Y. (1975). A note on Γ-minimax classification procedures. *Proceedings of the 40th Session of the International Statistical Institute*, Warsaw, Poland, 330-335.

[5] Gupta, S. S. and Huang, D. Y. (1975). On some parametric and nonparametric sequential subset selection procedures. In M. L. Puri (Ed.), *Statistical Inference and Related Topics*, Academic Press, New York, 101-128.

[6] Hodges, J. L. and Lehmann, E. L. (1952). The use of previous experience in reaching statistical decisions. *Ann. Math. Statist.*, 23, 396-407.

[7] Huang, D. Y. (1974). On some optimal subset selection procedures for model I and model II in treatments versus control problems. Dept. of Statist., Purdue Univ., Mimeo Ser. #373, W. Laf., Ind.

[8] Jackson, D. A., O'Donovan, T. M., Zimmer, W. J. and Deely, J. J. (1970). G_2-minimax estimates in the exponential family. *Biometrika* 57, 439-43.

[9] Lehmann, E. L. (1957a). A theory of some multiple decision problems, I. *Ann. Math. Statist.*, 28, 1-25.

[10] Lehmann, E. L. (1957b). A theory of some multiple decision problems, II. *Ann. Math. Statist.* 28, 547-572.

[11] Menges, G. (1966). On the Bayesification of the minimax principle. *Unternehmensforschung* 10, 81-91.

[12] Miller, R. G., Jr. (1966). *Simultaneous Statistical Inference.* McGraw-Hill, New York.

[13] Randles, R. H. and Hollander, M. (1971). Γ-minimax selection procedures in treatments versus control problems. *Ann. Math. Statist.* 42, 330-41.

[14] Robbins, H. (1964). The empirical Bayes approach to statistical decision problems. *Ann. Math. Statist.* 35, 1-20.

[15] Schaafsma, W. (1969). Minimax risk and unbiasedness for multiple decision problems of type I. *Ann. Math. Statist.* 40, 1648-1720.

[16] Solomon, D. L. (1972a). Λ-minimax estimation of a multivariate location parameter. *J. Amer. Statist. Assoc.* 67, 641-646.

[17] Solomon, D. L. (1972b). Λ-minimax estimation of a scale parameter. *J. Amer. Statist. Assoc.* 67, 647-648.

[18] Spjøtvol, E. (1972). On the optimality of some multiple comparison procedures. *Ann. Math. Statist.* 43, 398-411.

[19] Stäel von Holstein, C. A. S. (1971). Measurement of subjective probability. *Acta Psychologica* 34, 129-145.

[20] Watson, S. R. (1974). On Bayesian inference with incompletely specified prior distributions. *Biometrika* 61, 193-196.

MORE ON INCOMPLETE AND BOUNDEDLY COMPLETE FAMILIES OF DISTRIBUTIONS

By Wassily Hoeffding*
University of North Carolina at Chapel Hill

1. *Introduction*. Let $\mathcal{Q}$ be a family of distributions (probability measures) on a measurable space $(\mathcal{Y},\mathcal{B})$ and let Γ be a group of $\mathcal{B}$-measurable transformations of $\mathcal{Y}$. The family $\mathcal{Q}$ is said to be complete relative to Γ if no nontrivial Γ-invariant unbiased estimator of zero for $\mathcal{Q}$ exists. (A function is called Γ-invariant if it is invariant under all transformations in Γ.) The family $\mathcal{Q}$ is said to be boundedly complete relative to Γ if no bounded nontrivial Γ-invariant unbiased estimator of zero for $\mathcal{Q}$ exists.

Let $\mathcal{P}$ be a family of distributions on a measurable space $(\mathcal{X},\mathcal{A})$ and let $\mathcal{P}^{(n)} = \{P^n\colon P \in \mathcal{P}\}$ be the family of the n-fold product measures P^n on the measurable space $(\mathcal{X}^n,\mathcal{A}^{(n)})$ generated by $(\mathcal{X},\mathcal{A})$. The distributions P^n are invariant under the group Π_n of the n! permutations of the coordinates of the points in $\mathcal{X}^n$. It is known that if the family $\mathcal{P}$ is sufficiently rich (for instance, contains all distributions concentrated on finite subsets of $\mathcal{X}$) then $\mathcal{P}^{(n)}$ is complete relative to Π_n (Halmos [4]; Fraser [3]; Lehmann [6]; Bell, Blackwell and Breiman [1]).

Now consider a family, again denoted by $\mathcal{P}$, of distributions P on $(\mathcal{X},\mathcal{A})$ which satisfy the conditions

$$(1.1) \qquad \int u_i \, dP = c_i, \quad i = 1,\dots,k,$$

where $u_1,\dots,u_k$ are given functions and $c_1,\dots,c_k$ are given constants. In this case the family $\mathcal{P}^{(n)}$ is, in general, not complete relative to Π_n. Indeed, the statistic

---

*This research was supported by the National Science Foundation under Grant no. MPS75-07556.

$$(1.2)\quad g(x_1,\ldots,x_n) = \sum_{i=1}^{k} \sum_{j=1}^{n} \{u_i(x_j)-c_i\}h_i(x_1,\ldots,x_{j-1},x_{j+1},\ldots,x_n)$$

where $h_1,\ldots,h_k$ are arbitrary $A^{(n-1)}$-measurable, Π_{n-1}-invariant, $P^{(n-1)}$-integrable functions, is a Π_n-invariant estimator of zero. However, $P^{(n)}$ may be boundedly complete, as the following theorems, proved in [5], imply.

If A contains the one-point sets, let P_0 be the family of all distributions P concentrated on finite subsets of X which satisfy conditions (1.1). If μ is a σ-finite measure on (X,A), let $P_0(\mu)$ be the family of all distributions absolutely continuous with respect to μ whose densities $dP/d\mu$ are simple functions (finite linear combinations of indicator functions of sets in A) and which satisfy conditions (1.1).

THEOREM 1A. *If A contains the one-point sets and P is a convex family of distributions on (X,A) which satisfy conditions (1.1), such that $P_0 \subset P$, then every Π_n-invariant unbiased estimator $g(x_1,\ldots,x_n)$ of zero is of the form* (1.2).

THEOREM 2A. *If the conditions of Theorem 1A are satisfied and every nontrivial linear combination of $u_1,\ldots,u_k$ is unbounded then the family $P^{(n)}$ is boundedly complete relative to Π_n.*

(The assumption that the family P is convex is used in [5] only to prove that there are versions of the functions h_i that are integrable. Note that the families P_0, $P_0(\mu)$, and the family of all P which are absolutely continuous with respect to μ and satisfy conditions (1.1), are convex.)

Theorems 1B and 2B of [5] assert that if P is a convex family of distributions absolutely continuous with respect to a σ-finite measure μ which satisfies conditions (1.1), and $P_0(\mu) \subset P$, then conclusions analogous to those of Theorems 1A and 2A hold, except that $g(x_1,\ldots,x_n)$ is of the form (1.2) a.e. $(P^{(n)})$ and, in the second theorem, every nontrivial linear combination $u(x) = \sum a_i u_i(x)$ is assumed to be P-unbounded in the sense that for every real c there is a $P \in \mathcal{P}$ such that $P(|u(x)| > c) \neq 0$.

In this paper I shall deal with some extensions of these theorems. In Sections 2 and 3, straightforward extensions to some finite groups other than Π_n are briefly discussed. In Section 4 the generalization to the case where conditions (1.1) are replaced by $\int \int u(x,y) dP(x) dP(y) = c$ is considered, which has been investigated by N. I. Fisher [2].

2. *Distributions symmetric about* 0. Let $\mathcal{P}$ be a family of distributions which satisfy the conditions of one of the Theorems 1A, 1B, 2A, 2B with $X = R^1$ and A the Borel sets, and the additional condition that each P in $\mathcal{P}$ is symmetric about zero. In this case the distributions P^n are invariant under the group Γ_n which consists of all permutations of the components of the points $(x_1,\ldots,x_n) \in R^n$ and of all changes of signs of the components. In conditions (1.1) each $u_i(x)$ may be replaced by $\{u_i(x)+u_i(-x)\}/2$. Thus we may assume that $u_i(x) \equiv u_i(-x)$ for all i. Then $g(x_1,\ldots,x_n)$ in (1.2), where each h_i satisfies the additional condition of being Γ_{n-1}-invariant, is a Γ_n-invariant estimator of zero.

The four theorems quoted in the Introduction are true if the distributions in $\mathcal{P}$, $\mathcal{P}_0$ and $\mathcal{P}_0(\mu)$ satisfy the additional condition of being symmetric about zero, if the $u_i(x)$ are symmetric about zero, and if Π_n is replaced by Γ_n throughout. The proof is very simple. If P is the distribution of the random variable X, symmetric about zero, let P* denote the distribution of $|X|$. Conditions (1.1) and the condition that $g(x_1,\ldots,x_n)$ is a Γ_n-invariant unbiased estimator of zero can be expressed in terms of the distributions P*, and the problem is reduced to that of the theorems in the Introduction with X the set of the nonnegative numbers.

3. *Two-sample families*. For r = 1,2 let $\mathcal{P}_r$ be a family of distributions P on (X,A) which satisfy the conditions

$$\int u_{r,i}\, dP = c_{r,i}, \quad i = 1,\ldots,k_r. \tag{3.1}$$

Let

$$\mathcal{P}^{(m,n)} = \{P_1^m P_2^n : P_1 \in \mathcal{P}_1,\ P_2 \in \mathcal{P}_2\}.$$

The distributions in $\mathcal{P}^{(m,n)}$ are invariant under the group $\Pi_m\Pi_n$ of those permutations of the coordinates of the points in $\mathcal{X}^{m+n}$ which permute the first m coordinates among themselves and permute the remaining n coordinates among themselves.

To simplify notation (and with no loss of generality) let conditions (3.1) be satisfied with $c_{r,i} = 0$:

$$\int u_{r,i}\, dP = 0, \quad i = 1,\dots,k_r. \tag{3.2}$$

Then the statistic

(3.3) $g(x_1,\dots,x_m,y_1,\dots,y_n) =$

$$\sum_{i=1}^{k_1} \sum_{j=1}^{m} u_{1i}(x_j) h_{1i}(x_1,\dots,x_{j-1},x_{j+1},\dots,x_m,y_1,\dots,y_n)$$

$$+ \sum_{i=1}^{k_2} \sum_{j=1}^{n} u_{2i}(y_j) h_{2i}(x_1,\dots,x_m,y_1,\dots,y_{j-1},y_{j+1},\dots,y_n),$$

where each h_{1i} is $\Pi_{m-1}\Pi_n$-invariant and $\mathcal{P}^{(m-1,n)}$-integrable and each h_{2i} is $\Pi_m\Pi_{n-1}$-invariant and $\mathcal{P}^{(m,n-1)}$-integrable, is a $\Pi_m\Pi_n$-invariant unbiased estimator of zero for $\mathcal{P}^{(m,n)}$. It can be shown by the methods of [5] that the obvious analogs of the four theorems in the introduction are true. For instance, if $\mathcal{P}_{10}(\mathcal{P}_{20})$ denotes the family of all distributions concentrated on finite subsets of $\mathcal{X}$ which satisfy conditions (3.2) with r = 1 (r = 2), if $\mathcal{A}$ contains the one-point sets, and if $\mathcal{P}_1(\mathcal{P}_2)$ is a convex family of distributions which satisfy (3.2) with r = 1 (r = 2), and such that $\mathcal{P}_{r0} \subset \mathcal{P}_r$ (r = 1,2) then every $\Pi_m\Pi_n$-invariant estimator of zero is of the form (3.3). If, in addition, every nontrivial linear combination of $u_{r1},\dots,u_{rk_r}$ is unbounded (r=1,2) then the family $\mathcal{P}^{(m,n)}$ is boundedly complete relative to $\Pi_m\Pi_n$. Details of the proof are omitted.

4. *Families restricted by a nonlinear condition.* It is natural to replace the conditons $\int u_i \, dP = c_i$ $(i = 1,\ldots,k)$, which are linear in P, by one or more conditions of the form

$$\int_{X^s} u(x_1,\ldots,x_s)dP^s = c.$$

Here I will consider only a family $\mathcal{P}$ of distributions P on (X,A) which satisfy the single condition

$$\int_{X^2} u(x_1,x_2)dP^2 = 0. \tag{4.1}$$

For example, if $X = R^1$ and

$$u(x_1,x_2) = (x_1-x_2)^2 - 2\sigma^2, \tag{4.2}$$

condition (4.1) specifies the variance of the distribution P. For

$$u(x_1,x_2) = \sum_{i=1}^{k} \{u_i(x_1) - c_i\}\{u_i(x_2) - c_i\} \tag{4.3}$$

condition (4.1) is equivalent to the conditions (1.1). Some other interesting special cases of (4.1) will be discussed later. We may and shall assume that

$$u(x_1,x_2) \equiv u(x_2,x_1). \tag{4.4}$$

A Π_n-invariant unbiased estimator of zero (u.e.z.) for $\mathcal{P}^{(n)}$ $(n \geq 2)$ is

$$\begin{aligned} g(x_1,\ldots,x_n) = {} & u(x_1,x_2)h(x_3,\ldots,x_n) + \\ & u(x_1,x_3)h(x_2,x_4,\ldots,x_n)+\ldots+u(x_{n-1},x_n)h(x_1,\ldots,x_{n-2}), \end{aligned} \tag{4.5}$$

where h is any Π_{n-2}-invariant, $\mathcal{P}^{(n-2)}$-integrable function.

Even if $\mathcal{P}$ consists of all distributions P on (X,A) satisfying (4.1), a Π_n-invariant u.e.z., $g(x_1,\ldots,x_n)$, is not necessarily of the form (4.5). Whether it must be of this form depends on the function u. Thus if u is given by (4.3) then g is of the form (1.2), and cannot (in general) be expressed in the form (4.5). (Take, for instance, $u(x_1,x_2) = u_1(x_1)u_1(x_2)$ and $n = 2$.)

Families which satisfy condition (4.1) have been studied by N. I. Fisher in his Ph.D. dissertation [2]. One of Fisher's main results is as follows.

Let P be a convex family of distributions P on (X,A) satisfying (4.1), let A contain the one-point sets, and let $P \subset P_0$, where P_0 is the family of all distributions P concentrated on finite sets and satisfying (4.1). Define

$$L = \sum_{i=1}^{N-1} u_{iN} p_i, \quad Q = \sum_{i=1}^{N-1} \sum_{j=1}^{N-1} (u_{iN}u_{jN} - u_{NN}u_{ij}) p_i p_j,$$

where $u_{ij} = u(x_i, x_j)$.

THEOREM 3. _Suppose that for every_ $(x_1,\ldots,x_n) \in X^n$ _there exist_

$$N>n, \; x_{n+1},\ldots,x_N \text{ in } X, \; p_1>0,\ldots,p_{N-1}>0$$

such that

(a) $Q > 0$,

(b) $u_{NN} \neq 0$ _and_ $u_{NN}(Q^{\frac{1}{2}} - L) > 0$,

(c) $Q^{\frac{1}{2}}$, _considered as a function of_ $p_1,\ldots,p_{N-1}$, _is irrational_.

Then

(i) _Every_ s.u.e.z. _for_ $P^{(n)}$ _is of the form_ (4.5).

(ii) _If, in addition_, u( ) _is unbounded, then the family_ $P^{(n)}$ _is boundedly complete relative to_ Π_n.

A similar result, analogous to Theorems 1B and 2B, holds for dominated families satisfying condition (4.1).

To give just one example, the conditions of Theorem 3 are satisfied in the case (4.2).

The condition $u_{NN} \neq 0$ implies $u(x,x) \not\equiv 0$. Fisher shows that the conclusion of Theorem 3 also holds if $u(x,x) \equiv 0$. The conclusion of Theorem 3 also holds in some other cases, for instance if $u(x,y)$ is of the form

$$u(x,y) = v_1(x)v_2(y) + v_1(y)v_2(x).$$

In this case condition (4.1) is equivalent to

$$\int v_1 \, dP \cdot \int v_2 \, dP = 0.$$

In the case (4.3) it is easy to see that $Q \leq 0$, so that condition (a) is not satisfied.

To conclude we consider a special class of functions $u(x,y)$ not (in general) covered by the previously stated results. Let

$$u(x,y) = \sum_{i=1}^{\infty} c_i v_i(x) v_i(y), \tag{4.6}$$

where each $v_i(x)$ is bounded and the positive constants c_i are so chosen that $u(x,y)$ is bounded. In this case condition (4.1) is equivalent to the infinite set of conditions

$$\int v_i \, dP = 0, \quad i = 1,2,\ldots \tag{4.7}$$

Here are a few special cases.

(a) Let $X = R^1$, let P_0 be a given distribution with distribution function $F_0(x)$, and let, with I_A denoting the indicator function of set A,

$$v_i(x) = I_{(-\infty,r_i]}(x) - F_0(r_i),$$

where $\{r_i\}$ is a number sequence dense in R^1. Then the family P of all distributions satisfying condition (4.1) consists of the single distribution P_0.

(b) If $X = R^1$,

$$v_i(x) = I_{(-\infty,r_i]}(x) - I_{(-\infty,r_i]}(-x),$$

then the family of all distributions P on the Borel sets which satisfy (4.1) is the family of all distributions symmetric about 0. In this case the family $P^{(n)}$ is complete relative to the group Γ_n of Section 2.

(c) The family of all bivariate distributions with given marginal distributions can also be characterized in this way with $\{v_i\}$ suitably chosen. In this case the general form of an invariant unbiased estimator of zero is unknown.

References

[1] Bell, C. B., Blackwell, David, and Breiman, Leo (1960). On the completeness of order statistics. *Ann. Math. Statist.* 31, 794-797.

[2] Fisher, Nicholas I. (1976). The theory of unbiased estimation for some nonparametric families of probability measures. Inst. of Statist. (Univ. of North Carolina) Mimeo Ser. No. 1051.

[3] Fraser, D. A. S. (1954). Completeness of order statistics. *Canad. J. Math.* 6, 42-45.

[4] Halmos, Paul R. (1946). The theory of unbiased estimation. *Ann. Math. Statist.* 17, 34-43.

[5] Hoeffding, Wassily (1976). Some incomplete and boundedly complete families of distributions. Submitted to *Ann. Statist.*

[6] Lehmann, E. L. (1959). *Testing Statistical Hypotheses.* New York: John Wiley.

ROBUST COVARIANCES

By Peter J. Huber
Eidgenoessische Technische Hochschule, Zurich

1. *Introduction.* Covariance matrices and the associated ellipsoids are often used for describing the overall shape of distributions of points in p-dimensional Euclidean space. Important examples occur in principal component and factor analysis, and in discriminant analysis. But because of their high sensitivity to outliers, covariance matrices are not particularly well suited for this purpose, cf. the illustrations in Section 10.

Attempts to overcome this drawback and to replace covariances by outlier resistant alternatives - here to be called robust (pseudo-) covariances - have been around in the literature for some years (see in particular Gnanadesikan and Kettenring [2]).

Recently, Maronna [8] has considered M-estimates of multivariate location and scatter, has proved results on existence and uniqueness of solutions and has established consistency and asymptotic normality.

A first goal of the present paper is to systematize the affinely invariant case and to determine the most general form for an M-estimate of multivariate scatter. No judgment is implied whether invariant methods should be preferred to noninvariant ones.

Randomly placed outliers have the following effects on estimates: they may introduce a bias, they may increase the statistical variability of the estimate, and ultimately, if the number of outliers exceeds a critical fraction ε^* of the total sample, the estimate even may break down completely. For instance, the classical sample covariance has breakdown point $\varepsilon^*=0$: in a sample of arbitrary size, one single outlier will suffice to spoil the estimate.

Past experience with location estimates has shown that optimization with regard to the second effect - minimization of the statistical variability in worst possible cases - often leads to good robustness properties also with regard to the other two effects. We shall exploit this approach also here and shall determine least informative underlying distributions that make the estimation problem hardest.

2. *Maximum likelihood estimates.* Let $f(\underset{\sim}{x}) = f(|\underset{\sim}{x}|)$ be a fixed spherically symmetric probability density in R^p. We apply arbitrary non-degenerate affine transformations $\underset{\sim}{x} \rightarrow V(\underset{\sim}{x} - \underset{\sim}{\xi})$ and obtain a family of densities

$$f(\underset{\sim}{x};\underset{\sim}{\xi},V) = |\det V| f(|V(\underset{\sim}{x}-\underset{\sim}{\xi})|). \tag{2.1}$$

Let us assume that our data obeys an underlying model of the type (2.1); the problem is to estimate the vector $\underset{\sim}{\xi}$ and the matrix V from n observations of $\underset{\sim}{x}$.

Evidently, V is not uniquely identifiable (it can be multiplied by an arbitrary orthogonal matrix from the left), but V^TV is. We can enforce uniqueness of V by suitable side conditions, e.g. by requiring that it is positive definite symmetric, or that it is lower triangular with a positive diagonal. Mostly, we shall adopt the latter convention.

The maximum likelihood estimate of $\underset{\sim}{\xi}$, V is obtained by maximizing

$$\log(\det V) + \text{ave}\{\log f(|V(\underset{\sim}{x}-\underset{\sim}{\xi})|)\}, \tag{2.2}$$

where ave {...} denotes the average taken over the sample. A necessary condition for a maximum is that (2.2) remains stationary under infinitesimal variations of $\underset{\sim}{\xi}$ and V. So we let $\underset{\sim}{\xi}$ and V depend differentiably on a dummy parameter, take the derivative (denoted by superscribed dot) and obtain the condition:

$$\text{tr}(S) + \text{ave}\left\{\frac{f'(|\underset{\sim}{y}|)}{f(|\underset{\sim}{y}|)}\frac{\underset{\sim}{y}^T S\underset{\sim}{y}}{|\underset{\sim}{y}|}\right\} - \text{ave}\left\{\frac{f'(|\underset{\sim}{y}|)}{f(|\underset{\sim}{y}|)}\frac{\dot{\underset{\sim}{\xi}}^T V^T\underset{\sim}{y}}{|\underset{\sim}{y}|}\right\} = 0, \tag{2.3}$$

with the abbreviations

$$\underset{\sim}{y} = V(\underset{\sim}{x} - \underset{\sim}{\xi}), \tag{2.4}$$

$$S = \dot{V}V^{-1}. \tag{2.5}$$

Since this should hold for arbitrary infinitesimal variations $\dot{\xi}$, $\dot{V}$, (2.3) can be rewritten into the simultaneous matrix equations

$$\text{ave}\,\{w(|y|)y\} = 0, \tag{2.6}$$

$$\text{ave}\,\{w(|y|)yy^T - I\} = 0, \tag{2.7}$$

where I is the pxp identity matrix, and

$$w(|y|) = -\frac{f'(|y|)}{f(|y|)|y|}. \tag{2.8}$$

For example, if

$$f(|x|) = (2\pi)^{-p/2} \quad \exp\,(-|x|^2/2)$$

is the standard normal density, we have $w \equiv 1$, and (2.6), (2.7) can be written as

$$\xi = \text{ave}\,\{x\} \tag{2.9}$$

$$(V^TV)^{-1} = \text{ave}\,\{(x - \xi)(x - \xi)^T\} \tag{2.10}$$

In this case, $(V^TV)^{-1}$ is the ordinary covariance matrix of x (the sample one, if the average is taken over the sample, the true one, if the average is taken over the distribution).

More generally, if ξ and V are determined from equations of the form

$$\text{ave}\,\{w(|y|)y\} = 0 \tag{2.11}$$

$$\text{ave}\,\{u(|y|)\frac{yy^T}{|y|^2} - v(|y|)I\} = 0 \tag{2.12}$$

with $y = V(x - \xi)$, for any real valued functions w, u, v, we shall call

$$(V^TV)^{-1}$$

a <u>pseudo-covariance</u> matrix of x.

Note that (2.11) determines location ξ as a weighted mean

$$\xi = \frac{\text{ave}\,\{w(|y|)x\}}{\text{ave}\,\{w(|y|)\}} \tag{2.13}$$

with weights $w(|y|)$ depending on the sample.

Similarly, we can interpret the pseudo-covariance as a kind of weighted covariance

$$(2.14)\qquad (V^TV)^{-1} = \frac{\text{ave}\left\{\dfrac{u(|\tilde{y}|)(\tilde{x}-\tilde{\xi})(\tilde{x}-\tilde{\xi})^T}{|\tilde{y}|^2}\right\}}{\text{ave}\,\{v(|\tilde{y}|)\}}\,.$$

3. *Estimates determined by implicit equations.* This section shows that (2.12) in some sense is the most general form of an implicit matrix equation determining $(V^TV)^{-1}$.

In order to simplify the discussion, we assume that location is known in advance and fixed to be $\tilde{\xi} = 0$. Then we can write (2.12) in the form

$$(3.1)\qquad \text{ave}\,\{\Psi(V\tilde{x})\} = 0$$

with

$$(3.2)\qquad \Psi(\tilde{x}) = u(|\tilde{x}|)\,\frac{\tilde{x}\tilde{x}^T}{|\tilde{x}|^2} - v(|\tilde{x}|)I$$

Is this the most general form of Ψ?

Let us take any sufficiently smooth, but otherwise arbitrary function Ψ from R^p into the space of symmetric pxp matrices. This gives the proper number of equations for the $p(p + 1)/2$ unique components of $(V^TV)^{-1}$.

We shall determine a matrix V such that

$$(3.3)\qquad \text{ave}\,\{\Psi(V\tilde{x})\} = 0$$

where the average is taken with respect to a <u>fixed</u> (true or sample) distribution of $\tilde{x}$.

Let us assume that Ψ and the distribution of $\tilde{x}$ are such that (3.3) has at least one solution V, and that <u>all</u> solutions lead to the <u>same</u> pseudo-covariance matrix

$$(3.4)\qquad C_{\tilde{x}} = (V^TV)^{-1}$$

This uniqueness assumption implies at once that $C_{\tilde{x}}$ is invariant under linear transformations in the same sense as the usual covariance matrix:

(3.5) $$C_{B\underset{\sim}{x}} = BC_{\underset{\sim}{x}}B^T.$$

Now let S be an arbitrary orthogonal transformation and define

(3.6) $$\Psi_S(\underset{\sim}{x}) = S^T\,\Psi(S\underset{\sim}{x})S.$$

The transformed function Ψ_S determines a new pseudo-covariance $(W^TW)^{-1}$, namely the solution of

$$\text{ave}\,\{\Psi_S(W\underset{\sim}{x})\} = \text{ave}\{S^T\Psi(SW\underset{\sim}{x})S\} = 0.$$

Evidently, this is solved by $W = S^TV$, where V is any solution of (3.3), and thus

$$W^TW = V^TSS^TV = V^TV.$$

It follows that Ψ and Ψ_S determine the same pseudo-covariances.

We now form

(3.7) $$\bar{\Psi}(\underset{\sim}{x}) = \underset{S}{\text{ave}}\;\Psi_S(\underset{\sim}{x})$$

by averaging over the orthogonal group. Evidently, every solution of (3.3) still solves $\text{ave}\{\bar{\Psi}(V\underset{\sim}{x})\} = 0$, but of course the uniqueness postulated in (3.4) may be lost. We shall now show that $\bar{\Psi}$ is of the form (3.2).

Clearly, $\bar{\Psi}$ is invariant under orthogonal transformations in the sense that

(3.8) $$\bar{\Psi}_S(\underset{\sim}{x}) = S^T\,\bar{\Psi}(S\underset{\sim}{x})S = \bar{\Psi}(\underset{\sim}{x})$$

or

(3.9) $$\bar{\Psi}(S\underset{\sim}{x})S = S\bar{\Psi}(\underset{\sim}{x}).$$

Now let $\underset{\sim}{x} \neq 0$ be an arbitrary fixed vector, then (3.9) shows that the matrix $\bar{\Psi}(\underset{\sim}{x})$ commutes with all orthogonal matrices S which keep $\underset{\sim}{x}$ fixed. This implies that the restriction of $\bar{\Psi}(\underset{\sim}{x})$ to the subspace of R^p orthogonal to $\underset{\sim}{x}$ must be a multiple of the identity. Moreover, for every S which keeps $\underset{\sim}{x}$ fixed we have

$$S\bar{\Psi}(\underset{\sim}{x})\underset{\sim}{x} = \bar{\Psi}(\underset{\sim}{x})\underset{\sim}{x},$$

hence S also keeps $\bar{\Psi}(\underset{\sim}{x})\underset{\sim}{x}$ fixed, which therefore must be a multiple of $\underset{\sim}{x}$. It follows that $\bar{\Psi}(\underset{\sim}{x})$ can be written in the form

$$\bar{\Psi}(\underset{\sim}{x}) = u(\underset{\sim}{x}) \frac{\underset{\sim}{x}\underset{\sim}{x}^T}{|\underset{\sim}{x}|^2} - v(\underset{\sim}{x})I$$

with some scalar-valued functions u, v. Because of (3.8), u and v depend on $\underset{\sim}{x}$ only through $|\underset{\sim}{x}|$, and we conclude that $\bar{\Psi}$ is of the form (3.2).

Global uniqueness, as postulated in (3.4), may be an unrealistically severe requirement. But the arguments carry through in all essential respects with the much weaker local uniqueness requirement that there is a neighborhood of C_X which contains no other solutions besides C_X. For the symmetrized version (3.2), formulas (5.16 ff), together with the implicit function theorem, yield a simple set of sufficient conditions for local uniqueness.

4. *Breakdown properties.* Maronna [8] has pointed out the poor breakdown properties of covariance estimates in high dimensions; under his assumptions (implying $v \equiv 1$, u monotone increasing, $u(0) = 0$) he obtains $\varepsilon^* \leq 1/(p + 1)$. We shall now show that we cannot hope for much improvement from more general choices of u,v.

THEOREM. Assume that u, v satisfy some continuity conditions to be stated below. If $u \geq 0$, then the breakdown point is $\varepsilon^* \leq 1/p$, where p is the number of dimensions.

PROOF. Assume that a fraction $1-\varepsilon$ of the population is p-variate spherically symmetric and that the remaining fraction consists of two pointmasses $\varepsilon/2$ each at $(\pm t,0,0,\ldots,0)$. For reasons of symmetry, we then have $\underset{\sim}{\xi} = 0$, and V is diagonal, with diagonal vector $(\alpha,\beta,\ldots,\beta)$.

Continuity assumption. The functions u and v are continuous, and α, β depend continuously on t for $0 < t < \infty$, so long as $\varepsilon<\varepsilon^*$.

Actually, α and β satisfy the following two equations derived from (2.12)

$$(4.1)\quad (1-\varepsilon)\mathrm{ave}\,\{u(|\underset{\sim}{y}|)\frac{\alpha^2 x_1^2}{|\underset{\sim}{y}|^2} - v(|\underset{\sim}{y}|)\} + \varepsilon(u(s)-v(s)) = 0,$$

$$(4.2)\quad (1-\varepsilon)\mathrm{ave}\,\{u(|\underset{\sim}{y}|)\frac{\beta^2|\underset{\sim}{z}|^2}{|\underset{\sim}{y}|^2} - (p-1)v(|\underset{\sim}{y}|)\} - \varepsilon(p-1)v(s) = 0$$

with $\underset{\sim}{y} = (\alpha x_1, \beta x_2, \ldots, \beta x_p)$, $\underset{\sim}{z} = (0, x_2, x_3, \ldots, x_p)$ and $s = \alpha t$.

If we multiply (4.1) by (p-1) and subtract it from (4.2), we obtain a relation no longer containing v:

$$(4.3)\quad (1-\varepsilon)\mathrm{ave}\,\{u(|\underset{\sim}{y}|)\frac{\beta^2|\underset{\sim}{z}|^2-(p-1)\alpha^2 x_1^2}{|\underset{\sim}{y}|^2}\} - \varepsilon(p-1)u(s) = 0$$

If $\varepsilon < \varepsilon^*$, α and β must be bounded away from 0 and ∞. The continuity assumption now implies that $t \to s = \alpha t$ maps $(0,\infty)$ onto $(0,\infty)$. Hence, by a suitable choice of t, we can arrange that u(s) approximates $u_\infty = \sup u$ arbitrarily closely. The expression inside the curly parenthesis in (4.3) is bounded from above by u_∞, hence it follows that

$$\frac{\varepsilon}{1-\varepsilon} \leq \frac{1}{p-1},$$

or

$$\varepsilon \leq 1/p.$$

This holds whenever $\varepsilon < \varepsilon^*$, so we must have $\varepsilon^* \leq 1/p$.

Frank Hampel [4] has suggested that it should be possible to obtain a higher ε^* by allowing negative values of u. However, such a "superrobust" estimate, if it exists, would have rather strange properties. In particular, a contamination of the above type (two symmetric pointmasses in the x_1 - direction) would never be able to cause its breakdown by elongating the covariance ellipsoid beyond bounds in the x_1-direction; hence its behavior near breakdown would be extremely strange.

Sketch of the proof. By changing signs if necessary we can assume that $u_\infty > 0$ without loss of generality. Let $\varepsilon \uparrow \varepsilon^*$ and

choose $t = t(\varepsilon)$ so that $u(x)$ converges to u_∞.
The above-mentioned type of breakdown - infinite elongation in the x_1 direction - corresponds to $\alpha/\beta \to 0$. But then (4.1) and (4.3) yield the limiting relations

(4.4) $$\frac{\varepsilon^*}{1-\varepsilon^*} = \frac{\bar{v}}{u_\infty - v_\infty}$$

and

(4.5) $$\frac{\varepsilon^*}{1-\varepsilon^*} = \frac{1}{p-1}\,\frac{\bar{u}}{u_\infty}$$

respectively, with

$$\bar{u} = \lim \text{ ave } \{u(\beta|\underset{\sim}{z}|)\}$$

$$\bar{v} = \lim \text{ ave } \{v(\beta|\underset{\sim}{z}|)\}$$

$$v_\infty = \lim_{\varepsilon \uparrow \varepsilon^*} v(\alpha t).$$

Just as before, (4.5) implies $\varepsilon^* \leq 1/p$.

5. *A note on computation.* A simple minded and straightforward approach toward iterative computation of $\underset{\sim}{\xi}$, V is based on the fixed-point properties of formulas (2.13), (2.14). Start with some trial values for $\underset{\sim}{\xi}$ and $(V^T V)^{-1}$. Determine V and calculate $\underset{\sim}{y}$ by (2.4). Then find improved values for $\underset{\sim}{\xi}$ and $(V^T V)^{-1}$ from (2.13) and (2.14). It is fairly evident - and has been experimentally confirmed - that this procedure converges, provided the weights $w(|\underset{\sim}{y}|)$, $u(|\underset{\sim}{y}|)/|\underset{\sim}{y}|^2$ and $v(|\underset{\sim}{y}|)$ change little between iterations; a convergence proof is still outstanding.

Conceivably, the Newton-Raphson method should lead to faster convergence, even though, because of the large number of variables involved, it will not be practicable to solve the linearized versions of (2.11), (2.12) exactly.

The idea behind the following modified Newton-Raphson procedure is: assume that we have to find a $\underset{\sim}{\Theta}$ solving some implicit equation of the form

(5.1) $$\text{ave } \{\psi(\underset{\sim}{x};\ \underset{\sim}{\Theta})\} = 0$$

If $\underset{\sim}{\Theta}^{(m)}$ is a trial value, then a Taylor expansion gives

$$(5.2)\quad \text{ave}\,\{\psi(\underset{\sim}{x};\underset{\sim}{\Theta}^{(m)} + \Delta\underset{\sim}{\Theta})\} \cong \text{ave}\,\{\psi(\underset{\sim}{x};\underset{\sim}{\Theta}^{(m)})\} + \text{ave}\,\{\frac{\partial\psi}{\partial\Theta}(\underset{\sim}{x},\underset{\sim}{\Theta}^{(m)})\}\,\Delta\underset{\sim}{\Theta}$$

We put (5.2) equal to 0 and obtain the improved value

$$(5.3)\quad \underset{\sim}{\Theta}^{(m+1)} = \underset{\sim}{\Theta}^{(m)} + \Delta\underset{\sim}{\Theta}$$

with

$$(5.4)\quad \Delta\underset{\sim}{\Theta} = -(\text{ave}\,\{\frac{\partial\psi}{\partial\Theta}(\underset{\sim}{x};\underset{\sim}{\Theta}^{(m)})\})^{-1}\,\text{ave}\,\{\psi(\underset{\sim}{x};\underset{\sim}{\Theta}^{(m)})\}.$$

The precise value of the $\partial\psi/\partial\Theta$-term in (5.4) is not crucial. In the following, this term will be a rather complicated matrix, and we shall simplify it by replacing the average by an expectation, calculated at a convenient value "near" $\underset{\sim}{\Theta}^{(m)}$.

We begin with location. By expanding (2.11) with regard to $\underset{\sim}{\xi}$ we obtain

$$\text{ave}\,\{w(|V(\underset{\sim}{x}-\underset{\sim}{\xi}-\Delta\underset{\sim}{\xi})|)V(\underset{\sim}{x}-\underset{\sim}{\xi}-\Delta\underset{\sim}{\xi})\}$$

$$\cong \text{ave}\,\{w(|\underset{\sim}{y}|)\underset{\sim}{y}\} - \text{ave}\,\{w(|\underset{\sim}{y}|)V\Delta\underset{\sim}{\xi} + w'(|\underset{\sim}{y}|)\,\frac{\underset{\sim}{y}^T V\Delta\underset{\sim}{\xi}}{|\underset{\sim}{y}|}\,\underset{\sim}{y}\}$$

We now replace the last term in this approximate equation by its <u>conditional expected value</u> at the true $\underset{\sim}{\xi}$, V, given $|\underset{\sim}{y}|$. This amounts to averaging over the uniform distribution on the sphere $|\underset{\sim}{y}|$ = const. and yields

$$\cong \text{ave}\,\{w(|\underset{\sim}{y}|)\underset{\sim}{y}\} - \text{ave}\,\{w(|\underset{\sim}{y}|)V\Delta\underset{\sim}{\xi} + \frac{1}{p}\,w'(|\underset{\sim}{y}|)|\underset{\sim}{y}|V\Delta\underset{\sim}{\xi}\}.$$

We equate this to 0 and thus determine the correction as

$$(5.5)\quad \Delta\underset{\sim}{\xi} = \frac{\text{ave}\,\{w(|\underset{\sim}{y}|)(\underset{\sim}{x}-\underset{\sim}{\xi})\}}{\text{ave}\,\{w(|\underset{\sim}{y}|) + \frac{1}{p}\,w'(|\underset{\sim}{y}|)|\underset{\sim}{y}|\}}$$

In most cases, the second term in the denominator of (5.5) is small relative to the first; if it is neglected, we end up with the fixed-point approach derived from (2.13).

Now turning to covariances, we substitute $(I+S)\underset{\sim}{y}$ for $\underset{\sim}{y}$ in (2.12), where S is a small matrix, and expand into a Taylor series

(5.6) $$U_0 + U_1 + \ldots = 0,$$

where U_k collects the terms of order k in S. We neglect quadratic and higher order terms.

(5.7) $$U_0 = \text{ave}\,\{u(|\underset{\sim}{y}|)\frac{\underset{\sim}{y}\underset{\sim}{y}^T}{|\underset{\sim}{y}|^2} - v(|\underset{\sim}{y}|)I\}$$

(5.8) $$U_1 = \text{ave}\,\frac{\{u'(|\underset{\sim}{y}|)|\underset{\sim}{y}|-2u(|\underset{\sim}{y}|)}{|\underset{\sim}{y}|^4}(\underset{\sim}{y}^T S\underset{\sim}{y})\underset{\sim}{y}\underset{\sim}{y}^T + \frac{u(|\underset{\sim}{y}|)}{|\underset{\sim}{y}|^2}(S\underset{\sim}{y}\,\underset{\sim}{y}^T+\underset{\sim}{y}\,\underset{\sim}{y}^T S^T) - \frac{v'(|\underset{\sim}{y}|)}{|\underset{\sim}{y}|}(\underset{\sim}{y}^T S\underset{\sim}{y})I\}$$

Again, we approximate U_1 by its conditional expectation at the true $\underset{\sim}{\xi}$, V, given $|\underset{\sim}{y}|$, that is, we average over the uniform distribution on the sphere $|\underset{\sim}{y}|$ = const. In particular we obtain (for details see the appendix):

$$E_c\{(\underset{\sim}{y}^T S\underset{\sim}{y})\underset{\sim}{y}\underset{\sim}{y}^T\} = \frac{|\underset{\sim}{y}|^4}{p(p+2)}(tr(S)I+S+S^T),$$

$$E_c\{S\underset{\sim}{y}\underset{\sim}{y}^T+\underset{\sim}{y}\underset{\sim}{y}^T S^T\} = \frac{|\underset{\sim}{y}|^2}{p}(S+S^T),$$

$$E_c\{\underset{\sim}{y}^T S\underset{\sim}{y}\} = \frac{|\underset{\sim}{y}|^2}{p}\,tr(S)$$

Here, E_c stands for "conditional expectation, given $|\underset{\sim}{y}|$". Thus,

(5.9) $$E_c(U_1) = \text{ave}\,\frac{\{u'(|\underset{\sim}{y}|)|\underset{\sim}{y}|-2u(|\underset{\sim}{y}|)}{p(p+2)}(tr(S)I+S+S^T) + \frac{u(|\underset{\sim}{y}|)}{p}(S+S^T) - \frac{v'(|\underset{\sim}{y}|)|\underset{\sim}{y}|}{p}\,tr(S)I\}$$

or, if we separate diagonal and off-diagonal elements:

(5.10) $$E_c(\frac{1}{p}\,tr(U_1)) = \frac{1}{p}\,tr(S)\text{ave}\,\{\frac{1}{p}\,u'(|\underset{\sim}{y}|)|\underset{\sim}{y}| - v'(|\underset{\sim}{y}|)|\underset{\sim}{y}|\}$$

(5.11) $$E_c(U_{1_{jj}} - \frac{1}{p}\,tr(U_1)) = (S_{jj} - \frac{1}{p}\,tr(S))\,\frac{2}{p+2}\,\text{ave}\{\frac{1}{p}\,u'(|\underset{\sim}{y}|)|\underset{\sim}{y}| + u(|\underset{\sim}{y}|)\}$$

(5.12) $E_c(U_{1_{jk}}) = (S_{jk}+S_{kj}) \frac{1}{p+2} \text{ ave } \{\frac{1}{p} u'(|\underset{\sim}{y}|)|\underset{\sim}{y}| + u(|\underset{\sim}{y}|)\}, j \neq k$

Correspondingly, we have

$$(5.13) \qquad \frac{1}{p} \operatorname{tr}(U_0) = \text{ave } \{\frac{1}{p} u(|\underset{\sim}{y}|)-v(|\underset{\sim}{y}|)\}$$

$$(5.14) \qquad U_{0_{jj}} - \frac{1}{p} \operatorname{tr}(U_0) = \text{ave } \{u(|\underset{\sim}{y}|)(\frac{y_j^2}{|\underset{\sim}{y}|^2} - \frac{1}{p})\}$$

$$(5.15) \qquad U_{0_{jk}} = \text{ave } \{u(|\underset{\sim}{y}|) \frac{y_j y_k}{|\underset{\sim}{y}|^2}\}, \quad j \neq k$$

It turns out that the simplified equations (4.6), namely

$$U_0 + E_c(U_1) = 0$$

can be solved explicitly, which we shall do now. Evidently, only the symmetrized matrix $S + S^T$ enters into these equations. It is convenient to assume that S is lower triangular, then a lower triangular trial value V is improved to (I + S)V, which is also lower triangular, and where S is defined through

$$(5.16) \qquad \frac{1}{p} \operatorname{tr}(S) = - \frac{\text{ave } \{\frac{1}{p} u(|\underset{\sim}{y}|)-v(|\underset{\sim}{y}|)\}}{\text{ave } \{\frac{1}{p} u'(|\underset{\sim}{y}|)|\underset{\sim}{y}|-v'(|\underset{\sim}{y}|)|\underset{\sim}{y}|\}},$$

$$(5.17) \qquad S_{jj} - \frac{1}{p} \operatorname{tr}(S) = - \frac{p+2}{2} \frac{\text{ave } \{u(|\underset{\sim}{y}|)(\frac{y_j^2}{|\underset{\sim}{y}|^2} - \frac{1}{p})\}}{\text{ave } \{\frac{1}{p} u'(|\underset{\sim}{y}|)|\underset{\sim}{y}| + u(|\underset{\sim}{y}|)\}}$$

$$(5.18) \qquad S_{jk} = -(p+2) \frac{\text{ave } \{u(|\underset{\sim}{y}|) \frac{y_j y_k}{|\underset{\sim}{y}|^2}\}}{\text{ave } \{\frac{1}{p} u'(|\underset{\sim}{y}|)|\underset{\sim}{y}| + u(|\underset{\sim}{y}|)\}}, \ j > k,$$

$$S_{jk} = 0, \quad j < k.$$

6. *Large sample properties.* Assume that location $\underset{\sim}{\xi}$ and pseudo-covariance $(V^TV)^{-1}$ are estimated by solving (2.11), (2.12),

while the true underlying distribution belongs to a family (2.1). Without loss of generality we assume that it is spherically symmetric and suitably scaled, so that the true values of the parameters are $\xi^{(0)} = 0$ and $V^{(0)} = I$, i.e.

$$(6.1) \quad E(\Psi(\underset{\sim}{x})) = E(u(|\underset{\sim}{x}|)\frac{\underset{\sim}{x}\underset{\sim}{x}^T}{|\underset{\sim}{x}|^2} - v(|\underset{\sim}{x}|)I) = 0.$$

The more general case where the density is not affinely transformable into a spherically symmetric one presents difficulties.

Under mild regularity conditions, asymptotic existence, local uniqueness, consistency and asymptotic normality of the solutions $\underset{\sim}{\xi}$, $(V^TV)^{-1}$ can be established in a fairly straightforward way with the aid of the machinery developed by Huber [6] and Maronna [8]. However, a rigorous treatment is not intended here.

First, we note that the influence functions (Hampel [4]) of these estimates can be found by re-interpreting some of the formulas derived in the preceding section. From (5.5), we read off that the influence function for location is

$$(6.2) \qquad \frac{w(|\underset{\sim}{x}|)\underset{\sim}{x}}{E\{w(|\underset{\sim}{x}|) + \frac{1}{p}w'(|\underset{\sim}{x}|)|\underset{\sim}{x}|\}}$$

For V the influence function can be read off from (5.16) - (5.18). If we denote it by $S = IF(\underset{\sim}{x};\, V)$ we obtain

$$\frac{1}{p}\,tr(S) = -(\frac{1}{p}\,u(|\underset{\sim}{x}|)-v(|\underset{\sim}{x}|))\alpha$$

$$(6.3) \qquad S_{jj}-\frac{1}{p}\,tr(S) = -(u(|\underset{\sim}{x}|)(\frac{x_j^2}{|\underset{\sim}{x}|^2}-\frac{1}{p}))\,\frac{p+2}{2}\,\beta$$

$$S_{jk} = -u(|\underset{\sim}{x}|)\,\frac{x_jx_k}{|\underset{\sim}{x}|^2}\,(p+2)\beta, \quad j > k$$

$$S_{jk} = 0 \qquad\qquad , \quad j < k$$

with

$$(6.4)\qquad \begin{aligned}\alpha^{-1} &= E\{\tfrac{1}{p}\, u'(|\underset{\sim}{x}|)|\underset{\sim}{x}| - v'(|\underset{\sim}{x}|)|\underset{\sim}{x}|\},\\ \beta^{-1} &= E\{\tfrac{1}{p}\, u'(|\underset{\sim}{x}|)|\underset{\sim}{x}| + u(|\underset{\sim}{x}|)\}.\end{aligned}$$

For the pseudo-covariance $(V^T V)^{-1}$ itself, the influence function of course is $-(S+S^T)$.

The asymptotic variance and covariance properties of the estimates coincide with those of their influence functions and can easily be derived from (6.2) and (6.3). Note that for symmetry reasons location and covariance estimates are asymptotically uncorrelated and hence asymptotically independent.

Also, the location components $\hat{\xi}_j$ are asymptotically independent, with asymptotic variance

$$(6.5)\qquad n\,\mathrm{var}(\hat{\xi}_j) = \frac{\frac{1}{p}\, E((w(|\underset{\sim}{x}|)|\underset{\sim}{x}|)^2)}{\{E(w(|\underset{\sim}{x}|) + \frac{1}{p}\, w'(|\underset{\sim}{x}|)|\underset{\sim}{x}|)\}^2}$$

The asymptotic variances and covariances of the covariance estimates can be described as follows. We again assume that V is lower triangular.

$$(6.6)\qquad n\,\mathrm{var}(\tfrac{1}{p}\,\mathrm{tr}\hat{V}) = \frac{E((\frac{1}{p}\, u(|\underset{\sim}{x}|) - v(|\underset{\sim}{x}|))^2)}{\{E(\frac{1}{p}\, u'(|\underset{\sim}{x}|)|\underset{\sim}{x}| - v'(|\underset{\sim}{x}|)|\underset{\sim}{x}|)\}^2}$$

$$(6.7)\qquad n\,\mathrm{var}(\hat{V}_{jj} - \tfrac{1}{p}\,\mathrm{tr}\hat{V}) = \frac{(p-1)(p+2)}{2p^2}\,\lambda$$

$$(6.8)\qquad n\,E\{(\hat{V}_{jj} - \tfrac{1}{p}\,\mathrm{tr}\hat{V})(\hat{V}_{kk} - \tfrac{1}{p}\,\mathrm{tr}\,\hat{V})\} = \frac{p+2}{2p^2}\,\lambda, \quad j \neq k$$

$$(6.9)\qquad n\,\mathrm{var}(\hat{V}_{jk}) = \frac{p+2}{p}\,\lambda, \qquad j > k$$

with

$$(6.10)\qquad \lambda = \frac{E(u(|\underset{\sim}{x}|)^2)}{\{E(\frac{1}{p}\, u'(|\underset{\sim}{x}|)|\underset{\sim}{x}| + u(|\underset{\sim}{x}|))\}^2}\,.$$

All other asymptotic covariances between

$\frac{1}{p}\operatorname{tr}(\hat{V})$, $\hat{V}_{jj} - \frac{1}{p}\operatorname{tr}(\hat{V})$ and $\hat{V}_{jk}$ are 0.

7. *Least informative distributions: location.* Consider the family of distributions with density $f(\underset{\sim}{x};\underset{\sim}{\xi},I) = f(|\underset{\sim}{x}-\underset{\sim}{\xi}|)$, $\underset{\sim}{x}$, $\underset{\sim}{\xi} \in R^p$, where f belongs to some convex set $\mathcal{F}$ of densities. Assume that $\underset{\sim}{\xi}$ depends differentiably on a real parameter t. Then, Fisher information with respect to t is

$$
\begin{aligned}
I(f) &= E\{(\frac{d}{dt}\log f(\underset{\sim}{x};\underset{\sim}{\xi},I))^2\} \\
&= E\{(\frac{f'(|\underset{\sim}{x}|)}{f(|\underset{\sim}{x}|)}\frac{\dot{\underset{\sim}{\xi}}^T\underset{\sim}{x}}{|\underset{\sim}{x}|})^2\} \qquad (7.1)\\
&= E\{(\frac{f'(|\underset{\sim}{x}|)}{f(|\underset{\sim}{x}|)})^2\}\frac{|\dot{\underset{\sim}{\xi}}|^2}{p}.
\end{aligned}
$$

We now intend to find an $f_0 \in \mathcal{F}$ minimizing I(f). We know that in the one-dimensional location problem the maximum likelihood estimate based on the least informative f_0 possesses asymptotic minimax properties (Huber [5], [7]). This carries over, with some minor complications to the mentioned below, to the spherically symmetric p-dimensional case.

In our case, Fisher information I(f) is minimized by minimizing

$$E\{(\frac{f'(|\underset{\sim}{x}|)}{f(|\underset{\sim}{x}|)})^2\} = C_p\int_0^\infty(\frac{f'(r)}{f(r)})^2 r^{p-1} f(r)dr$$

where C_p denotes the surface of the unit sphere in R^p. This leads to the variational condition

$$\int_0^\infty\{-(\frac{f'}{f})^2 r^{p-1} - 2(\frac{f'}{f} r^{p-1})'\}\,\delta f\,dr \geq 0$$

under the side condition $\int r^{p-1}\,\delta f\,dr = 0$, or, for some Lagrange multiplier γ

$$4\gamma r^{p-1} - (\frac{f'}{f})^2 r^{p-1} - 2(\frac{f'}{f} r^{p-1})' = 0 \qquad (7.2)$$

on the set of r-values where f can be varied freely, with the equality sign to be replaced by ≥ 0 on the set where $\delta f \geq 0$.

With $u = \sqrt{f}$ we obtain the linear differential equation

$$u'' + \frac{p-1}{r} u' - \gamma u = 0, \tag{7.3}$$

valid on the set where f can be freely varied. The solutions of (7.3) can be described in terms of Bessel and Neumann functions, compare for instance [1], p. 227.

To give a specific example, let $p = 3$ and $\mathcal{F}$ be the set of spherically symmetric ε-contaminated normal distributions. Then (7.3) has the particular solution

$$u(r) = \frac{e^{-\sqrt{\gamma} r}}{r}. \tag{7.4}$$

Since f_0 and f_0'/f_0 should be continuous, we obtain after some calculations

$$\begin{aligned} f_0(r) &= a\, e^{-\frac{r^2}{2}}, && r \leq r_0 \\ &= b\, \frac{e^{-cr}}{r^2}, && r \geq r_0 \end{aligned} \tag{7.5}$$

with

$$a = (1-\varepsilon)(2\pi)^{-3/2}$$

$$b = (1-\varepsilon)(2\pi)^{-3/2}\, r_0^2\, e^{\frac{r_0^2}{2} - 2} \tag{7.6}$$

$$c = 2\sqrt{\gamma} = r_0 - \frac{2}{r_0}$$

and thus

$$\begin{aligned} -\frac{f_0'(r)}{f_0(r)} &= r, && r \leq r_0 \\ &= c + \frac{2}{r}, && r \geq r_0 \end{aligned} \tag{7.7}$$

Here, r_0 and ε are related by the requirement that f_0 be a probability density. In particular, we must have $c > 0$ and hence $r_0 > \sqrt{2}$; the limiting case $r_0 = \sqrt{2}$ corresponds to $\varepsilon = 1$.

It can be seen from (7.7) that the function $-\log f_0(|\underset{\sim}{x}|)$ is not convex. This causes some difficulties with proving uniqueness and consistency of the estimate when ε is large. For higher dimensions p, not even the qualitative behavior of f_0 is known, except when ε is small. However, it is probably not worthwhile in practice to work with such complicated minimax estimates; for our purposes - covariance estimation - any reasonably robust auxiliary estimate of location will do, just choose the weight function w so that the influence function (6.2) stays bounded.

Incidentally, it is possible to remove the assumption that f is spherically symmetric, both in this and in the following section. Note that Fisher information is a convex function of f, so by taking averages over the orthogonal group we obtain

$$I(\text{ave } f) \leq \text{ave } I(f),$$

where $\bar{f}$ = ave f is a spherically symmetric density. So, instead of minimizing I(f) for spherically symmetric f, we might minimize ave I(f) for more general f.

8. *Least informative distributions: covariance.* Consider the family of distributions

$$f(\underset{\sim}{x};\underset{\sim}{0},V) = |\det V| f(|V\underset{\sim}{x}|), \quad \underset{\sim}{x} \in R^p,$$

and assume that matrix V depends differentiably on a real parameter t. Then, Fisher information with respect to t at $V_0 = I$ is (because of symmetry it suffices to treat this special case):

$$I(f) = E\{(\frac{d}{dt} \log f(\underset{\sim}{x};\underset{\sim}{0},V))^2\}$$

(8.1)

$$= E\{(\text{tr}\dot{V} + \frac{f'(|\underset{\sim}{x}|)}{f(|\underset{\sim}{x}|)} \frac{\underset{\sim}{x}^T \dot{V} \underset{\sim}{x}}{|\underset{\sim}{x}|})^2\}.$$

In order to calculate this we first take the conditional expectation given $|\underset{\sim}{x}|$, i.e. we average over the uniform distribution on the spheres $|\underset{\sim}{x}|$ = const.

We shall use the following auxiliary results (see the appendix for proofs):

(8.2)
$$E\{\underset{\sim}{x}^T\dot{V}\underset{\sim}{x}\,|\,|\underset{\sim}{x}|\} = \lambda|\underset{\sim}{x}|^2$$
$$E\{(\underset{\sim}{x}^T\dot{V}\underset{\sim}{x})^2\,|\,|\underset{\sim}{x}|\} = (\lambda^2 + \frac{2\mu}{p+2})|\underset{\sim}{x}|^4 = \gamma|\underset{\sim}{x}|^4$$

with

(8.3)
$$\lambda = \frac{1}{p}\,\mathrm{tr}\,\dot{V},$$
$$\mu = \frac{1}{p}\,\Sigma\,\dot{V}_{jk}^2 - \lambda^2$$

(We may assume, without loss of generality, that $\dot{V}$ is symmetric). We put for short

(8.4)
$$u(r) = -\frac{f'(r)}{f(r)}\,r,$$

then (with C_p denoting the surface of the unit sphere in R^p)

(8.5)
$$E(u(|\underset{\sim}{x}|)) = -C_p \int \frac{f'(r)}{f(r)}\,r^p\,f(r)dr$$
$$= p\,C_p \int f\,r^{p-1}\,dr = p$$

and thus

(8.6)
$$I(f) = E\{(\frac{f'}{f}\,r)^2\gamma + 2\,\mathrm{tr}(\dot{V})(\frac{f'}{f}\,r)\lambda + (\mathrm{tr}\dot{V})^2\}$$
$$= C_p \int (u(r)^2\gamma - p^2\lambda^2)r^{p-1}\,f(r)dr$$
$$= \gamma\,C_p \int u(r)^2\,r^{p-1}\,f(r)dr - p^2\lambda^2.$$

In order to minimize $I(f)$ over $f \in \mathcal{F}$, it obviously suffices to minimize

(8.7)
$$J(f) = E(u(|\underset{\sim}{x}|)^2) = C_p \int_0^\infty u(r)^2\,r^{p-1}\,f\,dr$$
$$= C_p \int_0^\infty \frac{f'(r)^2}{f(r)}\,r^{p+1}\,dr.$$

A standard variational argument gives

(8.8)
$$\delta J(f) = C_p \int_0^\infty \{-u^2+2pu+2ru'\}r^{p-1}\,\delta f\,dr.$$

With the side condition $C_p \int r^{p-1} \delta f \, dr = 0$ we obtain that the u corresponding to the minimizing f_0 should satisfy

$$(8.9) \qquad 2ru' + 2pu - u^2 = c$$

for those r where f_0 can be varied freely, or

$$(8.10) \qquad -2ru' + (u-p)^2 = p^2 - c = \kappa^2.$$

For our purposes we only need the constant solutions corresponding to $u' = 0$. Thus

$$(8.11) \qquad u = p \pm \kappa$$

for some constant κ.

In particular, let

$$(8.12) \qquad \mathscr{F} = \{f | f(r) = (1-\varepsilon)\phi(r) + \varepsilon h(r),\ h \in \mathscr{M}_s\}$$

be the set of all spherically symmetric ε-contaminated normal densities, with

$$(8.13) \qquad \phi(r) = (2\pi)^{-p/2} e^{-r^2/2}$$

and $\mathscr{M}_s$ being the set of all spherically symmetric probability densities in R^p.

Then one easily verifies that J(f) and thus I(f) is minimized by choosing

$$(8.14) \qquad \begin{aligned} u(r) = -\frac{f_0'(r)}{f_0(r)} r &= a^2 && \text{for } 0 < r \leq a, \\ &= r^2 && \text{for } a \leq r \leq b, \\ &= b^2 && \text{for } b \leq r, \end{aligned}$$

and thus

$$(8.15) \qquad \begin{aligned} f_0(r) &= (1-\varepsilon)\phi(a)\left(\frac{a}{r}\right)^{a^2} && \text{for } 0 < r \leq a, \\ &= (1-\varepsilon)\phi(r) && \text{for } a \leq r \leq b, \\ &= (1-\varepsilon)\phi(b)\left(\frac{b}{r}\right)^{b^2} && \text{for } b \leq r. \end{aligned}$$

The constants a, b satisfy either

$$a = 0, \quad b^2 \geq 2p$$

or

$$0 < p - a^2 = b^2 - p = \kappa < p.$$

They have to be determined in such a way that the total mass of f_0 is 1. For actual computation, it is easier to begin with κ and to determine the corresponding ε: Choose $\kappa > 0$. Put

$$a = \sqrt{(p-\kappa)^+} \tag{8.16}$$
$$b = \sqrt{p+\kappa}$$

Determine ε from

$$\frac{1}{1-\varepsilon} = C_p\{\phi(a)\int_0^a (\frac{a}{r})^{a^2} r^{p-1}\, dr + \int_a^b \phi(r)r^{p-1}\, dr + \phi(b)\int_b^\infty (\frac{b}{r})^{b^2} r^{p-1}\, dr\}. \tag{8.17}$$

The maximum likelihood estimate of pseudo-covariance for f_0 can be described by (2.12) with u as in (8.14) and $v \equiv 1$. This estimate has the following minimax property: Let $\mathcal{F}_c \in \mathcal{F}$ be that subset for which it is a consistent estimate of the identity matrix. Then it minimizes the supremum over $\mathcal{F}_c$ of the asymptotic variances (6.6) to (6.10).

If $a > 0$, the least favorable density (8.15) is rather unrealistic in view of its singularity at the origin. In other words, one can legitimately object that the corresponding minimax estimate safeguards against an unlikely contingency. But there is also a counter argument: if one safeguards against outliers by censoring away a substantial amount of large values of u, without also censoring at the lower end, spurious tight clusters of points near the origin may cause breakdown by implosion.

It may be desirable to standardize the estimates such that they have the correct asymptotic values at the normal distribution.. This is best done by applying a correction factor τ^2 at the end. Note in particular, that with the u defined in (8.14) we have for standard normal observations $\underset{\sim}{x}$

$$E\{u(\tau|\underset{\sim}{x}|)\} = a^2\chi^2(p,a^2/\tau^2) + b^2(1-\chi^2(p,b^2/\tau^2)) \tag{8.18}$$

$$+ \tau^2 p(\chi^2(p+2, b^2/\tau^2) - \chi^2(p+2, a^2/\tau^2))$$

where $\chi^2(p,.)$ is the cumulative χ^2 distribution with p degrees of freedom. So we determine τ from $E\{u(\tau|\underset{\sim}{x}|)\} = p$ and then we multiply the pseudo-covariance $(V^T V)^{-1}$ found from (2.11), (2.12) by τ^2. Some numerical results are summarized in Section 10.

Note that these estimates are, essentially, ordinary covariances calculated from metrically Winsorized samples: observations for which $|V\underset{\sim}{x}| = |\underset{\sim}{y}| < a$ are moved radially outward to $|\underset{\sim}{y}| = a$, those for which $|\underset{\sim}{y}| > b$ are moved in to $|\underset{\sim}{y}| = b$, and V is adjusted until the thus modified transformed sample has unit covariance matrix.

Figure 1

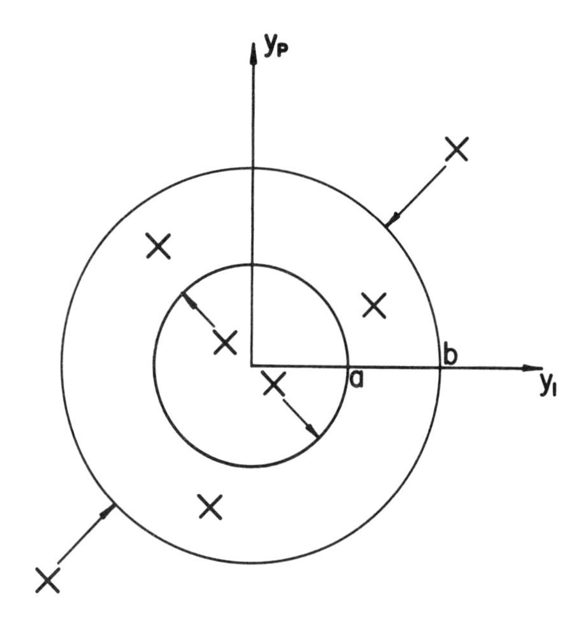

9. *Spherical symmetry? A note on models.* Apart from Section 4 we have been exclusively working within a family of models

which were either spherically symmetric or affinely transformable into spherically symmetric ones.

Such an assumption is reasonable if one is working with genuinely longtailed p-variate distributions. But for instance in the framework of the gross error models, the typical outliers will be generated by a process different from that of the main family and hence will tend to have quite a different covariance structure. For example, the main family may consist of a tight and narrow ellipsoid, with only a few principal axes significantly different from zero, while there is a diffuse and roughly spherical superimposed cloud of outliers. Or it is the outliers which show a structure and lie along some well defined lower dimensional subspaces, and so on. Of course, in an affinely invariant framework the two situations just mentioned are not really distinguishable.

But do we have the means to attack such separation problems directly, unless we have some definite prior information on the structures to be expected? I do not think so.

An estimate like that described in the previous section, just by being based on spherically symmetric models, may not only give an unprejudiced estimate of the overall shape of the principal part of a point cloud; it may also suggest an arbitrary, but objective separation into an inner ($|\underset{\sim}{y}| \leq a$), middle ($a < |\underset{\sim}{y}| < b$) and outer part ($|\underset{\sim}{y}| \geq b$).

From such a preliminary separation a more meaningful analysis of the composition of the point cloud might start off.

In any case, the empirical evidence with simulated data shows (see the next section) that these procedures are very well able to pick to main part of a point cloud under various types of contamination.

In my opinion, the restriction to affinely invariant estimates poses a much more serious problem. Section 4 suggests that in high dimensions no really robust, affinely invariant estimates may be available. Then one should presumably revert to coordinate dependent procedures. The practical turning point may lie between $p = 20$ to 50. Time series problems presumably are the prime examples where affine invariance makes little sense.

10. *Some quantitative results and empirical illustrations.* Table 1 collects some numerical results about the behavior of the least favorable distributions (8.15) and the correction factor τ^2 determined by (8.18). Note in particular that the mass above b is fairly constant over an extremely wide range of dimensions.

Figure 2 shows some experiments with simulated data. The composition of the samples is indicated at the lower edge of the pictures; for example 18N(1.0 1.0 0.9) 2N(9.0 9.0 -8.1) signifies that there are 18 observations from a bivariate normal distribution with covariance matrix

$$\begin{pmatrix} 1.0 & 0.9 \\ 0.9 & 1.0 \end{pmatrix}$$

and 2 observations with covariance matrix

$$\begin{pmatrix} 9.0 & -8.1 \\ -8.1 & 9.0 \end{pmatrix}$$

The mean vectors are all 0.

The solid lines show ellipses derived from the usual sample covariances, theoretically containing 80% of the total normal mass, the dotted lines derive from the estimate based on (8.14) with $\kappa = 2$ and correspond to the curve $|\underset{\sim}{y}| = b = 2$, which, asymptotically, also contains about 80% of the total mass if the underlying distribution is normal (79.7%, to be precise). Location is estimated simultaneously.

The pictures come in pairs, showing the identical sample with and without the contaminating points.

Appendix: Averages over the orthogonal group. Let $\underset{\sim}{x}$ be an arbitrary fixed p-vector and let $(x_1^S, \ldots, x_p^S)$ be the components of $S\underset{\sim}{x}$, where S is an orthogonal transformation. Then averaging over the orthogonal group gives

Table 1

			Mass of F_0		
	p		below a $a=\sqrt{(p-\kappa)^+}$	above b $b=\sqrt{p+\kappa}$	τ^2
$\varepsilon = 0.01$	1	4.1350	0	0.0332	1.0504
	2	5.2573	0	0.0363	1.0305
	3	6.0763	0	0.0380	1.0230
	5	7.3433	0	0.0401	1.0164
	10	9.6307	0.0000	0.0426	1.0105
	20	12.9066	0.0038	0.0440	1.0066
	50	19.7896	0.0133	0.0419	1.0030
	100	27.7370	0.0187	0.0395	1.0016
$\varepsilon = 0.05$	1	2.2834	0	0.1165	1.1980
	2	3.0469	0	0.1262	1.1165
	3	3.6045	0	0.1313	1.0873
	5	4.4751	0.0087	0.1367	1.0612
	10	6.2416	0.0454	0.1332	1.0328
	20	8.8237	0.0659	0.1263	1.0166
	50	13.9670	0.0810	0.1185	1.0067
	100	19.7634	0.0877	0.1141	1.0033
$\varepsilon = 0.10$	1	1.6086	0	0.1957	1.3812
	2	2.2020	0	0.2101	1.2161
	3	2.6635	0.0445	0.2141	1.1539
	5	3.4835	0.0912	0.2072	1.0908
	10	5.0051	0.1198	0.1965	1.0441
	20	7.1425	0.1352	0.1879	1.0216
	50	11.3576	0.1469	0.1797	1.0086
	100	16.0931	0.1523	0.1754	1.0043
$\varepsilon = 0.25$	1	0.8878	0.2135	0.3604	1.9470
	2	1.3748	0.2495	0.3406	1.3598
	3	1.7428	0.2582	0.3311	1.2189
	5	2.3157	0.2657	0.3216	1.1220
	10	3.3484	0.2730	0.3122	1.0577
	20	4.7888	0.2782	0.3059	1.0281
	50	7.6232	0.2829	0.3004	1.0110
	100	10.8052	0.2854	0.2977	1.0055

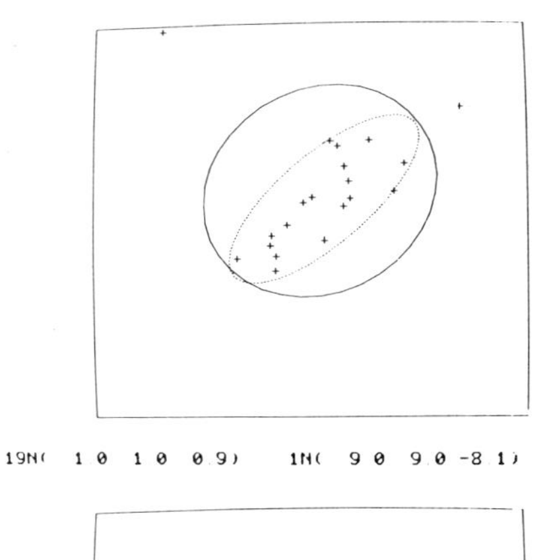

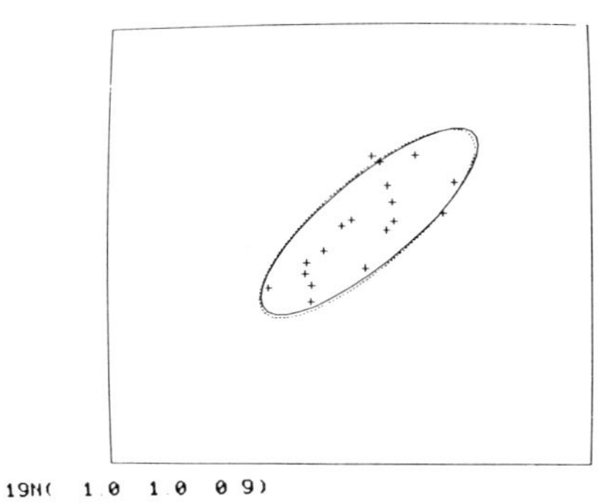

Figure 2a

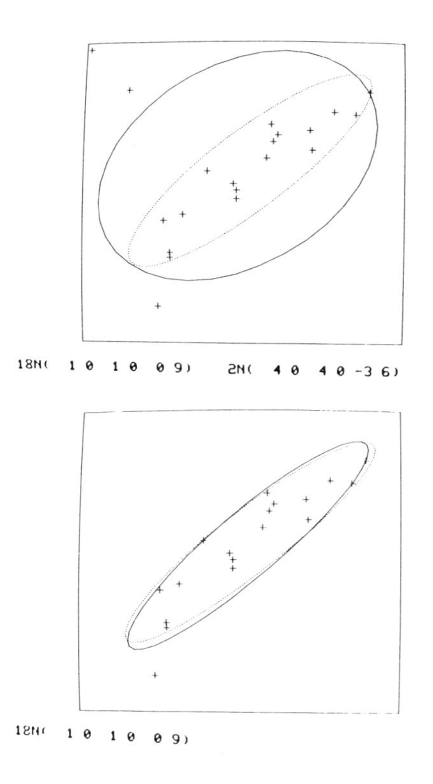

Figure 2b

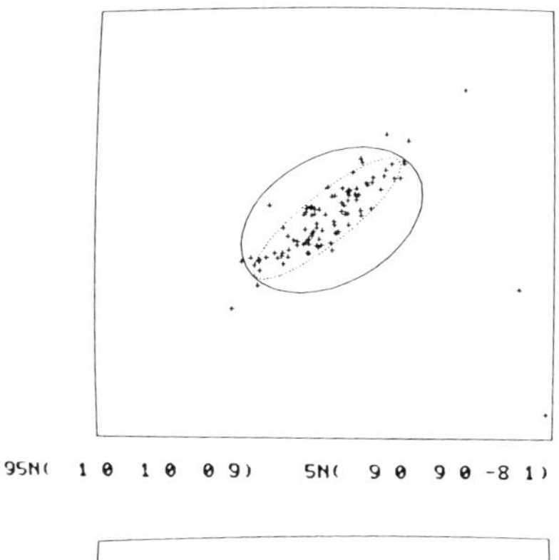

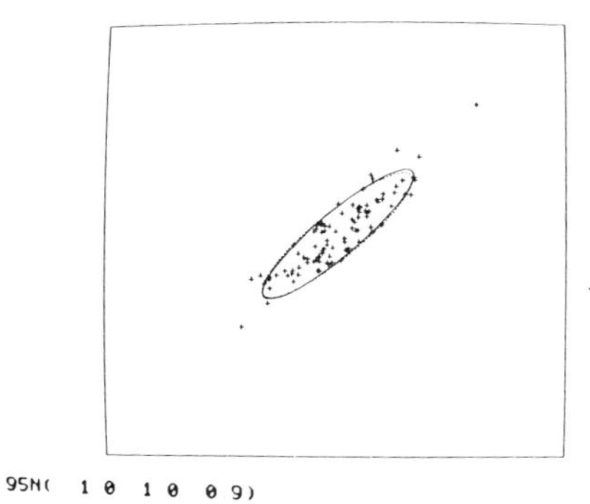

Figure 2c

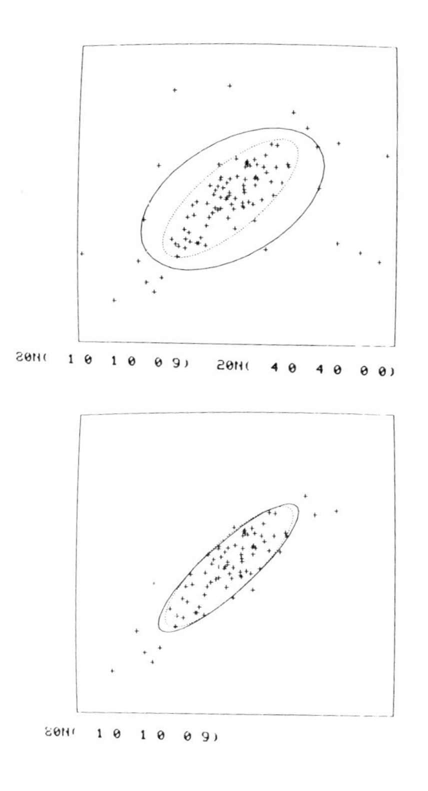

Figure 2d

$$(1)\qquad \operatorname*{ave}_S \{x_i^S x_j^S\} = \frac{1}{p} |\underset{\sim}{x}|^2 \delta_{ij},$$

$$(2)\qquad \operatorname*{ave}_S \{x_i^S x_j^S x_k^S x_l^S\} = \frac{1}{p(p+2)} |\underset{\sim}{x}|^4 (\delta_{ij}\delta_{kl} + \delta_{ik}\delta_{jl} + \delta_{il}\delta_{jk})$$

PROOF. For symmetry reasons, these averages are zero if one of the indices occurs an odd number of times, and their values remain unchanged under permutations of the indices. So we must have

$$(3)\qquad \operatorname*{ave}_S \{x_i^S x_j^S\} = \alpha |\underset{\sim}{x}|^2 \delta_{ij},$$

$$(4)\qquad \operatorname*{ave}_S \{x_i^S x_j^S x_k^S x_l^S\} = |\underset{\sim}{x}|^4 \{\beta(\delta_{ij}\delta_{kl} + \delta_{ik}\delta_{jl} + \delta_{il}\delta_{jk}) + \gamma\delta_{ijkl}\}$$

for some constants α, β, γ.

In order to determine these constants, we insert independent standard normal variables for the components of $\underset{\sim}{x}$ and take expectations on both sides of (3) and (4).

On the left hand side we can interchange the order of expectation and averaging. A comparison of coefficients yields (1) and (2).

We need the following immediate consequences of (1) and (2):

$$\operatorname{ave} \{\underset{\sim}{x}^T V \underset{\sim}{x}\} = \operatorname{ave} \{\sum_{ij} V_{ij} x_i x_j\} = \frac{1}{p} |\underset{\sim}{x}|^2 \operatorname{tr} V;$$

$$\operatorname{ave} \{(\underset{\sim}{x}^T V \underset{\sim}{x})^2\} = \operatorname{ave} \{\sum_{ijkl} V_{ij} V_{kl} x_i x_j x_k x_l\}$$

$$= \frac{|\underset{\sim}{x}|^4}{p(p+2)} \{(\operatorname{tr} V) + \sum_{ij} V_{ij}^2 + \sum_{ij} V_{ij} V_{ji}\},$$

$$\operatorname{ave} \{(\underset{\sim}{x}^T S \underset{\sim}{x}) \underset{\sim}{x}\underset{\sim}{x}^T\}_{kl} = \operatorname{ave} \{\sum_{ij} S_{ij} x_i x_j x_k x_l\}$$

$$= \frac{|\underset{\sim}{x}|^4}{p(p+2)} \{\operatorname{tr}(S)\, \delta_{kl} + S_{kl} + S_{lk}\}.$$

Acknowledgement. The search for the most general form of a covariance M-estimate was stimulated by F. Hampel in a seminar talk, and the result (3.2) has been found independently also by B. Stahel.

References

[1] Courant, R., and D. Hilbert (1937). *Methoden der Mathematischen Physik*, Vol. II. Springer, Berlin.

[2] Gnanadesikan, R., and J. R. Kettenring (1972). Robust estimates, residuals, and outlier detection with multiresponse data. *Biometrics* 28, 81-124.

[3] Hampel, F. R. (1974). The influence curve and its role in robust estimation. *J. Amer. Statist. Assoc.* 69, 383-393.

[4] Hampel, F. R. (1975). Beyond location parameters: Robust concepts and methods. 40th Session International Statistical Institute, Warsaw.

[5] Huber, P. J. (1964). Robust estimation of a location parameter. *Ann. Math. Statist.* 35, 73-101.

[6] Huber, P. J. (1967). The behavior of a maximum likelihood estimates under nonstandard conditions. *Proc. Fifth Berkeley Symp. Math. Statist. Prob.* 1, 221-233.

[7] Huber, P. J. (1972). Robust statistics: A review. *Ann. Math. Statist.* 43, 1041-1067.

[8] Maronna, R. A. (1976). Robust M-estimators of location and scatter. *Ann. Statist.* 4, 51-67.

ASYMPTOTICALLY MINIMAX ESTIMATION OF CONCAVE AND CONVEX DISTRIBUTION FUNCTIONS. II

By J. Kiefer* and J. Wolfowitz**

Cornell University and University of Illinois

0. *Introduction.* The present paper (II) is a sequel to our paper (I) of the same name, which appeared in the *Zeitschrift fuer Wahrscheinlichkeitstheorie und Verwandte Gebiete*, 34 (1976), 73-85. The notation and definitions of I will be adopted in toto, and familiarity with I is required for the understanding of II. To make II self-contained would require an intolerable amount of repetition from I. We therefore number the following section as the fifth (of the combined paper).

The purpose of the present paper is to prove Theorem 3 below. This result may be considered as essentially a generalization of Theorem 1 (this is not literally true). Conditions (3.2) are now replaced by (5.3) below, and the conclusions are very little different. Conditions (5.3) are still more restrictive than we would like, although a fairly formidable amount of argument is nevertheless required. To prove the theorem under essentially weaker conditions would require even more argument or a completely new idea or both.

Let $X_1,\ldots,X_n$ be independent chance variables with the common concave (respectively, convex) distribution function F, let F_n be their empiric distribution function, and let C_n be the least concave majorant (resp., greatest convex minorant) of F_n. The conclusions of both Theorems 1 and 3 may be stated approximately and simply as follows:

$$\sup_x |C_n(x) - F_n(x)| = o_p(n^{-\frac{1}{2}}).$$

*Research under NSF Grant MCS 75-22481.

**Research supported by the U. S. Air Force under Grant AFOSR-76-2877, monitored by the Office of Scientific Research.

The precise results are stated in Theorems 1 and 3, which explain the role of our result (Theorem 2) on estimation in the case described in the title.

In Section 5 we proceed as we did in I, and state our results and arguments for concave distribution functions. These apply, mutatis mutandis, to convex distribution functions.

To make things easier for the reader, we now restate Theorem 2 from I, with the additions described in the text of I. By concave (resp. convex) d.f. we mean a d.f. for which that property holds on the smallest interval $[\alpha_0(F), \alpha_1(F)]$ of F-probability one.

THEOREM 2. <u>Let</u> $\mathfrak{F}$ <u>be the set of all continuous concave</u> (resp. convex) d.f.'s <u>on</u> R^1. <u>Then</u>

$$\lim_{n\to\infty} \frac{\sup_{F\in\mathfrak{F}} r_n(F;F_n)}{\inf_{g_n} \sup_{F\in\mathfrak{F}} r_n(F;g_n)} = 1 \tag{4.5}$$

Because of Marshall's lemma (I, Lemma 3), F_n may be replaced by C_n in (4.5) above. This result also holds with any or all of the following modifications of $\mathfrak{F}$:(1)F can be permitted to have a jump at the left (resp., at the right) end of the interval of support; (2) finiteness of $\alpha_1(F)$ (resp., $\alpha_0(F)$), boundedness of α_0 and/or α_1 with known bounds, or exact knowledge of α_0 and/or α_1 can be assumed.

A usual treatment of the concave case is that in life testing, where $\alpha_0(F)$ is assumed to be zero. The obvious modifications in the definition of C_n, required under these variations, are discussed in Remark 1 of Section 3 of I.

5. *Estimating concave* F *under conditions* (5.3). In this section we weaken assumption (3.2) and conclusion (3.18). It is trivial that $||F_n-C_n|| \leq n^{-\frac{1}{2}}(\log n)^c$, $c > 0$, with high probability, and we shall only treat here cases where an exponent $< -\frac{1}{2}$ is attainable.

The outline of the proof is the same as for Theorem 1. The

principal change in detail is a more complex definition of $L^{(k)}$ (and, hence, $L_n^{(k)}$) in order to improve the estimate of Lemma 6. This redefinition in turn necessitates changes in the proof of Lemma 4. However, the form of $L^{(k)}$ here will be such that we will be able to refer to the estimates of Section 3 for much of the calculation.

The change in the definition of $L^{(k)}$ is made in order to handle behavior in a neighborhood of any point where $-f'(x)/f^2(x)$ approaches 0 or $+\infty$, corresponding to $\beta = 0$ or $\gamma = +\infty$ in (3.2). Near such a point, we use "local" analogues of β and γ to obtain a better bound. (The global values used in the simpler proof of Theorem 1 do not suffice to treat most of the examples listed at the end of this section, including those with $\alpha_1 = +\infty$.) At the same time, the probability assigned to intervals $[a_j^{(k)}, a_{j+1}^{(k)}]$ near such a point will depend on detailed behavior near the point, and cannot be taken as k^{-1} as it could under (3.2).

We introduce the concepts used in the replacement (5.3) of (3.2).

We define x' to be an <u>exceptional point</u> for F in $[\alpha_0(F), \alpha_1(F)]$ if f or f' is not both bounded and bounded away from 0 as $x \to x'$. In this definition, if x' is α_0 or α_1, the approach $x \to x'$ is one-sided. If $\alpha_1 = +\infty$, the value $x' = \alpha_1$ is always exceptional.

The development is made simpler if f or f' does not oscillate too wildly or approach 0 or ∞ too rapidly at an exceptional point. For $0 < \rho + F(x') < 1$, let x'_ρ satisfy $F(x'_\rho) = F(x') + \rho$. Here ρ can be positive or negative unless x' is α_0 or α_1. Write $I(x',\rho)$ for the open interval with end points x'_ρ and $x'_{2\rho}$. We shall say that F is <u>regular</u> at the isolated exceptional point x' (or that x' is a regular exceptional point) if

$$\text{(5.1)}\quad \begin{aligned} &\lim\sup_{\rho\to 0}\sup_{\xi',\xi''\in I(x',\rho)} f'(\xi')/f'(\xi'') < \infty,\\ &\lim\sup_{\rho\to 0}\sup_{\xi',\xi''\in I(x',\rho)} f(\xi')/f(\xi'') < \infty,\\ &\lim\inf_{\rho\to 0}\inf_{\xi',\xi''\in I(x',\rho)} f'(\xi')/f'(\xi'') > 0,\\ &\lim\inf_{\rho\to 0}\inf_{\xi',\xi''\in I(x',\rho)} f(\xi')/f(\xi'') > 0. \end{aligned}$$

The last two conditions are trivially satisfied except at $x' = \alpha_0$ or α_1; there is some redundancy in (5.1) at those points, which we shall not take the space to discuss.

In order to phrase our final simplifying regularity condition, we define, for ε small and positive and x' an isolated exceptional point,

$$\text{(5.2)}\quad \begin{aligned} \beta^+(\varepsilon;x',F) &= -f'(x'_\varepsilon)/f^2(x'_\varepsilon),\\ \beta^-(\varepsilon;x',F) &= -f'(x'_{-\varepsilon})/f^2(x'_{-\varepsilon}). \end{aligned}$$

Half of the definition is omitted if $x' = \alpha_0$ or α_1.

In place of (3.2) we now assume

(5.3)
- $\alpha_0(F) = 0$;
- F is continuous;
- F has only finitely many exceptional points, all regular;
- F is concave on $(0,\alpha_1(F))$, and is twice continuously differentiable there except possibly at the exceptional points;
- for each exceptional point x', there are values $\mu^+(x',F) > -1$ and $\mu^-(x',F) > -1$ such that, for $* = +$ or $-$, $\beta^*(\varepsilon;x',F) = \varepsilon^{\mu^*(x',F)+q(\varepsilon)}$ as $\varepsilon \downarrow 0$, where $q(\varepsilon) = o(1)$ as $\varepsilon \downarrow 0$.

We define $\bar\mu(F)$ to be the maximum of zero and of the quantities $\mu^+(x',F)$ and $\mu^-(x',F)$ over all exceptional points. Note that $\bar\mu(F) = 0$ for F satisfying (3.2). The possible values of the quantities μ^*, and the need for the inequality in the last condition of (5.3), will be seen later. Our result, weakening (3.2) but obtaining a weaker conclusion in place of (3.18) if $\bar\mu(F) > 0$, is

THEOREM 3. <u>If</u> F <u>satisfies</u> (5.3), <u>there is a function</u> τ_F <u>with</u> $\lim_{n\to\infty} \tau_F(n) = 0$ <u>such that</u>

$$(5.4) \quad P_F\{\|C_n - F_n\| > n^{-(\bar{\mu}(F)+2)/(2\bar{\mu}(F)+3)+\tau_F(n)}\} < 2n^{-2},$$

<u>so that, for every function</u> h <u>with</u> $\lim_{n\to\infty} h(n) = 0$,

$$(5.5) \quad P_F[h(n)n^{(\bar{\mu}(F)+2)/(2\bar{\mu}(F)+3)-\tau_F(n)}\|C_n - F_n\| = 0] = 1.$$

We note that the case $\bar{\mu}(F) = 0$ essentially includes Theorem 1, if one bothers to relate the function τ_F to the o(1) term in the last condition of (5.3). Before turning to the proof, we remark on the way in which the μ^* enter into (5.4), to aid in understanding our approach.

First, if the μ^* are all < 0, a smaller order than $n^{-2/3+\tau}$ is indeed achievable for the maximum of $|C_n - F_n|$ in a neighborhood (shrinking as $n \to \infty$) of each x'; however, at nonexceptional points the proof of Theorem 1 shows that the best order obtainable by our approach is $n^{-2/3+\tau}$, and this accounts for the presence of $\bar{\mu}$ rather than max u* in this case. The development in the neighborhood of x' in the case of negative μ^* (Part II of the proof) is thus aimed at achieving order $n^{-2/3+\tau}$ rather than the best order there.

Second, the proof is divided into four parts to help in simplifying the notation and understanding the detailed calculations by isolating the various modifications needed in the proof of Theorem 1. The first part contains the simplest modifications. It is also the longest since, having gone through those modifications fully in the case of a single exceptional x' and simplest μ^* as treated there, we can limit ourselves in the other three parts to describing the additional changes.

Third, the order of our development is to find a partition of $[\alpha_0, \alpha_1]$, with associated (varying) interval probabilities that are as small as possible, yet yield the analogue of Lemma 4, to see what this yields in Lemma 6 and Lemma 2, and to assemble the bounds as in the paragraph of proof below the statement of Theorem 1.

Throughout the proof we write c_i for positive constants and $\Lambda(n)$ or $\Lambda_i(n)$ for positive functions of the form $n^{o(1)}$. The symbol $o(1)$ always denotes a function which approaches 0 as $n \to \infty$; if it depends on other variables, the approach is uniform. The symbol $\Omega(1)$ denotes a positive function that is bounded away from 0 and ∞. The precise meaning of these symbols can vary even within the same equation; in contrast, Λ is sometimes used with a subscript when it is necessary to keep track of detailed relationships.

PART I. In this part of the proof we assume _there is a single exceptional point_ $x' = \alpha_0(F) = 0$, with $0 \leq \mu^+(x',F) = \bar{\mu}$. As will be discussed in Part III, the choice of x' is inessential and is made to simplify notation; the treatment of the behavior to the _left_ of an x' involves slight additional modification, described in Part III. (Part II treats a single $x' = \alpha_0$ when $\mu^+ < 0$, and Part IV treats the case of several exceptional points.) We shall now prove Theorem 3 under the above restriction.

In our proof, it will be necessary to vary the probabilities of intervals $[a_j^{(k)}, a_{j+1}^{(k)}]$. We alter the definitions of Section 2 as follows: For each k we consider a vector $(p_0^{(k)},\ldots,p_{k-1}^{(k)})$ of positive values which sum to 1, and determine the $a_j^{(k)}$ by $a_0^{(k)} = \alpha_0, a_k^{(k)} = \alpha_1$, and, in place of the rest of (2.6),

$$F(a_{j+1}^{(k)}) - F(a_j^{(k)}) = p_j^{(k)}, \; 0 \leq j \leq k-1. \tag{5.6}$$

We define

$$\sigma_j^{(k)} = \sum_{i=0}^{j} p_i^{(k)}, \; a_*^{(k)} = F^{-1}(p_0^{(k)}/2).$$

The function $L^{(k)}$ again satisfies (2.7) (for the $a_j^{(k)}$ of (5.6)), and, in addition,

(5.7) $$L^{(k)}(a_*^{(k)}) = F(a_*^{(k)}) = p_0^{(k)}/2.$$

The remainder of the definition of $L^{(k)}$, replacing that just above Lemma 4, is that $L^{(k)}$ is linear on each of the intervals $[a_*^{(k)}, a_1^{(k)}]$ and $[a_j^{(k)}, a_{j+1}^{(k)}]$, $1 \le j \le k-1$, and that

(5.8) $$L^{(k)}(x) = F(x) \quad \text{for} \quad 0 \le x \le a_*^{(k)}.$$

The use of this last modification will become apparent in the discussion of Lemma 6.

The function $L_n^{(k)}$ is defined by the following analogue of (2.8), in terms of the $L^{(k)}$ just given:

(5.9) $$L_n^{(k)}(x) = F_n(a_j^{(k)})+(p_j^{(k)})^{-1}[F_n(a_{j+1}^{(k)})-F_n(a_j^{(k)})][L^{(k)}(x) - F(a_j^{(k)})]$$

$$\text{for } a_j^{(k)} \le x \le a_{j+1}^{(k)},\ 0 \le j \le k.$$

Note that for $j = 0$ this is $F_n(a_1^{(k)})L^{(k)}(x)/p_0^{(k)}$; although the definition of $L^{(k)}$ is broken into two parts at $a_*^{(k)}$, the random multiplier $F_n(a_1^{(k)})$ is the same for the whole interval. Once more, $L_n^{(k)}$ is nondecreasing, and (2.9) holds for $0 \le j \le k$.

We define $\gamma_j^{(k)}$ and $\beta_j^{(k)}$, for $1 \le j \le k-1$, to be the values obtained if the supremum and infimum in (3.1) are taken over $(a_j^{(k)}, a_{j+1}^{(k)})$ instead of over $(0,\alpha_1(F))$. Similarly, $\gamma_0^{(k)}$ and $\beta_0^{(k)}$ are defined by computing these extrema over $(a_*^{(k)}, a_1^{(k)})$.

We begin by looking at the modifications needed in Lemma 4. For $j \ge 0$, the definition of $T_{n,j}$ is as given in (3.7) for the present $a_j^{(k_n)}$; for $\Delta_{n,j}$ the definition in (3.7) will still be used for $j \ge 1$, but in considering the event $B_{n,0}$ we will now use

(5.10) $$\Delta_{n,*} = a_1^{(k_n)} - a_*^{(k_n)}.$$

By differentiating (5.9) to the right and left of each

$a_j^{(k_n)}$, $1 \leq j \leq k_n - 1$, we see that the event A_n of (3.4) is now $(\cap_{j=1}^{k_n-2} B_{n,j}) \cap B_{n,*}$, where the $B_{n,j}$ are as defined in (3.8) and

$$(5.11) \qquad B_{n,*} = \{T_{n,0}/2\Delta_{n,*} \geq T_{n,1}/\Delta_{n,1}\}.$$

(The form of $L^{(k_n)}$ on $[0, a_1^{(k_n)}]$ insures that $L_n^{(k_n)}$ is concave there.)

For $1 \leq j \leq k_n - 2$, we now obtain, in place of (3.9) - (3.10), that $B_{n,j}$ follows from

$$(5.12) \qquad |T_{n,i} - p_i^{(k_n)}| \leq \delta_{n,j} p_i^{(k_n)} \quad \text{for } i = j,\ j+1$$

and

$$(5.13) \qquad p_j^{(k_n)} \Delta_{n,j+1} / p_{j+1}^{(k_n)} \Delta_{n,j} \geq 1 + 3\delta_{n,j},$$

where $0 < \delta_{n,j} < 1/3$. The relation (3.11) becomes

$$(5.14) \qquad \Delta_{n,j+1} = p_{j+1}^{(k_n)}/f(a_{j+1}^{(k_n)}) - (p_{j+1}^{(k_n)})^2 f'(\xi)/2f^3(\xi);$$

and $\Delta_{n,j} \leq p_{j+1}^{(k_n)}/f(a_{j+1}^{(k_n)})$ yields, in place of (3.12),

$$(5.15) \qquad \Delta_{n,j+1} p_j^{(k_n)} / \Delta_{n,j} p_{j+1}^{(k_n)} \geq 1 + \beta_{j+1}^{(k_n)} p_{j+1}^{(k_n)}/2.$$

Similarly, $B_{n,*}$ follows from (5.12) for $i = 0$ and, replacing (5.13),

$$(5.16) \qquad p_0^{(k_n)} \Delta_{n,1} / 2p_1^{(k_n)} \Delta_{n,*} \geq 1 + 3\delta_{n,0},$$

the line after (5.14) becomes $2\Delta_{n,*} \leq p_1^{(k_n)}/f(a_1^{(k_n)})$, and we thus also obtain, in place of (5.15),

$$(5.17) \qquad \Delta_{n,1} p_0^{(k_n)} / 2\Delta_{n,*} p_1^{(k_n)} \geq 1 + \beta_1^{(k_n)} p_1^{(k_n)}/2.$$

Thus, defining $\bar{\delta}_0 = \beta_1^{(k_n)} p_1^{(k_n)}/6$, $\bar{\delta}_{k_n-1} = \beta_{k_n-1} p_{k_n-1}/6$, and

$\bar{\delta}_i = \min_{j=i,i-1}\beta_{j+1}^{(k_n)} p_{j+1}^{(k_n)}/6$ for $1 \leq i \leq k_n-2$, we conclude that, if all $\bar{\delta}_i < 1/3$, the event A_n follows from

$$(5.18)\qquad |T_{n,i}-p_i^{(k_n)}| \leq \bar{\delta}_i p_i^{(k_n)}, \quad 0 \leq i \leq k_n-1.$$

If $\max_{0\leq i\leq k_n-1}\bar{\delta}_i$ and $\max_{0\leq i\leq k_n-1} p_i^{(k_n)}$ approach 0, we can thus use Lemma 1 and obtain, in place of (3.7),

$$(5.19)\ ,\qquad 1-P\{A_n\} \leq 2\sum_{i=0}^{k_n-1} e^{-np_i^{(k_n)}\bar{\delta}_i^2/3}$$

for large n.

In order to make the right side of (5.19) $\leq n^{-2}$, we shall choose the $p_i^{(k_n)}$ to be about as small as possible, so as to obtain the best advantage in the analogue of Lemma 6. In the calculations that follow, it will turn out that the first term ($i = 0$) of (5.19) already establishes the power of n attainable in Lemma 6, and the remaining terms are then merely chosen so as not to destroy that order.

We now establish that, for appropriate Λ_i, the choices $k_n = \Lambda_1(n)n^{1/3}$ and

$$(5.20)\qquad p_j^{(k_n)} = \begin{cases} n^{-1/(3+2\bar{\mu})}j^{-2\mu^+/(3+2\mu^+)}\Lambda_2(n) & \text{for } j \leq \Lambda_3(n)n^{1/3}, \\ n^{-1/3}\Lambda_4(n) & \text{otherwise} \end{cases}$$

make $1-P\{A_n\} \leq n^{-2}$. This calculation will occupy us through the paragraph of (5.26).

First, for $j \leq \Lambda_3(n)n^{1/3} = m_n$ (say) we obtain, since $3/(3+2\mu^+) > 0$,

$$(5.21)\qquad \sigma_j^{(k_n)} = n^{-1/(3+2\mu^+)}j^{3/(3+2\mu^+)}\Omega(1)\Lambda_2(n).$$

Hence, supposing m_n an integer (a matter of definition of Λ_3),

$$p_{m_n}^{(k_n)} = n^{-1/3}[\Lambda_3(n)]^{-2\mu^+/(3+2\mu^+)}\Lambda_2(n),$$

(5.22)
$$\sigma_{m_n}^{(k_n)} = [\Lambda_3(n)]^{3/(3+2\mu^+)}\Omega(1)\Lambda_2(n).$$

Whatever the choice of Λ_2 (to be determined below, from (5.25)), we choose Λ_3 so as to make $\sigma_{m_n}^{(k_n)} = 1/\log n$. (Any other sequence $\sigma_{m_n}^{(k_n)} = \Lambda(n) \to 0$ would work as well; Λ_2 can be chosen for large n to depend on this sequence and on q, but not circularly on Λ_3.) This step is not even necessary for the present case of a single $x' = 0$, but we have introduced it here to illustrate the procedure used in Part IV, of "localizing" the behavior of F and choice of $p_j^{(k_n)}$ in each of a number of small non-overlapping intervals containing an exceptional point as endpoint.

The function Λ_4 (and hence Λ_1) is determined so that $n^{-1/3}\Lambda_4(n)$ is as close as possible to $p_{m_n}^{(k_n)}$; exact equality may be impossible because $p_j^{(k_n)}$ is constant for $j \geq m_n$ and $\sum_0^{k_n-1} p_j^{(k_n)} = 1$. However, this choice makes $p_{m_n+1}^{(k_n)}/p_{m_n}^{(k_n)} \to 1$, so that $p_j^{(k_n)}$ varies "smoothly" on j at $j = m_n$. Thus, $p_j^{(k_n)}/p_{j+1}^{(k_n)} = \Omega(1)$ uniformly for $0 \leq j \leq k_n-1$. It follows from (5.3) and the fact that $\sigma_{m_n}^{(k_n)} \to 0$ that, for $1 \leq j \leq m_n$,

(5.23)
$$\beta_j^{(k_n)} = (\sigma_j^{(k_n)})^{\mu^+ + q(\sigma_j^{(k_n)})},$$

and hence that the exponent $-np_i^{(k_n)}\bar{\delta}_i^2/3$ of (5.19) is, uniformly for $0 \leq i \leq m_n-1$,

$$(5.24) \quad -n(p_{i+1}^{(k_n)})^3(\beta_{i+1}^{(k_n)})^2\Omega(1) =$$

$$- \Omega(1)[\Lambda_2(n)]^{3+2\mu^+ +o(1)} n^{o(1)}(1+i)^{o(1)},$$

where the three expressions o(1), obtained from q and (5.21), are not the same. The function Λ_2 is chosen large enough so that (5.24) will be $\le -(\log n)^2$ for $0 \le i \le m_n-1$.

Since $\beta_{m_n}^{(k_n)} = (\log n)^{-\mu^+ +o(1)}$ (possibly unbounded if $\mu^+ =0$) and $-f'/f^2$ is bounded, we have, uniformly for $j \ge m_n$,

$$(5.25) \quad \beta_j^{(k_n)}/\beta_{m_n}^{(k_n)} \le (\log n)^{\mu^+ +o(1)}.$$

Consequently, from (5.25), uniformly for $i \ge m_n$,

$$(5.26) \quad -np_i^{(k_n)}\frac{\bar{\delta}_i^2}{3} = -n(p_{m_n}^{(k_n)})^3(\beta_{i+1}^{(k_n)})^2\Omega(1)$$

$$\le - (\log n)^2(\beta_{i+1}^{(k_n)}/\beta_{m_n}^{(k_n)})^2\Omega(1)$$

$$\le - (\log n)^{2+2\mu^+ +o(1)}$$

From this and the end of the previous paragraph we conclude that each of the k_n terms in the sum (5.19) is $\le n^{-3}$ for n sufficiently large, uniformly in i, provided that $\max_i \bar{\delta}_i \to 0$ and $\max_i p_i^{(k_n)} \to 0$, and these are easily verified, completing the proof of the assertion containing (5.20).

Next, we turn to the analogue of Lemma 6. The proof of that lemma demonstrates that, for $1 \le j \le k-1$,

$$(5.27) \quad \sup_{a_j^{(k)} \le x \le a_{j+1}^{(k)}} |F(x)-L^{(k)}(x)| \le \gamma_j^{(k)}(p_j^{(k)})^2/2.$$

Similarly,

$$(5.28)\qquad \sup_{a_o^{(k)} \le x \le a_1^{(k)}} |F(x) - L^{(k)}(x)|$$
$$= \sup_{a_*^{(k)} \le x \le a_1^{(k)}} |F(x)-L^{(k)}(x)|$$
$$\le \gamma_*^{(k)}(p_1^{(k)})^2/8.$$

From (5.1) and (5.2), we have $\gamma_*^{(k)} = \beta_*^{(k)}\Omega(1)$ (which is where (5.8) is used), and $\gamma_j^{(k)} = \beta_j^{(k)}\Omega(1)$ uniformly in j. Consequently,

$$(5.29)\qquad ||F-L^{(k)}|| \le \Omega(1)\max_{j=*,1,\ldots,k_n-1}\beta_j^{(k)}(p_j^{(k)})^2.$$

While using k_n in (5.29) we recall the remark just above (5.23) (which, with the choice of $p_j^{(k_n)}$, allows us to limit the range of j to values $\le m_n$ in what follows) and the calculations of p_{j+1}/p_j and β_{j+1}/β_j which in like manner apply for p_1/p_* and β_1/β_*. From (5.20), (5.21), and (5.23) we obtain, for a suitable Λ_5,

$$||F-L^{(k_n)}|| \le \Omega(1)\max_{1\le j\le m_n}\beta_j^{(k_n)}(p_j^{(k_n)})^2$$
$$(5.30)\qquad = \max_{1\le j\le m_n} n^{-(2+\mu^+)/(3+2\mu^+)} j^{-\mu^+/(3+2\mu^+)}\Lambda_5(n)$$
$$= n^{-(2+\mu^+)/(3+2\mu^+)} .$$

(We explain briefly why it is sufficient to maximize over $1 \le j \le m_n$ in (5.30). Since 0 is the only exceptional point so far, f'/f^2 and thus $\beta_j^{(k)}$ is bounded and bounded away from 0 when j is large. Specifically, by (5.22) $\sigma_{m_n}^{(k_n)} = n^{o(1)}$ and hence $\beta_j^{(k_n)} = n^{o(1)}$ for $j \ge m_n$. Hence, for $j \ge m_n$,

$$\beta_j^{(k_n)}(p_j^{(k_n)})^2 = n^{o(1)} \times [n^{-1/3}\Lambda_4(n)]^2 = n^{-2/3+o(1)}$$

$$\leq n^{o(1)} \times [\text{bound on } \beta_{m_n}^{(k_n)}(p_{m_n}^{(k_n)})^2 \text{ used in (5.30)}]).$$

Finally, for the choice (5.20) of $p_j^{(k_n)}$ it is easily checked that, for an appropriate Λ_6, the proof of Lemma 2 is modified (treating intervals of varying probabilities and the modification (5.8)) to yield, with $\bar{p}^{(k_n)} = \max_j p_j^{(k_n)}$,

$$(5.31) \qquad P_F\{||F_n - F - L_n^{(k_n)} + L^{(k_n)}|| > \Lambda_6(n)[n^{-1}\bar{p}^{(k_n)}]^{\frac{1}{2}}\} < n^{-2}.$$

In the present case $\bar{p}^{(k_n)} = p_1^{(k_n)}$ and hence

$$\Lambda_6(n)[n^{-1}\bar{p}^{(k_n)}]^{1/2} = n^{-(2+\mu^+)/(3+2\mu^+)}\Lambda_7(n).$$

The final conclusions of the last three paragraphs, when substituted for the roles of Lemma 4, Lemma 6, and Lemma 2 in the paragraph of proof just below (3.19), yield the proof of Theorem 3 in our special case of Part I.

PART II. We again suppose <u>there is a single exceptional point</u> $x' = \alpha_0(F) = 0$, but now $-1 \leq \mu^+(x',F) < 0$. We shall now indicate as succinctly as possible the way in which the proof of Part I must be altered to prove Theorem 3 under the present restriction.

This is the case in which, as mentioned earlier, one can achieve a "local" improvement in the order of $|C_n - F_n|$ near x', but it is pointless to do so because the behavior at nonexceptional points makes -2/3 the most negative possible power of n (for our method of proof) in (5.4). This fact means that, rather than using the analogue of Lemma 4 to find the smallest order of $p_j^{(k_n)}$ that will make (5.19) small enough, and having that determine the best order obtainable in Lemmas 6 and 2, we could instead now let the order $n^{-2/3+\tau}$ in Lemmas 6 and 2 determine the $p_j^{(k_n)}$ and then check Lemma 4. Since then less is "wasted" in making the deviation obtained from Lemmas 6 and 2 smaller than needed near x', it might be thought that the argument could apply for a larger range of μ^+ values than does the determination based on Lemma 4. Formally this is true except for one crucial point. In order for the argument of the analogue of Lemma 4 to make sense, $\beta_j^{(k_n)} p_j^{(k_n)}$ must not be unbounded, as we see from (5.15) and the fact that $\delta_{n,i}$ must be < 1/3 (or, with alterations, < 1). But it turns out that the minimum μ^+ value for which this is so is -1, whichever way of determining the $p_j^{(k_n)}$ is used. So the extra applicability of the second method is illusory, and for the sake of uniformity we use the method of Part I.

Thus, the $p_j^{(k_n)}$ are determined so as to yield (5.26). It turns out that such a determination is then again given by (5.20), with possible changes of the Λ_i. The proof of Part I is now used, with minor exceptions which we shall list.

The development through (5.19) remains intact.

Since the exponent of j in (5.21) is still > 0, that formula is still correct.

The only other place in which the sign of μ^+ enters is in the development leading to (5.26). The exponent of the lower bound $(\log n)^{\mu^+ + o(1)}$ is now negative, and at the end of the previous paragraph Λ_2 must consequently be chosen to make (5.25) $\leq -(\log n)^{2-2\mu^+}$. The result (5.26) (in particular, the inequality) is then valid.

Once more $\max_i p_i^{(k_n)} \to 0$ is clear, and for the essential domain $1 \leq j \leq m_n$ we compute, from (5.20), (5.21), and (5.23),

$$\beta_j^{(k_n)} p_j^{(k_n)}$$

(5.32)
$$= \Omega(1) n^{-[\mu^+ + 1 + o(1)]/(3+2\mu^+)} j^{(\mu^+ + 3[o(1)])/(3+2\mu^+)} \times [\Lambda_2(n)]^{1+\mu^+ o(1)},$$

where o(1) has the same meaning in all three appearances; namely, it is $q(\sigma_j)$. Thus, $\max_i \bar{\delta}_i \to 0$, provided $\mu^+ > -1$. (We have written out (5.32) in order to indicate the difficulty when $\mu^+ \leq -1$. A more careful analysis might take care of the value $\mu^+ = -1$ for some q, but a completely different device is needed when $\mu^+ < -1$.) This completes the proof in our special case of Part II.

PART III. We now suppose that <u>there is a single exceptional</u> $x' = \alpha_1(F)$ with $-1 < \mu^-(x', F) < \infty$. In order to use the previous developments with a minimum of rewriting, it is convenient, <u>for this Part III only</u>, to denote the interval endpoints by $\alpha_0(F) = b_k^{(k)} < b_{k-1}^{(k)} < \ldots < b_1^{(k)} < b_*^{(k)} < b_0^{(k)} = \alpha_1(F)$, with

(5.33)
$$F(b_j^{(k)}) - F(b_{j+1}^{(k)}) = \pi_j^{(k)}, \quad 0 \leq j \leq k-1,$$
$$F(b_0^{(k)}) - F(b_*^{(k)}) = \pi_0^{(k)}/2,$$

where $(\pi_0^{(k)}, \ldots, \pi_{k-1}^{(k)})$ is again a vector of positive values summing to 1. Essentially all of Parts I and II proceeds for the present case with minor notational changes, upon replacing $[a_j^{(k)}, a_{j+1}^{(k)}]$ and $p_j^{(k)}$ there by $[b_{j+1}^{(k)}, b_j^{(k)}]$ and $\pi_j^{(k)}$ (with $b_*^{(k)}$ for $a_*^{(k)}$, etc.) in the present setting. We let $\Delta_{m,j}$, $\beta_j^{(k)}$, and $\gamma_j^{(k)}$ refer to the interval $(b_{j+1}^{(k)}, b_j^{(k)})$ here. The one major change is caused by the fact that, because of the monotonicity of f that led to the line just below (5.14), the right side of the relation (5.15) refers to the interval <u>to the right</u> of $a_j^{(k_n)}$. Thus,

$\beta_*^{(k_n)} p_*^{(k_n)}$ did not occur in the analogue of Lemma 4 in Part I, but in the present case the quantity $\beta_*^{(k_n)} \pi_o^{(k_n)}$ occurs. The analogue of the development (5.14), (5.16), and (5.17) is now, with $b_1^{(k_n)} < \xi < b_*^{(k_n)}$,

$$\Delta_{n,*} = (\pi_o^{(k_n)}/2)f(b_1^{(k_n)}) - (\pi_o^{(k_n)}/2)^2 f'(\xi)/2f^3(\xi),$$

$$2\Delta_{n,*}\pi_1^{(k_n)}/\Delta_{n,1}\pi_o^{(k_n)} \geq 1 + 3\delta_{n,o},$$

$$\Delta_{n,1} \leq \pi_1^{(k_n)}/f(b_1^{(k_n)}), \tag{5.34}$$

$$2\Delta_{n,*}\pi_1^{(k_n)}/\Delta_{n,1}\pi_o^{(k_n)} \geq 1+\beta_*^{(k_n)}\pi_o^{(k_n)}/4.$$

This and the previously argued regularity of p_j/p_{j+1} and β_j/β_{j+1} make the previous arguments apply with obvious changes.

We note here the second role of the form (5.8); without that adjustment of the piecewise linear $L^{(k)}$ of Section 3, we would have $\beta_o^{(k_n)}$ in place of $\beta_*^{(k_n)}$ in (5.34), and this would be infinite if $\mu^- < 0$. The other role of (5.8), as in Part I, is to replace $\gamma_o^{(k_n)}$ by $\gamma_*^{(k_n)}$ in the analogue of Lemma 6.

PART IV. We <u>now treat the general case assumed in Theorem 3</u>. For any exceptional point x' interior to $[\alpha_o,\alpha_1)$, we use the construction of Parts I and II with $[0,F^{-1}(1/\log n)]$ replaced by the interval $[x',F^{-1}(F(x') + 1/\log n)]$, and that of Part III with $[F^{-1}(1 - 1/\log n), \alpha_1(F)]$ replaced by $[F^{-1}(F(x') - 1/\log n), x']$, to obtain on each of these a set of $n^{1/3}\Lambda(n)$ subintervals and corresponding probabilities that yield n^{-3} for the probability of (5.4) when $\sup|F_n - C_n|$ is computed only on either of these two intervals. The construction is such that, near each of the interval endpoints other than x', the subinterval has probability $\Lambda(n)n^{-1/3}$. Thus, if x" is the next exceptional point to the right of x', for large n the interval $J = [F^{-1}(F(x')+1/\log n), F^{-1}(F(x")-1/\log n)]$ can be subdivided into subintervals of varying probability uniformly of the form $n^{-1/3}\Lambda(n)$ in such a way as to maintain the smoothness of p's and β's for neighboring intervals throughout $[0,\alpha_1(F))$. As in the paragraph containing (5.26), we obtain $n^{-2/3+o(1)}$ for the supremum on J of $|F_n - C_n|$ that is exceeded with probability $< n^{-1/3}$. Adding the contributions to (5.4) from the various J's and sequences of shrinking intervals with exceptional endpoints, we obtain Theorem 3.

REMARKS. The remarks of the previous section apply to the present setting (5.3) as well as to (3.2). We now give examples of F's that exhibit the various values of μ^*.

(a) <u>The point</u> x' = 0. We go through only one calculation in detail, and merely state the results in other cases. Suppose that, for x sufficiently small,

$$F(x) = x - x^{r+1}/(r+1) \tag{5.35}$$

for some $r > 0$. Then $f(x) = 1-x^r$ and $f'(x) = -rx^{r-1}$, so that $-f'(x)/f^2(x) \sim rx^{r-1}$ for x small. Since $F^{-1}(\varepsilon) \sim \varepsilon$ as $\varepsilon \downarrow 0$, we obtain $\mu^+ = r-1$. In similar fashion it can be checked that (again with $r > 0$ and the behavior studied being near $x = 0$)

$$\begin{aligned} F(x) &= (\log x^{-1})^{-r} \Longrightarrow \mu^+ = -r^{-1}-1, \\ F(x) &= x(\log x^{-1})^{r} \Longrightarrow \mu^+ = -1, \\ F(x) &= x^{r'}(\log x^{-1})^{r''} \Longrightarrow \mu^+ = -1, \end{aligned} \tag{5.36}$$

the last with r" arbitrary and $0 < r' < 1$. To summarize, <u>all real values are possible for</u> μ^+.

(b) <u>The cases</u> $0 < x' < \alpha_1$ <u>and</u> $x' = \alpha_1 < \infty$ <u>with</u> $f(\alpha_1) > 0$. These two cases exhibit the same behavior, and we consider the latter. If $\alpha_1 = 1$, an analogue of (5.35) is (with $r > 0$, $c > 0$, and stated behavior near $x = 1$)

$$F(x) = (x+c)/(1+c) - (1-x)^{r+1}/(r+1) \Longrightarrow \mu^- = r-1. \tag{5.37}$$

The values $\mu^- < -1$ cannot occur in this case if $f(\alpha_1) > 0$ (or, similarly, for $0 < x' < \alpha_1$), as a simple argument shows. Suppose $\alpha_1 = 1$. Then, writing $H = -(\mu^- + 1)/2 > 0$, we have, for u sufficiently near 1, from (5.3),

$$-f'(u)/f(u) > f(u)[1-F(u)]^{-(H+1)}. \tag{5.38}$$

Integration, with $u_1 < u_2 < 1$, yields

$$\log \frac{f(u_1)}{f(u_2)} > H^{-1}\{[1-F(u_2)]^{-H} - [1-F(u_1)]^{-H}\}. \tag{5.39}$$

Letting $u_2 \to 1$, the left side remains bounded and the right side is unbounded, yielding a contradiction. If $\mu^- = -1$ and $q \leq 0$, a similar proof of impossibility can be given.

(c) <u>The case</u> $\alpha_1 < \infty$, $f(\alpha_1) = 0$. Here μ^- is completely determined. Suppose $\alpha_1 = 1$, and, if $\mu^- < -1$, rewrite (5.39) as $f(u_2) < ce^{-H^{-1}[1-F(u_2)]^{-H}}$. Integrating from $F^{-1}(1-\varepsilon)$ to 1, we obtain

$$\varepsilon < \int_{F^{-1}(1-\varepsilon)}^{1} ce^{-H^{-1}[1-F(u_2)]^{-H}} du_2 < \int_{F^{-1}(1-\varepsilon)}^{1} ce^{-H^{-1}\varepsilon^{-H}} du_2 \tag{5.40}$$

$$< ce^{-H^{-1}\varepsilon^{-H}},$$

giving a contradiction as $\varepsilon \downarrow 0$. Thus, we cannot have $\mu^- < -1$. Similarly, if $\mu^- > -1$, write $H = (\mu^- + 1)/2 > 0$, and (5.3) yields, for u near 1,

$$-f'(u)/f(u) < f(u)[1-F(u)]^{-1+H}, \tag{5.41}$$

which, after integration from u_1 to u_2, gives

$$\log[f(u_1)/f(u_2)] < H^{-1}\{[1-F(u_1)]^{H} - [1-F(u_2)]^{H}\}, \tag{5.42}$$

again producing a contradiction as $u_2 \uparrow 1$. We conclude that $\mu^- = -1$ in this case.

(d) <u>The case</u> $\alpha_1 = +\infty$. Here $\mu^- \leq -1$ again. In fact, one cannot have $-f'(x)/f^2(x) < c[1-F(x)]^{-1}$ for all $x > x_0$ (say) and constant $c < 1$. To see this, integrate $-f'/f$ as before to obtain, after slight manipulation, $f(x)/[1-F(x)]^c > c_1$ for $x > x_0$, where $c_1 > 0$. Integration from x_0 to $+\infty$ yields a contradiction. In the other direction, if $1-F(x) = x^{-r}$ for $r > 1$ as $x \to \infty$ we obtain $\mu^- = -1$ and $\varepsilon^{q(\varepsilon)}$ constant. Unlike the situation in (c), we can now have $\mu < -1$: if $F(x) = 1-(\log x)^{-r}$ for $r > 0$ and large x, we obtain $\mu^- = -1-r^{-1}$.

Unfortunately, the interesting case $\alpha_1 = +\infty$, $\mu^- = -1$ requires further research.

SEQUENTIAL DECISION ABOUT A NORMAL MEAN

By T. L. Lai*, Herbert Robbins* and D. Siegmund**

Columbia University

1. *Introduction*. Let $x_1, x_2, \ldots$ be independent and normally distributed with mean μ and variance σ^2. If σ is known, one can test the compound hypothesis H_0: $\mu < 0$ against H_1: $\mu > 0$ by a symmetric sequential probability ratio test of the surrogate simple hypothesis $\mu = -\mu_0$ against $\mu = \mu_0$, where μ_0 is a fixed positive number (cf. Wald, 1947). Such a test involves a constant $c > 0$ and the stopping rule

$$N = \text{first } n \geq 1 \text{ such that } |s_n| \geq c\sigma^2/\mu_0, \tag{1.1}$$

where $s_n = x_1 + \ldots + x_n$, H_0 or H_1 being accepted according as $s_N < 0$ or $s_N > 0$. Let $b = c/\mu_0$ for notational convenience. Wald's approximations for the error probability and expected sample size of this test are

$$P_{\mu,\sigma}\{s_N < 0\} = P_{-\mu,\sigma}\{s_N > 0\} \leq 1/(1+e^{2\mu b}) \quad (\mu \geq 0) \tag{1.2}$$

and

$$\begin{aligned} E_{\mu,\sigma}(N) &\cong \frac{b\sigma^2}{|\mu|}\left(\frac{e^{2|\mu|b} - 1}{e^{2|\mu|b} + 1}\right) \quad && (\mu \neq 0) \\ &\cong b^2\sigma^2 && (\mu = 0). \end{aligned} \tag{1.3}$$

It is customary to regard the inequality in (1.2) as an approximate equality. It follows from (1.2) that for any given $\epsilon > 0$ the probability of error can be made $\leq \epsilon$ for all $|\mu| \geq \mu_0$ by choosing b appropriately.

The purpose of this note is to present a simple modification of (1.1) which defines an analogous test of H_0: $\mu < 0$ against H_1: $\mu > 0$ when σ is unknown.

* Research supported by National Science Foundation Grant NSF-MPS-7518471 at Columbia University.

** Research supported by ONR Grant N00014-75C-0560 at Columbia University.

Before proceeding we note that the usual sequential t test (cf. Ghosh, [2], p. 300) is not "analogous" to the test described above. It is a test about μ/σ in the sense that both its error probability and expected sample size depend on the pair (μ,σ) only through their ratio. (In the case where σ is known such a test may be obtained by replacing σ^2 by σ in (1.1).) By way of contrast, for the test given above the error probability and expected sample size are genuine functions of the two parameters μ and σ. Previous investigations of sequential tests having these properties have been made by Stein [9], Baker [1], Hall [3], and Weiss and Wolfowitz [11].

Let $\bar{x}_n = s_n/n$, and for any $k_1, k_2 \geq 0$ let

$$v_n^2 = \sum_{i=1}^{n} (x_i - \bar{x}_n)^2/h_n, \quad \text{where} \tag{1.4}$$
$$h_n = h_n(k_1,k_2) = \max(1, n - k_1 n^{\frac{1}{2}} - k_2).$$

For $b > 0$, by analogy with (1.1), define

$$T = \text{first } n \geq 5 \text{ such that } |s_n| \geq b\, v_n^2. \tag{1.5}$$

To test H_0: $\mu < 0$ against H_1: $\mu > 0$ we propose to terminate sampling with the stopping rule T and accept H_0 or H_1 according as $s_T < 0$ or $s_T > 0$. In this paper we show by simulation for $(k_1,k_2) = (1,1.5)$ and $(1.4,0)$ and a range of values of μ, σ, and b that the error probability of this test is moderately well approximated by the left hand side of (1.2). Thus by an appropriate choice of b and (k_1,k_2) *we can make the probability of error less than an arbitrary preassigned* ϵ *for all* $|\mu| \geq \mu_0$ *and all* $\sigma \geq 0$, even though the value of σ is unknown. We also give a useful heuristic approximation for the expected sample size based on asymptotic considerations.

Remark. Replacing v_n^2 by v_n in (1.5) yields a test for which the error probability and expected sample size depend only on μ/σ, as in the sequential t test. Results paralleling those presented below hold for this test, and indeed the corresponding heuristic approximations seem to be somewhat more accurate in this case.

2. *Probability of error.* For the stopping rule (1.1) the inequality (1.2) becomes an equality as $\sigma \to \infty$, or equivalently as $\mu \to 0$, $b \to \infty$ in such a way that μb remains fixed. Standard weak convergence arguments show that $P_{\mu,\sigma}\{s_T < 0\}$ converges to the same limit. Nevertheless, for most interesting values of μ, σ, and b the right hand side of (1.2) is quite a bit larger than the left. We have attempted to determine through simulation a value of (k_1, k_2) in (1.4) so that for T defined by (1.5)

$$(2.1) \qquad P_{\mu,\sigma}\{s_T < 0\} \tilde{=} P_{\mu,\sigma}\{s_N < 0\}$$

for a range of interesting values for μ, σ, and b. The probability on the right in (2.1) has been evaluated according to the approximation given by Siegmund [7], which is known to be very accurate over the range of parameter values selected. To estimate the left hand side of (2.1) by simulation, a slight modification of the variance reducing technique discussed by Siegmund [8] was used.

In Tables 1 and 2 we have put $\alpha = P_{\mu,\sigma}\{s_N < 0\}$ and $\hat{\alpha}$ = the estimated value of $P_{\mu,\sigma}\{s_T < 0\}$. The $\pm$ figures are one estimated standard error of $\hat{\alpha}$.

Table 1: Error Probabilities: $\sigma = 1$, $b = 5$

μ	$\hat{\alpha}$		α	Wald Approximation to α
	$k_1=1$, $k_2=1.5$	$k_1=1.4$, $k_2=0$		
.075	.2848 $\pm$ .0026	.2867 $\pm$ .0019	.3025	.3208
.125	.1811 $\pm$ .0031	.1781 $\pm$ .0019	.1985	.2227
.25	.0535 $\pm$ .0022	.0544 $\pm$ .0015	.0579	.0759
.50	.0062 $\pm$ .0010	.0088 $\pm$ .0008	.0037	.0067

For Table 2 we have set $b = 6$, for which the Wald approximation to $P_{\mu,\sigma}\{s_N < 0\}$ is .0474 at $\mu = .25$, and for $\mu = .25$ we have studied the error probability as a function of σ. It is helpful to keep in mind that changing σ to $\sigma\lambda$ is equivalent to changing the pair (μ, b) to $(\mu\lambda^{-1}, b\lambda)$ with σ held fixed.

Table 2: Error Probabilities: $\mu = .25$, $b = 6$

σ	$\hat{\alpha}$		α
	$k_1=1$, $k_2=1.5$	$k_1=1.4$, $k_2=0$	
.5	.0160 ± .0012	.0197 ± .0010	.0271
.67	.0287 ± .0019	.0311 ± .0013	.0312
1.00	.0367 ± .0019	.0375 ± .0014	.0359
1.25	.0371 ± .0018	.0427 ± .0016	.0379
1.67	.0400 ± .0019	.0422 ± .0016	.0401
2.50	.0428 ± .0019	.0406 ± .0012	.0424
3.33	.0432 ± .0016	.0401 ± .0011	.0437
∞	.0474	.0474	.0474

Remarks. (a) For practical purposes there is probably no important difference between the error probabilities attained by our two choices of (k_1,k_2). However, the apparent lack of monotonicity of the error probability as a function of σ for the choice $k_1 = 1.4$, $k_2 = 0$ given in Table 2 makes it seem somewhat less satisfactory. A second objection to this choice is that the error probability seems to be more sensitive to the (arbitrary) minimum sample size in (1.5). For example, when the number 5 in (1.5) was replaced by 4, for $b = 5$, $\mu = .25$, $\sigma = 1$ we obtained $\hat{\alpha}=$ .0561 in the case $k_1 = 1$, $k_2 = 1.5$ and $\hat{\alpha} = .0633$ in the case $k_1 =$ 1.4, $k_2 = 0$.

(b) It is disappointing that even in the row $\sigma = 3.33$ of Table 2 (equivalently $\sigma = 1$, $\mu = 20$, $b = .075$) one does not see unambiguous evidence of the convergence of $P_{\mu,\sigma}\{s_T < 0\}$ and α to the limiting value .0474. As we shall see in the next section these parameter values correspond to an expected sample size of about 270, which is in the range for which large sample theory might be expected to apply if it is to be useful.

(c) In the last row of Table 1 both $\hat{\alpha}$ and α are quite small, but the ratio of $\hat{\alpha}$ to α is about 2 to 1. This considerable relative discrepancy is presumably unavoidable to a certain extent. For example, for fixed $\mu > 0$ as $b \to \infty$ $P_{\mu,\sigma}\{s_T < 0\} \geq P_{\mu,\sigma}\{s_5 < -bv_5^2\} =$ $\exp(-o(b))$, whereas by (1.2) $\alpha \leq \exp(-2\mu b)$. This phenomenon persists even if the minimum sample size 5 in (1.5) is replaced by an arbitrary number n_0 which may depend on b but diverges more slowly than b. Hopefully, we have confined this difficulty to points in the parameter space at which the error probabilities are so small that a considerable relative change is not very important.

3. *Expected sample size.* Wald's approximation (1.3) to $E_{\mu,\sigma}(N)$ is based on regarding s_N as approximately a two-valued random variable, assuming the values $b\sigma^2$ and $-b\sigma^2$ with probabilities $P_{\mu,\sigma}\{s_N > 0\}$ and $P_{\mu,\sigma}\{s_N < 0\}$ respectively. For $\mu > 0$, treating (1.2) as an approximate equality yields

$E_{\mu,\sigma}(s_N) \cong b\sigma^2(\frac{e^{2\mu b} - 1}{e^{2\mu b} + 1})$, which together with Wald's identity

$E_{\mu,\sigma}(s_N) = \mu\, E_{\mu,\sigma}(N)$ gives (1.3). It may be shown that as $\sigma \to \infty$

$$E_{\mu,\sigma}(T) \sim E_{\mu,\sigma}(N) \sim \frac{b\sigma^2}{|\mu|}\left(\frac{e^{2|\mu|b} - 1}{e^{2|\mu|b} + 1}\right), \tag{3.1}$$

but this approximation underestimates $E_{\mu,\sigma}(N)$ and is unsatisfactory theoretically as an approximation for $E_{\mu,\sigma}(T)$ because it gives no idea of the price we pay for not knowing σ.

A different approximation for $E_{\mu,\sigma}(T)$ is suggested by the arguments of Pollak and Siegmund [6] and Lai and Siegmund [5]. Suppose that $\mu > 0$, $\sigma > 0$ are fixed and $b \to \infty$. By writing

$$v_n^{-2}\, s_n = h_n\, f(n^{-1}\sum_1^n x_i, n^{-1}\sum_1^n x_i^2), \tag{3.2}$$

where $f(x,y) = x/(y-x^2)$, and expanding f in a Taylor series about the point $(\mu, \sigma^2 + \mu^2)$, we obtain

$$v_n^{-2}s_n = h_n\mu/\sigma^2 + n^{-1}h_n \sum_{i=1}^n \{x_i-\mu - \frac{\mu}{\sigma^2}[(x_i-\mu)^2-\sigma^2]\}/\sigma^2+\ldots \tag{3.3}$$

Neglecting higher order terms we see that the stochastic component on the right hand side of (3.3) is a sum of independent identically distributed mean zero random variables, and hence to a first approximation the behavior of $v_n^{-2}s_n$ will be governed by the deterministic part $h_n\mu/\sigma^2$. For fixed positive μ and σ, as $b \to \infty$ $P_{\mu,\sigma}\{s_T < 0\} \to 0$ very rapidly and hence with overwhelming probability $v_T^{-2}s_T \cong b$. Making this argument precise shows that as $b \to \infty$

$$(3.4) \qquad P_{\mu,\sigma}\{b^{-1}T \to \sigma^2/\mu\} = 1, \ b^{-1}E_{\mu,\sigma}(T) \to \sigma^2/\mu,$$

However, the second part of (3.4) can be improved. Let S_n denote the sum in (3.3) and recall the definition (1.4) of h_n. Continuing to neglect higher order terms, we obtain

$$(3.5) \quad \sigma^2 b \cong \sigma^2 E_{\mu,\sigma}(v_T^{-2}s_T) \cong \mu\, E_{\mu,\sigma}(T-k_1T^{1/2}-k_2) + E_{\mu,\sigma}(S_T-k_1S_T/T^{1/2}-k_2S_T/T).$$

By Wald's identity, $E_{\mu,\sigma}(S_T) = 0$. By (3.4), the central limit theorem for randomly stopped sums, and a uniform integrability argument, $E_{\mu,\sigma}(S_T/T^{1/2}) \to 0$ as $b \to \infty$. Also $E_{\mu,\sigma}(S_T/T) \to 0$, and by (3.4) and uniform integrability we see that $E_{\mu,\sigma}(T^{1/2}) \sim \sigma(b/\mu)^{1/2}$. Hence by (3.5), for large b

$$(3.6) \qquad E_{\mu,\sigma}(T) \cong \sigma^2 b/\mu + k_1\sigma(b/\mu)^{1/2} + k_2 \qquad (\mu > 0).$$

Analyzing the quadratic terms in the Taylor series for f as well as the excess over the boundary $v_T^{-2}s_T - b$ gives another constant and terms tending to 0 on the right hand side of (3.6) (cf. Lai and Siegmund [5]). Since this constant is difficult to compute exactly and is small except when μ/σ^2 is small, we have chosen to ignore it.

A comparison of the arguments leading to (3.1) and (3.6) shows that the approximation given by (3.6) will be poor unless $bP_{\mu,\sigma}\{s_T < 0\} \cong 0$. If we agree that we have been successful in choosing k_1 and k_2 so that $P_{\mu,\sigma}\{s_T < 0\} \cong \alpha$, where as in Section 2, $\alpha = P_{\mu,\sigma}\{s_N < 0\}$ $(\mu > 0)$, then according to Wald's argument at the beginning of this section

$$E_{\mu,\sigma}(v_T^{-2}s_T) \cong b(1-2\alpha).$$

Together with (3.5) and the subsequent argument this suggests replacing b by $b(1-2\alpha)$ in (3.6) to give the approximation

$$E_{\mu,\sigma}(T) \cong \frac{\sigma^2 b}{\mu}(1-2\alpha) + k_1\sigma\left(\frac{b}{\mu}(1-2\alpha)\right)^{1/2} + k_2. \tag{3.7}$$

Tables 3 and 4 give Monte Carlo estimates for $E_{\mu,\sigma}(T)$ together with the approximation (3.7) and the value of $E_{\mu,\sigma}(N)$.

Table 3: Expected Sample Size: $\sigma = 1$, $b = 5$

μ	Approximation to $E_{\mu,\sigma}(T)$ from (3.7)		Monte Carlo		$E_{\mu,\sigma}(N)$
	$k_1=1, k_2=1.5$	$k_1=1.4, k_2=0$	$k_1=1, k_2=1.5$	$k_1=1.4, k_2=0$	
.075	33	34	37	39	29.7
.125	31	31	33	33	27.2
.25	23	24	24	24	20.0
.50	15	14	15	14	11.4

Table 4: Expected Sample Size: $\mu = .25$, $b = 6$

σ	Approximation to $E_{\mu,\sigma}(T)$ from (3.7)		Monte Carlo		$E_{\mu,\sigma}(N)$
	$k_1=1, k_2=1.5$	$k_1=1.4, k_2=0$	$k_1=1, k_2=1.5$	$k_1=1.4, k_2=0$	
.5	9.6	9.0	11	10	7.05
.67	15	14	15	15	10.9
1.00	28	29	29	29	24.7
1.25	42	43	42	43	37.6
1.67	71	72	73	73	65.1
2.50	150	154	152	157	142.8
3.33	260	265	264	275	250.7

It is difficult to state a simple general conclusion, but these figures seem to indicate that our modification for σ unknown of the sequential test defined by (1.1) achieves approximately the same operating characteristic at a 10-20% increase in the expected sample size.

The approximation (3.7) appears to give reasonable numerical results over the range of values of μ, σ, b tested. However, it does not seem to have any "natural" theoretical justification. It would be interesting to try to obtain an expansion of the form

$$(3.8)\quad E_{\mu,\sigma}(T) = \frac{b\sigma^2}{\mu}\left(\frac{e^{2\mu b}-1}{e^{2\mu b}+1}\right) + K\sigma + o(\sigma) \quad (\sigma \to \infty,\ \mu > 0),$$

but we have not been able to do this. A particular difficulty is to find the rate of convergence of $P_{\mu,\sigma}\{s_T < 0\}$ to $1/(1+e^{2\mu b})$ as $\sigma \to \infty$. An expansion corresponding to (3.8) is easy to obtain for the stopping rule N and gives good numerical results in terms of the moments of s_{τ_+}, where $\tau_+ = \inf\{n: s_n > 0\}$. Arguments similar to those used by Siegmund [7] suggest that

$$(3.9)\quad E_{\mu,\sigma}(N) = \left(\frac{e^{2\mu b}-1}{e^{2\mu b}+1}\right)\frac{\sigma^2 b}{\mu} + \frac{\sigma E_{0,1}(s_{\tau_+}^2)}{2\mu E_{0,1}(s_{\tau_+})}\left[1 + \frac{4\mu b\, e^{2\mu b}}{e^{4\mu b}-1}\right]\} + o(\sigma)$$

as $\sigma \to \infty$. Here $\mu \geq 0$ and the right hand side of (3.9) for $\mu = 0$ is to be evaluated in the limit as $\mu \to 0$. Using classical fluctuation results of Spitzer, one can compute $E_{0,1}(s_{\tau_+}^2)/E_{0,1}(s_{\tau_+})$ to be approximately 1.168 (cf. Lai [4]). With this value the approximation (3.9) gives results which are typically too small by only about 2-3%.

References

[1] Baker, A. (1950). Properties of some tests in sequential analysis. *Biometrika* 37, 334-46.

[2] Ghosh, B. K. (1970). *Sequential Tests of Statistical Hypotheses*. Addison-Wesley, Reading, Mass.

[3] Hall, W. J. (1962). Some sequential analogues of Stein's two-stage test. *Biometrika* 49, 367-78.

[4] Lai, T. L. (1976). Asymptotic moments of random walks with applications to ladder variables and renewal theory. *Ann. Prob.* 4, 51-66.

[5] Lai, T. L. and Siegmund, D. (1976). A non-linear renewal theory with applications to sequential analysis II (in preparation).

[6] Pollak, M. and Siegmund, D. (1975). Approximations to the expected sample size of certain sequential tests. *Ann. Statist.* 3, 1267-1282.

[7] Siegmund, D. (1975). Error probabilities and average sample number of the sequential probability ratio test. *Jour. Roy. Statist. Soc.* B, 37, 394-401.

[8] Siegmund, D. (1976). Importance sampling in the Monte Carlo study of sequential tests. *Ann. Statist.*, 4.

[9] Stein, C. (1945). A two-sample test for a linear hypothesis whose power is independent of the variance. *Ann. Math. Statist.* 16, 243-258.

[10] Wald, A. (1947). *Sequential Analysis.* John Wiley and Sons, New York.

[11] Weiss, L. and Wolfowitz, J. (1972). An asymptotically efficient sequential equivalent of the t-test. *Jour. Roy. Statist. Soc.* B, 34, 456-460.

A REDUCTION THEOREM FOR CERTAIN SEQUENTIAL EXPERIMENTS

By L. Le Cam*

University of California, Berkeley

1. *Introduction.* It has been shown in [1] and [2] that asymptotic optimality results can often be obtained by first taking the limits of the experiments under consideration and then applying the optimality results known to be valid for the limit situation. This method is particularly effective and simple when the limit is the standard Gaussian shift experiment for which many optimality results are well-known.

The experiments considered in [1] are of the "non-sequential" type. However, whenever the stopping rule is fixed, a "sequential experiment" is just an experiment in the sense of [1].

The distinguishing feature is that, in the sequential case, one wishes also to give prescriptions on the selection of the stopping rule.

In the present paper we consider a restricted, but fairly wide, class of sequential experiments, relate them to non sequential analogues and show that in some cases the selection of the stopping rule can be approximated by a similar problem on a suitable limit stochastic process. For instance in regular cases where the non sequential limits are Gaussian experiments, the problem is reduced to a problem involving stopping times of Brownian motion.

Some applications to testing and estimation are briefly mentioned. In particular, we recover some results of Ibragimov and Has'minskii [3] and the observation that for standard estimation problems sequential procedures are often no better than their non-sequential counterparts. We also recover some results related to those of A. C. Singh [4] on the sequential generalization of

*This research was prepared with the support of National Science Foundation Grant No. MPS 75-10376 and U. S. Army Research Office Grant No. ARO-DA-AG 29-76-G-0167.

Neyman's $C(\alpha)$ tests. The results are also related to the method used by Chernoff in [5].

The specifics of the notation and definitions form the bulk of Section 2. This same Section 2 includes Propositions concerning the non-sequential limiting distributions. Section 3 describes the main features of the reduction process with the passage to stopping times of a limiting stochastic process. Section 4 gives some indications on the applicability of the results to the independent, or independent identically distributed, case.

Section 5 describes briefly some applications to estimation and testing.

2. *Notation and main assumptions.* The sequential experiments considered here can be described by systems

$$S = \{X,\{A_n\},\{P_\theta\},\Theta,N\}$$

in which the letters have the following meaning. The sequence $\{A_n\}$, $n = 0,1,2,\ldots$ is an increasing (i.e. $A_n \subset A_{n+1}$) sequence of σ-fields on the set X. It is assumed that A_0 is the trivial σ-field $A_0 = \{\phi,X\}$. The variable N is an integer valued function defined on X and such that $A_k = \{N = k\} \in A_k$.

The set Θ is an arbitrary set representing the possible values of the parameters. For each $\theta \in \Theta$ and each integer n one is given a probability measure $P[\theta,n;.]$ on the σ-field A_n. The measures are assumed to be <u>coherent</u> in the sense that if $A \in A_m$ with $m < n$ then $P[\theta,m;A] = P[\theta,n;A]$.

The entire coherent sequence $\{P[\theta,n;.];\ n = 0,1,2,\ldots,\ \theta \in \Theta\}$ will be called an experiment and noted $E = \{P_\theta;\ \theta \in \Theta\}$. The experiment $E(n)$ obtainable from E is formed by taking the measures $E(n) = \{P[\theta,n;.];\ \theta \in \Theta\}$ on A_n.

It has not been assumed so far that the coherent sequence $\{P[\theta,n;.],\ n = 0,1,2,\ldots\}$ defines a σ-additive measure on the σ-field A_∞ generated by the union $\bigcup_n A_n$ of the A_n. This assumption can always be thrown in at low cost, since under all circumstances, one can force it to be satisfied by completing X in an appropriate manner.

Thus, the arguments given below will use this assumption without mentioning it.

The stopping variable N defines a field obtainable from the partition $\{A_k;\ k = 0,1,2,\ldots\}$ by taking on each A_k the trace of the corresponding σ-field A_k. According to the above mentioned assumption we shall call $\mathcal{B}(N)$ the σ-field generated in this manner and let $Q(\theta,N)$ or $Q[\theta,N;.]$ be the restriction to $\mathcal{B}(N)$ of the measure P_θ. The family $E(N) = \{Q(\theta,N);\ \theta \in \Theta\}$ is then the experiment induced by the stopping variable N.

To arrive at asymptotic problems, let ν be an index whose values are the integers $\nu = 1,2,\ldots$. Instead of just one system $S = \{X,\{A_n\},\{P_\theta\};\Theta,N\}$ consider a sequence $S_\nu = \{X_\nu,A_{n,\nu},P_{\theta,\nu},\Theta_\nu, N_\nu\}$, $\nu = 1,2,\ldots$ of such systems.

Note that here absolutely <u>everything</u> is allowed to change as ν changes. The corresponding experiments become E_ν, $E_\nu(N_\nu)$, etc. However it will be convenient to <u>assume that the set Θ_ν is a certain set Θ independent of ν</u>. This assumption can readily be modified. It is used here to avoid complicating notation and language unduly.

In addition to the sequence $\{S_\nu;\ \nu = 1,2,\ldots\}$ we shall consider a particular sequence $\{n_\nu\}$ of integers with $n_\nu \to \infty$ as $\nu \to \infty$. The assumptions listed below refer to that particular sequence and to a particular point θ_0 selected in Θ.

Let H be a finite dimensional Euclidean space with a norm noted $||v||$ for $v \in H$. Let ξ be a map from Θ to H and let β be a real valued function defined on the image $\xi(\Theta)$ of Θ in H. Assume $\xi(\theta_0) = 0$ and let H' be the dual of H.

DEFINITION 1. The sequence $\{E_\nu,n_\nu\}$ is asymptotically of Koopman-Darmois type for the maps (ξ,β) if for every integer b the following two conditions are satisfied:

i) the sequences $\{P_\nu(\theta_0,bn_\nu)\}$ and $\{P_\nu(\theta,bn_\nu)\}$ are contiguous,

ii) there are H' valued statistics $S_{n,\nu}$ such that $S_{n,\nu}$ is $A_{n,\nu}$ measurable and such that the Radon-Nikodym densities satisfy the requirement that

$$\sup_{n\leq bn_\nu} \left|\frac{dP_\nu(\theta,n)}{dP_\nu(\theta_0,n)} - \exp\{\xi(\theta)S_{n,\nu} - \frac{n}{n_\nu}\beta(\xi(\theta))\}\right|$$

tends to zero in probability as $\nu \to \infty$.

The preceding definition requires some comments. First let us note that there is no loss of generality in assuming that the linear span of $\xi(\Theta)$ is the space H itself. We shall proceed under this assumption below. Second, note that the requirement that $S_{n,\nu}$ be $A_{n,\nu}$ measurable is also a cheap requirement. If there are vector valued random variables satisfying the other requirements there are certainly $A_{n,\nu}$ measurable variables with the required property. Third, the occurrence of the supremum $\sup_{n\leq bn_\nu}$ in the definition seems to be rather strong. However, note that under the contiguity assumption (i) the likelihood ratios $\frac{dP_\nu(\theta,n)}{dP_\nu(\theta_0,n)}$, $n = 0,1,2,\ldots$ form a sequence, which, for each ν, is very close to a martingale sequence. If the exponentials $\exp\{\xi S_{n,\nu} - \frac{n}{n_\nu}\beta(\xi)\}$ were also subjected to similar martingale like restrictions one could dispense with the $\sup_{n\leq bn_\nu}$ and just require that the difference at sequences $m_\nu \leq bn_\nu$ tend to zero. This will be so, for instance, if the $S_{n,\nu}$ are truncated in an appropriate way. The existence of appropriate truncations can be proved, as in [6], if $\xi(\Theta)$ is a closed convex cone with a finite number of extreme edges.

In the common, independent identically distributed case we shall show in Section 4 that one can dispense not only with the "sup" but also with the arbitrary integer b.

Finally, we have assumed here that the real valued function

β depends on θ only through the value $\xi(\theta)$. It is easily seen that this is also a very minor requirement. If there existed a real valued function of θ, say $\beta'(\theta)$ which could replace $\beta[\xi(\theta)]$ then there would also be a function β of $\xi(\theta)$.

In Definition 1 the dependence of the measures P_θ occurs mostly through the values of $\xi(\theta)$. For many problems one cannot keep Θ_ν fixed, independent of ν, as assumed here. One would then have maps ξ_ν from a set Θ_ν to the Euclidean space H. In such situations, Definition 1 could be modified accordingly. One would assume for instance that some $\theta_{0,\nu}$ has been fixed, that $\xi_\nu(\theta_{0,\nu})=0$ and that the sets $\xi_\nu(\Theta_\nu)$ are increasing subsets of H. The conditions (i) and (ii) of Definition 1 would be stated with sequences $\{\theta_\nu\}$ such that $\xi_\nu(\theta_\nu)$ remains equal to a fixed $\xi \in H$.

Alternatively, one could assume that the sets $\xi_\nu(\theta_\nu)$ converge, in the Hausdorff distance sense, to a certain limit and state the conditions of Definition 1 for sequences $\{\theta_\nu\}$ such that $\xi_\nu(\theta_\nu)$ converges in H. While the situations covered by this second type of assumption are very frequent, the use of arbitrary convergent sequences $\{\xi_\nu(\theta_\nu)\}$ introduces some uniformity requirements which will not be particularly needed below. Thus we shall describe only the case where Θ and ξ are fixed, independently of ν. However we shall, throughout, impose a fairly severe restriction on $\xi(\Theta)$ as follows.

ASSUMPTION A. *The closure in* H *of the set* $\xi(\Theta)$ *contains an open subset of* H.

There are many interesting cases which are ruled out by (A). They could be dealt with under suitable requirements. The nice feature of Assumption (A) is that it entails automatically a wealth of specific features, which, otherwise, would need special hypotheses.

A particular case of Definition 1 will be the one receiving most attention below. It is covered by the following Definition.

DEFINITION 2. The sequence $\{\xi_\nu, n_\nu\}$ will be called asymptotically standard Gaussian if it satisfies Definition 1 for a function ξ which obeys assumption (A) and for a function β such that $2\beta(\xi)$ is the square Euclidean norm $2\beta(\xi) = ||\xi||^2$.

To proceed construct stochastic processes $\{S_\nu(t), t \in [0,\infty)\}$ by taking $S_\nu(t) = S_{n,\nu}$ if $n \leq n_\nu t < n+1$.

Under the conditions of Definition 1 the sequences of distributions $L\{S_\nu(t)|\theta_0\}$ are always relatively compact sequences since $\xi(\Theta)$ spans all of H. Furthermore, if Assumption (A) holds, the sequence $L\{S_\nu(t)|\theta_0\}$ must converge to the distribution of a certain random vector $S(t)$ such that

$$E \exp\{\xi S(t)\} = \exp\{t\beta(\xi)\}$$

for all $\xi \in \xi(\Theta)$. In particular $\exp\{t\beta(\xi)\}$ is the Laplace transform of some distribution. This being true for all $t \in (0,\infty)$, there is a certain process $\{S(t): t \in [0,\infty)\}$ which is a process with independent increments satisfying the equality $E \exp\{\xi S(t)\} = \exp\{t\beta(\xi)\}$ for all t. The distribution of this process is well determined by the function β. We can and will assume that the paths of S are continuous from the right.

PROPOSITION 1. *Assume that the sequence* $\{E_\nu, n_\nu\}$ *satisfies the conditions of Definition 1 for a given function* β. *Suppose that Assumption (A) is satisfied and let* $\{S(t); t \quad [0,\infty)\}$ *be the process with homogeneous independent increments such that*

$$E \exp\{\xi S(t)\} = \exp\{t\beta(\xi)\}$$

for all $t \in [0,\infty)$ *and* $\xi \in \xi(\Theta)$.

Then, for every finite set $\{t_i; i = 1,2,\ldots,r\}$ *the joint distributions* $L\{S_\nu(t_i); i = 1,2,\ldots,r|\theta_0\}$ *converge to the corresponding joint distributions of the process* S.

(In particular, when Definition 2 holds the limit process is the standard H-valued Brownian motion process).

PROOF. We already know that individual distributions such as $L\{S(t)|\theta_0\}$ converge. Consider sequences of integers $m_{\nu,i}$, $i = 1, 2,\ldots,r$ such that $\frac{m_{\nu,i}}{n} \to t_i$ with $t_1 \leq t_2 \leq \ldots \leq t_r$. On the product $H' \times H'$ of H^ν by itself let $M_{\nu,i}(\theta)$ be the joint distribution $M_{\nu,i}(\theta) = L\{S_\nu(\frac{m_{\nu,i-1}}{n_\nu}), S_\nu(\frac{m_{\nu,i}}{n_\nu})|\theta\}$. Let $M_i(\theta)$ be a cluster point of this sequence of distributions. Denote by pairs (x,y) points of the product $H' \times H'$.

The contiguity and asymptotic sufficiency conditions included in Definition 1 imply immediately that $M_i(\theta)$ must have the form

$$M_i[dx,dy;\theta] = \exp\{\xi y - t_i\beta(\xi)\}M_i(dx,dy)$$

where, for short, $\xi = \xi(\theta)$ and $M_i = M_i(\theta_0)$.

The joint distribution M_i has marginals M_i' and M_i'' respectively. One may also write M_i in the form $M_i(dx,dy) = K(dx|y)M_i''(dy)$ for a certain Markov kernel K. For every measurable $A \subset H'$ one must have

$$M_i'[A|\theta] = \int K(A|y)\exp\{\xi y - t_i\beta(\xi)\}M_i''(dy).$$

Since $\xi = \xi(\theta)$ runs through a set which is dense in an open subset of H, this identity determines K uniquely. Thus, the joint limiting distribution of $S_\nu(\frac{m_{\nu,1}}{n_\nu})$, $S_\nu(\frac{m_{\nu,2}}{n_\nu})$ is uniquely determined. Repeating the argument with the conditional distribution $L[S(t_1), S(t_2)|S(t_3)]$ one sees that joint distributions of triplets are also uniquely determined. Proceeding recursively it follows that the sequence $L\{S_\nu(\frac{m_{\nu,i}}{n_\nu}); i = 1,2,\dots,r|\theta_0\}$ has only one cluster point. Thus, it converges as claimed. The limit distributions are entirely determined by the given function β.

Now consider a process $\{X(t); t \in [0,\infty)\}$ with independent increments such that for all $\xi \in \xi(\theta)$ and all t the identity $E \exp\{\xi X(t)\} = \exp\{t\beta(\xi)\}$ holds. Let P_0 be the distribution of $X(t_i); i = 1,2,\dots,r$ for this process. Let P_ξ be the probability measure whose density with respect to P_0 is $\exp\{\xi X(t_r)-t_r\beta(\xi)\}$. This is a Radon-Nikodym density taken on the σ-field generated by $X(t_i); i = 1,2,\dots,r$. On the σ-field B_j generated by $X(t_i); i = 1,2,\dots,j$ the corresponding density is

$$E^{\mathcal{B}_j}\exp\{\xi X(t_r)-t_r\beta(\xi)\} = E^{\mathcal{B}_j}\exp\{\xi[X(t_r)-X(t_j)]+\xi X(t_j)-t_r\beta(\xi)\}$$
$$= \exp\{\xi X(t_j)-t_j\beta(\xi)\}.$$

Thus the P_ξ satisfy the conditions required by Definition 1. This shows that the limit distributions of the processes S_ν must be those of the independent increment process and completes the proof of the proposition.

REMARK 1. The above argument shows also that one can obtain, under the conditions of Definition 1, any process with homogeneous independent increments whose Laplace transform exists in an open set. In the Gaussian case the measures P_ξ are those of a Brownian process with linear drift. In the general case the P_ξ are not usually obtainable by such a shift.

REMARK 2. The argument of Proposition 1 shows that if $\{t_\nu\}$ is any non-random sequence converging to t, the distributions $L[S_\nu(t_\nu)]$ converge to $L[S(t)]$. This result can be strengthened to some extent as follows. Let m_ν and m'_ν be two integers such that $m_\nu \leq m'_\nu$. Assume that $m_\nu n_\nu^{-1}$ and $m'_\nu n_\nu^{-1}$ both converge to the same limit t. The experiments $E_\nu(m_\nu)$ and $E_\nu(m'_\nu)$, corresponding respectively to the σ-fields $A_{m_\nu,\nu}$ and $A_{m'_\nu,\nu}$, converge to the same limit. Thus on every finite set $\Theta_0 \subset \Theta$ the deficiency of $E_\nu(m_\nu)$ with respect to $E_\nu(m'_\nu)$ converges to zero. According to [1] (Section 5, Proposition 5) this entails that for the finite set Θ_0, the insufficiencies converge to zero. It follows that the differences

$$\frac{dP[\theta,m_\nu]}{dP[\theta_0,m_\nu]} - \frac{dP[\theta,m'_\nu]}{dP[\theta_0,m'_\nu]}$$

converge to zero in L_1-norm. Therefore, under the conditions of Definition 1 the differences $S_\nu(\frac{m_\nu}{n_\nu}) - S_\nu(\frac{m'_\nu}{n_\nu})$ converge to zero in probability. Unfortunately the same conclusion does not appear to be valid if m_ν and m'_ν are replaced by stopping times of the sequences $\{A_{n,\nu}\}$.

3. *A reduction procedure.* In this section we shall consider a sequence $\{S_\nu\}$; $\nu = 1,2,\ldots$ of systems $S_\nu = \{X_\nu, \{A_{n,\nu}\}; P_{\theta,\nu}, \Theta, N_\nu\}$ as described in Section 2. Also, we suppose given a particular sequence of integers $\{n_\nu; \nu = 1,2,\ldots\}$ such that $n_\nu \to \infty$. The measure available on the σ-field $A_{n,\nu}$ for the parameter point θ is $P_\nu[\theta,n]$. The measure induced similarly by the stopping variable N_ν is $Q_\nu[\theta,N_\nu]$.

The following assumption will be used throughout.

ASSUMPTION B. For each $\theta \in \Theta$, the sequence of distributions $L\{\frac{N_\nu}{n_\nu}|\theta\}$ is a relatively compact sequence on $[0,\infty)$.

LEMMA 1. Assume that (B) holds and that for each fixed integer b the sequences $\{P_\nu[\theta,bn_\nu]\}$ and $\{P_\nu[\theta_0,bn_\nu]\}$ are contiguous. Then the sequences $\{Q_\nu[\theta,N_\nu]\}$ and $\{Q_\nu[\theta_0,N_\nu]\}$ are also contiguous.

PROOF. Let $B(N_\nu)$ be the σ-field induced by the stopping time N_ν. Suppose that there would be an $\varepsilon > 0$ and sets $B_\nu \in B(N_\nu)$ such that $\lim_\nu Q_\nu[\theta_0,N_\nu;B_\nu] = 0$ but $\lim_\nu \sup Q_\nu[\theta,N_\nu;B_\nu] > 2\varepsilon$. According to Assumption (B) there is some integer b such that $Q_\nu[\theta,N_\nu;N_\nu > bn_\nu] < \varepsilon$ for all ν. Let A_ν be the intersection of B_ν with the set $\{N_\nu \leq bn_\nu\}$. This is an element of the σ-field $A_{bn_\nu,\nu}$. For this set we have

$$\lim_\nu P_\nu[\theta_0,bn_\nu;A_\nu] = 0$$

and

$$\lim_\nu \inf P_\nu[\theta,bn_\nu;A_\nu] \geq \varepsilon > 0.$$

This contradicts the contiguity assumption.

The desired result follows upon interchanging the respective roles of θ and θ_0.

Suppose now that the sequence $\{E_\nu,n_\nu\}$ satisfies the requirements of Definition 1 for suitable functions ξ and β and for random vectors $S_{n,\nu}$. Consider the joint distribution M_ν of $(S_{N_\nu,\nu},\frac{N_\nu}{n_\nu})$ under θ_0. When Assumption (B) holds, the sequence $\{M_\nu\}$

is a relatively compact sequence on the product of H' by the positive interval $[0,\infty)$. Thus the following assumption could be made to hold by passing to subsequences.

ASSUMPTION C. *The sequence* $M_\nu = L[S_{N_\nu,\nu}; \frac{N_\nu}{n_\nu} | \theta_0]$ *converges on* $H' \times [0,\infty)$ *to a limit* F.

LEMMA 2. *Let the sequence* $\{E_\nu, n_\nu\}$ *satisfy the conditions of Definition 1. Let Assumptions* (A), (B) *and* (C) *be satisfied.*

Then the sequential experiments $E_\nu(N_\nu) = \{Q_\nu[\theta, N_\nu]; \theta \in \Theta\}$ *converge, pointwise on* Θ *to a certain limit* $F = \{F_\theta; \theta \in \Theta\}$. *The measures* F_θ *constituting* F *can be taken of the form*

$$F_\theta(dx, d\tau) = \exp\{\xi(\theta)x - \tau\beta(\xi(\theta))\}F(dx, d\tau)$$

on the space $H' \times [0,\infty)$.

PROOF. If $A_{n,\nu} = \{N_\nu = n\}$, then, on the particular set $A_{n,\nu}$ the Radon-Nikodym densities of $E_\nu(N_\nu)$ have the form

$$\frac{dQ_\nu[\theta, N_\nu]}{dQ_\nu[\theta_0, N_\nu]} = \frac{dP_\nu[\theta, n]}{dP_\nu[\theta_0, n]} = \exp\{\xi(\theta)S_{n,\nu} - \frac{n}{n_\nu}\beta[\xi(\theta)]\} + \varepsilon_\nu.$$

In other words

$$\frac{dQ_\nu[\theta, N_\nu]}{dQ_\nu[\theta_0, N_\nu]} = \exp\{\xi S_{N_\nu,\nu} - \frac{N_\nu}{n_\nu}\beta(\xi)\} + \varepsilon_\nu.$$

It follows from this that the pairs $(S_{N_\nu,\nu}, \frac{N_\nu}{n_\nu})$ form a "distinguished" sequence of asymptotically sufficient statistics (see [1], p. 50). Since the sequences of measures are contiguous the result follows by application of the remarks in [1] (p. 38).

So far we have considered only non randomized stopping variables. One could just as well use randomized stopping times. This can be done by replacing each X_ν space by the product of X_ν by a countable product of intervals [0,1] each carrying the Lebesgue measure. This leads to the following.

PROPOSITION 2. *Let Assumptions* (A), (B), (C) *be satisfied*

<u>for a sequence</u> (E_ν, n_ν) <u>which fulfills the requirements of Definition 1.</u>

<u>Then there are randomized stopping variables</u> $\{M_\nu\}$ <u>of the sequence</u> $\{S_{n,\nu};\ n = 0,1,2,\ldots\}$ <u>such that the sequential experiments</u> $E_\nu(M_\nu)$ <u>have the same limit as</u> $E_\nu(N_\nu)$.

PROOF. Consider first distributions when the value of the parameter is the point $\theta_0 \in \Theta$. The stopping variable N_ν can also be described by a sequence $\{u_0, u_1, u_2, \ldots\}$ of indicator variables such that $u_j = 1$ if $N_\nu = j$ and zero otherwise. Thus u_0 has a certain probability p_0 of being equal to unity. Since p_0 is $A_{0,\nu}$ measurable, it is a constant. Construct indicator variables $\{v_0, v_1, \ldots\}$ recursively as follows. To start, v_0 is a binomial variable with $\Pr\{v_0 = 1\} = \Pr\{u_0 = 1\} = p_0$. If $v_0, v_1, \ldots, v_{n-1}$ have to be constructed, we consider the conditional probability p_n that $u_n = 1$ given $S_{0,\nu}, S_{1,\nu}, \ldots, S_{n,\nu}$ and let $v_n = 1$ with this probability and zero otherwise.

Proceeding recursively one sees that

$$L(S_{0,\nu}, v_0) = L(S_{0,\nu}, u_0),$$

$$L(S_{0,\nu}, S_{1,\nu}; u_0, u_1) = L(S_{0,\nu}, S_{1,\nu}, v_0, v_1)$$

and so forth.

Now the v_j determine a stopping time M_ν and the above equalities say that

$$L[S_{M_\nu,\nu}, M_\nu] = L(S_{N_\nu,\nu}, N_\nu).$$

This is of course possible under the distributions induced by θ_0. To pass to other values of θ, fix a number b and let

$$N_\nu(b) = N_\nu \wedge (bn_\nu) \quad \text{and} \quad M_\nu(b) = M_\nu \wedge (bn_\nu).$$

Both stopping times $N_\nu(b)$ and $M_\nu(b)$ satisfy the compactness requirement (B). If b is a continuity point of the limit distribution F of Assumption (C) the limiting distribution, under θ_0, of the pair $(S_{M_\nu(b),\nu}, M_\nu(b))$ is the same as that of $(S_{N_\nu(b),\nu}, N_\nu(b))$. Thus the sequential experiments $E_\nu(M_\nu(b))$ have the same limit as the experiments $E_\nu(N_\nu(b))$.

A particular feature, by contiguity, is that the distributions $L\{\frac{N_\nu(b)}{n_\nu}|\theta\}$ and $L\{\frac{M_\nu(b)}{n_\nu}|\theta\}$ tend to the same limit, which can be obtained from the expression given in Lemma 2, from the marginal distribution of τ in $F_\xi(dx,d\tau)$. Since this marginal distribution is entirely on $[0,\infty)$, letting b tend to infinity one concludes that $\{L[M_\nu|\theta]\}$ is also a relatively compact sequence. Since $L[S_{M_\nu,\nu},M_\nu|\theta] = L[S_{N_\nu,\nu},N_\nu|\theta_0]$, the desired result follows immediately from Lemma 2.

Returning to the processes S_ν defined in Section 2 by the relation $S_\nu(t) = S_{n,\nu}$ if $n \leq n_\nu t < n+1$, what Proposition 2 means is that it will be enough to consider (randomized) stopping times τ_ν of the processes S_ν.

Unfortunately it appears difficult to conclude that the limiting distribution F can be realized as the joint distribution $L[S(\tau),\tau]$ where τ is a randomized stopping time of the limiting process S itself. To describe this it is convenient to introduce a definition as follows.

DEFINITION 3. The sequence $\{\tau_\nu\}$ of stopping times subject to (C) will be called <u>regular</u> if there is a stopping time randomized or not of the process $\{S(t);\ t \in [0,\infty)\}$ such that $L[S(\tau),\tau] = F = \lim_\nu L[S_\nu(\tau_\nu),\tau_\nu]$.

It is abundantly clear that most of the stopping times used in statistical practice are regular in this sense. However it is perfectly easy to construct "irregular" sequences.

Consider a pair (X_ν,Y_ν) of random variables and a stopping variable τ_ν whose conditional distribution given (X_ν,Y_ν) is a function of X_ν only. It may happen that the joint distribution $L(X_\nu,Y_\nu,\tau_\nu)$ converges to a limit $L(X,Y,\tau)$ with X, Y independent Gaussian variables, but with a τ whose conditional distribution given (X,Y) is a function of Y only.

We have an embryo of a proof to the effect that when the process S is the Gaussian process, reduction to stopping times of

S can be achieved for all stopping times τ_ν which are well behaved in a sense which requires mostly that

$$\lim_{\varepsilon \downarrow 0} \lim_{\nu} L[S_\nu(\tau_\nu + \varepsilon)] = \lim_{\nu} L[S_\nu(\tau_\nu)].$$

However, since the proof, if correct, is rather cumbersome we shall restrict ourselves to a statement valid in a very special situation.

PROPOSITION 3. <u>Assume that the sequence $\{E_\nu, n_\nu\}$ is of asymptotic Koopman-Darmois type in the sense of Definition 1. Assume also that there exists particular choices of the vectors $S_{n,\nu}$ of Definition 1 which are such that the successive differences $(S_{n+1,\nu} - S_{n,\nu})$, $n = 0,1,2,\ldots$ are independent. Then, all sequences $\tau_\nu = \frac{N_\nu}{n_\nu}$ of stopping times which satisfy Assumption (C) are automatically regular, in the sense of Definition 3.</u>

NOTE. The limit processes S are assumed to have right continuous paths. Since we use randomized stopping times we may as well assume that the σ-fields used to define such times are completed.

PROOF. According to Proposition 2, we can limit ourselves to randomized stopping times of the sequences $\{S_{n,\nu}: n = 0,1,2,\ldots\}$. However, in the present case, to say that τ_ν is a stopping time of $\{S_{n,\nu}; n = 0,1,2,\ldots\}$ amounts to saying that the events $\{\tau_\nu \leq s_1 \leq s\}$ are independent of the differences $\{[S_\nu(t) - S_\nu(s)], t > s\}$.

This independence is preserved by passages to the limit.

To summarize, under (A), (B), (C), sequential problems relating to systems which satisfy Definition 1 have for limits analogous problems in the process with independent increments $\{S(t); t \in [0,\infty)\}$ at least whenever the stopping times τ_ν are regular and certainly when the successive differences $[S_{n+1,\nu} - S_{n,\nu}]$ are themselves independent.

4. *Independent observations.* In this section we shall retain the structure described in Section 2 but specialize it to the

case where passing from the σ-field $A_{n,\nu}$ to the next σ-field $A_{n+1,\nu}$ is effected by taking one observation on a variable independent of those which generate $A_{n,\nu}$. This can be stated more explicitly in many ways, one of which is as follows.

ASSUMPTION D. The space (X_ν, A_ν) is a countable product of spaces $(X_{j,\nu}, B_{j,\nu})$. The σ-field $A_{n,\nu}$ is the σ-field generated by the coordinate σ-fields $B_{j,\nu}$, $j \leq n$. The measure $P_\nu[\theta,n]$ on $A_{n,\nu}$ is the direct product of measures $p_\nu(\theta,j)$, $j \leq n$, with $p_\nu(\theta,j)$ defined on the σ-field $B_{j,\nu}$.

To state the results, we shall assume given particular functions ξ and β and a process $\{S(t);\ t \in [0,\infty)\}$ with independent increments as specified in Section 2. Let $P_{\xi,t}$ be the distribution which has density $\exp\{\xi S(t)-\beta(\xi)\}$ with respect to the distribution $P_{0,t}$ of $S(t)$. Let F_t be the experiment $F_t = \{P_{\xi(\theta),t};\ \theta \in \Theta\}$.

PROPOSITION 4. Assume that the sequence (E_ν, n_ν) is obtained from independent observations according to Assumption (D). If Assumption (A) is satisfied in order that the sequence (E_ν, n_ν) be of Koopman-Darmois type for the functions ξ and β in the sense of Definition 1, it is necessary and sufficient that for every sequence $\{m_\nu\}$ of integers such that $\frac{m_\nu}{n_\nu} \to t$ the experiments $E_\nu(m_\nu)$ converge weakly to the experiment F_t.

PROOF. Let $\theta_0, \theta_1, \ldots, \theta_k$ be elements of Θ such that $\xi(\theta_0) = 0$, and such that the set $\{\xi_j;\ j = 1,\ldots,k\}$ with $\xi_j = \xi(\theta_j)$ be a basis of H. Define random vectors $S_{n,\nu}$ by the relations

$$\xi_j\ S_{n,\nu} = \log \frac{dP_\nu(\theta_j,n)}{dP_\nu(\theta_0,n)} + \frac{n}{n_\nu}\ \beta(\xi_j).$$

Let $\xi = \xi(\theta)$ be another element of $\xi(\Theta)$. The fact that the limit experiments F_t are homogeneous, together with the weak convergence of $E_\nu(m_\nu)$ to F_t implies (is in fact equivalent) to the convergence in distribution of the logarithms of likelihood ratios for the points $\theta_0, \theta_1, \ldots, \theta_k, \theta$.

It follows that for any sequence $\{m_\nu\}$ for which $\frac{m_\nu}{n_\nu} \to t$ the difference

$$\log \frac{dP_\nu(\theta, m_\nu)}{dP_\nu(\theta_0, m_\nu)} - \xi(\theta) S_{m_\nu,\nu} + \frac{m_\nu}{n_\nu} \beta[\xi(\theta)]$$

will tend in probability to zero.

Let $X_{\nu,j}(\theta)$ be the variable

$$X_{\nu,j}(\theta) = \log \frac{dp_\nu(\theta, j)}{dp_\nu(\theta_0, j)}.$$

By Assumption (D), these variables are independent and the logarithms of likelihood ratios have the form

$$\log \frac{dP_\nu(\theta, m_\nu)}{dP_\nu(\theta_0, m_\mu)} = \sum_{j \leq m_\mu} X_{\nu,j}(\theta).$$

Take a fixed integer b. The variables $S_{bn_\nu,\nu}$ are sums of independent variables analogous to the $X_{\nu,j}(\theta)$. Since $\sum_{j \leq bn_\nu} X_{\nu,j}(\theta) - S_{bn_\nu,\nu}$ tends to zero in probability, there are constants, say $C_{n,\nu}$, such that

$$\sup_{n \leq bn_\nu} \left| \sum_{j \leq n} X_{\nu,j}(\theta) - S_{n,\nu} - C_{n,\nu} \right| \to 0.$$

However we already know that for every sequence $m_\nu \leq bn_\nu$, $\sum_{j \leq m_\nu} X_{\nu,j}(\theta) - S_{m_\nu,\nu} \to 0$. Thus $\sup_{n \leq bn_\nu} |C_{n,\nu}| \to 0$ and the variables $S_{n,\nu}$ so defined satisfy all the conditions of Definition 1. The reverse implication being trivial, the desired equivalence is proved.

REMARK. Let $\{Z_{\nu,j};\ j = 0,1,2,\ldots\}$ be independent vectors such that $Z_{\nu,j}$ is a function on the coordinate space $X_{j,\nu}$. Let $S^*_{n,\nu} = \sum_{j \leq n} Z_{\nu,j}$. If the variables $S^*_{n,\nu}$ also satisfy the conditions of Definition 1 then $\sup_{n \leq bn_\nu} |S^*_{n,\nu} - S_{n,\nu}|$ must tend to zero in probability. However, for these $S^*_{n,\nu}$ to satisfy the requirements it will be sufficient that for every b, every b, every sequence $m_\nu \le bn_\nu$ the differences $S^*_{m_\nu,\nu} - S_{m_\nu,\nu}$ converge in probability to zero.

It is clear that, even under Assumption (D) one cannot expect to have the asymptotic behavior described by Definition 1 unless the various $p_\nu(\theta,j)$, $j = 1,2,\ldots$ are somewhat alike. One particular case, which is probably the one which has received most attention in sequential analysis, is the case where the $p_\nu(\theta,j)$ are in fact copies of one and the same $p_\nu(\theta)$ for all $j = 1,2,\ldots$. This $p_\nu(\theta)$ still depends on ν, of course.

This can be referred to as the "independent identically distributed case." To make the statement precise, the independent identically distributed case satisfies Assumption (D) with spaces $\{X_{j,\nu}, \mathcal{B}_{j,\nu}, p_\nu(\theta,j)\}$ which are copies of each other for each given ν.

This leads to the following statement.

COROLLARY. Let Assumption (D) be satisfied for spaces $\{X_{j,\nu}, \mathcal{B}_{j,\nu}, p_\nu(\theta,j)\}$ which are, for each ν and θ, exact copies of one another. Then the conditions of Definition 1 hold if and only if the experiment $E_\nu(n_\nu)$ converges to the experiment F_1.

PROOF. One can first check that for each integer b, the experiment $E_\nu(bn_\nu)$ converge to F_b. Similarly if m_ν is the integer part of $\frac{jn_\nu}{m}$ for a fixed integer m, the experiments $E_\nu(m_\nu)$ will converge to $F_{j/m}$. The rest follows easily.

Examples of situations which come under the purview of the above Corollary can be obtained by taking independent identically distributed observations and looking at "nearby alternatives" as is common practice in many asymptotic arguments. Specifically suppose that G is an open subset of H which contains the origin as interior point. Fix a probability space $\{X_0, \mathcal{B}_0, \mu\}$ and assume that for each $s \in G$ one has a density $f(x,s)$ with respect to μ. Assume that the process $s \to \sqrt{f(x,s)}$ is differentiable in quadratic mean at the point $s = 0$ and, changing coordinate systems if necessary, assume the "Fisher information matrix" at $s = 0$ is the identity matrix. For every $s \in G$ let $\theta(s) = \sqrt{\nu}s$. The image of G by this map is a certain set Θ_ν. On any fixed subset $\Theta \subset \Theta_\nu$ the corresponding experiments E_ν satisfy the conditions of Definition 2 for the sequence $n_\nu = \nu$. The independence and identity of distribution assumptions of Corollary 2 are also satisfied.

In this situation, one can take for vectors $S_{n,\nu}$ either the vectors described in Proposition 4 or the vectors which are successive sums of the derivatives in quadratic mean at $s = 0$. It should be clear that such an example is a very special case of situations covered by the Corollary since the latter allows among other things fairly arbitrary variations of the densities $f(x,s)$ as ν changes.

5. *Some applications to tests and estimates.* The purpose of the previous sections was to show that sequential experiments on systems $\{E_\nu, n_\nu\}$ covered by Definition 1 have for limits sequential experiments on the stochastic processes $\{S(t);\ t \in [0,\infty)\}$. The passages to the limit considered here are only the weak passages in the sense of [1]. However, according to Theorem 1 of [2], weak convergences are already enough to provide information on lower bounds for risk functions. The results of [2] say in particular that if the sequential experiments F_ν converge to a limit F and if r is a function which is not a possible risk function for F, then $r+\varepsilon$ is not possible either for F_ν if ν is sufficiently large and $\varepsilon > 0$ is sufficiently small.

In the present case we have to consider not only the losses coming from wrong terminal decisions, but also the cost of sampling. In the case of our systems S_ν this would usually be introduced by taking a cost proportional to $E[\frac{N_\nu}{n_\nu}]$. However, just as exact expected square deviations may be misleading, so can these expectations. Extraordinarily large values of N_ν which occur with negligibly small probabilities can increase EN_ν enormously, without introducing any difference in the practical properties of the sequential scheme. For this reason, we shall proceed here as if the cost of sampling was measured by the expectation $E[\tau|\theta]$ where τ has the limiting distribution $\lim_{\nu} \mathcal{L}[\frac{N_\nu}{n_\nu}|\theta]$.

To give some examples of possible arguments we shall consider the following situation.

ASSUMPTION E. *The systems*

$$S_\nu = \{X_\nu, \{A_{n,\nu}\}, P_{\theta,\nu}, \Theta, N_\nu\}$$

satisfy the conditions of Definition 2 for a given sequence $\{n_\nu\}$. *The assumptions* (B) *and* (C) *are satisfied. The stopping variables* $\tau_\nu = \frac{N_\nu}{n_\nu}$ *form a regular sequence. Furthermore the range* $\xi(\Theta)$ *is the entire space* $H = R^k$.

Under this assumption, let us consider the problem of estimating the parameter ξ $(= \xi(\theta))$, the loss being the usual quadratic loss $||\hat{\xi}-\xi||^2$.

Here the limiting process S is the Brownian process with values in H' and the sequential experiment F_ν converge to a limit F given by measures F_ξ such that

$$F_\xi(dx,d\tau) = \exp\{\xi x - \frac{\tau}{2}||\xi||^2\}F_0(dx,d\tau).$$

LEMMA 3. *For the experiment* F *and for the square loss function, the minimax risk is at least equal to*

$$\lim_{\sigma\to 0} \inf_{\theta} E[\frac{k}{\tau+\sigma^2}|\theta] \geq k\{\sup_{\theta} E(\tau|\theta)\}^{-1}.$$

PROOF. This follows immediately from the computation of Bayes risks for Gaussian prior distributions of ξ.

Letting $a_0 = \{\sup_{\theta} E(\tau|\theta)\}^{-1}$ we can state the following corollary.

COROLLARY. *Let Assumption* (E) *be satisfied. For each* ν *and each number* $c \geq 1$ *let* $r(\nu,c)$ *be the minimax risk obtainable for the sequential experiment* F_ν *and the loss function* $\min\{||\hat{\xi}-\xi||^2, c\}$.

If a_1 *is a number* $a_1 < a$, *there is a* ν_0 *and a* c *such that for all* $\nu \geq \nu_0$ *one has* $r(\nu,c) \geq ka_1$.

This follows from [2], Theorem 1.

The form of the bounds given in Lemma 3 suggests that in fact, if τ is not degenerate, the minimax risks will be strictly bigger than what would be obtainable by non-random stopping. However, at this time we do not have a proof yielding strict inequalities. In the form given above the statement resembles a result of Ibragimov and Has'minskii [3].

To pass to a more complicated example retain Assumption (E) and consider the problem of testing the hypothesis (H_0) that ξ belongs to a certain hyperplane $H_0 \subset H$ against the alternative that ξ is in a specified half-space bounded by H_0. One can then take an orthogonal system of coordinates in H_0 and complete it by a unit vector u orthogonal to H_0. We shall assume that the half space under consideration is the set of points whose projections on u are positive.

Let us suppose that it is desired to have tests based on the sequential systems S_ν such that i) their power $\beta_\nu(\xi)$ converges to a given $\alpha \in (0,1)$ for all $\xi \in H_0$, ii) subject to limitations on the expectations $E(\tau|\theta)$ of the limiting distributions, the limit power is maximized on the affine variety $H_c = H_0 + cu$, for a given $c > 0$.

Consider then, for the limiting Brownian process S, the problem of testing a particular $\xi_0 \in H_0$ against the simple alternative $\xi_0 + cu$.

For this simple hypothesis against simple alternative problems, one can use the sequential probability ratio test (SPRT). If $P_{\xi,t} = L[S(t)|\xi]$, the logarithms of likelihood ratios have the form

$$Z(t) = \log \frac{dP_{\xi_0+cu,t}}{dP_{\xi_0,t}} = cuS(t) - \frac{c^2}{2}t.$$

Thus an SPRT test will be given by two numbers b_0, b_1, and a stopping time τ which is the first time t at which an inequality of the type $b_0 + \frac{ct}{2} \leq uS(t) \leq b_1 + \frac{ct}{2}$ fails. At this time τ one decides in favor or against H_0 according to the obvious rule.

The coefficients (b_0, b_1) can be adjusted to obtain a level α and a power β as desired. This being done the basic result of Wald and Wolfowitz [7] says that the sequential test in question minimizes both $E(\tau|\xi_0)$ and $E[\tau|\xi_0 + cu]$.

Since the test in question depends only on the projection uS(t) (which is a standard Brownian motion with linear drift) it is also a similar test of H_0 against H_c, with constant power β on H_c.

Here an asymptotic optimality statement could be formulated in the following way. Let τ be the stopping time of Brownian motion described above, for a test with level α on H_0 and power β on H_c. Let $\tau_\nu = \frac{N_\nu}{n_\nu}$ be any regular sequence of stopping time yielding a sequence $\{S_\nu, \nu = 1,2,\ldots\}$. Suppose that a test available on F_ν has limit level not exceeding α on H_0 and power at least β on H_c. Then, if $L[\tau_\nu|\xi] \to L(\tau'|\xi)$ one has $E[\tau'|\xi_0] \geq E[\tau|\xi_0]$ and $E[\tau'|\xi_0 + cu] \geq E[\tau|\xi_0 + cu]$. At this point it is tempting to mimick the solution for Brownian motion on the processes S_ν themselves, taking for stopping times τ_ν the first time when the inequalities $b_0 + \frac{ct}{2} \leq uS_\nu(t) \leq b_1 + \frac{ct}{2}$ fail.

It is easily seen that <u>under Assumption (E) such a stopping rule will precisely achieve the desired limiting behavior whenever Assumption (D) is satisfied and the $S_{n,\nu}$ are successive sums, as described in the proof of Proposition 4</u>. Indeed, in such circumstances the processes S_ν converge to the Brownian process S in the Prohorov sense on every finite interval $0 \leq t \leq c$. This is obvious here. For other considerations see [8].

However, even under circumstances essentially covered by (D) and (E) one is often led to use statistics which are not successive sums. This occurs for instance in A. Singh's generalization of Neyman's $C(\alpha)$ tests [4].

In the situation studied by Singh the reduction to the local problem described by our Assumption (E) is effected by using suitably consistent estimates $\hat{\eta}_\nu$ of a nuisance parameter, say η. (Locally the range of the nuisance parameter is our H_0.) One then estimates for each n the quadratic forms which occur in the Gaussian approximations around the estimated value $\hat{\eta}_\nu$.

In this case, instead of the statistics $uS_{n,\nu}$ one uses approximations to them, say $T_{n,\nu}$.

The above argument shows that sequential tests with required level and power cannot be asymptotically better than the test derived from the Brownian motion argument.

Whether the stopping rule where N_ν is the first integer n violating the inequalities

$$b_0 + \frac{c}{2}\,\frac{n}{n_\nu} \leq T_{n,\nu} \leq b_1 + \frac{c}{2}\,\frac{n}{n_\nu}$$

achieves the desired asymptotic behavior depends on the behavior of the processes $\{T_{n,\nu};\ n = 0,1,2,\ldots\}$.

Constructing processes $\{W_\nu(t);\ t \in [0,\infty)\}$ by the relation $W_\nu(t) = T_{n,\nu}$ if $n \leq tn_\nu < n+1$, one sees immediately that the asymptotic behavior will be optimal whenever the W_ν converge in the Prohorov, or Skorohod sense to the limit Wiener process $\{uS(t);\ t \in [0,\infty)\}$.

It should be apparent that one can handle in a similar manner sequences $\{E_\nu, n_\nu\}$ where the experiments $E_\nu(m_\nu)$, for non-random stopping times m_ν, are obtained from the general asymptotically Gaussian sequences of [1] (Section 12). This affords a vast generalization which however has to rely on the "regularity" of the stopping times $\tau_\nu = \frac{N_\nu}{n_\nu}$. We shall return to it elsewhere.

References

[1] L. Le Cam. (1974). *Notes on asymptotic methods in statistical decision theory.* Chapter I. Centre de Recherches Mathématiques, Université de Montréal.

[2] L. Le Cam. (1976). On a theorem of J. Hájek. To be published in the *Hájek Memorial* volume. Academia, Praha.

[3] I. A. Ibragimov and R. Z. Has'minskii. (1974). On sequential estimation. *Theor. Probability Appl.* 19, 233-244.

[4] Avinash C. Singh. (1975). *Sequential Version of* C(α) *Tests of Composite Hypotheses.* Ph.D. Thesis, University of California, Berkeley.

[5] H. Chernoff. (1961). Sequential tests for the mean of a normal distribution. *Proc. Fourth Berkeley Symp. Math. Statist. Prob.* 1, 79-91.

[6] L. Le Cam. (1960). Locally asymptotically normal families of distributions. *University of California Publications in Statistics* 3, 37-98.

[7] A. Wald and J. Wolfowitz. (1948). Optimum character of the sequential probability ratio test. *Ann. Math. Statist.* 19, 326-339.

[8] W. J. Hall and R. M. Loynes. (1975). Weak convergence of processes related to likelihood ratios. University of Rochester Technical Report.

ROBUST DESIGNS FOR REGRESSION PROBLEMS*

By Michael B. Marcus and Jerome Sacks
Northwestern University

1. *Introduction.* Consider the regression design setup given by

$$Y(x_i) = f(x_i) + \varepsilon_i \quad i = 1,\ldots,n$$

where the errors, $\{\varepsilon_i\}$, are i.i.d with mean 0 and constant variance σ^2, $x_i \in [-1,1]$, f is a function from a class $\mathscr{F}_0$. $\mathscr{F}_0$ is commonly a linear combination of specified functions $f_0,\ldots,f_k$ (e.g. $f_j = x^j$). The regression problem is concerned with inference about the (unknown) coefficients of these specified f_j's, and the associated design problem is to choose the x_i's in an optimal way for this inference. Numerous papers by Kiefer and others have been addressed to this problem. Box and Draper [1] seem to have been the first to expose some of the dangers inherent in a strict formulation of $\mathscr{F}_0$ which ignores the possibility that the "true" f may only be approximated by an element of $\mathscr{F}_0$, e.g., in estimation there may result a large bias term. A careful description of some problems in this context has been given by Kiefer [3] in the case where the class of possible f's, $\mathscr{F}$, is a finite dimensional space containing $\mathscr{F}_0$.

In a related direction Huber [2] formulated a problem where $\mathscr{F}_0$ = {linear functions} and $\mathscr{F} = \{f(x) = a + bx + g(x)$ with $\int_{-1/2}^{1/2} g^2(x)dx \leq \varepsilon\}$. Huber confines himself to the use of the standard least squares estimates based on the model $\mathscr{F}_0$ and finds the design which minimizes the loss

$$\sup_{f\in\mathscr{F}} E\int_{-1/2}^{1/2} (\hat{a} + \hat{b}x - f(x))^2 dx.$$

*This work was partially supported by grants from the National Science Foundation.

An unfortunate consequence of this formulation is that it leads to the restriction that the designs must be absolutely continuous, otherwise the loss above is infinite. This means that no implementable design can have finite loss. As for the restriction to the standard estimates see Remark 3 in Section 4 and the last paragraph of Section 4.

In the spirit of Huber's formulation, we take $\mathscr{F}_0$ to be the linear functions and take $\mathscr{F}$ to be a class of "nearly" linear functions f of the form $f(x) = a + bx + g(x)$ where $|g(x)| \leq \varphi(x)$, $\varphi(0) = 0$, with φ a given function. The canonical example is to take $a = f(0)$, $b = f'(0)$ and $\varphi(x) = mx^2$ where m is a bound on f". The further condition, f" continuous, is not used. In fact this condition leads to what appears to be a much more difficult problem. We restrict the estimates to be linear, but not necessarily the standard least squares estimates based on the model $\mathscr{F}_0$, and look for designs and estimates to minimize the mean square error

$$\sup_f E((\hat{a}-a)^2 + \theta^2(\hat{b}-b)^2)$$

where $\hat{a}$, $\hat{b}$ denote the estimates and θ is specified. (θ can be related to the length L of the interval on which the problem is originally stated, in which case $\theta = 1/L$, or θ can serve the purpose of weighting the mean square errors differently.)

We are able to solve this problem in a number of cases. In Theorem 3.4 we show that if $\varphi(x) \geq mx$ then the (unique) optimal design is on the points $\{-1,0,1\}$. In Theorem 3.2 we conclude that when φ is convex there is a wide range of cases for which a best design can be found on 2 points $\{-z,z\}$, (z is determined by (3.3)).

When $\varphi(x) = mx^2$ and $\rho = \frac{\sigma^2}{nm^2} \leq \theta^4$ then $z = \rho^{1/4}$. This should cover many cases of interest. As one would expect, small ρ leads to designs concentrated close to 0. Unfortunately, our results which pertain to this example (Theorems 3.2 and 3.3) do not cover all possible choices of ρ, θ (see the last paragraph of Section

3).

In Section 4 we compare some of the optimal solutions with the ones based on the standard estimates and it appears that there isn't much difference in the minimum loss. See, however, Remark 3 in Section 4 which suggests that if the standard estimates are not used with their best design there may result a severe loss whereas the use of a non-optimal design together with the best estimates for that design may not result in severe loss.

Section 2 contains the formulation of the problem and some preliminary results. Section 3 contains the results we can get on optimal solutions. Section 4 treats the problem using the standard estimates and contains some other remarks.

2. *Formulation of the problem.* Let $\mathcal{G}$ be a class of functions on [-1,1], which will be specified later, and let $\mathcal{F}$ be the class of functions f which are of the form $f(x) = a + bx + g(x)$ where $g \in \mathcal{G}$. An observation of $f \in \mathcal{F}$ at a point $x \in [-1,1]$ will be assumed to have mean f(x) and constant variance σ^2. Let $Y(x_1),\ldots,Y(x_n)$ denote n observations taken at points $x_1,\ldots,x_n$ ($x_i = x_j$, $i \neq j$ is possible). We assume that the observations are uncorrelated.

If ξ is a discrete probability measure on [-1,1] satisfying $\xi(x) = j(x)/n$ where j(x) is an integer then ξ is called an exact design for n observations. We denote by Σ_n the class of such designs. Eventually, as is customary, we will simplify the problem by considering designs as arbitrary probability measures with finite support. To this end let $\mathcal{D}'_k = \{\xi \mid \text{Card(supp of } \xi) \leq k\}$ and let $\mathcal{D}' = \bigcup_{k=1}^{\infty} \mathcal{D}'_k$.

Let $\xi \in \Sigma_n$ and take linear estimates of a,b defined by

$$\hat{a} = \int_{-1}^{1} Y(x)A(x)d\xi(x) \tag{2.1}$$

$$\hat{b} = \int_{-1}^{1} Y(x)\tilde{B}(x)d\xi(x) \tag{2.2}$$

where A, $\tilde{B}$ are real valued functions on $[-1,1]$. Let $0 < \theta < \infty$ and consider the "mean square error"

$$E(\hat{a}-a)^2 + \theta^2 E(\hat{b}-b)^2 \tag{2.3}$$

The "variance" term in (2.3), $E(\hat{a}-E\hat{a})^2 + \theta^2 E(\hat{b}-E\hat{b})^2$, is equal to

$$\frac{\sigma^2}{n} \int_{-1}^{1} (A^2(x) + \theta^2 \tilde{B}^2(x))\xi(dx) \tag{2.4}$$

and the "bias" term $(a-E(\hat{a}))^2 + \theta^2(b-E(\hat{b}))^2$ is determined by the equations

$$a-E(\hat{a}) = a\left(\int_{-1}^{1} A(x)d\xi(x)-1\right)+b \int_{-1}^{1} A(x)d\xi(x) + \int_{-1}^{1} A(x)g(x)d\xi(x). \tag{2.5}$$

$$\theta(b-E(\hat{b})) = \theta a\int_{-1}^{1} \tilde{B}(x)d\xi(x) + \theta b\left(\int_{-1}^{1} \tilde{B}(x)d\xi(x)-1\right) + \theta\int_{-1}^{1} \tilde{B}(x)g(x)d\xi(x). \tag{2.6}$$

If no restrictions are placed on a,b then the bias term is unbounded unless the coefficients of a and b are 0 in (2.5) and (2.6). Therefore we require that

$$\int_{-1}^{1} A(x)d\xi(x) = 1, \quad \int_{-1}^{1} A(x)xd\xi(x) = 0 \tag{2.7}$$

$$\int_{-1}^{1} B(x)d\xi(x) = 0, \quad \int_{-1}^{1} B(x)xd\xi(x) = \theta \tag{2.8}$$

where we set $B(x) = \theta\tilde{B}(x)$.

Let $\rho = \sigma^2/n$ and define

$$J(A,B,\xi,g) = \left(\int_{-1}^{1} A(x)g(x)d\xi(x)\right)^2+\left(\int_{-1}^{1} B(x)g(x)d\xi(x)\right)^2 + \rho \int_{-1}^{1} (A^2(x) + B^2(x))d\xi(x). \tag{2.9}$$

Note that $J(A,B,\xi,g)$ is equal to (2.3) with conditions (2.7) and (2.8) imposed. Conditions (2.7) and (2.8) and $J(A,B,\xi,g)$ all are well defined for $\xi \in \mathscr{D}'$ and, from now on, we shall not restrict ξ to be an exact design.

Our objective is to find A, B and ξ satisfying (2.7) and (2.8) that minimize $\sup_{g\in\mathscr{G}} J(A,B,\xi,g)$. We begin with some lemmas.

LEMMA 2.1. *If $A^2 + B^2$ is not constant on the support of ξ then there is an A_0, B_0, ξ_0 for which $J(A_0,B_0,\xi_0,g) < J(A,B,\xi,g)$ for all $g \in \mathscr{G}$.*

PROOF. Let $d\xi_0 = c(A^2 + B^2)^{1/2}d\xi$ where c is the constant that makes ξ a probability measure. Set $A_0 = A/c(A^2 + B^2)^{1/2}$ and $B_0 = B/c(A^2 + B^2)^{1/2}$ and define them to be zero if the denominator is zero. Note that $A_0 d\xi_0 = Ad\xi$ and $B_0 d\xi_0 = Bd\xi$. Consider (2.7), (2.8) and (2.9) with A_0,B_0,ξ_0 replacing A,B.ξ. Everything is unchanged with the possible exception of the last term in (2.9). Since

$$\int_{-1}^{1} (A_0^2+B_0^2)d\xi_0 = \frac{1}{c^2}\int_{-1}^{1} d\xi_0 = [\int_{-1}^{1} (A^2+B^2)^{1/2}d\xi]^2$$
$$\leq \int_{-1}^{1} (A^2 + B^2)d\xi,$$

with strict inequality unless $A^2 + B^2$ is constant on the support of ξ, the lemma is proved.

The next lemma is an invariance result which enables us to simplify conditions (2.7) and (2.8) and restrict our attention to functions A,B and designs ξ supported on [0,1].

LEMMA 2.2. *Suppose that for $g \in \mathscr{G}$, $Tg(x) = g(-x)$ also is contained in $\mathscr{G}$. If*

$$(2.10) \qquad J^* = \inf_{\xi\in\mathscr{D}_k'} \inf_{A,B} \sup_{g\in\mathscr{G}} J(A,B,\xi,g) = \sup_{g\in\mathscr{G}} J(A_0,B_0,\xi_0,g)$$

for specific functions A_0,B_0 and design $\xi_0 \in \mathscr{D}_k'$ satisfying (2.7) and (2.8) then there exists a $\bar{\xi} \in \mathscr{D}_{2k}'$ and functions $\bar{A}$, $\bar{B}$ such

<u>that</u> $\bar{A}(x) = \bar{A}(-x)$, $\bar{B}(x) = -\bar{B}(-x)$, $\bar{\xi}(x) = \bar{\xi}(-x)$, $\bar{A}$, $\bar{B}$, $\bar{\xi}$ <u>satisfy</u> (2.7) <u>and</u> (2.8) <u>and</u>

$$J^* = \sup_{g\in\mathcal{G}} J(\bar{A},\bar{B},\bar{\xi},g) = \sup_{g\in\mathcal{G}} J(A_0,B_0,\xi_0, g). \tag{2.11}$$

PROOF. By Lemma 2.1 we have that

$$\int_{-1}^{1} (A_0^2 + B_0^2)d\xi_0 = [\int_{-1}^{1} (A_0^2 + B_0^2)^{1/2}d\xi_0]^2. \tag{2.12}$$

Let $\xi_1(x) = \xi_0(-x)$, $A_1(x) = A_0(-x)$, $B_1(x) = -B_0(-x)$. For each $g\in\mathcal{G}$, $J(A_1,B_1,\xi_1,Tg) = J(A_0,B_0,\xi_0,g)$ and it follows that

$$J^* = \sup_{g\in\mathcal{G}} J(A_0,B_0,\xi_0,g) = \sup_{g\in\mathcal{G}} J(A_1,B_1,\xi_1,g). \tag{2.13}$$

Let $u_i(x) = A_i(x)\xi_i(x)$, $v_i(x) = B_i(x)\xi_i(x)$, $i = 0,1$. The u_i and v_i are defined on $S = \{x|x \in \text{supp}(\xi_0) \text{ or } -x \in \text{supp}(\xi_0)\}$. Using (2.12) we get $J(A_i,B_i,\xi_i,g) = \tilde{J}(u_i,v_i,g)$ where (for any functions u,v)

$$\tilde{J}(u,v,g) = (\sum_{x\in S} g(x)u(x))^2 + (\sum_{x\in S} g(x)v(x))^2$$
$$+ \rho (\sum_{x\in S} (u^2(x) + v^2(x))^{1/2})^2.$$

$\tilde{J}(u,v,g)$ is convex in the pair (u,v). (To see this for the last term note that $(u^2(x) + v^2(x))^{1/2} = ||q(x)||$ for the vector $q(x) = (u(x), v(x))$. Then use the triangle inequality, sum on x, etc.) Clearly $\sup_{g\in\mathcal{G}} \tilde{J}(u,v,g)$ is also convex in (u,v) and (2.7) and (2.8) remain valid for convex combinations of (u,v). Let $\bar{u} = (u_0+u_1)/2$, $\bar{v} = (v_0+v_1)/2$ and note that $\bar{u}(x) = \bar{u}(-x)$ and $\bar{v}(x) = -\bar{v}(-x)$. We have

$$\sup_{g\in\mathcal{G}} \tilde{J}(\bar{u},\bar{v},g) \leq J^*$$

by convexity and (2.13). Define $\bar{\xi}(x) = \bar{c}(\bar{u}^2(x) + \bar{v}^2(x))^{1/2}$, where $\bar{c}$ is the constant that makes $\bar{\xi}$ a probability measure, $\bar{A}(x) = \bar{u}(x)/\bar{\xi}(x)$ and $\bar{B}(x) = \bar{v}(x)/\bar{\xi}(x)$. These are the values $\bar{A}$, $\bar{B}$, $\bar{\xi}$ that satisfy (2.11).

Let $\varphi(x)$ be a positive bounded even function on $[-1,1]$,

$\varphi(0) = 0$, $\varphi(1) = 1$. The normalization $\varphi(1) = 1$ is only for convenience since we can replace 1 by m and take $\tilde{\varphi}(x) = \varphi(x)/m$ and change ρ to $\tilde{\rho} = \sigma^2/nm^2$. (Generally one would want $\varphi(x)$ increasing on [0,1].) For such a function φ we take $\mathcal{G} = \mathcal{G}_\varphi = \{g \mid |g(x)| \leq \varphi(x)\}$.

LEMMA 2.3. <u>For $\mathcal{G}_\varphi$ as defined above (2.10) is satisfied.</u>

PROOF. This follows easily if A and B can be restricted to be bounded. Note that

$$J^* \leq \max[1,\theta^2] + \rho[1 + \theta^2] \tag{2.14}$$

(the bound on the right can be achieved by a design that places mass 1 at x = 1.) By Lemma 2.1 we can restrict our attention to functions A and B for which $A^2 + B^2 = \text{Const}$. These two observations imply that A and B can be restricted to be bounded.

LEMMA 2.4. <u>If g maximizes $J(A,B,\xi,g)$ then $|g(x)| = \varphi(x)$ on the support of ξ.</u>

PROOF. $J(A,B,\xi,g)$ is a convex quadratic function of g(x) for fixed x. Therefore its maximum for $-\varphi(x) \leq g(x) \leq \varphi(x)$ is at the end points.

Let $\bar{A}$, $\bar{B}$, $\bar{\xi}$ be the functions and design obtained in Lemma 2.2. We have

$$J(\bar{A},\bar{B},\bar{\xi},g) = \left(\int_0^1 \bar{A}(x)h_1(x)d\bar{\xi}(x)\right)^2 + \left(\int_0^1 \bar{B}(x)h_2(x)d\bar{\xi}(x)\right)^2 + \rho\int_0^1 (\bar{A}(x)^2 + \bar{B}^2(x))d\bar{\xi}(x) \tag{2.15}$$

where $h_1(x) = (g(x) + g(-x))/2$, $h_2(x) = (g(x)-g(-x))/2$ and $\xi = 2\bar{\xi}$. Also, conditions (2.7) and (2.8) reduce to

$$I_1(\bar{A}) = \int_0^1 \bar{A}(x)\xi(dx) = 1,\ I_2(\bar{B}) = \int_0^1 \bar{B}(x)d\xi(x) = \theta. \tag{2.16}$$

By Lemma 2.4 we see that $h_1(x)$, $h_2(x)$ can assume only the three values 0, $\pm\varphi(x)$ and $h_1(x) = 0$ if and only if $h_2(x) \neq 0$.

Let E be a subset of [0,1]. Define

$$J_0(\bar{A},\bar{B},\xi,E) = \left(\int_0^1 |\bar{A}| I_E \varphi \, d\xi\right)^2 + \left(\int_0^1 |\bar{B}| I_{E^c} \varphi \, d\xi\right)^2 + \rho \int_0^1 (\bar{A}^2 + \bar{B}^2) d\xi$$

where I_E is the indicator function of E and E^c denotes the complement of E. Because of the special relationship between $h_1(x)$ and $h_2(x)$ we have

$$\sup_{E \subset [0,1]} J_0(\bar{A},\bar{B},\xi,E) = \sup_{g \in \mathcal{G}} J(\bar{A},\bar{B},\xi,g) = J^*$$

Finally, let $\mathcal{D}_N$ be the class of probability measures on [0,1] with support on at most N points. Consider

$$J^{**} = \min_{\xi \in \mathcal{D}_N} \min_{A,B} \max_{E \subset [0,1]} J_0(A,B,\xi,E) \tag{2.17}$$

where A, B are now functions on [0,1] satisfying $I_1(A) = 1$ and $I_2(B) = \theta$ (see (2.16)).

The following lemma gives us further simplification in (2.17).

LEMMA 2.5. *In* (2.17) *we can take* A, B $\geq$ 0.

PROOF. If $A(x_0) < \theta$ for some $x_0 \in [0,1]$ define $\tilde{A}(x) = \frac{A(x)}{1-A(x_0)\xi(x_0)}$ for $x \neq 0$, $\tilde{A}(x_0) = 0$, where $\xi(x_0)$ is the mass at x_0. Clearly $J_0(\tilde{A},B,\xi,E) < J_0(A,B,\xi,E)$ and (2.16) is satisfied for $\tilde{A}(x)$. A similar argument holds for B.

This leads us to consider the following problem: Let $\mathcal{D}_N$ be the class of N point designs on [0,1] and let $A \geq 0$, $B \geq 0$ satisfy

$$\int_0^1 A(x) d\xi(x) = 1, \int_0^1 B(x) x d\xi(x) = \theta. \tag{2.18}$$

Find the solution for

$$J^{**} = \min_{\xi \in \mathcal{D}_N} \min_{A,B} \max_{E \subset [0,1]} J_0(A,B,\xi,E) \tag{2.19}$$

where

$$(2.20)\qquad J_0(A,B,\xi,E) = \left(\int_0^1 AI_E\varphi\, d\xi\right)^2 + \left(\int_0^1 BI_{E^c}\varphi\, d\xi\right)^2 + \rho\left[\int_0^1 (A^2+B^2)d\xi\right]$$

where φ is the function that determines $\mathcal{G}$. Suppose, for some k the ξ which minimizes (2.19) remains the same for all $N \geq k$. Then by Lemma 2.2 this ξ leads to a solution of the problem of minimizing J^* (see (2.10)). It will be seen in the next section that our solutions have designs which are supported on small sets. Therefore they are also solutions of the original problem.

We proceed to study the min-max problem (2.19). First we introduce some new notation. Let $\emptyset$ be the empty set. Define (for $J_0(A,B,\xi,E)$ as given in (2.20))

$$(2.21)\qquad J_1(A,B,\xi) = J_0(A,B,\xi,\emptyset)$$

$$(2.22)\qquad J_2(A,B,\xi) = J_0(A,B,\xi,[0,1])$$

$$(2.23)\qquad J_3(A,B,\xi) = \max(J_1(A,B,\xi),\ J_2(A,B,\xi))$$

$$(2.24)\qquad J_0(A,B,\xi) = \max_{E\subset[0,1]} J_0(A,B,\xi,E).$$

Also we define $J_i = J_i(A,B,\xi)$ and $J_i(\xi) = \min_{A,B} J_i(A,B,\xi)$, $i = 0,\ldots,3$. It is clear that

$$(2.25)\qquad J_0 \geq J_3 \geq J_1 \text{ and } J_0 \geq J_3 \geq J_2.$$

The next result is from Sacks and Ylvisaker [4]. It is obtained by a variational argument. We cite it without proof.

THEOREM 2.1. _For_ ξ _fixed_, $J_2(A,B,\xi)$ _is minimized by functions_ A, B _satisfying_

$$(2.26)\qquad A = \frac{\mu}{\rho}(\alpha-\varphi(x))_+$$

on the support of ξ, $((f(x))_+ = \max(f(x),0))$, _where_ $\mu = \int_0^1 A\varphi\, d\xi$ _and_ α _and_ μ _are determined by_

$$(2.27)\qquad \int_0^1 (\alpha-\varphi)_+\varphi\, d\xi = \rho$$

(2.28)
$$\mu = \frac{\rho}{\int_0^1 (\alpha-\varphi)_+ d\xi}$$

and

(2.29)
$$B(x) = \frac{\theta x}{\int_0^1 x^2 d\xi} .$$

In this case

(2.30)
$$J_2(\xi) = \alpha\mu + \rho\ \theta^2 / \int_0^1 x^2 d\xi .$$

For fixed ξ, $J_1(A,B,\xi)$ is minimized by

(2.31)
$$A(x) = 1,\ B(x) = \frac{\lambda}{\rho}(\beta-\varphi)_+$$

on the support of ξ where $\lambda = \int_0^1 B\,\varphi d\xi$ and β and λ are determined by

(2.32)
$$\int_0^1 (\beta x-\varphi)_+ \varphi\ d\xi = \rho$$

and

(2.33)
$$\lambda = \frac{\rho\theta}{\int_0^1 (\beta x-\varphi)_+ x d\xi} .$$

In this case

$$J_1(\xi) = \beta\lambda\theta + \rho .$$

The next theorem gives a simple but useful condition for finding $\inf_{\xi\in\mathscr{D}_N} J_0(\xi)$. (For notation, see the remarks following (2.24)).

THEOREM 2.2. *Suppose* $\min_{\xi\in\mathscr{D}_N} J_3(\xi)$ *is achieved at* A^*, B^*, ξ^* *and* ξ *either has support on one point or has support on two points one of which is* $\{0\}$, *then* $\inf_{\xi\in\mathscr{D}_N} J_0(\xi)$ *is achieved at* A^*, B^*, ξ^*.

PROOF. We have

$$J_3(A^*,B^*,\xi^*) = \min_{\xi}\min_{A,B} J_3(A,B,\xi) \leq \min_{\xi}\min_{A,B} \sup_{E\subset[0,1]} J_0(A,B,\xi,E)$$

$$\leq \sup_{E\subset[0,1]} J_0(A^*,B^*,\xi^*,E).$$

Since there is at most one point in the support of E not = 0 and $\varphi(0) = 0$ we get

$$\sup_{E\subset[0,1]} J_0(A^*,B^*,\xi^*,E) = J_3(A^*,B^*,\xi^*).$$

REMARK 1. Display the dependence of J_3 (see (2.23)) on φ by writing $J_3(A,B,\xi,\varphi)$. Suppose $\varphi_0 \leq \varphi$ on $[0,1]$, $\varphi_0(0) = 0$, $\varphi_0(1)=1$ and suppose that $\min_{\xi,A,B} J_3(A,B,\xi,\varphi_0)$ is achieved at A^*, B^*, ξ^* with supp $(\xi^*) \subset \{0,1\}$. It is then easy to see that the same A^*,B^*,ξ^* minimizes $J_3(A,B,\xi,\varphi)$ because $\varphi_0 = \varphi$ on supp (ξ^*). We use this observation to reduce the case of $\varphi(x) \geq x$ to the case $\varphi(x) = x$.

3. *Solutions*. In this section we consider the problem for J_1, J_2 and J_3 for a wide class of functions φ. We noted at (2.25) that these are all lower bounds for J_0 and Theorem 2.2 indicates why consideration of J_1, J_2, J_3 may give results about J_0. It turns out that this fortuitous situation occurs in a number of cases. Under certain extra conditions the solutions for J_1 and J_2 are also solutions for J_3.

We first consider J_1. We use Theorem 2.1 with t = 0. In this case A = 1 and we have the problem of minimizing

$$(\int_0^1 B\varphi\, d\xi)^2 + \rho \int_0^1 (1 + B^2)d\xi \tag{3.1}$$

under the side condition $\int_0^1 Bx d\xi = \theta$. Refer to (2.31), (2.32) and (2.33) for the explicit form of the function B(x).

THEOREM 3.1. <u>Let</u> $\varphi(x) \geq x$, $\varphi(0) = 0$, $\varphi(1) = 1$. <u>Then the</u> B <u>and</u> ξ <u>which minimize</u> (3.1) <u>are given by</u> $\xi(1) = 1$ <u>and</u> $B(1) = \theta$. <u>In this case</u>

$$\inf_{A,B,\xi} J_1(A,B,\xi) = \theta^2 + \rho(1 + \theta^2). \tag{3.2}$$

If $\theta \geq 1$ *and* $k \geq 2$ *(see (2.10)) these values for* B, ξ *and* $A(1) = 1$ *minimize* J_0.

PROOF. By Remark 1 we need only consider $\varphi(x) = x$. Employing the method of Lemma 2.1 we see that for the optimal B and ξ, $1 + B^2$ and consequently B^2 is constant on the support of ξ. By (2.31) we see that B^2 is convex and strictly increasing. Thus ξ has only one point in its support, call it t. The side condition is $B(t) = \theta/t$ and from (3.1) we get

$$\inf J_1(A,B,\xi) = \inf_{0 \leq t \leq 1} \left(\theta^2 + \rho \left(1 + \frac{\theta^2}{t^2}\right) \right)$$

This is minimized at $t = 1$.

When $\theta \geq 1$ we have $\theta^2 = (\int Bx d\xi)^2 \geq 1 = (\int Ax d\xi)^2$ since $A=1$ in this case. It follows that J_3 is minimized by these values of A, B, ξ. It follows from Theorem 2.2 that these values also minimize J_0.

Clearly all concave $\varphi(x)$, $\varphi(0) = 0$, $\varphi(1) = 1$ are covered by Theorem 3.1. The next result will cover all convex φ with $\varphi(0) = 0$ and $\varphi(1) = 1$ except for linear φ's which are covered above.

THEOREM 3.2. *Suppose that* $\varphi(x)$ *is continuous, strictly increasing and that* $\varphi(0) = 0$, $\varphi(1) = 1$. *Assume also that* $(\beta x - \varphi)_+$ *is unimodal on* $[0,1]$ *for each* $\beta > 0$, *(i.e. it achieves a unique maximum* C^* *on* $[0,1]$ *and* $(\beta x - \varphi)_+ = C$ *has at most two solutions for each* $0 < C < C^*$). *Then, the* B *and* ξ *which minimize* J_1 *are such that* ξ *has one point of support,* z, *which is where*

$$J_1 = \inf_{\xi} J_1(\xi) = \inf_{0 \leq x \leq 1} \left[\theta^2 \frac{\varphi^2(x)}{x^2} + \rho\left(1 + \frac{\theta^2}{x^2}\right)\right] \tag{3.3}$$

is attained, and $B(z) = \theta/z$. *If* $\theta \geq z$ *and* $k \geq 2$ *(see (2.10)) these values for* B, ξ *and* $A = 1$ *minimize* J_0.

Note that the problem for J_0 is now solved if $\theta \geq 1$.

PROOF. By Lemma 2.1 and (2.31) we see that $B^2 = \text{Const.}$ on at most two points. We will show that the support of B is on only one of these points. In order to do this we first prove the following lemma.

LEMMA 3.1: *Let* B, ξ *minimize* (3.1). *Let* η *be any other design such that* $\int_0^1 Bx d\eta = \theta$. *Then*

$$\int_0^1 B^2 d(\xi-\eta) \geq 0. \tag{3.4}$$

PROOF. Let

$$I(t) = \left(\int_0^1 B\varphi d[(1-t)\xi + t\eta]\right)^2 + \rho\left(\int_0^1 (1 + B^2) d[(1-t)\xi + t\eta]\right).$$

Since B, ξ are optimal

$$I'(0) = 2\left(\int_0^1 B\varphi\, d\xi\right) \int_0^1 B\varphi\, d(\eta-\xi) + \rho \int_0^1 B^2 d(\eta-\xi) \geq 0. \tag{3.5}$$

Write $B^2 = B \frac{\lambda}{\rho}(\beta x-\varphi)$. Using (2.31), (2.32) and (2.33) we get

$$2\lambda B\varphi = 2\lambda\beta x B - 2\rho B^2. \tag{3.6}$$

Substituting (3.6) into (3.5) and noting that $\int_0^1 Bx d(\eta-\xi) = 0$ by assumption on η, we get (3.4).

PROOF OF THEOREM 3.2 CONTINUED. Now assume that the support of the optimal design is on two points $x_0 \neq x_1$. Let ξ be the mass at x_0, $1 - \xi$ the mass at x_1 $(0 < \xi < 1)$. By our assumptions on $(\beta x-\varphi)_+$, $B(x_0) = B(x_1) = C < C^*$ and $0 < x_0 < x_1 < 1$.

The side condition $B(x_0)x_0\xi + B(x_1)x_1(1-\xi) = \theta$ implies, since $B(x_0) = B(x_1)$, that $B(x_0)x_0 < \theta$ and $B(x_1)x_1 > \theta$. Since $(\beta x-\varphi)_+$ is continuous we can choose $x_0' > x_0$, $x_1' < x_1$ so that $B(x_0')x_0' < \theta$, $B(x_1')x_1' > \theta$ and η, $0 < \eta < 1$, such that

$$B(x_0')x_0'\eta + B(x_1')x_1'(1-\eta) = \theta.$$

But by the conditions on $(\beta x-\varphi)_+$, $B(x_0') > C$ and $B(x_1') > C$. Therefore

$$B^2(x_0')\eta + B^2(x_1')(1-\eta) > C^2$$

which contradicts Lemma 3.1. Therefore the theorem is proved.

A more direct proof can be given when $\varphi(x)$ is convex but our method and Lemma 3.1 may be useful in other contexts.

REMARK 2. The z of Theorem 3.2 is the point at which $(\beta x-\varphi)_+$ achieves its maximum on [0,1]. To see this relate (2.32) (which gives β) to the equation that determines z.

We now determine the solution of J_2. By Theorem 2.1 we get $B = \theta x/\int_0^1 x^2 d\xi$ and have the problem of minimizing

$$(3.7) \qquad \left(\int_0^1 A\varphi\, d\xi\right)^2 + \rho\left[\int_0^1 A^2 d\xi + \frac{\theta^2}{\int_0^1 x^2 d\xi}\right]$$

under the side condition $\int_0^1 A d\xi = 1$. Refer to (2.26), (2.27) and (2.28) for the explicit form of the function A.

THEOREM 3.3. *Let φ be a continuous strictly increasing function such that $\varphi(x)/x^2$ is non-increasing,* $\varphi(0) = 0$, $\varphi(1) = 1$. *Let* c *be the value of* t, $0 < t \leq 1$, *that minimizes*

$$(3.8) \qquad H(t) = \frac{\rho(\rho+t)}{\rho+t-t^2} + \frac{\rho\theta^2}{t}.$$

Then

$$H(c) = \inf_{\xi\in\mathcal{D}_k} \inf_{A,B} J_2(A,B,\xi)$$

and the minimizing A,B,ξ *are given by*

$$(3.9) \qquad \xi(0) = 1 - c, \ \xi(1) = c$$

$$(3.10) \qquad B(0) = 0, \ B(1) = \frac{\theta}{c}$$

$$(3.11) \qquad A(0) = \frac{\rho+c}{\rho+c-c^2}, \ A(1) = \frac{\rho}{\rho+c-c^2}.$$

If

$$(3.12) \qquad \frac{\rho c}{\rho+c-c^2} = \int_0^1 A\varphi\, d\xi \geq \theta$$

and $k \geq 3$ (*see* (2.10)) *then this triple* A,B,ξ *minimizes* $J_0(A,B,\xi)$.

Note that $\int_0^1 A\varphi\, d\xi \leq 1$ so (3.12) can only be satisfied for

$\theta < 1$. Of course, for $\theta \geq 1$ the problem is solved by Theorem 3.2. The typical values of ρ and θ for which Theorem 3.3 gives a solution to the J_0 problem are such that θ is quite small compared to ρ.

PROOF. We will show that the optimal design is supported on $\{0\}$ and $\{1\}$. The relations (3.9), (3.10) and (3.11) are just the solutions to (3.7) subject to this condition. By (2.27) we get

$$\alpha = (\rho + \int_0^{a-} \varphi^2 d\xi)/\int_0^{a-} \varphi\, d\xi \tag{3.13}$$

where $a = \sup\{x | (\alpha - \varphi(x))_+ > 0\}$ or 1, whichever is smaller. Using (2.28) and (3.13) in (2.30) we see that the problem of minimizing (3.7) reduces to finding the design that minimizes

$$I(\xi) = \alpha\mu + \rho\int_0^1 B^2 d\xi = \frac{\rho\theta^2}{m_2} + \frac{\rho(\rho + \int_0^{a-} \varphi^2 d\xi)}{(\int_0^{a-} d\xi)(\rho + \int_0^{a-} \varphi^2 d\xi) - (\int_0^{a-} \varphi\, d\xi)^2} \tag{3.14}$$

and where

$$m_2 = \int_0^1 x^2 d\xi. \tag{3.15}$$

Let ξ minimize (3.14). We will show that $\mathrm{supp}(\xi) \subset \{0,1\}$. First note that ξ cannot have support on any values x such that $a \leq x < 1$ because a design that moved such mass at x to 1 would be an improvement since everything would be unchanged in (3.14) except m_2 which would become larger. Therefore assume that ξ has support at an $x_0 \in (0,a)$ (it can have support at other points in $(0,a)$ as well). We define a design ξ' as follows: Let $x_0 < x_0' < a$ and suppose x_0' is not in the support of ξ.

Let $\xi'(x_0') = \dfrac{\varphi(x_0)}{\varphi(x_0')}\xi(x_0)$, let $\xi'(0) = \xi(0) + \xi(x_0) - \xi'(x_0')$

and let $\xi'(x_i) = \xi(x_i)$ for any other points in the support of ξ. Since φ is strictly increasing $\xi'(x_0') < \xi(x_0)$. It is easy to check that the following hold and that they imply that $I(\xi') < I(\xi)$:

$$\int_0^{a-} d\xi = \int_0^{a-} d\xi', \int_0^{a-} \varphi(x)d\xi = \int_0^{a-} \varphi(x)d\xi',$$

$$\int_0^{a-} \varphi^2(x)d\xi < \int_0^{a-} \varphi^2(x)d\xi', \int_0^1 x^2 d\xi < \int_0^1 x^2 d\xi'.$$

(It is helpful to write the last term in (3.14) as

$\rho[1-\frac{(\int_0^{a-}\varphi\, d\xi)^2}{\rho + \int_0^{a-}\varphi^2 d\xi}]^{-1}$). We have proved that the minimizing design has support on {0} and {1}.

The statement in (3.12) is just the observation that $\int_0^1 B\varphi d\xi \leq \int_0^1 A\varphi\, d\xi$ and so in this case $\inf_{\xi\in\mathcal{D}_k} J_2(\xi) = \inf_{\xi\in\mathcal{D}_k} J_3(\xi)$ for $k \geq 3$ ($k \geq 4$ is obvious but $k \geq 3$ is all that is needed because {0} is in the support of the best ξ.) Theorem 3.3. is proved.

THEOREM 3.4. *Let* $\varphi(x)$, $x \in [0,1]$ *be a bounded function with* $\varphi(x) \geq x$, $\varphi(0) = 0$, $\varphi(1) = 1$. *For these functions* φ, $J_0(A,B,\xi)$ *is minimized by a design* ξ *which is supported on* {0} *and* {1}. *If* $\theta \geq 1$ *the best design is given in Theorem* 3.1. *If* $\theta < 1$ *the best design* ξ *puts mass* c *at* {1} *if the value* c *which minimizes* (3.8) *satisfies* (3.12). *In this case* A *and* B *are given in Theorem* 3.3. *Otherwise* ξ *puts mass*

$$d = \frac{\sqrt{2}\theta}{1 + (\sqrt{2}-1)\theta} \tag{3.16}$$

at {1} *and*

$$B(0) = 0,\ B(1) = A(1) = \frac{\theta}{d},\ A(0) = \frac{1-\theta}{1-d}\ . \tag{3.17}$$

and

$$\inf_\xi J_3(\xi) = \inf_\xi J_0(\xi) = \theta^2 + \rho(1 + (\sqrt{2}-1)\theta)^2. \tag{3.18}$$

As in Theorem 3.3 $k \geq 3$ (*see* (2.10)).

PROOF. By Remark 1 we need only consider the case $\varphi(x) = x$. The case $\theta \geq 1$ is dealt with in Theorem 3.1. Assume $\theta < 1$ and

A,B,ξ minimize J_3. We first show that

$$(3.19) \qquad \int_0^1 Ax\,d\xi < \int_0^1 Bx\,d\xi$$

is not possible. Assume (3.19), then the variational argument of Theorem 2.1 shows that the minimizing A and B satisfy $A = 1$, $B = \frac{\theta x}{m_2}$ on the support of ξ and

$$(3.20) \qquad \inf_{\xi} J_3(\xi) = \theta^2 + \rho(1 + \frac{\theta^2}{m_2})$$

where $m_2 = \int_0^1 x^2 d\xi$. If $\xi(1) = 1$ then $\int_0^1 Ax\,d\xi = 1 > \int_0^1 Bx\,d\xi = \theta$ which contradicts (3.19) but if $\xi(1) < 1$ we can change ξ and reduce (3.20).

Therefore, if $\theta < 1$ we can assume that $\int_0^1 Ax\,d\xi \geq \theta$. If $\int_0^1 Ax\,d\xi > \theta$, following the argument of Theorem 2.1, we get the same solutions for A and B as in considering the problem for $J_2(A,B,\xi)$. The convexity of $A^2 + B^2$ and Lemma 2.1 imply that the minimizing design is supported on at most two points. Exactly as in the proof of Theorem 3.3 we see that these points must be {0} and {1}. The best design on [0,1] is given in Theorem 3.3 and when $\int_0^1 Ax\,d\xi > \theta$ (see (3.12) it minimizes $J_0(A,B,\xi)$.

The final possibility is that $\int_0^1 Ax\,d\xi = \theta$. This gives us the problem of minimizing

$$(3.21) \qquad \tilde{J}(A,B,\xi) = \theta^2 + \rho \int_0^1 (A^2 + B^2)d\xi$$

subject to the conditions

$$(3.22) \qquad \int_0^1 Ax\,d\xi = \int_0^1 Bx\,d\xi = \theta, \ \int_0^1 A\,d\xi = 1, \ A \geq 0, \ B \geq 0$$

Using a variational argument similar to the one used in

Theorem 2.1 it follows that, for a fixed design ξ, the minimizing A and B have the form

$$(3.23) \qquad A(x) = (a-bx)_+, \; B(x) = \frac{\theta x}{m_2}.$$

By Lemma 2.1 and the convexity of $A^2 + B^2$ we see that ξ has at most two points in its support. Suppose that $A > 0$ at both points. Then $\int_0^1 (a-bx)d\xi = 1$ and $\int_0^1 (a-bx)x d\xi = \theta$ imply

$$a = 1 + bm_1, \; b = \frac{m_1-\theta}{m_2-m_1^2}$$

where $m_1 = \int_0^1 x d\xi$ and $m_2 = \int_0^1 x^2 d\xi$.

$$\int_0^1 A^2 d\xi = 1 + \frac{(m_1-\theta)^2}{(m_2-m_1^2)}$$

and consequently

$$(3.24) \quad \tilde{J}(\xi) = \inf_{A,B} \tilde{J}(A,B,\xi) = \theta^2 + \rho(1 + \frac{(m_1-\theta)^2}{m_2-m_1^2} + \frac{\theta^2}{m_2}).$$

For fixed m_1 we minimize the right side of (3.24) by taking $\xi(1)= m_1$ and $\xi(0) = 1-m_1$. Then the corresponding

$$(3.25) \qquad a = 1 + \frac{m_1-\theta}{1-m_1}, \; b = \frac{m_1-\theta}{m_1(1-m_1)}$$

and

$$(3.26) \qquad \tilde{J}(\xi) = \theta^2 + \rho(1 + \frac{(m_1-\theta)^2}{m_1(1-m_1)} + \frac{\theta^2}{m_1}).$$

The function A corresponding to these values of a and b satisfies $A(0), A(1) > 0$.

We now show that a two point design at $0 \le u < v \le 1$ with $A(u) = 1/\xi(u)$ and $A(v) = 0$ can not be optimal. Since

$\int_0^1 Ax d\xi = \theta$ then $u = \theta$. Let $J'(\xi)$ be the value of $\inf_B \tilde{J}(A,B,\xi)$ for this choice of A. Then

$$J'(\xi) = \theta^2 + \rho\left(\frac{1}{\xi} + \frac{\theta^2}{\theta^2\xi + (1-\xi)}\right) \tag{3.27}$$

Here $\xi = \xi(u)$ and use the fact that the best one can do is to take $v = 1$. The value of ξ which minimizes is either $\xi = 1$ or is given by

$$\frac{1}{\xi} = t + \sqrt{t(1-t)} \tag{3.28}$$

where $t = 1 - \theta^2$. If $\xi = 1$ this is not a two point design so we need not be concerned with this case. If ξ is given by (3.28) the value of (3.27) is

$$\theta^2 + \rho(1 + 2\theta(1-\theta^2)^{1/2}). \tag{3.29}$$

In (3.26) with $m_1 = \theta$ we get $\tilde{J}(\xi) = \theta^2 + \rho(1 + \theta)$ which is smaller than (3.29) unless $2\theta(1-\theta^2)^{1/2} < \theta$, i.e., unless $t < 1/4$. But such a value for t implies that $\xi > 1$ in (3.28) which is a contradiction.

We have shown that, when ξ is an optimal design, $A > 0$ on all the points in the support of ξ. Therefore, we minimize (3.26) with respect to m_1 and use (3.25) to get (3.16), (3.17) and (3.18).

We have not been able to complete the solution of J_3 for φ convex. As above we would have to consider the case when $\int_0^1 A\varphi\, d\xi = \int_0^1 B\varphi\, d\xi$. In the example $\varphi(x) = x^2$ calculations with a computer indicate that there are cases where the optimal design is supported on $\{0,x\}$ and cases where the optimal design is supported on $\{y,1\}$. (The optimality in this second situation does not follow from Theorem 2.2 but by a separate argument). We can only conjecture that this description is complete.

4. *Standard estimates*. We compare the results of Section 3

with the best design when the "usual" least squares estimates are used i.e.,

$$A = 1,\ B = \frac{\theta x}{m_2} \tag{4.1}$$

where $m_2 = \int_0^1 x^2 d\xi$.

Let

$$L(\varphi,\xi) = \max\left(\left(\int_0^1 \varphi\, d\xi\right)^2, \frac{\theta^2}{m_2^2}\left(\int_0^1 x\varphi\, d\xi\right)^2\right) + \rho\left(1 + \frac{\theta^2}{m_2}\right). \tag{4.2}$$

We consider the case $\varphi(x) = x$ and show that the ξ which minimizes (4.2) is concentrated on $\{0,1\}$. It follows from Remark 1 that the best ξ for $\varphi(x) = x$ will also be the best for any φ satisfying $\varphi(x) \geq x$, $\varphi(0) = 0$, $\varphi(1) = 1$. When $\varphi(x) = x$, (4.2) becomes

$$L(x,\xi) = \max(m_1^2,\theta^2) + \rho\left(1 + \frac{\theta^2}{m_2}\right) \tag{4.3}$$

and it is obvious that, if m_1 is fixed, the best design ξ puts mass m_1 at 1, and the rest at 0. It follows that the ξ which minimizes (4.3) puts mass $\xi(1)$ at 1 where $\xi(1)$ is the value of m that minimizes

$$H(m) = \max(m^2,\theta^2) + \rho\left(1 + \frac{\theta^2}{m}\right),\ 0 < m \leq 1 \tag{4.4}$$

THEOREM 4.1. <u>If</u> $\varphi(x) \geq x$, $\varphi(0) = 0$, $\varphi(1) = 1$, <u>then the optimal designs using the estimates given by</u> (4.1) <u>and the corresponding</u> $L^* = \inf_\xi L(x,\xi)$ <u>are as follows</u>:

(4.5) <u>If</u> $\theta \geq 1$ <u>then</u> $\xi(1) = 1$, $L^* = \theta^2 + \rho(1 + \theta^2)$

(4.6) <u>If</u> $\theta < 1$, $\rho > 2/\theta^2$, <u>then</u> $\xi(1) = 1$, $L^* = \theta^2 + \rho(1 + \theta^2)$

(4.7) <u>If</u> $\theta<1$, $2\theta<\rho<2/\theta^2$, <u>then</u> $\xi(1)=(\frac{\rho\theta^2}{2})^{1/3}$, $L^*=\rho+3(\frac{\rho\theta^2}{2})^{2/3}$

(4.8) *If* $\theta < 1$ *and* $\rho < 2\theta$, *then* $\xi(1) = \theta$, $L^* = \theta^2 + \rho(1+\theta)$.

Following Lemma 2.1 we can make a simple modification of the A, B, and ξ of Theorem 4.1 and get A_0, B_0, ξ_0 with $J_3(A_0, B_0, \xi_0) < L^*$ when $\theta < 1$ and $\rho < 2/\theta^2$ (the cases given by (4.7) and (4.8)). For the (4.8) case we obtain

$$\xi_0(1) = \frac{\sqrt{2}\theta}{1 + (\sqrt{2}-1)\theta}$$

$$A_0(0) = 1 + (\sqrt{2}-1)\theta$$

$$A_0(1) = B_0(1) = \frac{1}{\sqrt{2}} + \left(1 - \frac{1}{\sqrt{2}}\right)\theta$$

$$B_0(0) = 0$$

and this provides $J_3(A_0, B_0, \xi_0) = \theta^2 + \rho(1 + (\sqrt{2}-1)\theta)^2$.

A_0, B_0, ξ_0 is the optimal triple found in Theorem 3.4 at (3.16) and (3.17).

When (4.7) pertains this method leads to the following: Let $\xi = \xi(1) = \left(\frac{\rho\theta^2}{2}\right)^{1/3}$. Then

$$\xi_0(1) = \frac{\sqrt{\theta^2 + \xi^2}}{1 - \xi + \sqrt{\xi^2 + \theta^2}}$$

$$A_0(0) = 1 - \xi + \sqrt{\xi^2 + \theta^2} \tag{4.8a}$$

$$A_0(1) = \xi + \frac{\xi(1-\xi)}{\sqrt{\xi^2 + \theta^2}}$$

$$B_0(1) = \theta + \frac{\theta(1-\xi)}{\sqrt{\xi^2 + \theta^2}}$$

and

$$J_3(A_0, B_0, \xi_0) = \xi^2 + \rho\, A_0^2(0)$$

Contrary to the previous case, this is not the optimal solution given in Theorem 3.4.

REMARK 3. Despite the differences that turn up in the A's

and B's the values of $\inf_{\xi} J_3(A,B,\xi)$ for the three A,B,ξ's discussed (the one of Theorem 4.1, the improvement just discussed and the optimal one given by Theorem 3.4) are nearly the same. In fact there does not seem to be much difference between the values of $\inf_{\xi} J_3(A,B,\xi)$ corresponding to the best design using the standard ((4.1)) estimates and the optimal (Theorem 3.4) solution. It is worth noting however that if one starts with a wrong design then the standard estimates may lead to bad results but the best A, B (in Theorem 2.1) for the wrong design seems to behave decently. For example, in the context of Theorem 4.1, if we use $\xi(0) = \xi(1) = 1/2$ then here are some values of $\inf_{\xi} J_3(A,B,\xi)$.

	Stand. Est	Best A,B from Theorem 2.1
$\rho = .25,\ \theta = .5$	.63	.69
$\rho = .1,\ \theta = .5$	.40	.45
$\rho = .1,\ \theta = .2$	.36	.20

Note that for $\theta = 1/2$ and the values of ρ that we are looking at Theorem 4.1 gives $\xi(1) = 1/2$ so the standard estimates are taken with their best design in the first two cases. When $\rho = .1$, $\theta = .2$ the standard estimate has no opportunity to cut down the bias whereas the optimal estimate does. In this third case the smallest $J_3 \sim .15$. In fact matters would look even more striking if ρ became smaller and such cases would be typical, e.g. if we started with a regression on the interval $[-5,5]$ and $\theta = 1$, $\rho = .1$ then, in transforming the problem to $[-1,1]$ we would get $\theta = .2$ and the new ρ would be .004.

For $\varphi(x) = x^2$ the best design using the estimators of (4.1) is obtained by minimizing

$$(4.9)\qquad L(x^2,\xi) = \max(m_2^2, \theta^2 m_3^2/m_2^2) + \rho\left(1 + \frac{\theta^2}{m_2}\right)$$

$(m_j = \int_0^1 x^j d\xi)$. Since $(\int x^3 d\xi)^2 \geq (\int x^2 d\xi)^3$ we can take η to put mass 1 at $\sqrt{m_2}$ and obtain $L(x^2,\eta) \leq L(x^2,\xi)$. (The same argument will work for $\varphi(x) = x^p$, $p \geq 2$). It follows that the ξ which minimizes (4.9) is concentrated at one point and we then want to minimize

$$H(t) = \max(t^4, \theta^2 t^2) + \rho(1 + \frac{\theta^2}{t^2}). \tag{4.10}$$

Doing this gives

THEOREM 4.2. *If $\varphi(x) = x^2$ then the optimal designs using the estimators (4.1) are all supported at one point x. Let* $L^* = \inf_{\xi} L(x^2, \xi)$.

If $\theta < 1$

$$\rho \leq \theta^4, \; x = \rho^{1/4}, \; L^* = 2\rho^{1/2}\theta^2 + \rho \tag{4.11}$$

$$\theta^4 < \rho \leq 2\theta^4, \; x = \theta, \; L^* = \theta^4 + 2\rho \tag{4.12}$$

$$2\theta^4 < \rho \leq \frac{2}{\theta^2}, \; x = (\frac{\rho\theta^2}{2})^{1/6}, \; L^* = 3(\frac{\rho\theta^2}{2})^{2/3} + \rho \tag{4.13}$$

$$\frac{2}{\theta^2} < \rho, \; x = 1, \; L^* = 1 + \rho[1 + \theta^2]. \tag{4.14}$$

If $\theta \geq 1$

$$\text{all } \rho, \; x = 1, \; L^* = \theta^2 + \rho[1 + \theta^2] \tag{4.15}$$

Results like Theorem 4.2 can be obtained for other φ's such as $\varphi(x) = x^p$.

From Theorem 3.2 we see that the optimal solution there is the one found here in Theorem 4.2 when (4.11) and (4.15) pertain. As for the other cases see the comment at the end of Section 3 which indicates the presence of optimal designs on two points so that use of the standard estimates won't lead to optimal results.

Finally we note that some of the methods here can be applied to the problem considered by Huber [2] (see the introduction above) but without the restriction to the standard estimates. The problem becomes one of finding A,B,p to minimize

$$\text{Max}(\int_0^{1/2} A^2p^2dx, \int_0^{1/2} B^2p^2dx) + \rho \int_0^{1/2} (A^2+B^2)p\, dx$$

subject to

$$\int_0^{1/2} Ap\, dx = 1/2, \int_0^{1/2} Bp\sqrt{12}x\, dx = 1/2,$$

where p is the density of the design measure. Lemma 2.1 is useful here and gives immediate but not substantial improvements over the solutions found by Huber. The minimax problem can be solved (at least computationally) but in parallel with the comments relating to the case $\varphi(x) \geq x$, there is no great decrease in loss obtained by using the optimal solution instead of the solution in [2]. We omit the details.

References

[1] Box, G. E. P. and Draper, N. R. (1959). A basis for the selection of a response surface design. *J. Amer. Statist. Assoc.* 54, 622-654.

[2] Huber, P. (1975). Robustness and Designs. In Srivastava, J. N. (Ed.) A *Survey of Statistical Design and Linear Models*, 287-303. North Holland, Amsterdam.

[3] Kiefer, J. (1973). Optimal designs for fitting biased multi-response surfaces. In Krishnaiah, P. R. (Ed.) *Multivariate Analysis* III, 287-297. Academic Press, New York.

[4] Sacks, J. and Ylvisaker, D., manuscript.

LARGE SAMPLE PROPERTIES OF NEAREST NEIGHBOR DENSITY FUNCTION ESTIMATORS

By David S. Moore* and James W. Yackel
Purdue University

1. *Introduction.* Let $X_1, X_2, \ldots$ be iid random variables having unknown density function f with respect to Lebesgue measure λ on Euclidean p-space R^p. We wish to estimate f(z) for a given z. Let $\{k(n)\}$ be a sequence of positive integers satisfying

$$k(n) \to \infty \text{ and } k(n)/n \to 0 \quad \text{as } n \to \infty. \tag{1.1}$$

Define R(n) as the distance from z to the k(n)th closest of $X_1, \ldots, X_n$, distance being measured in a norm $||\cdot||$ on R^p which generates the usual topology. Denote by S(r) the "sphere"

$$S(r) = \{x \text{ in } R^p\colon ||x-z|| \leq r\}.$$

A nearest neighbor estimator of f(z) is

$$g_n(z) = \frac{k(n)/n}{\lambda\{S(R(n))\}} .$$

Note that $g_n(z)$ is simply empiric measure divided by Lebesgue measure for the region S(R(n)). This estimator is essentially due to Fix and Hodges [2], and was explicitly introduced and studied by Loftsgaarden and Quesenberry [5]. These and subsequent authors used the Euclidean norm, but for $p > 1$ other norms may be useful (e.g., squares about z rather than spheres are obtained from the "maximum component" norm), and proofs are unaffected by this generality. We have suppressed the dependence on z of R(n) and S(r), since in this paper we consider only results for a fixed z in R^p.

Loftsgaarden and Quesenberry proved consistency in probability of $g_n(z)$. Wagner [8] established almost sure (a.s.)

*Research of this author was sponsored by the Air Force Office of Scientific Research, Air Force Systems Command, USAF, under Grant No. 72-2350 B.

consistency under a condition equivalent to $k(n)/\log n \to \infty$. For the case $p = 1$, Moore and Henrichon [6] proved a.s. uniform consistency when $k(n)/\log n \to \infty$. (They state only convergence in probability, but an application of the Borel-Cantelli lemma shows that their proof yields a.s. convergence.) The local control of the estimation process which is a feature of the nearest neighbor estimator has been popular with practitioners, who have used g_n in discrimination and pattern recognition problems.

Sections 2 and 3 of this paper are devoted to g_n. In Section 2 we establish a.s. consistency under the condition $k(n)/\log\log n \to \infty$. Since $R(n)$ is a sample $k(n)/n$-tile, the study by Kiefer [3] of sample p_n-tiles for $p_n \to 0$ provides the tools needed for our result. What is more, it follows from Kiefer's work that $k(n)/\log\log n \to \infty$ is the weakest condition on $\{k(n)\}$ satisfying (1.1) which guarantees a.s. convergence of $g_n(z)$ to $f(z)$. Section 3 proves asymptotic normality of $g_n(z)$. The proof uses the standard device of restating an event defined in terms of an order statistic as an event given in terms of a binomial random variable. The limiting distribution derived in Section 3 is required in the more general study in Section 4.

Recently, the authors [7] observed that g_n could be viewed as the uniform kernel case of the general nearest neighbor density function estimator defined by

$$f_n(z) = \frac{1}{nR(n)^p} \sum_{i=1}^{n} K\left(\frac{z-X_i}{R(n)}\right)$$

where

$$\text{(1.2)} \qquad \begin{aligned} &K(u) \text{ is a bounded density on } R^p \\ &K(u) = 0 \quad \text{for } ||u|| > 1 \end{aligned}$$

Here the norm $||\cdot||$ must satisfy the additional restriction that

$$\lambda\{S(r)\} = cr^p \quad \text{where } c = \lambda\{S(1)\}.$$

This is the case for, e.g., the usual Euclidean norm and the maximum-component norm.

The estimator f_n is the analog of the Rosenblatt-Parzen

class of bandwidth estimators defined by

$$\hat{f}_n(z) = \frac{1}{nr(n)^p} \sum_{i=1}^{n} K\left(\frac{z-X_i}{r(n)}\right)$$

where $\{r(n)\}$ is a sequence of positive bandwidths satisfying $r(n) \to 0$ and $nr(n)^p \to \infty$ as $n \to \infty$. In our earlier paper we showed, roughly speaking, that any consistency theorem (in probability or almost sure, pointwise or uniform) true for $\hat{f}_n$ remains true for f_n having the same kernel K and $k(n) \sim \alpha nr(n)^p$ for some $\alpha > 0$. This allows the large literature on consistency of $\hat{f}_n$ to be restated for f_n. See [7] for details and qualifications. Here we mention only that either by this consistency-equivalence result or directly from Kiefer's work it follows that the uniform kernel case of $\hat{f}_n$ is a.s. consistent when $nr(n)^p/\log\log n \to \infty$. This is the analog of the result of Section 2 below, and is similarly best possible and stronger than known results for general kernels. Thus Section 2 sets a goal for work on a.s. consistency of $\hat{f}_n$ or f_n, and the results of [7] show that attaining this goal for either of $\hat{f}_n$ or f_n is sufficient to reach it for both.

Sections 4 and 5 concern the general nearest neighbor estimator f_n. Section 4 establishes asymptotic normality. It is noteworthy that f_n does not have the same asymptotic variance as the matching bandwidth estimator $\hat{f}_n$. The nearest neighbor method is more efficient than the bandwidth method when $f(z)$ is small, as intuition might suggest. Section 5 shows that weak consistency of f_n implies mean consistency, a supplement to the consistency results in [7].

2. *Almost sure consistency, uniform kernel case.* We make use of a lemma which extracts a very small portion of Theorem 6 of Kiefer [3].

LEMMA 1. Let Z_n be a sample α_n-tile from n iid random variables uniformly distributed on (0,1). If $\alpha_n \to 0$ and $n\alpha_n/\log\log n \to \infty$, then $Z_n/\alpha_n \to 1$ a.s.

Lemma 1 is applied to $g_n(z)$ by noting that if

$$H(r) = P[||X-z|| \leq r] = \int_{S(r)} f(x)dx \tag{2.1}$$

then $H(R(n))$ is the sample $k(n)/n$-tile from n iid uniform random variables. Here is our a.s. consistency result.

THEOREM 1. *Let* f *be continuous at* z, *and let* $\{k(n)\}$ *satisfy* (1.1) *and* $k(n)/\log\log n \to \infty$. *Then* $g_n(z) \to f(z)$ a.s.

PROOF. Lemma 1 states that

$$\frac{k(n)/n}{H(R(n))} \to 1 \quad \text{a.s.} \tag{2.2}$$

from which it follows that

$$H(R(n)) \to 0 \quad \text{a.s.} \tag{2.3}$$

We claim that

$$R(n) \to r_0 = \inf\{r: H(r) > 0\} \quad \text{a.s.} \tag{2.4}$$

For clearly $R(n) \geq r_0$ a.s. for each n, and if for some $\varepsilon > 0$, $R(n) \geq r_0 + \varepsilon$ for a sequence of n at a sample point ω, then $H(R(n)) \geq H(r_0 + \varepsilon) > 0$ for these n at ω. By (2.3), this can occur only on a set of ω having probability zero.

Applying the mean value theorem for integrals to (2.1), there exist λ_n satisfying

$$\inf_{S(R(n))} f(x) \leq \lambda_n \leq \sup_{S(R(n))} f(x)$$

such that

$$H(R(n)) = \lambda_n \, \lambda\{S(R(n))\}. \tag{2.5}$$

With (2.2), (2.5) implies that $g_n(z)/\lambda_n \to 1$ a.s. If $f(z) > 0$, then by (2.4), $R(n) \to 0$ a.s. and by continuity of f at z, $\lambda_n \to f(z)$ a.s. If $f(z) = 0$, then (2.3), (2.4) and (2.5) imply that $\lambda_n \to 0$ a.s. In either case, $g_n(z) \to f(z)$ a.s.

The proof of Theorem 1 amounts to observing that

$$g_n(z) \sim \frac{k(n)/n}{H(R(n))} f(z) = a_n f(z)$$

and applying (2.2) to a_n. From Kiefer's Theorem 6 it follows also that if $k(n)/\log\log n \to v$, $0 < v < \infty$, then $\underline{\lim}\, a_n$ and

$\overline{\lim}\, a_n$ are unequal, finite and positive. If $k(n)/\log\log n \to 0$, then $\overline{\lim}\, a_n = \infty$ and $\underline{\lim}\, a_n = 0$. Thus $g_n(z)$ is not a.s. consistent if $k(n)$ increases more slowly than is assumed in Theorem 1. Of course, $g_n(z)$ remains weakly consistent as long as (1.1) holds.

3. *Asymptotic normality, uniform kernel case.* Although the proof in this section is straightforward, both it and the proof of Section 4 require the assumption

$$(3.1) \qquad (k(n))^{\frac{1}{2}}|f(z_n)-f(z)| \to 0(P) \text{ when } ||z_n-z|| \leq R(n).$$

The assumption (3.1) connects $\{k(n)\}$ and the local behavior of f at the point z. It can be restated in more explicit form for specific norms $||\cdot||$. In particular, when $f(z) > 0$ and either the Euclidean norm or the maximum-component norm is used,

$$(3.2) \qquad \frac{k(n)/n}{cR(n)^p} \to f(z)(P) \qquad c = \lambda\{S(1)\}$$

(This is just weak consistency as proved in [5]), so then $R(n) = 0_p\{(k(n)/n)^{1/p}\}$ and (3.1) is implied by

$$(3.3) \qquad (k(n))^{\frac{1}{2}}|f(z_n)-f(z)| \to 0 \text{ when } ||z_n-z|| = 0\{(\frac{k(n)}{n})^{1/p}\}.$$

If the p first partial derivatives exist and are bounded near z, (3.3) in turn is satisfied when

$$k(n) = o\{n^{2/(p+2)}\}.$$

THEOREM 2. Let f be continuous at z, $f(z) > 0$, $\{k(n)\}$ satisfy (1.1), and let (3.1) hold. Then

$$\mathcal{L}\{(k(n))^{\frac{1}{2}}(g_n(z)-f(z))\} \to N(0,f^2(z))$$

PROOF. As in the proof of Theorem 1, we can write

$$g_n(z) = \frac{k(n)/n}{H(R(n))} f(z_n) \qquad \text{for some } z_n \text{ in } S(R(n)).$$

Now (1.1) and $f(z) > 0$ are sufficient for $R(n) \to 0(P)$ and $H(R(n))/(k(n)/n) \to 1$ (P) (see [5]). Therefore from

$$\frac{(k(n))^{\frac{1}{2}}}{f(z)}(g_n(z)-f(z)) = (k(n))^{\frac{1}{2}}\left(\frac{k(n)/n}{H(R(n))} - 1\right) + (k(n))^{\frac{1}{2}}\left(\frac{f(z_n)}{f(z)} - 1\right)\frac{k(n)/n}{H(R(n))}$$

and (3.1), we need only show that

$$\mathcal{L}\left\{(k(n))^{\frac{1}{2}}\left(\frac{k(n)/n}{H(R(n))} - 1\right)\right\} \to N(0,1).$$

Since $H(R(n))$ is the $k(n)$th order statistic of n iid uniform random variables $U_1,\ldots,U_n$ on $(0,1)$,

$$\begin{aligned} P_n(a) &= P\left[(k(n))^{\frac{1}{2}}\left(\frac{k(n)/n}{H(R(n))} - 1\right) \leq a\right] \\ &= P\left[H(R(n)) \geq \frac{k(n)/n}{1+ak^{-\frac{1}{2}}}\right] \\ &= P[B_n < k(n)] \end{aligned}$$

where B_n is the number of $U_1,\ldots,U_n$ falling below $\pi_n = (k(n)/n)/(1+ak(n)^{-\frac{1}{2}})$ and has the binomial (n,π_n) distribution. By (1.1), $\pi_n \to 0$ and $n\pi_n \to \infty$, so that B_n is asymptotically normal. Writing

$$P_n(a) = P\left[\frac{B_n-n\pi_n}{\sigma_n} < \frac{k(n)-n\pi_n}{\sigma_n}\right]$$

where $\sigma_n = [n\pi_n(1-\pi_n)]^{\frac{1}{2}}$, and computing

$$\begin{aligned} \frac{k(n)-n\pi_n}{\sigma_n} &\sim \frac{k(n)-n\pi_n}{(n\pi_n)^{\frac{1}{2}}} \\ &= a\left(\frac{k^{\frac{1}{2}}}{a+k^{\frac{1}{2}}}\right)^{\frac{1}{2}} \to a \end{aligned}$$

we obtain $P_n(a) \to \Phi(a)$, Φ being the standard normal df. This completes the proof.

4. *Asymptotic normality, general case.* Recall that in order to formulate the general nearest neighbor estimator f_n, we require that the norm $||\cdot||$ satisfy

$$\lambda\{S(r)\} = cr^p \quad \text{where } c = \lambda\{S(1)\}. \tag{4.1}$$

In this case, (3.1) is equivalent to the more usable condition (3.3).

THEOREM 3. Let f be continuous at z, $f(z) > 0$, $K(u)$ satisfy (1.2) and $\{k(n)\}$ satisfy (1.1). Let also (3.3) and (4.1) hold. Then

$$\mathcal{L}\{(k(n))^{\frac{1}{2}}(f_n(z)-f(z))\} \to N(0,cf^2(z)\int K^2(u)du)$$

The proof will be divided into several parts. First note that in

$$f_n(z) = \frac{1}{nR(n)^p} \sum_{i=1}^{n} K(\frac{z-X_i}{R(n)})$$

there are exactly $k(n)-1$ nonzero summands by (1.2), corresponding to the first $k(n)-1$ order statistics of $||X_i-z||$. Denote by $Y_1,\ldots,Y_{k(n)-1}$ the subsequence of $X_1,\ldots,X_n$ defined by

$$Y_1 = X_{i_1} \qquad i_1 = \min\{i: ||X_i-z|| < R(n)\}$$

$$Y_j = X_{i_j} \qquad i_j = \min\{i > i_{j-1}: ||X_i-z|| < R(n)\}$$

and let $K_{n,i} = K(\frac{z-Y_i}{R(n)})$ be the nonzero summands in f_n. Then the conditional distribution of $Y_1,\ldots,Y_{k(n)-1}$ given $R(n) = r$ is that of $k(n)-1$ independent observations each having the density function

$$f(y)/P(S(r)) \quad \text{for } y \text{ in } S(r)$$

where

$$P(S(r)) = \int_{S(r)} f(x)dx.$$

Therefore the conditional distribution of $K_1,\ldots,K_{k(n)-1}$ given $R(n) = r$ is that of $k(n)-1$ iid random variables having mean

$$E(r) = E[K_i|R(n) = r] = \int_{S(r)} K(\frac{z-y}{r})\frac{f(y)}{P(S(r))}\, dy$$

and variance

$$\sigma^2(r) = \int_{S(r)} K^2(\frac{z-y}{r})\frac{f(y)}{P(S(r))}\, dy - E^2(r).$$

By the (vector) change of variables $u = (z-y)/r$ and the mean value theorem for integrals, we can write

$$(4.2)\qquad E(r) = \frac{\lambda_{1,r}}{P(S(r))} r^p \int_{S(1)} K(u)du = \frac{\lambda_{1,r}}{P(S(r))} r^p$$

$$= \frac{\lambda_{1,r}}{c\lambda_{2,r}}$$

and

$$(4.3)\qquad \sigma^2(r) = \frac{\lambda_{3,r}}{c\lambda_{2,r}} \int_{S(1)} K^2(u)du - \left(\frac{\lambda_{1,r}}{c\lambda_{2,r}}\right)^2$$

where

$$\inf_{S(r)} f(x) \leq \lambda_{i,r} \leq \sup_{S(r)} f(x).$$

We first consider the normalized sum

$$Z_n = \sum_{i=1}^{k(n)-1} \frac{K_i - E(R(n))}{(k(n))^{\frac{1}{2}}\sigma(R(n))}$$

LEMMA 2. _Under the conditions of Theorem_ 3, _if_ $K(u)$ _is not constant on_ $S(1)$, _then_

$$\mathcal{L}\{Z_n\} \to N(0,1).$$

PROOF. If $F_n(x|r)$ is the conditional df of Z_n given $R(n) = r$, then by the remarks above, the Berry-Esseen theorem applies to give

$$(4.4)\qquad |F_n(x|r) - \Phi(x)| \leq \frac{3M^3}{\sigma(r)(k(n))^{\frac{1}{2}}}$$

where $M = \sup|K(u)| < \infty$. Since (1.1) implies that $R(n) \to 0(P)$,

$$(4.5)\qquad \sigma^2(R(n)) \to \sigma^2 = c^{-1} \int_{S(1)} K^2(u)du - c^{-2} \qquad (P)$$

and $\sigma^2 > 0$ when $K(u)$ is not the uniform pdf on $S(1)$. Then if G_n is the df of $R(n)$ and $\delta > 0$,

$$|P[Z_n \leq x] - \Phi(x)| \leq \int |F_n(x|r) - \Phi(x)| dG_n(x)$$

$$\leq \frac{3M^2}{\delta(k(n))^{\frac{1}{2}}} P[\sigma(R(n)) > \delta] + 2P[\sigma(R(n)) \leq \delta]$$

and this with (4.5) establishes that $\mathcal{L}\{Z_n\} \to N(0,1)$.

PROOF OF THEOREM 3. We write

$$(4.6)\quad (k(n))^{\frac{1}{2}}(f_n(z)-f(z)) = \frac{k(n)\sigma(R(n))}{nR(n)^p} Z_n + (k(n))^{\frac{1}{2}}\left[\frac{k(n)E(R(n))}{nR(n)^p} - f(z)\right]$$

Since by (3.2) and (4.5),

$$\frac{k(n)\sigma(R(n))}{nR(n)^p} \to cf(z)\sigma \qquad (P)$$

The first term on the right in (4.6) has

$$N(0,cf^2(z)\int K^2(u)du - f^2(z))$$

as its limit in law by Lemma 2. The second term on the right of (4.6) can be written as

$$\frac{k(n)/n}{R(n)^p}(k(n))^{\frac{1}{2}}(E(R(n))-c^{-1})+(k(n))^{\frac{1}{2}}\left(\frac{k(n)/n}{cR(n)^p} - f(z)\right)$$

$$= \frac{k(n)/n}{cR(n)^p}(k(n))^{\frac{1}{2}}\left(\frac{f(z_{1,n})}{f(z_{2,n})} - 1\right) + (k(n))^{\frac{1}{2}}(g_n(z)-f(z))$$

for large n, by (4.2) and continuity of f at z. Here $z_{i,n}$ lie in $S(R(n))$. By (3.1) and (3.2) applied to the first term and Theorem 2 applied to the second, this last expression has $N(0,f^2(z))$ as its limiting distribution. Moreover, it is asymptotically independent of Z_n. To see this, it is sufficient to show that

$$P[Z_n \leq a|k^{\frac{1}{2}}(g_n-f) \geq b] = P\left[Z_n \leq a|R(n) \geq \left(\frac{k/nc}{f+bk^{-\frac{1}{2}}}\right)^{1/p}\right]$$

converges to $\Phi(a)$ for any b. That this is true follows from the argument used to prove Lemma 2. Theorem 3 now follows from (4.6).

Note that the bandwidth estimator $\hat{f}_n$ using the same kernel $K(u)$ and $r(n) = (k(n)/n)^{1/p}$ does not have the same limiting distribution as does f_n. Cacoullos [1] shows (under conditions which ask more of f and less of $\{k(n)\}$ than those of Theorem 3) that

$$\mathcal{L}\{(nr(n)^p)^{\frac{1}{2}}(\hat{f}_n(z)-f(z))\} \to N(0,f(z)\int K^2(u)du).$$

Comparison of asymptotic variances shows that $\hat{f}_n(z)$ is less efficient at points z where f(z) is small, that is, where use of the fixed radii {r(n)} may result in few observations.

5. *Mean consistency.* Pointwise weak consistency results for f_n are available both by direct proof (for g_n) and by the consistency-equivalence results of [7]. It is easy to show that under quite general conditions, weak consistency of f_n implies mean consistency. This we now do.

THEOREM 4. If K(u) is bounded, f is bounded in a neighborhood of z, and {k(n)} satisfies (1.1), then $f_n(z) \to f(z)(P)$ implies that $E[|f_n(z)-f(z)|] \to 0$.

PROOF. We must show (Loeve (1963), p. 163) that

$$\lim_{a\to\infty} \int_{\{|f_n|>a\}} |f_n|\, dP \to 0$$

uniformly in n. Let M denote an arbitrary positive constant. Since K is bounded,

$$|f_n| \leq M \frac{k(n)/n}{R(n)^p}$$

and hence if $c(n) = (Mk(n)/an)^{1/p}$,

$$P(n,a) = \int_{\{|f_n|>a\}} |f_n|\, dP \leq M \frac{k(n)}{n} \int_{\{R(n)<c(n)\}} R(n)^{-p}\, dP.$$

But R(n) is the k(n)th order statistic from n observations on the df H(r) (see (2.1)). So

$$P(n,a) \leq M \frac{k}{n} k\binom{n}{k} \int_0^{c_n} r^{-p} H^{k-1}(r)[1-H(r)]^{n-k} dH(r)$$

$$\leq M \frac{k^2}{n} \binom{n}{k} \int_0^{c_n} r^{-p} H^{k-1}(r)\, dH(r)$$

From $H(r) = \lambda(r)cr^p$ for $\inf f \leq \lambda(r) \leq \sup f$, we see

$$P(n,a) \leq M \frac{k^2}{n} \binom{n}{k} \int_0^{c_n} H^{k-2}(r)\, dH(r)$$

$$= M \frac{k^2}{n} \binom{n}{k} \frac{1}{k-1} H^{k-1}(c_n)$$

$$\leq M \left(\frac{k}{n}\right)^k \binom{n}{k} \left(\frac{M}{a}\right)^{k-1}$$

after again substituting $H(r) = \lambda(r)cr^p$ in $H(c_n)$. Thus we must show that given $\varepsilon > 0$, there is a $\delta > 0$ such that

$$\left(\frac{k}{n}\right)^k \binom{n}{k} \delta^{k-1} < \varepsilon \quad \text{for all } n.$$

That this is true follows easily from

$$\left(\frac{k}{n}\right)^k \binom{n}{k} \delta^{k-1} < \delta^{k-1} \frac{k^k}{k!} \sim \frac{\delta^{k-1}}{e^k (2\pi k)^{\frac{1}{2}}} \quad \text{as } k \to \infty.$$

References

[1] Cacoullos, T. (1966). Estimation of a multivariate density. *Ann. Inst. Statist. Math.* 18, 178-189.

[2] Fix, E. and Hodges, J. L. (1951). Nonparametric discrimination: consistency properties. USAF Sch. Aviation Medicine, Rep. 4, Proj. 21-49-004.

[3] Kiefer, J. (1972). Iterated logarithm analogues for sample quantiles when $p_n \downarrow 0$. *Proc. Sixth Berkeley Symp. Math. Statist. Prob.* 1, 227-244.

[4] Loève, M. (1963). *Probability Theory*, 3rd. Ed., D. Van Nostrand Company, Princeton, N. J.

[5] Loftsgaarden, D. O. and Quesenberry, C. P. (1965). A nonparametric estimate of a multivariate density function. *Ann. Math. Statist.* 36, 1049-1051.

[6] Moore, D. S. and Henrichon, E. G. (1969). Uniform consistency of some estimates of a density function. *Ann. Math. Statist.* 40, 1499-1502.

[7] Moore, D. S. and Yackel, J. W. (1976). Consistency properties of nearest neighbor density function estimators. *Ann. Statist.* to appear.

[8] Wagner, T. J. (1973). Strong consistency of a nonparametric estimate of a density function. *IEEE Trans. Systems, Man and Cybernetics* 3, 289-290.

A STOCHASTIC VERSION OF WEAK MAJORIZATION, WITH APPLICATIONS

By S. E. Nevius*, F. Proschan*, and J. Sethuraman**
The Florida State University

1. *Introduction.* In Proschan and Sethuraman [10], a basic theorem concerning the preservation of a Schur function under a certain integral transformation was derived. A stochastic version of majorization was introduced in Nevius, Proschan, and Sethuraman [7] and multivariate applications of this new concept were developed. (Definitions from [7] and [10] will generally not be repeated.) In the present paper we introduce a stochastic version of weak majorization (see Section 2 for new definitions), develop its properties, and obtain multivariate applications.

2. *A stochastic version of weak majorization.* The concept of weak majorization has been mentioned by several authors (Mitrinović [6]; Marshall, Walkup, and Wets [5]; Beckenbach and Bellman [2]), but has not as yet been applied to statistical problems.

The deterministic definition of weak majorization will first be stated. Given a vector $\underline{x} = (x_1,\ldots,x_n)$, let $x_{[1]} \geq \cdots \geq x_{[n]}$ denote a decreasing rearrangement of $x_1,\ldots,x_n$. $\underline{x}$ is said to weakly majorize $\underline{x}'$ if

$$\sum_{i=1}^{j} x_{[i]} \geq \sum_{i=1}^{j} x'_{[i]}, \; j = 1,\ldots,n;$$

in symbols, $\underline{x} \geq \geq^m \underline{x}'$.

The following lemma will enable us to derive an important characterization of weak stochastic majorization.

*Research sponsored by the Air Force Office of Scientific Research, AFSC, USAF, under Grant No. AFOSR-74-2581.

**Research supported by the United States Army Research Office, Durham, under Grant No. DAHC04-74-G-0188.

LEMMA 2.1. *If $\underline{x} \geq \geq^m \underline{x}'$, then there exists a vector $\underline{x}^*$ such that $\underline{x} \geq^m \underline{x}^*$ and $\underline{x}^* \geq \underline{x}'$.*

PROOF. Without loss of generality, assume $\underline{x} = (x_1,\ldots,x_n)$ and $\underline{x}' = (x_1',\ldots,x_n')$, where $x_1 \geq x_2 \geq \ldots \geq x_n$, and $x_1' \geq \ldots \geq x_n'$. Let i_0 be the largest j, $1 \leq j \leq n$, such that $\sum_{i=1}^{j} x_i = \sum_{i=1}^{j} x_i'$; however, if $\sum_{i=1}^{j} x_i > \sum_{i=1}^{j} x_i'$ for all j, $1 \leq j \leq n$, set $i_0 = 0$. If $i_0 = n$, let $\underline{x}^* = \underline{x}'$.

If $i_0 < n$, let $d_1 = \min\{\sum_{i=1}^{j} x_i - \sum_{i=1}^{j} x_i', j = i_0 + 1,\ldots,n\}$ and $i_1 = \max\{j, i_0 + 1 \leq j \leq n$, such that $\sum_{i=1}^{j} x_i - \sum_{i=1}^{j} x_i' = d_1\}$. Set $x_{i_0+1}^{(1)} = x_{i_0+1}' + d_1$, $x_i^{(1)} = x_i'$ for $i \neq i_0 + 1$. If $i_1 = n$, let $\underline{x}^* = (x_1^{(1)},\ldots,x_n^{(1)})$.

If $i_1 < n$, let $d_2 = \min\{\sum_{i=1}^{j} x_i - \sum_{i=1}^{j} x_i^{(1)}, j = i_1 + 1,\ldots,n\}$ and $i_2 = \max\{j, i_1 + 1 \leq j \leq n$, such that $\sum_{i=1}^{j} x_i - \sum_{i=1}^{j} x_i^{(1)} = d_2\}$. Set $x_{i_1+1}^{(2)} = x_{i_1+1}^{(1)} + d_2$ and $x_i^{(2)} = x_i^{(1)}$ for $i \neq i_1 + 1$. If $i_2 = n$, let $\underline{x}^* = (x_1^{(2)},\ldots,x_n^{(2)})$.

Continuing in this manner, we get in a finite number, say m, of steps that $i_m = n$ and $\underline{x}^* = \underline{x}^{(m)}$.

By construction, we have $\underline{x}^* \geq \underline{x}'$ and $\underline{x} \geq^m \underline{x}^*$.

We may now prove the following characterization of weak majorization.

THEOREM 2.2. *$\underline{x} \geq \geq^m \underline{x}'$ if and only if $f(\underline{x}) \geq f(\underline{x}')$ for all nondecreasing Schur-convex functions.*

PROOF. Suppose $f(\underline{x}) \geq f(\underline{x}')$ for all nondecreasing Schur-convex functions f. Consider the nondecreasing Schur-convex functions

$f_j(x) = \sum_{i=1}^{j} x_{[i]}$, $j = 1,\ldots,n$. By assumption, $f_j(\underline{x}) \geq f_j(\underline{x}')$, $j = 1,\ldots,n$. Thus $\underline{x} \geq \geq^m \underline{x}'$ by definition.

Conversely, suppose $\underline{x} \geq \geq^m \underline{x}'$. By Lemma 2.1, there exists a vector $\underline{x}^*$ such that $\underline{x} \geq^m \underline{x}^*$ and $\underline{x}^* \geq \underline{x}'$. Let f be a nondecreasing Schur-convex function. Since f is Schur-convex and $\underline{x} \geq^m \underline{x}^*$, we have by the definition of Schur-convexity that $f(\underline{x}) \geq f(\underline{x}^*)$. Since f is nondecreasing and $\underline{x}^* \geq \underline{x}'$, then $f(\underline{x}^*) \geq f(\underline{x}')$. Thus $f(\underline{x}) \geq f(\underline{x}')$.

This equivalent version of weak majorization leads us to the following stochastic version. Throughout, let $\underline{X}$ and $\underline{X}'$ have probability measures P and P', respectively, and distribution functions G and G', respectively.

DEFINITION. (a) $\underline{X}$ is said to *stochastically weakly majorize* $\underline{X}'$ if

$$f(\underline{X}) \geq^{st} f(\underline{X}')$$

for every nondecreasing Borel-measurable Schur-convex function f; in symbols, $\underline{X} \geq \geq^{st.m.} \underline{X}'$.

(b) We say $P \geq \geq^{st.m.} P'$ if $\underline{X} \geq \geq^{st.m.} \underline{X}'$.

(c) We say $G \geq \geq^{st.m.} G'$ if $\underline{X} \geq \geq^{st.m.} \underline{X}'$.

We need the following definition to state Theorem 2.3 which characterizes stochastic weak majorization.

DEFINITION. A subset S of R_n is said to be *increasing* if $\underline{x} \in S$ and $\underline{x}' \geq \underline{x}$ imply $\underline{x}' \in S$.

THEOREM 2.3. *The following statements are equivalent.*

(i) $\underline{X} \geq \geq^{st.m.} \underline{X}'$.

(ii) $Ef(\underline{X}) \geq Ef(\underline{X}')$ *for every nondecreasing Schur-convex function* f *for which both these expectations exist*.

(iii) $Ef(\underline{X}) \geq Ef(\underline{X}')$ *for every bounded nondecreasing Schur-convex function* f.

(iv) $P(A) \geq P'(A)$ *for all measurable increasing Schur-convex sets*.

PROOF. The proof is similar to the proof of Theorem 2.2 of [7]; the proof of the implication (ii) $\Rightarrow$ (i) requires the additional observation that a nondecreasing function of a nondecreasing Schur-convex function is itself nondecreasing and Schur-convex, while the proof of (iv) $\Rightarrow$ (ii) uses the fact that $\{x: f(\underline{x}) \geq t\}$ is an _increasing_ Schur-convex set for every nondecreasing Schur-convex function f.

Next, we relate stochastic weak majorization to deterministic weak majorization in terms of a stochastic comparison of vectors. See Veinott [11] for a characterization of the following stochastic ordering of vectors.

DEFINITION. A random vector $\underline{X}$ is _stochastically larger_ than a random vector $\underline{X}'$ if $h(\underline{X}) \geq^{st} h(\underline{X})$ for all nondecreasing functions h; in symbols, $\underline{X} \geq^{st} \underline{X}'$.

An additional characterization of stochastic weak majorization may be stated in terms of the stochastic comparison of the vectors of partial sums of the reverse order statistics $x_{[1]} \geq \ldots \geq x_{[n]}$. Define the map T from R_n into R_n by $T(\underline{x}) = (y_1, \ldots, y_n)$, where $y_i = \sum_{j=1}^{i} x_{[j]}$, $i = 1, \ldots, n$. Thus, T gives the partial sums of the reverse order statistics of $x_1, \ldots, x_n$. Let $TR_n = C$. The following characterization of nondecreasing Schur-convex functions will prove useful in the characterization of stochastic weak majorization developed in Theorem 2.5 below.

LEMMA 2.4. _For any permutation-invariant function f on R_n, define the function g on C by putting $g(\underline{y}) = f(\underline{x})$ whenever $\underline{y} = T\underline{x}$. This defines a 1-1 correspondence, $f \leftrightarrow g$, between permutation-invariant functions on R_n and functions on C. Moreover, f is a nondecreasing Schur-convex function if and only if g is nondecreasing._

PROOF. The first statement is easily verified. The second statement is an immediate consequence of the fact that $\underline{x} \geq \geq^{m} \underline{x}'$ if and only if $y_i \geq y_i'$, $i = 1, \ldots, n$, where $\underline{y} = T\underline{x}$ and $\underline{y}' = T\underline{x}'$.

THEOREM 2.5. _Let $\underline{X}$ and $\underline{X}'$ be random vectors in R_n. Set_

$\underline{Y} = T\underline{X}$ and $\underline{Y}' = T\underline{X}'$. Then $\underline{X} \geq\geq^{\text{st.m.}} \underline{X}'$ if and only if $\underline{Y} \geq^{\text{st}} \underline{Y}'$.

PROOF. This theorem is an immediate consequence of Lemma 2.4.

As a direct consequence of Theorem 2.5, we obtain:

COROLLARY 2.6. *If* $\underline{X} \geq\geq^{\text{st.m.}} \underline{X}'$, *then* $\sum_{i=1}^{j} X_{[i]} \geq^{\text{st}} \sum_{i=1}^{j} X'_{[i]}$ *for* $j = 1,\ldots,n$.

Note that Corollary 2.6 also follows directly from the definition of stochastic weak majorization since $\sum_{i=1}^{j} x_{[i]}$ is a nondecreasing Schur-convex function of $\underline{x}$ for $j = 1,\ldots,n$. To see that the converse does not hold, consider the following example. Let $\underline{X}$ take on the values (4,1) and (3,1), each with probability $\frac{1}{2}$, and let $\underline{X}'$ take on the values (4,0) and (3,2), each with probability $\frac{1}{2}$. Then $X_{[1]} =^{\text{st}} X'_{[1]}$ and $X_{[1]} + X_{[2]} =^{\text{st}} X'_{[1]} + X'_{[2]}$. Hence $\sum_{i=1}^{j} X_{[i]} \geq^{\text{st}} \sum_{i=1}^{j} X'_{[i]}$, $j = 1,2$. Consider the nondecreasing Schur-convex function $f(\underline{x}) = x_1^3 + x_2^3$. Note that $P(f(\underline{X}) \leq 28) > P(f(\underline{X}') \leq 28)$. Thus it is *not* true that $\underline{X} \geq\geq^{\text{st.m.}} \underline{X}'$. (A version of this example appears in Marshall and Olkin [3].)

In the deterministic case, we know that $\underline{x} \geq \underline{x}'$ implies that $\underline{x} \geq\geq^{m} \underline{x}'$. Theorem 2.7 gives a similar result for the stochastic case.

THEOREM 2.7. *If* $\underline{X} \geq^{\text{st}} \underline{X}'$, *then* $\underline{X} \geq\geq^{\text{st.m.}} \underline{X}'$.

PROOF. Every nondecreasing Schur-convex function of $\underline{x}$ is itself a nondecreasing function of $\underline{x}$. Hence the result holds.

To see that the converse does not hold even in the deterministic case, let $\underline{X} = (5,3,3)$ with probability one and $\underline{X}' = (4,4,2)$ with probability one. Then $\underline{X} \geq\geq^{\text{st.m.}} \underline{X}'$, but $\underline{X}$ is *not* stochastically larger than $\underline{X}'$.

The following statement is equivalent to the definition of weak majorization in the deterministic case.

EQUIVALENT DEFINITION. A vector $\underline{x}$ *weakly majorizes* a vector

$\underline{x}'$ if $\underline{x} \geq \geq^m \underline{z}$ for every $\underline{z}$ such that $\underline{x}' \geq \geq^m \underline{z}$.

We have the following stochastic analogue of this definition.

THEOREM 2.8. *If* $\underline{X} \geq \geq^{st.m.} \underline{X}'$, *then* $P(\underline{X} \geq \geq^m \underline{z}) \geq P(\underline{X}' \geq \geq^m \underline{z})$ *for each vector* $\underline{z}$.

PROOF. The result follows by choosing the bounded nondecreasing Schur-convex function $g(\underline{x}) = I_{[\underline{x} \geq \geq^m \underline{z}]}$ and applying Theorem 2.3.

Intuitively, Theorem 2.8 states that if $\underline{X} \geq \geq^{st.m.} \underline{X}'$, then $\underline{X}$ is more likely to weakly majorize any vector $\underline{z}$ than is $\underline{X}'$. The converse, however, is not true. To see this, consider the example following Corollary 2.8 of [7], where $\underline{X}$ takes on the values $(1,0,0)$, $(\frac{1}{2},\frac{1}{2},0)$, $(\frac{2}{3},\frac{1}{6},\frac{1}{6})$, and $(\frac{1}{2},\frac{1}{3},\frac{1}{6})$ with probabilities .1, .1, .1, and .7, respectively, and $\underline{X}'$ takes on these values with probabilities 0, .2, .2, and .6, respectively. The fact that $P[\underline{X} \geq \geq^m \underline{z}] \geq P[\underline{X}' \geq \geq^m \underline{z}]$ for all $\underline{z}$ follows from the easily verified fact that a vector $\underline{z}$ is weakly majorized by both $(\frac{1}{2},\frac{1}{2},0)$ and $(\frac{2}{3},\frac{1}{6},\frac{1}{6})$ only if it is weakly majorized by $(\frac{1}{2},\frac{1}{3},\frac{1}{6})$. Since $f(\underline{X})$ is not stochastically larger than $f(\underline{X}')$ for the nondecreasing Schur-convex function $f(\underline{x}) = x_1^3 + x_3^3 + x_3^3$, it follows that $\underline{X}$ does not stochastically weakly majorize $\underline{X}'$.

3. *Preservation of stochastic weak majorization under operations.* The applicability of our new notion of stochastic weak majorization is greatly expanded by the degree to which it is preserved under various standard mathematical, probabilistic, and statistical operations. In this section we display operations which preserve stochastic weak majorization.

First, we show that *stochastic weak majorization is preserved under mixtures of distributions*.

THEOREM 3.1. *Let* $\underline{X}$ *and* $\underline{X}'$ *be two random vectors and let* U *be a random variable such that the conditional distribution of* $\underline{X}$ *given* $U = u$ *stochastically weakly majorizes the conditional distribution of* $\underline{X}'$ *given* $U = u$ *for each* u. *Then* $\underline{X} \geq \geq^{st.m.} \underline{X}'$.

The proof of Theorem 3.1 is essentially the same as the proof of the analogous theorem for stochastic majorization in [7].

Another preservation result involving sums of independent random vectors may be shown utilizing the following two lemmas. Corresponding to a random vector $\underline{X}$, denote by $\underline{X}^*$ the vector of reverse order statistics $(X_{[1]}, X_{[2]}, \ldots, X_{[n]})$.

LEMMA 3.2. $\underline{X} \geq \geq^{st.m.} \underline{X}'$ *if and only if* $\underline{X}^* \geq \geq^{st.m.} \underline{X}'^*$.

The proof utilizes the fact that a Schur-convex function is necessarily permutation-invariant. The details are omitted.

Lemma 3.3 below is proven in Pledger and Proschan [9]. It is a special case of a result of Esary and Proschan (unpublished) which is utilized later as Lemma 3.6.

LEMMA 3.3. *If* $\underline{Z} \geq^{st} \underline{Z}'$, $\underline{W} \geq^{st} \underline{W}'$, $\underline{Z}$ *and* $\underline{W}$ *are independent, and* $\underline{Z}'$ *and* $\underline{W}'$ *are independent, then* $\underline{Z} + \underline{W} \geq^{st} \underline{Z}' + \underline{W}'$.

We may now state the following preservation result for sums.

THEOREM 3.4. *If* $\underline{X} \geq \geq^{st.m.} \underline{X}'$, $\underline{Y} \geq \geq^{st.m.} \underline{Y}'$, $\underline{X}$ *and* $\underline{Y}$ *are independent, and* $\underline{X}'$ *and* $\underline{Y}'$ *are independent, then* $\underline{X}^* + \underline{Y}^* \geq \geq^{st.m.} \underline{X}'^* + \underline{Y}'^*$.

PROOF. By Theorem 2.5, it suffices to show $(X_{[1]} + Y_{[1]}, X_{[1]} + X_{[2]} + Y_{[1]} + Y_{[2]}, \ldots, \sum_{i=1}^n X_{[i]} + \sum_{i=1}^n Y_{[i]}) \geq^{st} (X'_{[1]} + Y'_{[1]}, X'_{[1]} + X'_{[2]} + Y'_{[1]} + Y'_{[2]}, \ldots, \sum_{i=1}^n X'_{[i]} + \sum_{i=1}^n Y'_{[i]})$. But this follows directly from Lemma 3.3.

To see that the result does *not* hold even in the deterministic case without rearranging vectors before adding, let $\underline{X} = (6,2)$, $\underline{Y} = (3.6)$, $\underline{X}' = \underline{Y}' = (1,5)$ with probability one. Then $\underline{X} \geq \geq^{st.m.} \underline{X}'$, $\underline{Y} \geq \geq^{st.m.} \underline{Y}'$, but $\underline{X} + \underline{Y}$ does *not* stochastically weakly majorize $\underline{X}' + \underline{Y}'$.

Next, we show that stochastic weak majorization is also preserved under multiplication by an increasing function of $\sum_{i=1}^n X_i$.

THEOREM 3.5. *Let* $\underline{X} \geq \geq^{st.m.} \underline{X}'$ *and* f *be a nonnegative nondecreasing Borel-measurable function. Then*

$$f(\sum_{i=1}^{n} X_i)\underline{X} \geq \geq^{st.m.} f(\sum_{i=1}^{n} X_i')\underline{X}'.$$

PROOF. Let g be a nondecreasing Schur-convex function. Then $g(f(\sum_{i=1}^{n} x_i)\underline{x})$ is a nondecreasing Schur-convex function of $\underline{x}$. The desired result follows.

In proving the next theorem, which gives the most important and useful operation preserving stochastic majorization, we require the following lemmas. The first lemma, originally proven by Esary and Proschan (unpublished), is presented in [9].

LEMMA 3.6. *Let $\underline{X}_1,\ldots,\underline{X}_k$ be mutually independent random vectors, and similarly $\underline{X}_1',\ldots,\underline{X}_k'$ be mutually independent. Suppose $\underline{X}_i \geq^{st} \underline{X}_i'$ for $1 \leq i \leq k$. Then $(\underline{X}_1,\ldots,\underline{X}_k) \geq^{st} (\underline{X}_1',\ldots,\underline{X}_k')$.*

We also require the following deterministic lemma.

LEMMA 3.7 (*Deterministic*). *f is a nondecreasing Schur-convex function if and only if*

$$\underline{x} \geq \geq^{m} \underline{x}' \Rightarrow f(\underline{x}) \geq f(\underline{x}'). \tag{3.1}$$

PROOF. Let f be a nondecreasing Schur-convex function and $\underline{x} \geq \geq^{m} \underline{x}'$. By Lemma 2.1, $f(\underline{x}) \geq f(\underline{x}')$.

Conversely, suppose (3.1) holds. Since $\underline{x} \geq^{m} \underline{x}'$ implies $\underline{x} \geq \geq^{m} \underline{x}'$, it follows that $\underline{x} \geq^{m} \underline{x}' \Rightarrow f(x) \geq f(\underline{x}')$. Thus f is Schur-convex by definition. Since $\underline{x} \geq \underline{x}'$ implies $\underline{x} \geq \geq^{m} \underline{x}'$, we have $f(\underline{x}) \geq f(\underline{x}')$. Thus f is nondecreasing.

We may now prove the following preservation theorem.

THEOREM 3.8. *Let $\phi(\lambda,x)$ be TP_2 in $\lambda > 0$ and $x \geq 0$, and satisfy the semigroup property. (See [10] for definitions.) Let $\underline{X}_{\underline{\lambda}} = (X_{\lambda_1},\ldots,X_{\lambda_n})$ be a random vector of independent components, where X_{λ_i} is a nonnegative random variable with density $\phi(\lambda_i,x)$, wrt Lebesgue or count measure. If $\underline{\lambda} \geq \geq^{m} \underline{\lambda}'$, then $\underline{X}_{\underline{\lambda}} \geq \geq^{st.m.} \underline{X}_{\underline{\lambda}'}$.*

PROOF. Let g be a nondecreasing bounded Schur-convex function. Define $h(\underline{\lambda}) = Eg(X_{\lambda_1},\ldots,X_{\lambda_n})$. We want to show $\underline{\lambda} \geq \geq^{m} \underline{\lambda}'$ implies $h(\underline{\lambda}) \geq h(\underline{\lambda}')$. Thus, by Lemma 3.7 it suffices to show that h is a nondecreasing Schur-convex function of $\underline{\lambda}$. By Theorem 3.3 of [7], h is Schur-convex.

To show that h is nondecreasing, let $\underline{\lambda} \geq \underline{\lambda}'$. Then for each set of corresponding components X_{λ_i} and $X_{\lambda_i'}$ of $\underline{X}_{\lambda}$ and $\underline{X}_{\lambda'}$, respectively, we have $X_{\lambda_i} \geq^{st} X_{\lambda_i'}$. Since g is nondecreasing, we have by Lemma 3.6 that $g(\underline{X}_{\lambda}) \geq^{st} g(\underline{X}_{\lambda'})$. Thus $h(\underline{\lambda}) \geq h(\underline{\lambda}')$.

Thus a <u>deterministic property (weak majorization) of the parameter vector λ is transformed into a corresponding stochastic property (stochastic weak majorization) of the random vector</u> $\underline{X}_{\lambda}$. This result will be exploited in Section 4 to obtain useful multivariate applications of stochastic weak majorization.

An immediate corollary of Theorem 3.8 is an application to stochastic processes.

COROLLARY 3.9. *Let* $\{X(t), 0 \leq t < \infty\}$ *be a stochastic process with stationary, independent, and nonnegative increments. Let the density* $\phi(\lambda,x)$ *of* $X(t+\lambda) - X(t)$ *be* TP_2 *in* $\lambda > 0$ *and* $x \geq 0$. *Let* $0 = t_0 < t_1 < \ldots < t_n$, $0 = t_0' < t_1' < \ldots < t_n'$, $\lambda_i = t_i - t_{i-1}$, *and* $\lambda_i' = t_i' - t_{i-1}'$. *Then* $\underline{\lambda} \geq \geq^m \underline{\lambda}'$ *implies that*

$$(X(t_1) - X(t_0),\ldots,X(t_n) - X(t_{n-1})) \geq \geq^{st.m.}$$
$$(X(t_1') - X(t_0'),\ldots,X(t_n') - X(t_{n-1}')).$$

The proof is analogous to the proof of Corollary 3.4 in [7].

Theorem 3.8 gives sufficient conditions on the distribution of a random vector $\underline{X}_{\lambda}$ to state that $\underline{\lambda} \geq \geq^m \underline{\lambda}'$ implies that $\underline{X}_{\lambda} \geq \geq^{st.m.} \underline{X}_{\lambda'}$. We now consider prior distributions on the parameter vector $\underline{\lambda}$ with probability distributions G and G', where $G \geq \geq^{st.m.} G'$. The next theorem shows that stochastic weak majorization is preserved under a mixing operation.

THEOREM 3.10. *Let* $\{\underline{X}_{\lambda}\}$ *be a family of random vectors indexed by* $\underline{\lambda} \in \Theta \subset R_n$ *such that* $\underline{\lambda} \geq \geq^m \underline{\lambda}'$ *implies* $\underline{X}_{\lambda} \geq \geq^{st.m.} \underline{X}_{\lambda'}$. *Let G and G' be two probability measures on* Θ *such that* $G \geq \geq^{st.m.} G'$.

Let $Q(A) = \int_\Theta P(\underline{X}_{\underline{\lambda}} \in A)dG(\underline{\lambda})$ *and* $Q'(A) = \int_\theta P(\underline{X}_{\underline{\lambda}} \in A)dG'(\underline{\lambda})$ *for all Borel sets* A *in* R_n. *Then* $Q \geq \geq^{st.m.} Q'$.

PROOF. Let f be a bounded nondecreasing Schur-convex function. Let $\underline{X}$ and $\underline{X}'$ have probability measures Q and Q', respectively. Then

$$Ef(\underline{X}) = \int_\Theta Ef(\underline{X}_{\underline{\lambda}})dG_1(\lambda).$$

By assumption and Theorem 2.3, $\underline{\lambda} \geq \geq^m \underline{\lambda}'$ implies $Ef(\underline{X}_{\underline{\lambda}}) \geq Ef(\underline{X}_{\underline{\lambda}'})$. By Lemma 3.7, $Ef(\underline{X}_{\underline{\lambda}})$ is a nondecreasing Schur-convex function of $\underline{\lambda}$. Since $G_1 \geq \geq^{st.m.} G_2$, it follows by Theorem 2.3 that $\int_\Theta Ef(\underline{X}_{\underline{\lambda}})dG_1(\underline{\lambda}) \geq \int_\Theta Ef(\underline{X}_{\underline{\lambda}})dG_2(\underline{\lambda})$; that is, $Ef(\underline{X}) \geq Ef(\underline{X}')$. Thus $\underline{X} \geq \geq^{st.m.} \underline{X}'$.

A type of preservation theorem under conditioning may be proven for certain types of discrete random vectors. We will use the following definitions from Barlow and Proschan [1].

DEFINITION. A random variable T is *stochastically increasing* in a random variable S if $P(T > t|S = s)$ is increasing in s for all t. We write $T \uparrow$ st in S.

DEFINITION. A random vector $\underline{T}$ is *stochastically increasing* in S if $f(\underline{T}) \uparrow$ st in S for all nondecreasing functions $f(\underline{t})$. We write $\underline{T} \uparrow$ st in S.

Let $\underline{X}_N$ have the conditional probability measure of the random vector $\underline{X}$ given that $\sum_{i=1}^n X_i = N$. The following preservation theorem may now be stated.

THEOREM 3.11. *Let* $\underline{X}$ *and* $\underline{X}'$ *be discrete random vectors and* $\underline{X} \uparrow$ st *in* $\sum_{i=1}^n X_i$. *Let* $P(\sum_1^n X_i = N) > 0$ *and* $P(\sum_1^n X_i = N') > 0$. *If* $\underline{X} \geq^{st.m.} \underline{X}'$ *and* $N \geq N'$, *then* $\underline{X}_N \geq \geq^{st.m.} \underline{X}'_{N'}$.

PROOF. Since $\underline{X} \uparrow$ st in $\sum_{i=1}^n X_i$, we have $\underline{X}_N \geq^{st} \underline{X}_{N'}$, and thus by Theorem 2.7, $\underline{X}_N \geq \geq^{st.m.} \underline{X}_{N'}$. Since $\underline{X} \geq^{st.m.} \underline{X}'$, we have by

Corollary 2.10 of [7] that $\underline{X}_{N'} \geq^{st.m.} \underline{X}'_{N'}$ and hence $\underline{X}_{N'} \geq \geq^{st.m.} \underline{X}'_{N'}$. Thus $\underline{X}_N \geq \geq^{st.m.} \underline{X}'_{N'}$.

Next, we give a sufficient condition for $\underline{X} \uparrow$ st in $\sum_{i=1}^{n} X_i$ which will provide applications of Theorem 3.11 in the next section.

THEOREM 3.12. Let $\underline{X}$ be a random vector such that $\underline{X}_{N+1}$ has the same distribution as $\underline{X}_N + \Delta(\underline{X}_N)$, where $\Delta(\underline{X}_N)$ is a nonnegative random vector whose distribution may depend on $\underline{X}_N$. Then $\underline{X} \uparrow$ st in $\sum_{i=1}^{n} X_i$.

PROOF. Note that $\underline{X}_N + \Delta(\underline{X}_N) \geq \underline{X}_N$ a.s. Thus $\underline{X}_{N+1} \geq^{st} \underline{X}_N$. That is, $\underline{X} \uparrow$ st in $\sum_{i=1}^{n} X_i$.

Let $\underline{X}_{\mathit{N}}$ represent a random vector whose conditional distribution given $\mathit{N} = N$ is that of $\underline{X}_N$ as defined above. In other words, the distribution of N is a mixing distribution for the parameter N. We may now state the following theorem.

THEOREM 3.13. Let $\underline{X}$ and $\underline{X}'$ be discrete random vectors with $\underline{X} \uparrow$ st in $\sum_{i=1}^{n} X_i$. If $\underline{X} \geq^{st.m.} \underline{X}'$ and $\mathit{N} \geq^{st} \mathit{N}'$, then $\underline{X}_{\mathit{N}} \geq \geq^{st.m.} \underline{X}'_{\mathit{N}'}$.

PROOF. Let f be a nondecreasing Schur-convex function. Then

$$\begin{aligned} Ef(\underline{X}_{\mathit{N}}) &= EE[f(\underline{X}_{\mathit{N}})|\mathit{N}] \\ &\geq EE[f(\underline{X}'_{\mathit{N}})|\mathit{N}] \text{ [by Corollary 2.10 of [7]]} \\ &\geq EE[f(\underline{X}'_{\mathit{N}'})|\mathit{N}'] \text{ [by Theorem 3.11 and} \\ &\qquad \text{the hypothesis that } \mathit{N} \geq^{st} \mathit{N}'\text{].} \end{aligned}$$

We may combine Theorems 3.10 and 3.13 to obtain the following result.

THEOREM 3.14. Let $\{\underline{X}_{\underline{\lambda}}\}$ be a family of discrete random vectors indexed by $\underline{\lambda} \in \Theta \subset R_n$ such that $\underline{\lambda} \geq \geq^{m} \underline{\lambda}'$ implies $\underline{X}_{\underline{\lambda}} \geq \geq^{st.m.} \underline{X}_{\underline{\lambda}'}$ and $\underline{X}_{\underline{\lambda}} \uparrow$ st in $\sum_{i=1}^{n} X_{i,\underline{\lambda}}$. Let $\underline{X}_{\underline{\lambda},N}$ denote a random vector with the conditional distribution of $\underline{X}_{\lambda}$ given that $\sum_{i=1}^{n} X_{i,\underline{\lambda}} = N$, and let $\underline{X}_{\underline{\lambda},N}$ denote the resultant mixture as in Theorem 3.13. Finally, let $\underline{X}_{\underline{\lambda},N}$ have probability measure

$$\mu(A) = \int P(\underline{X}_{\underline{\lambda},N} \in A)dH(\underline{\lambda})$$

for all Borel sets A in R_n, where $\underline{\Lambda}$ is a random vector taking values in Θ a.s. and with distribution H.

If $N \geq^{st} N'$ and $\underline{\Lambda} \geq\geq^{st.m.} \underline{\Lambda}'$, then $\underline{X}_{\underline{\Lambda},N} \geq\geq^{st.m.} \underline{X}'_{\underline{\Lambda},N'}$.

PROOF. Let g be a bounded nondecreasing Schur-convex function. Then

$$\begin{aligned} Eg(\underline{X}_{\underline{\Lambda},N}) &= EE[g(\underline{X}_{\underline{\Lambda},N})|N] \\ &\geq EE[g(\underline{X}_{\underline{\Lambda}',N})|N] \text{ [by Theorem 3.10]} \\ &= Eg(\underline{X}_{\underline{\Lambda}',N}) \\ &= EE[g(\underline{X}_{\underline{\Lambda}',N})|\underline{\Lambda}'] \\ &\geq EE[g(\underline{X}_{\underline{\Lambda}',N'})|\underline{\Lambda}'] \text{ [by Theorem 3.13]} \\ &= Eg(\underline{X}_{\underline{\Lambda}',N'}). \end{aligned}$$

By Theorem 2.3, $\underline{X}_{\underline{\Lambda},N} \geq\geq^{st.m.} \underline{X}_{\underline{\Lambda}'N'}$.

4. *Applications of stochastic weak majorization.* In this section we present applications of stochastic weak majorization to obtain general inequalities for several well known multivariate distributions.

APPLICATION 4.1. Let $\underline{X}_{\lambda}$ be a random vector of independent components, where X_{λ_i} has a density of the form given in (a), (b), or (c) below. Let $\underline{\lambda} \geq\geq^{m} \underline{\lambda}'$. Then $\underline{X}_{\underline{\lambda}} \geq\geq^{st.m.} \underline{X}_{\underline{\lambda}'}$.

(a) Poisson. $f_{\lambda}(x) = \frac{(\lambda\theta)^x}{x!} e^{-\lambda\theta}$, $x = 0,1,\ldots,\lambda > 0$, and fixed $\theta > 0$.

(b) Binomial. $f_{\lambda}(x) = \binom{\lambda}{x}p^x(1-p)^{\lambda-x}$, $x = 0,1,\ldots,n$; $n=1,2,\ldots$; and $0 < p < 1$.

(c) Gamma. $f_\lambda(x) = \frac{\theta^\lambda x^{\lambda-1}}{\Gamma(\lambda)} e^{-\theta x}$, $x \geq 0$, $\lambda > 0$, and fixed $\theta > 0$.

The result follows from Theorem 3.8 by noting that $\phi(\lambda,x) = f_\lambda(x)$ is TP_2 and satisfies the semigroup property in each of the cases (a), (b), and (c).

Application 4.1 is next used to obtain stochastic comparisons for the multivariate negative binomial distribution by applying Theorem 3.1.

APPLICATION 4.2. Let $\underline{Y}_{\underline{\lambda}} = (Y_{1,\underline{\lambda}},\ldots,Y_{n,\underline{\lambda}})$ have a multivariate negative binomial distribution with density

$$f_{\underline{\lambda}}(\underline{y}) = \frac{\Gamma(N+\sum_{i=1}^n y_i)}{\Gamma(N)} \prod_{i=1}^n \frac{\lambda_i^{y_i}}{y_i!} (1+\sum_{i=1}^n \lambda_i)^{-N-\sum_{i=1}^n y_i},$$

where $y_i = 0,1,\ldots,\lambda_i > 0$, $i = 1,\ldots,n$, $N > 0$. If $\underline{\lambda} \geq \geq^m \underline{\lambda}'$, then $\underline{Y}_{\underline{\lambda}} \geq \geq^{st.m.} \underline{Y}_{\underline{\lambda}'}$.

PROOF. The result follows from Application 4.1(a), Theorem 3.1, and the fact utilized in [7] that the multivariate negative binomial distribution is a mixture of independent Poisson random variables with densities 4.1(a) under a gamma distribution for θ with density $g(\theta) = \frac{\theta^{N-1}}{\Gamma(N)} e^{-\theta}$.

Using the fact that $g(\underline{x}) = -I_{[\max(x_1,\ldots,x_n)\leq r]}(\underline{x})$ is a nondecreasing Schur-convex function, we obtain the following application.

APPLICATION 4.3. Let $\underline{Y}_\lambda$ have any one of the multivariate distributions given in Applications 4.1 or 4.2. If $\underline{\lambda} \geq \geq^m \underline{\lambda}'$, then

$$P(Y_{1,\underline{\lambda}} \leq r,\ldots,Y_{n,\underline{\lambda}} \leq r) \leq P(Y_{1,\underline{\lambda}'} \leq r,\ldots,Y_{n,\underline{\lambda}'} \leq r).$$

Application 4.3 generalizes a similar majorization theorem for the multinomial distribution proven by Olkin [8].

Applications of stochastic weak majorization to the

multinomial and multivariate hypergeometric distributions with random sample size are possible using the preservation results of Section 3. First, we present the following result.

THEOREM 4.4. *Let* $\underline{X}_{\underline{p},N}$ *be a multinomial random vector with parameters* N *and* p. *Let* $\underline{p} \geq^{m} \underline{p}'$ *and* $N \geq N'$. *Then* $\underline{X}_{N,\underline{p}} \geq \geq^{st.m.} \underline{X}_{N',\underline{p}'}$.

PROOF. Recall that the multinomial may be derived by conditioning a vector of independent Poisson random variables $\underline{X}$ on $\sum_{i=1}^{n} X_i$. By Theorem 3.12 and Application 4.1 of [7], the hypotheses of Theorem 3.11 are satisfied. Thus we may apply Theorem 3.11 to conclude $\underline{X}_{\underline{p},N} \geq \geq^{st.m.} \underline{X}_{\underline{p}',N'}$.

An analagous theorem holds for the multivariate hypergeometric distribution.

THEOREM 4.5. *Let* $\underline{X}_{N,\underline{\lambda}}$ *be a multivariate hypergeometric random vector with density*

$$f(\underline{y}) = \frac{\prod_{i=1}^{n} \binom{\lambda_i}{y_i}}{\binom{\sum_{i=1}^{n} \lambda_i}{N}},$$

where $y_1,\ldots,y_n$ *are integers satisfying* $0 \leq y_i \leq \lambda_i$, $i=1,\ldots,n$, $\sum_{i=1}^{n} y_i = N$; $\lambda_1,\ldots,\lambda_n$ *are positive integers; and integer* N *satisfies* $0 < N \leq \sum_{i=1}^{n} \lambda_i$. *If* $\underline{\lambda} \geq^{m} \underline{\lambda}'$ *and* $N \geq N'$, *then* $\underline{X}_{N,\underline{\lambda}} \geq \geq^{st.m.} \underline{X}_{N',\underline{\lambda}'}$.

PROOF. Recall that the multivariate hypergeometric distribution may be derived by conditioning a vector of independent binomial random vectors $\underline{X}$ on $\sum_{i=1}^{n} X_i$. By Theorem 3.12 and Application 4.1 of [7], the hypotheses of Theorem 3.11 are satisfied. Thus we may apply Theorem 3.11 as in the proof of Theorem 4.4. to obtain the desired result.

Similar results for the multinomial and multivariate hypergeometric with random sample size are possible by utilizing Theorem 3.13. The following theorem is for the multinomial; an analogous result holds for the multivariate hypergeometric.

THEOREM 4.6. Let N be a random variable taking values in $\{1,2,\ldots\}$ and let the conditional distribution of $\underline{X}_{\underline{p},N}$ given $N = r$ be multinomial with parameters r and $\underline{p}$. If $N \geq^{st} N'$ and $\underline{p} \geq^{m} \underline{p}'$, then $\underline{X}_{\underline{p},N} \geq\geq^{st.m.} \underline{X}_{\underline{p}',N'}$.

PROOF. The result follows directly from Theorem 3.13 by similar arguments as in the proof of Theorem 4.4.

We may also apply Theorem 3.14 to get a stochastic weak majorization result for the multinomial with both N and $\underline{p}$ random. An analogous result holds for the multivariate hypergeometric.

THEOREM 4.7. Let $\underline{X}_{\underline{p},N}$ be as defined in Theorem 4.6. Let $\underline{\Lambda}$ and $\underline{\Lambda}'$ be random vectors with distributions G and G', respectively, and which take values in $\{\underline{p}\colon\ p_i > 0,\ i = 1,\ldots,n,\ \sum_{i=1}^{n} p_i = 1\}$ a.s. Let $\underline{X}_{\underline{\Lambda},N}$ have probability measure

$$\mu(A) = \int P(\underline{X}_{\underline{p},N} \in A)dG(\underline{p})$$

for all Borel sets A in R_n; similarly for $\underline{X}_{\underline{\Lambda}',N'}$. If $N \geq^{st} N'$ and $\underline{\Lambda} \geq^{st.m.} \underline{\Lambda}'$, then $\underline{X}_{\underline{\Lambda},N} \geq\geq^{st.m.} \underline{X}_{\underline{\Lambda}',N'}$.

PROOF. The result follows directly from Theorem 3.14.

References

[1] Barlow, R. E. and Proschan, F. (1975). *Statistical Theory of Reliability and Life Testing: Probability Models*. Holt, Rinehart, and Winston, New York.

[2] Beckenbach, E. F. and Bellman, R. L. (1965). *Inequalities*. Springer-Verlag, New York.

[3] Marshall, A. W. and Olkin, I. (in preparation). *Majorization and Schur Functions*.

[4] Marshall, A. W., Olkin, I., and Proschan, F. (1967). Monotonicity of ratios of means and other applications of majorization. In O. Shiska (Ed.), *Inequalities*. Academic Press, New York, 177-190.

[5] Marshall, A. W., Walkup, D. W., and Wets, R. J. (1967). Order-preserving functions; applications to majorization and order statistics. *Pacific J. Math.* 23, 569-584.

[6] Mitrinović, D. S. (1970). *Analytic Inequalities*. Springer-Verlag, New York.

[7] Nevius, S. E., Proschan, F., and Sethuraman, J. Schur functions in statistics. II: Stochastic majorization. *Ann. Statist.*, to appear.

[8] Olkin, I. (1972). Monotonicity properties of Dirichlet integrals with applications to the multinomial distribution and the analysis of variance. *Biometrika* 59, 303-307.

[9] Pledger, G. and Proschan, F. (1973). Stochastic comparisons of random processes, with applications in reliability. *J. Appl. Prob.* 10, 572-585.

[10] Proschan, F. and Sethuraman, J. Schur functions in statistics, I: The preservation theorem. *Ann. Statist.*, to appear.

[11] Veinott, A. F., Jr. (1965). Optimal policy in a dynamic, single product, nonstationary inventory model with several demand classes. *Operations Research* 13, 761-778.

SYNERGISTIC EFFECTS AND THE CORRESPONDING OPTIMAL VERSION OF THE MULTIPLE COMPARISON PROBLEM

By Jerzy Neyman*
University of California, Berkeley

1. *Introduction.* The sources of most interesting theoretical problems of relatively young mathematical disciplines are, frequently, "substantive" problems, that is problems arising in studies of nature. One of the important substantive problems of modern times is the detection of noxious synergisms. Here the interests of the U. S. Food and Drug Administration, of the Environmental Protection Agency, and the National Cancer Institute come to one's mind. Briefly and roughly, the substantive problem is to find out: (i) whether any of the agents A,B,...,C expected to penetrate our environment is likely to be deleterious to public health, and (ii) whether the combined effect of two or more of these agents, say of A and B, or of A and B and C, etc. is deleterious. In the affirmative case, the agencies with regulatory powers would take steps to prevent the proliferation of the noxious agents in our environment.

As just described, the problem of detecting deleterious synergisms is of particular interest to regulatory agencies. However, obviously, from the point of view of pure Science, beneficial synergisms are of equal interest.

The purpose of the present paper is to sketch the history of the problem of mathematical statistics generated by the above substantive problem. As I see it, this history begins with the remarkable insight of Fisher [2], in the middle 1920's and encompasses our present efforts in Berkeley begun in the Spring of 1975, with quite a few individuals involved. Although a

*This investigation was jointly supported by the National Institutes of Health, research grant No. ES01299-13, and the U. S. Energy Research and Development Agency in the Statistical Laboratory of the University of California, Berkeley.

substantial number of interesting results have been attained, and here I am indebted to Henry Scheffé for important references, the theoretical optimality problem required by the substantive problem of synergistic effects continues to be unsolved. Here, the connection with confidence regions and with multiple comparison problems is important.

2. *Early times*. In statistical parlance, "synergism" means "interaction." The honor of being the first to become aware of the phenomenon of synergisms, strongly enough to invent the term "interactions", seems to belong to R. A. Fisher. The following passage, quoted from Yates [13], indicates Fisher's conviction [2] that a properly organized experiment should present Nature not just one question, but a whole carefully thought out questionnaire:

> No aphorism is more frequently repeated in connection with (agricultural) field trials than that we must ask Nature few questions, or ideally, one question at a time. The writer is convinced that this view is wholly mistaken. Nature, he suggests, will best respond to a logical and carefully thought out questionnaire...

Subsequent pages of Yates' paper indicate what the intended questionnaire might look like. Discussing agricultural trials, which at the time were the main concern of both Fisher and Yates, Yates visualized factorial experiments. With three treatments designated by the letters $\underline{n}$, $\underline{p}$, and $\underline{k}$, compared with a control designated by (1), the type of questionnaire embodied in a factorial experiment included the following differences:

$$\begin{aligned} &\underline{n} - (1), \\ &\underline{np} - \underline{p}, \\ &\underline{nk} - \underline{k}, \\ &\underline{npk} - \underline{pk}. \end{aligned} \tag{2.1}$$

Successively, these differences represent the effect of treatment $\underline{n}$ applied alone, the effect of the same treatment $\underline{n}$ applied in the presence of $\underline{p}$, in the presence of $\underline{k}$ and in the presence of both $\underline{p}$ and $\underline{k}$. The average of these four differences is labelled the "main effect" of $\underline{n}$, denoted by N. The differences exemplified by

$$(\underline{np} - \underline{p}) - [\underline{n} - (1)], \tag{2.2}$$

perhaps multiplied by a norming factor, are described as interactions. It is such interactions that are fundamental in the current concern with synergistic effects of various agents.

Clearly, the theoretical-statistical counterpart of a questionnaire of the above type represents a particular case of what we now call the multiple comparison problem. It would be too much to expect that in the 1920's or 1930's this problem could have been properly formulated. Thus, after quoting Fisher's passage about the questionnaire, Yates appears to have drifted away from the idea and his subsequent discussion was directed towards a much less ambitious goal: the summary assessment of particular treatments concerned, considered one at a time.

The mathematical assumptions used were those of the analysis of variance: normal distribution of all the observed variables, with unknown population means and a common, though unknown, variance. Yates' suggestion was that "in the absence of consistent interactions ... over a series of experiments and of any knowledge of when they are likely to occur, the average response to $\underline{n}$ (that is, the "main effect" N) ... and the average depression with $\underline{k}$ ... are the best estimates to adopt in assessing the advantages of these fertilizers."

Here it is important to notice that, if the "true values" of all the interactions are really zero, then the factorial design ensures a considerable saving in allowing several treatments to be assessed in a single experiment. However, the "true" interactions could be different from zero without this fact being detected and could vitiate the assessments of the treatments. The question to consider is how frequently this could occur. The concept of power of a test was introduced [6] only two years before the presentation of the paper by Yates and at the time did not penetrate the thinking of the statistical community. In particular, neither Yates nor Fisher had the idea that, if the state of nature happens to be unfavorable, then the methodology described could lead to most undesirable consequences occuring with an embarrassing frequency. Simple approximate calculations of power I performed showed that this is a real possibility, and Table 1, reproduced from [5] illustrates the situation.

Table 1: Results of sampling experiment: 'main' effect of a and interactions a×b and a×c in 30 hypothetical experiments

	'Main' effect a	S.D.(a) -	Interaction a×b	Interaction a×c	S.D. (interaction)	
'True' values	1.755	0.4082	1.545	1.545	0.8165	
No.			Sample values			
1	1.64 SS	0.38	0.44	2.44 SS	0.76	
2	2.15 SS	.34	1.49 S	0.69	.69	
3	1.50 SS	.43	2.36 S	2.13 S	.87	
4	1.61 SS	.36	1.86	2.38 SS	.73	
5	1.50 SS	.48	2.03	2.13 S	.96	
6	1.90 SS	.31	2.83 SS	0.56	.62	
7	0.89	.42	1.86 S	0.88	.83	
8	1.77 SS	.42	1.80	2.20 S	.83	
9	1.66 SS	.30	0.44	1.11	.60	*
10	1.34 SS	.43	1.94 S	0.94	.87	
11	1.87 SS	.36	3.29 SS	1.62 S	.73	
12	1.08	.52	1.20	0.86	1.04	
13	2.01 SS	.25	-0.09	2.05 SS	0.50	
14	2.34 SS	.38	0.72	1.12	.76	*
15	2.66 SS	.40	1.72	1.55	.80	*
16	2.62 SS	.49	1.99	0.75	.99	*
17	1.38 SS	.36	0.50	2.40 SS	.73	
18	1.58 SS	.49	0.51	2.26 S	.99	
19	1.82 SS	.38	0.75	1.61	.76	*
20	1.69 SS	.49	1.03	1.60	.99	*
21	1.93 SS	.38	1.57	2.32 S	.76	
22	2.16 SS	.28	2.18 SS	-0.52	.55	
23	0.45	.30	2.18 SS	1.04	.60	
24	1.68 SS	.38	1.40	-0.41	.76	*
25	1.68 SS	.38	1.08	1.60	.76	*
26	1.76 SS	.36	0.93	2.99 SS	.73	
27	1.45 SS	.40	1.62	2.04 S	.80	
28	1.47 SS	.28	2.00 SS	1.25 S	.55	
29	2.29 SS	.38	1.13	1.36	.76	*
30	2.44 SS	.45	0.98	2.86 SS	.90	

Notes: (i) S means significant at 5%; (ii) SS means significant at 1%; (iii) asterisks in the last column mark those experiments in which the main effect of a proved highly significant while the interactions were not significant even at 5%.

The table was modeled on an actual experiment described by Yates using his estimate of error variance, his number of replicates, etc. It refers to three hypothetical treatments, a, b, and c. The "true" state of nature in terms of responses of controls and of the several combinations of treatments was as follows:

	(1) = control	a	b,c,bc	ab,ac,abc
(2.3)	102	98	91	109

The numbers were selected so as to average 100. Obviously, the treatment a applied by itself is deleterious,

(2.4) $$a - (1) = -4.$$

However, in the presence of either b, or c, or bc, the effect of a is positive and large, e.g.

(2.5) $$ab - b = +18.$$

There is therefore a strong interaction (or synergism),

(2.6) $$[ab - b] - [a - (1)] = +22.$$

One might think that interactions of this magnitude would be easily detectable, but they are not. On the other hand, the main effect of a, equal to +12.5 is easily detectable. In consequence, if Nature behaves frivolously, with its true state represented by numbers similar to (2.3), a large proportion of factorial experiments as described by Yates would produce just one significant effect, namely the main effect of a, which would be positive and highly significant (at 1%). The recommended conclusion would be that the only treatment worth talking about is treatment a and that it is beneficial, contrary to the true state of nature. Table 1 shows that in the 30 Monte Carlo experiments I performed, highly misleading outcomes occurred nine times.

As mentioned, it is not surprising that in 1935 the planning of an experiment and of its evaluation was not combined with the examination of power of the tests to be used. During the 40-odd years that elapsed since that time the concept of power became one of the basic subjects taught in relatively elementary university courses. Also, a variety of relevant tables became available. Many books were published on experimental design, including factorial experiments. I looked at some of them but failed to see any kind of warning that the reliance on main effects, even if highly significant, may be dangerous. This seems a little surprising.

All the above refers to the situation where the response variables are normally distributed with the same though unknown variance. Currently, many experiments are performed with binary response variables, e.g. death or survival, etc. Table 2 is due to Traxler [9] who explored factorial experiments conducted not long ago by an important scientist. The purpose of the experiments was to discover the most advantageous combination of different therapeutic treatments to be used against a particular disease of man. Here, then, the possibility of synergistic effects has been explicitly considered. The response variable was of binary type: presence or absence of an improvement in a patient's health. In consequence, the variance of the response variable could not be assumed constant, and this could make a difference.

The hypothetical "state of nature" contemplated by Traxler is characterized by the following probabilities:

	Treatment	(control)	<u>a</u>	<u>b</u>,<u>c</u>,<u>bc</u>	<u>ab</u>,<u>ac</u>,<u>abc</u>
(2.7)	Success probability	0.6	0.4	0.2	0.8

It will be seen that treatment <u>a</u> applied singly is deleterious: with this treatment the probability of "success", that is the probability of an improvement in a patient's health, is only 0.4, while that corresponding to "control", that is with no treatment at all, is 0.6. Treatments <u>b</u> and <u>c</u> and also their combination <u>bc</u> are even more deleterious. On the other hand, the combinations of treatments ab, ac and abc are beneficial; with these combinations, the probability of improving the health of the patient is 0.8. The reader will realize that the question of how realistic all these assumptions are is not the subject of the present discussion. This discussion is concerned solely with the effectiveness of the experimental design combined with statistical analysis. With particular reference to treatment a Traxler explored the situation empirically.

Table 2: Summary of 100 Monte Carlo simulations of binary factorial experiments, reprinted from [9], p. 294, by courtesy of Marcel Dekker, Inc.

	Percent of simulated experiments which resulted in:					
	"Very wrong conclusion" regarding treatment a		Failure to find interaction significant at 5% level			
			a × b		a × c	
No. of Patients per treament category	χ^2method	logistic	χ^2method	logistic	χ^2method	logistic
8	19	21	65	70	68	71
16	4	4	43	48	49	51
32	0	1	9	17	8	12

Table 2 summarizes the 100 Monte Carlo experiments performed by Traxler. Here "very wrong conclusion" regarding treatment a means the outcome of the experiment with the main effect of a being positive and highly significant, and with no other significant effects. The first line of the table assumes only 8 patients per experimental category, the number used by the experimentalist. The other two lines illustrate the effect of doubling and of redoubling of the number of patients.

It is seen that, if the true state of nature happens to be similar to that visualized by Traxler, and if the analysis of data is performed in the traditional manner in terms of main effects, etc., then about 20% of experiments, with only 8 patients per category, would lead to results deserving the description disastrous: instead of providing the patients an 80% chance of improvement in their health (using combinations of treatments ab, or ac or abc), they would be given treatment a alone, with the probability of improvement of only 0.4, one-half of that possible and even less than that without any treatment at all! In the same conditions, the strong synergisms of the treatments would be missed very frequently.

Clearly, Tables 1 and 2 indicate the necessity of an intense effort at a more satisfactory design and evaluation of experiments involving synergistic effects.

3. *Bartlett's priority in treating a questionnaire relating to binary variables.* The first attempt at a questionnaire treatment of a binary data problem appears to be due to M.S. Bartlett. The relevant publication [1], under the telling title, "Contingency table interactions," appeared in 1935, next to the paper by Yates discussed above. Bartlett is explicit in acknowledging the inspiration of Fisher.

The paper, developing a modification of the chi-square technique, includes a numerical example concerned with survival or death of certain plants subjected to certain possibly interacting treatments. As in the case of Yates, power considerations are missing. The discussion of the experiment with the survival of plants ends with the conclusion: "In any case it (the chi-square) reveals no evidence of any interaction." The question of how large this interaction could have been with little chance of being detected is not considered. Also, effects of particular treatments and of their interactions are treated, not simultaneously, but one by one.

4. *Enter Hotelling and confidence regions.* Although not specifically stated by Fisher, his idea of a questionnaire to be submitted to Nature tends to imply the appropriateness of some kind of list of answers, all covered by a single overall probability statement as to their overall reliability. As illustrated above, the treatments of the problem both by Yates, presuming normality of the observable variables, and by Bartlett, referring to

binaries, did not involve the assessment of the overall reliability of all the findings.

The priority with regard to summary assessment of reliability of all the findings in response to a questionnaire appears to belong to Harold Hotelling. Here two papers have to be mentioned. In a paper of 1929, coauthored with Working [12], a particular problem was treated. Next in a paper of 1931, by Hotelling alone [3], primarily concerned with the distribution of Hotelling's T-statistic, the last paragraph reads as follows:

> To means of a single variate it is customary to attach a 'probable error', with the assumption that the difference between the true and calculated values is almost certainly less than a certain multiple of the probable error. A more precise way to follow out this assumption would be to adopt some definite level of probability, say P = .05, of a greater discrepancy, and to determine from a table of Student's distribution the corresponding value of t, which will depend on n; adding and subtracting the product of this value of t by the estimated standard error would give upper and lower limits between which the true values may with the given degree of confidence be said to lie. With T and exactly analogous procedure may be followed, resulting in the determination of an ellipse or ellipsoid centered at the point $\xi_1, \xi_2, \ldots, \xi_p$. Confidence corresponding to the adopted probability P may then be placed in the proposition that the set of true values is represented by a point within this boundary.

Here, then if a screening experiment is performed, yielding observations on controls, on responses to several treatments applied singly and on responses to the various combinations of these treatments, all supposed to be normally distributed, the procedure invented by Hotelling provides the possibility of filling out the relevant questionnaire with a preassigned probability that all the answers are correct.

Although Hotelling's work fits the situation involved in the search for synergisms, his motivations were very different: "coefficients of racial likeness", problem of testing a prediction of relativity theory, a comparison of prices of a list of commodities at two epochs, etc. The specific subject of Hotelling's study was a set of p possibly correlated normal variables "each of which is measured for N individuals ..."

5. *Developments after World War II.* The most important development in mathematical statistics that occurred since World War II is, undoubtedly, the book Statistical Decision Functions by Abraham Wald [11]. One might say that the appearance of this book altered fundamentally the whole structure of the theory of statistics. However, just because of all-inclusiveness and of great generality confined within its 179 pages, the monograph could not deal with specific, particular problems like experimentation to discover synergisms.

New theoretical-statistical developments particularly relevant to the substantive problem of synergisms appear to have occurred in 1953 at the hands of Scheffé [7] and of Tukey [10], more than a quarter of a century since Fisher's advocacy of a questionnaire.

Scheffé's paper appears as a direct descendant of Hotelling's paper [3], but the results of Scheffé are much more general. Rather than being concerned with a confidence region for several population means of correlated normal variables, Scheffé's confidence regions relate simultaneously to all "contrasts", linear combinations of such means.

Tukey's procedure is based on a different distributional result: the Studentized range of a sample from a normal population. According to Scheffé [8, p. 83], Tukey's procedure, allowing for the estimation of only a few selected contrasts, may result in shorter individual confidence intervals, jointly corresponding to the preassigned confidence coefficient.

The works of Scheffé and Tukey marked the beginning of an epoch of strong interest in the multiple decision problem. An excellent critical examination of many studies is made available by Rupert Miller [4] in a nice monograph published ten years ago. Here the concepts of power and, more generally, of performance characteristics of particular procedures, play an important role.

This being a half-century since Fisher's pronouncement of the questionnaire principle, one might expect that by now all the relevant theoretical problems would be solved and the appropriate techniques would be developed and embedded in manuals. The current theoretical research interests could then be devoted to novel domains, not dreamed about in the 1920's. Remarkably, this is not the case. The June 1976 issue of the JASA contains two articles relating to multiple comparisons and/or simultaneous confidence intervals, and the March issue of the same journal contains seven of them!

6. *What is the outstanding theoretical-statistical problem related to screening of agents?* Let us consider a laboratory routinely engaged in screening what we shall describe as "agents", say $A_1, A_2, \ldots, A_n$. These may be food additives or cosmetic preparations expected to be put on the market, etc. However they may also be environmental pollutants, perhaps stemming from industrial establishments or from motor cars. Some of these pollutants may be radioactive. In each case there would be the presumption that exposures of the living organisms to the agents studied, either singly or in various combinations, may be deleterious to health. Let X denote some conventionally defined measure of the state of health of the experimental organism used in the laboratory. Conceivably, this may mean life span or blood pressure, etc. Whatever the definition of X, its values corresponding to particular members of the population studied will vary and it is unavoidable to consider X as a random variable with its distribution, say F, depending upon the environment, whether "normal", that is in the absence of the agents A or in the presence of any one of them singly or in some combinations. In the past it was customary to assume that F is normal with unknown means and variance, the latter being the same for all conditions contemplated. This may have been satisfactory in many cases. However, when units of observations are experimental animals, the situation is likely to be more complicated. What to assume about the nature of the

function F in all these conditions is a problem by itself. However, this problem is not the subject of the present discussion and we shall assume that it is realistically solved.

Now, it is necessary to consider a measure of the possible health effects of the agents studied. Here it is convenient to select a few, perhaps one or two conventional measures on "goodness" of health conditions in the population studied. One possibility is to consider the expectation of X. Another possibility is the variance of X or its square root, the SD. If X designates, say, the life span of the experimental mouse, then the finding of an experiment might be expressed in terms of "mean life shortening", etc. The suggestion of using the SD of X is related to the possibility that the population of experimental mice is really a mixture of strains that react differently to the agents studied. Here, then, the effect of an agent A may be an increase in the SD.

Obviously, the definition of "goodness" of health conditions must be chosen in consultations with knowledgeable biologists. From now on, we shall assume that this problem is solved and will denote by θ_0 the measure corresponding to the control conditions, and by θ_i that influenced by the ith of the n agents screened and, in addition, of a certain number of their combinations that may be considered relevant. Here, then, $i = 1,2,..,n,...,N$. For each i, the quantity of interest may be the difference $\theta_i - \theta_0$ or some other selected function, such as $100[\theta_i-\theta_0]/\theta_0$, etc. In the following only the differences $\theta_i - \theta_0$ will be discussed.

Still another symbol has to be defined. This is a conventional limit $\Delta > 0$, between changes in the goodness of health conditions that are "trivial" and those that are not. Here Tukey's word "allowance" comes to my mind. How large should Δ be in any particular case is again a matter for discussion between experimentalist-biologists and statisticians, particularly with reference to the numbers of animals to be used and the cost of experimentation.

With all the above choices, of the F's, of θ and of Δ, determined, the proper laboratory's report on screening of $A_1, A_2, \ldots, A_n$ agents and of their selected combinations (totalling N), might look as follows:

The screening of agents $A_1, A_2, \ldots, A_n$ indicated that
(i) the first k of these agents appear harmless; the deleterious health effects (if any) of these agents taken singly or of their combinations (specify, e.g. combinations of any two of them) cannot be more than "trivial" (specify, e.g. $\theta_1 - \theta_0 \leq \Delta$). The probability of at least one error in this multiple assertion does not exceed α (e.g. $\alpha = 0.01$, or $\alpha = 0.05$, or $\alpha = 0.10$, etc.).
(ii) Without risking an error with probability greater than α, no such statements can be made regarding the remaining n-k agents, $A_{k+1}, A_{k+2}, \ldots, A_n$.

I emphasize the difference between the above statement (i) and the frequently seen statement, say (i*), "no significant effects were found". Obviously, as illustrated by Tables 1 and 2, the statement (i*) is not good enough. In common parlance, statement (i) may be viewed as the result of testing the hypothesis symbolized by $\theta_i - \theta_0 > \Delta$ for a number of values of i, and of its rejection with the joint level of significance α. On the other hand, statement (i*) is comparable to the joint failure to reject the hypothesis symbolized by $\theta_i = \theta_0$.

The above statement might represent a preliminary report on screening the agents A. Its consequence might be to absolve the first k agents from suspicion of being deleterious singly and in the specified combinations. A subsequent report, relating the remaining n-k agents may conceivably specify that some of them taken singly are "innocent", but in particular combinations, possibly involving some of the first k agents, are deleterious. Further studies may reveal that, say, A_k is an "initiator" of carcinogenesis and A_{k+1} the "promoter", etc. The general problem of screening agents is very complex indeed.

Obviously, the development of an optimal methodology to formulate the reports of the above kind would require quite a bit of preliminary research. Some of it would have to be theoretical, but there would be also a substantial experimental part, beginning with the choice of distribution functions F. There are here many important ramifications, such as the estimation of the number of animals to be used, the decision about the number of agents to be studied in a single trial, the determination of a realistic value of Δ, etc.

7. *Concluding remarks.* As mentioned earlier, the problem of a reliable needed methodology to reach conclusions of the type described in the preceeding section was initiated in our Statistical Laboratory in the Spring of 1975. As of now, an interesting paper by Robert Bohrer has been submitted for publication. Another paper is now in the process of being typed. It was prepared jointly by Robert Bohrer (apparently the leader), by V. M. Joshi, and by Ray Faith, supported by graduate students W. Chow and C. Wu.

I am looking forward to seeing the results obtained.

References

[1] M. S. Bartlett (1935). Contingency table interactions. *J. Royal Stat. Soc. Supplement*, Vol. 2, pp. 248-252.

[2] R. A. Fisher (1926). The arrangement of field experiments. *J. of Min. Agric.* 33, 505-515.

[3] H. Hotelling (1931). The generalization of Student's ratio. *Ann. Math. Stat.*, Vol. 2, pp. 360-378.

[4] R. G. Miller, Jr. (1966). Simultaneous Statistical Inference. McGraw-Hill, New York.

[5] J. Neyman (1935). Contribution to the discussion of the paper by F. Yates, 'Complex experiments'. *J. Royal Stat. Soc. Supplement*, Vol. 2, pp. 235-243. See also: A selection of early statistical papers of J. Neyman, Univ. of California Press, Berkeley (1967), pp. 225-232.

[6] J. Neyman and E. S. Pearson (1933). On the problem of the most efficient tests of statistical hypotheses. *Philos. Trans. Royal Soc. of London*, Ser. A, Vol. 231, pp. 289-337. See also: Joint Statistical Papers of J. Neyman and E. S.

Pearson, Univ. of California Press, Berkeley (1967), pp. 140-185.

[7] H. Scheffé (1953). A method for judging all contrasts in the analysis of variance. *Biometrika*, Vol. 40, pp. 87-104.

[8] H. Scheffé (1959). The analysis of variance, Wiley, New York.

[9] R. M. Traxler (1976). A snag in the history of factorial experiments, in On the history of statistics and probability, D. B. Owen, ed., Marcel Dekker, Inc., New York, pp. 281-296.

[10] J. W. Tukey (1953). The problem of multiple comparisons. Dittoed manuscript of 396 pages, Princeton University.

[11] A. Wald (1950). Statistical Decision Functions. Wiley, New York.

[12] H. Working and H. Hotelling (1929). Application of the theory of error to the interpretation of trends. *JASA Supplement*, Vol. 24, pp. 73-85.

[13] F. Yates (1935). Complex experiments. *J. Royal Stat. Soc. Supplement*, Vol. 2, pp. 181-247.

ESTIMATING COVARIANCES IN A MULTIVARIATE NORMAL DISTRIBUTION

By I. Olkin and J. B. Selliah
Stanford University
University of Sri Lanka

1. *Introduction*. This paper is concerned with estimating certain functions of the covariance matrix $\Sigma = (\sigma_{ij})$ in a multivariate normal distribution. We consider the following problems:

(i) Estimation of tr $A\ \Sigma$ when the mean vector is known and the matrix A is symmetric. In particular, we may obtain estimators of σ_{ii} and σ_{ij} by a suitable choice of the matrix A.

(ii) Estimation of det Σ when the mean vector is known.

(iii) Estimation of a variance when the mean is unknown.

In case (i) we show that tr $AV/(n+2)$, where V is the sample cross product matrix, is admissible for squared error loss functions, (perhaps more aptly called quadratic loss or normalized squared error loss), and is minimax for a particular loss function. We also show that the natural estimator of det Σ is minimax.

Stein and James [10] have shown that the natural estimator of Σ is not a minimax estimator for the loss function

$$L[d(V),\Sigma] = \text{tr}\ \Sigma^{-1}d(V) - \log\det \Sigma^{-1}d(V) - p,$$

and have explicitly obtained a minimax estimator of Σ. Eaton [2] discussed this and other problems in Covariance Estimation. Efron and Morris [3] show that for $p \geq 2$

$$(n-p-1)V^{-1} + \frac{(p^2+p-2)I}{\text{tr}\ V}$$

is a uniformly better estimator of Σ^{-1} than the best unbiased estimator $(n-p-1)V^{-1}$. The loss function used is

$$L[d(V),\Sigma] = \frac{\text{tr}[(d(V)-\Sigma^{-1})^2 V]}{n\ \text{tr}\ \Sigma^{-1}}\ .$$

We use the loss function

$$L[d(V),\Sigma] = \mathrm{tr}[\Sigma^{-1}d(V)-I]^2,$$

and obtain a minimax estimator (not the natural estimator) of Σ. We also obtain a minimax estimator of Σ^{-1} for the loss function

$$L(D(V),\Sigma) = \mathrm{tr}(\Sigma\, D(V)-I)^2.$$

In the univariate case, Stein [9] has shown that $s^* = \sum_1^n (x_i-\bar{x})^2/(n+1)$ is not an admissible estimator of the variance σ^2 for squared error loss functions when the mean is unknown. Shorrock and Zidek (1974) obtain the generalizations of this result for the problem of estimating Σ. We show that under certain restrictions on the bias, s^* is both admissible and minimax.

2. *Preliminaries*. Let $x_1, x_2, \ldots, x_n$ be n independent observations from a p-variate normal distribution with known mean vector μ and covariance matrix Σ. The sample cross product matrix $V = \sum_{i=1}^n (x_i-\mu)'(x_i-\mu)$ is a sufficient statistic for Σ and has the Wishart distribution $\mathcal{W}(\Sigma,p,n)$ with density function given by

$$p(V) = c(p,n)(\det \Sigma)^{-n/2}(\det V)^{(n-p-1)/2}\mathrm{etr}(-\tfrac{1}{2}\Sigma^{-1}V), \tag{2.1}$$

where

$$[c(p,n)]^{-1} = 2^{np/2}\,\pi^{p(p-1)/4}\prod_{i=1}^p \Gamma\left(\frac{n-i+1}{2}\right). \tag{2.2}$$

We state some well-known theorems (suitably extended) which we use in proving either admissibility or minimax property of an estimator. If t is an estimator of $\theta \in \Theta$, its bias function is defined by $b_t(\theta) \equiv E(t)-\theta$.

THEOREM 2.1. (Hodges and Lehmann [4]). <u>Suppose that for all estimators</u> f <u>of</u> θ, <u>the inequality</u>

$$\mathrm{Var}(f) \geq [1 + b_f'(\theta)]^2/E_\theta[\{\tfrac{\partial}{\partial\theta}\log p_\theta(x)\}^2] \tag{2.3}$$

<u>is satisfied, where</u> $b'(\theta)$ <u>denotes the derivative of</u> b <u>with respect to</u> θ. <u>If</u> g(x) <u>is an estimator of</u> θ <u>for which</u> (2.3) <u>becomes an</u>

equality and if for every bias function $b(\theta)$,

$$(2.4)\quad b^2(\theta) + \frac{[1 + b'(\theta)]^2}{E_\theta\{\frac{\partial}{\partial\theta} \log p_\theta(x)\}^2} \le b_g^2(\theta) + \frac{[1 + b_g'(\theta)]^2}{E_\theta\{\frac{\partial}{\partial\theta} \log p_\theta(x)\}^2},$$

for all $\theta \varepsilon \Theta$,

implies that $b(\theta) \equiv b_g(\theta)$, then $g(x)$ is admissible for squared error loss functions.

The basic idea of this procedure is to find a lower bound, $\ell(b,\theta)$ for the variance of the estimator of θ, and to see whether the bound is attained for some estimator $g(x)$. If the bound is attained by $g(x)$, and if

$$b^2(\theta) + \ell(b,\theta) \le b_g^2(\theta) + \ell(b_g,\theta), \text{ for all } \theta \varepsilon \Theta,$$

implies that $b(\theta) \equiv b_g(\theta)$, then $g(x)$ is an admissible estimator of θ. In Theorem 2.1, the lower bound used is the Cramér-Rao bound; however, bounds such as the Bhattacharya, Chapman-Robbins, or Kiefer bounds could also be used. In the multiparameter case we use a multiparameter version of the Cramér-Rao bound; this bound involves the information matrix.

We also extend Karlin's Theorem [5] on admissibility as follows:

THEOREM 2.2. If in a subspace $\Omega = (\underline{\omega}_\Omega, \bar{\omega}_\Omega)$ of the parameter space Θ, the probability density function with respect to a measure μ is $p(x,\omega) = B(\omega) \exp[\omega t(x)]$, then $t(x)/(\lambda+1)$ is an admissible estimator of $E_\omega t(x)$ for squared error loss, when the following two conditions are satisfied:

$$\text{(i)}\quad \int_c^b [B(\omega)]^{-\lambda}\, d\omega \to \infty \text{ as } b \to \bar{\omega}_\Omega,$$

$$\text{(ii)}\quad \int_a^c [B(\omega)]^{-\lambda}\, d\omega \to \infty \text{ as } a \to \underline{\omega}_\Omega,$$

where c is an interior point of $(\underline{\omega}_\Omega, \bar{\omega}_\Omega)$, and $\underline{\omega}_\Omega$, $\bar{\omega}_\Omega$ are the lower and upper bounds of Ω, respectively.

The extension follows from the facts that (a) if there is a

"unique admissible" estimator in a subspace Ω of the parameter space Θ, then it is also admissible in Θ, and (b) if there is an estimator $g(x)$ whose risk is less than or equal to that of $t(x)/(\lambda+1)$ in this subspace, then $g(x) = t(x)/(\lambda+1)$ a.e.

The principle of invariance (Kiefer [6]) is also used to obtain minimax estimators. (The solvable group version has been proved by Kiefer [7].)

THEOREM 2.3. *If d is a minimax estimator among the class of estimators invariant under a solvable group of transformations leaving the problem invariant, then it is a minimax estimator among all estimators.*

3. *Estimation of covariances.* We first discuss the estimation of tr $A\Sigma$ where A is a symmetric matrix. Results on the estimation of a single covariance follow as a particular case. We also discuss estimating the generalized variance.

Estimation of tr $A\Sigma$. Any symmetric matrix A can be written as

$$A = B \begin{pmatrix} I_k & 0 & 0 \\ 0 & -I_\ell & 0 \\ 0 & 0 & 0 \end{pmatrix} B',$$

where B is nonsingular, k and ℓ are the number of positive and negative characteristic roots of A, respectively. Let $S = B'VB$ and $\Psi = B'\Sigma B$. Then S has a Wishart distribution $\mathscr{W}(\Psi,p,n)$. The problem now is to estimate

$$\operatorname{tr} A\Sigma = \operatorname{tr} \begin{pmatrix} I_k & 0 & 0 \\ 0 & -I_\ell & 0 \\ 0 & 0 & 0 \end{pmatrix} \Psi = \sum_1^k \psi_{ii} - \sum_{k+1}^{k+\ell} \psi_{ii}.$$

THEOREM 3.1. *If A is a symmetric* $p \times p$ *matrix, then* $\operatorname{tr}AV/(n+2)$ *is an admissible estimator for quadratic loss and is minimax for the loss function*

$$L[d(V),\Sigma] = [d(V)-\operatorname{tr} A\Sigma]^2/\{2n \operatorname{tr} A\Sigma A\Sigma + 4(\operatorname{tr} A\Sigma)^2\}.$$

PROOF. Case (i). $k > 0$, $\ell > 0$. Consider the subspace

$$\Omega = \left\{ \Psi^{-1} = \begin{pmatrix} (1-t)I_k & 0 & 0 \\ 0 & (1+t)I_\ell & 0 \\ 0 & 0 & I \end{pmatrix} \right\} .$$

In this subspace, the probability density function is

$$p_\Omega(S)=c(1-t)^{\frac{nk}{2}}(1+t)^{\frac{n\ell}{2}}|S|^{\frac{n-p-1}{2}}\exp\{-\tfrac{1}{2}[(1-t)\text{tr } S_{11}+(1+t)\text{tr } S_{22}$$
$$\text{tr } S_{33}]\},$$

where

$$S = \begin{pmatrix} S_{11} & S_{12} & S_{13} \\ S_{12}' & S_{22} & S_{23} \\ S_{13}' & S_{23}' & S_{33} \end{pmatrix} ,$$

S_{11} is order $k \times k$ and S_{22} is order $\ell \times \ell$.

Now

$$p_\Omega(S) = c(1-t)^{\frac{nk}{2}}(1+t)^{\frac{n\ell}{2}} e^{\frac{1}{2}(\text{tr } S_{11}-\text{tr } S_{22})t} f(S).$$

In applying Karlin's Theorem, if we make the correspondence

$\omega \leftrightarrow t/2$, $B(\omega) \leftrightarrow (1-t)^{\frac{nk}{2}}(1+t)^{\frac{n\ell}{2}}$, $\lambda = \frac{2}{n}$, the natural parameter space being $-1 < t < 1$, it is clear that $n(\text{tr } S_{11}-\text{tr } S_{22})/(n+2)$ is an admissible estimate of $E[\text{tr } S_{11}-\text{tr } S_{22}]$. Now

$$\text{tr } S_{11}-\text{tr } S_{22} = \text{tr}\begin{pmatrix} I_k & 0 & 0 \\ 0 & -I_\ell & 0 \\ 0 & 0 & 0 \end{pmatrix} S = \text{tr}\begin{pmatrix} I_k & 0 & 0 \\ 0 & -I_\ell & 0 \\ 0 & 0 & 0 \end{pmatrix} B'VB$$

$$= \text{tr } B\begin{pmatrix} I_k & 0 & 0 \\ 0 & -I_\ell & 0 \\ 0 & 0 & 0 \end{pmatrix} B'V = \text{tr}AV.$$

Hence $\text{tr } AV/(n+2)$ is an admissible estimator of $\text{tr } A\Sigma$ for quadratic loss.

Case (ii). $k > 0$, $\ell = 0$. Consider the subspace

$$\Omega = \left\{ \Psi^{-1} = \begin{pmatrix} (1-t)I_k & 0 \\ 0 & I \end{pmatrix} \right\} ,$$

the natural parameter space being $(-\infty,1)$ and the proof is similar to case (i).

Case (iii). $k = 0$, $\ell > 0$. Consider the subspace

$$\Omega = \left\{ \Psi^{-1} = \begin{pmatrix} (1+t)I_\ell & 0 \\ 0 & I \end{pmatrix} \right\},$$

the natural parameter space being $(-1,\infty)$ and the proof is again similar to case (i).

In all cases, since the risk is constant for the given loss function, the estimator is also minimax.

Estimation of a single covariance. Our result here is as follows.

THEOREM 3.2. <u>The statistic</u> $v_{ij}/(n+2)$ <u>is an admissible and minimax estimator of</u> σ_{ij} <u>for the loss function</u>

$$L[d(V),\Sigma] = [d(V)-\sigma_{ij}]^2/g(\Sigma),$$

<u>where</u>

$$g(\Sigma) = n(\sigma_{ii}\sigma_{jj}+\sigma_{ij}^2) + 4\sigma_{ij}^2.$$

PROOF. Case (i) $i = j$. Let $A = (a_{ij})$ where $a_{ii} = 1$ and all other elements of A are zero. The result follows from Theorem 3.1.

Case (ii) $i \neq j$. Let $A = (a_{ij})$ where $a_{ij} = a_{ji} = \frac{1}{2}$ and all other elements of A are zero. The result follows from Theorem 3.1.

REMARK. We note that for $i = j$, $g(\Sigma)$ can be chosen to be just σ_{ii}^2. That $v_{ii}/(n+2)$ is a minimax estimator for the given loss function can also be proved by using the sequence of a priori distributions in which the σ_{ij}'s are independently distributed with marginal distributions

$$p(\sigma_{jk}=0) = 1 \text{ for } j \neq k,\ p(\sigma_{kk} = 1) = 1 \text{ for } k \neq i,$$

$$p(\sigma_{ii}) = \frac{C(\theta)}{\sigma_{ii}^{(M+1)/M}} \exp\left(\frac{-\theta}{2\sigma_{ii}}\right), \qquad \theta > 0.$$

By computing the Bayes estimator and showing that its risk tends to that of $v_{ii}/(n+2)$ as $M \to \infty$, we see that it is minimax.

COROLLARY 3.3. *Let* $a = (a_1, a_2, \ldots, a_p)$, $b = (b_1, b_2, \ldots, b_p)$, *then* $aVb'/(n+2)$ *is an admissible estimator of* $a\Sigma b'$ *for quadratic loss.*

PROOF. $V^* = \binom{a}{b} V(a', b')$ has a Wishart distribution $\mathcal{W}(\Sigma^*, 2, n)$, where $\Sigma^* = \binom{a}{b} \Sigma (a' \cdot b')$.

Estimation of the generalized variance. If $\mathcal{L}(V) = \mathcal{W}(\Sigma, p, n)$, then $E \det V = [\prod_1^p (n-i+1)] \det \Sigma$. We now show that $(\det V)/\prod_1^p (n-i+3)$ is a minimax estimator of $\det \Sigma$ for the loss function

$$L[d(V), \Sigma] = \left(\frac{d(V)}{\det \Sigma} - 1\right)^2 .$$

To see this, consider the group of lower triangular transformations $V \to AVA'$, and the induced group of transformations in the parameter space $\Sigma \to A\Sigma A'$. This group is solvable. Since $\det A\Sigma A' = (\det A)(\det \Sigma)(\det A)$, invariance of an estimator d under transformation A implies

$$d(AVA') = (\det A) d(V) (\det A) = (\det A)^2 d(V),$$

for all $V > 0$. In particular, when $V = I$,

$$d(AA') = (\det A)^2 d(I) = k(\det A)^2, \tag{3.5}$$

where k is a constant. For any $V > 0$, there is a lower triangular matrix T, such that $V = TT'$, so that by letting $A = T$ in (3.5), we obtain

$$d(V) = d(TT') = k(\det T)^2 = k(\det V).$$

The risk of an estimator $k(\det V)$ is given by

$$\frac{\{k(\det V) - (\det \Sigma)\}^2}{(\det \Sigma)^2} = k^2 \frac{c(p,n)}{c(p,n+4)} - 2k \frac{c(p,n)}{c(p,n+2)} + 1,$$

and is minimized by $k = \dfrac{c(p,n+4)}{c(p,n+2)} = \prod_{i=1}^{p} (n-i+3)^{-1}$. Thus $(\det V)/\prod_1^p (n-i+3)$ is a minimax estimator of $\det \Sigma$ among the class of estimators invariant with respect to the group of lower triangular transformations. Since this group is solvable, the result follows by Theorem 2.3.

A minimax estimator of the covariance matrix. In this section we obtain a minimax estimator of Σ for the loss function

$$L[d(V),\Sigma] = \mathrm{tr}(\Sigma^{-1}d(V)-I)^2,$$

and in the course of the proof, observe that the natural estimator is not minimax. The problem is invariant under the transformation $V \to AVA'$, $\Sigma \to A\Sigma A'$, $d(V) \to A\,d(V)A'$, where A is a lower triangular matrix. The condition of invariance of an estimator d under the transformation by the matrix A is that

$$d(AVA') = A\,d(V)A',$$

for all $V > 0$. In particular, for $V = I$, we have that $d(AA') = A\,d(I)A'$. When A is a diagonal matrix, say D_ϵ, with diagonal elements $\epsilon_i = \pm 1$, then $AA' = I$, so that $d(I) = D_\epsilon\,d(I)D_\epsilon$, for all D_ϵ. This implies that $d(I)$ is diagonal. For every $V > 0$, there is a lower triangular matrix T, such that $V = TT'$, and hence

$$d(V) = TD_\delta T',$$

where $D_\delta \equiv \mathrm{diag}(\delta_1,\delta_2,\ldots,\delta_p)$. The group of lower triangular matrices operates transitively on the parametric space, so that the risk of an invariant procedure d is constant. Thus we need compute the risk only for $\Sigma = I$, which is given by

$$\begin{aligned} R[d(V)] &= E\,\mathrm{tr}[d(V)-I]^2 \\ &= E\,\mathrm{tr}(TD_\delta T'\;TD_\delta T')-2E\,\mathrm{tr}\,TD_\delta T' + p \\ &= E\,\mathrm{tr}[D_\delta(T'T)D_\delta(T'T)]-2E\,\mathrm{tr}\,D_\delta(T'T)+p. \end{aligned}$$

If we set $T'T = M$, then

$$m_{ii} = \sum_{j\geq i} t_{ji}^2,\quad m_{i\alpha} = \sum_{j\geq\alpha} t_{ji}\,t_{j\alpha} \text{ for } \alpha > i.$$

The elements of T are independently distributed with $\mathcal{L}(t_{ii}^2) = \mathcal{L}(\chi^2_{n-i+1})$, $\mathcal{L}(t_{ij}) = N(0,1)$, $i > j$, so that

$$Em_{ii} = E\,\chi^2_{n-i+1} + E\,\chi^2_{p-i} = n+p-2i+1,$$

$$Em_{ii}^2 = (n+p-2i+1)(n+p-2i+3),$$

$$Em_{i\alpha}^2 = E\, t_{\alpha\alpha}^2\, t_{\alpha i}^2 + \sum_{j>\alpha} (t_{j\alpha}^2 t_{ji}^2) + \sum E(t_{j\alpha}\, t_{ji}\, t_{k\alpha}\, t_{ki})$$

$$= (n-\alpha+1) + p-\alpha+0 = n+p-2\alpha+1, \text{ for } \alpha > i.$$

Thus

$$E \operatorname{tr} D_\delta M = \sum_i E\, \delta_i\, m_{ii} = \sum_i \delta_i\, (n+p-2i+1),$$

$$E \operatorname{tr}(D_\delta M\, D_\delta M) = \sum_{i,\alpha} E\, \delta_i \delta_\alpha\, m_{i\alpha}^2$$

$$= \sum_i \delta_i^2 (n+p-2i+1)(n+p-2i+3) + 2 \sum_{\alpha>i} \delta_i \delta_\alpha (n+p-2\alpha+1),$$

and

$$R[d(V)] = \sum_j \delta_j^2 (n+p-2j+1)(n+p-2j+3) + 2 \sum_{j<\alpha} \delta_j \delta_\alpha \{n+p-2\alpha+1\}$$
$$-2 \sum_j \delta_j (n+p-2j+1) + p.$$

In order to minimize $R[d(V)]$, differentiate $R[d(V)]$ with respect to the δ's and equate the derivatives to zero. The minimizing δ's, denoted δ^*, are given by the solution to the linear equations

$$(\delta_1^*, \ldots, \delta_p^*) B = (n+p-1,\ n+p-3, \ldots, n-p+1),$$

where $B = (b_{ij})$ is a symmetric matrix with $b_{ii} = (n+p-2i+1)(n+p-2i+3)$, and $b_{ij} = n+p-2j+1$ for $j > i$.

The matrix B can be written as the sum of two positive definite matrices, and hence is nonsingular. Further, an iterative procedure shows that the solution is nonnegative. Thus $TD_{\delta^*}T'$ is a minimax estimator among estimators invariant with respect to lower triangular transformations. Since this group is solvable it follows by Theorem 2.3, that $TD_{\delta^*}T'$ is also minimax among all estimators. Clearly the natural estimator kV where k is a constant, is not minimax.

For the case $p = 2$, the minimizing solution can be explicitly obtained:

$$\delta_1^* = \frac{(n+1)^2-(n-1)}{[(n+1)^2(n+2)-(n-1)]}, \quad \delta_2^* = \frac{(n+1)(n+2)}{[(n+1)^2(n+3)-(n-1)]}.$$

Estimating the inverse of the covariance matrix. Consider estimating Σ^{-1} for the loss function

$$L[D(V),\Sigma] = \mathrm{tr}[\Sigma\, D(V)-I]^2.$$

The problem is invariant under the transformation $V \to AVA'$, $\Sigma \to A\Sigma A'$, $D(V) \to (A')^{-1}D(V)A^{-1}$, where A is an arbitrary non-singular p×p matrix. The condition of invariance of an estimator D under the transformation by the matrix A is

$$D(AVA') = (A')^{-1}D(V)A^{-1}.$$

We restrict ourselves to the group of transformations where A is lower triangular; setting V = I we have that

$$D(AA') = (A')^{-1}D(I)A^{-1}.$$

When $A = D_\epsilon$ is diagonal with ±1 on the diagonal, it follows that

$$D(I) = D_\epsilon^{-1}D(I)D_\epsilon^{-1},$$

which implies that d(I) is diagonal. Since there is a lower triangular matrix T such that $V = TT'$, we deduce that

$$D(V) = (T')^{-1}\ \mathrm{diag}(\delta_1,\delta_2,\ldots,\delta_p)T^{-1}.$$

As in the estimation of Σ, we need to compute the risk only for $\Sigma = I$. The minimax invariant estimate is obtained by finding those values of δ which minimize the risk and this will be a minimax estimate among all estimators by Theorem 2.3.

For the case p = 2,

$$D(V) = \begin{pmatrix} \frac{1}{t_{11}} & -\frac{t_{21}}{t_{11}t_{22}} \\ 0 & \frac{1}{t_{22}} \end{pmatrix} \begin{pmatrix} \delta_1 & 0 \\ 0 & \delta_2 \end{pmatrix} \begin{pmatrix} \frac{1}{t_{11}} & 0 \\ -\frac{t_{21}}{t_{11}t_{22}} & \frac{1}{t_{22}} \end{pmatrix},$$

$$D(V)-I = \begin{pmatrix} \frac{\delta_1}{t_{11}^2} + \frac{\delta_2 t_{21}^2}{t_{11}^2 t_{22}^2} - 1 & -\frac{\delta_2 t_{21}}{t_{11}t_{22}^2} \\ -\frac{\delta_2 t_{21}}{t_{11}t_{22}^2} & \frac{\delta_2}{t_{22}^2} - 1 \end{pmatrix},$$

$$\text{risk} = E\left(\frac{\delta_1}{t_{11}^2} + \frac{\delta_2 t_{21}^2}{t_{11}^2 t_{22}^2} - 1\right)^2 + \frac{2\,\delta_2^2 t_{21}^2}{t_{11}^2 t_{22}^4} + \left(\frac{\delta_2}{t_{22}^2} - 1\right)^2$$

$$= \delta_1^2 \frac{1}{(n-2)(n-4)} + \frac{\delta_2^2(n-1)}{(n-2)(n-4)(n-5)} - \frac{2\delta_1}{n-2} - \frac{2\delta_2(n-1)}{(n-2)(n-3)}$$

$$+ \frac{2\delta_1\delta_2}{(n-2)(n-4)(n-3)} + 2.$$

The minimizing δ's are given by

$$\delta_1^* = \frac{(n-4)(n-1)[n^2-7n+14]}{[(n-3)^2(n-1)-(n-5)]},$$

$$\delta_2^* = \frac{(n-2)(n-3)(n-4)(n-5)}{[(n-3)^2(n-1)-(n-5)]}.$$

It is clear that the natural estimator is not minimax.

4. *Estimation of the variance when the mean is unknown in a univariate normal distribution.* Suppose $x_1, x_2, \ldots, x_n$ are independently and identically distributed, each having a $N(\mu, \sigma^2)$ distribution. It has been shown by Stein (1964) that $s^* = \sum_{i=1}^{n} (x_i - \bar{x})^2/(n+1)$ is not an admissible estimator of σ^2 for squared error loss functions. However, it is admissible among some subclasses of estimators, namely, those for which the bias is a multiplicative or additive function of μ and σ.

THEOREM 4.1. The estimator s^* is an admissible estimator of σ^2 among estimators whose bias function $b(\sigma^2, \mu)$ is of the form $b(\sigma^2, \mu) = c(\mu)d(\sigma^2)$ or $b(\sigma^2, \mu) = c(\mu) + d(\sigma^2)$.

PROOF. Since the Cramér-Rao bound is not attained by s^* we use the first Bhattacharya bound,

$$\text{Var}\, t \geq \frac{b^2(\theta,\mu)}{\theta^2} + (1 + b_\theta,\ b_{\mu\mu}) \begin{pmatrix} n/2 & n \\ n & 2n^2 \end{pmatrix}^{-1} \begin{pmatrix} 1 + b_\theta \\ b_{\mu\mu} \end{pmatrix},$$

where $\theta \equiv \sigma^2$, $b_\theta = \frac{\partial b}{\partial \theta}$ and $b_{\mu\mu} = \frac{\partial^2 b}{\partial \mu^2}$. The bias of s^* is $b_{s^*}(\theta) =$

$-2\theta/(n+1)$, and its risk is $2/(n+1)$. We must show that the inequality

$$\frac{b^2(\theta,\mu)}{\theta^2} + (1+b_\theta,\ b_{\mu\mu})\begin{pmatrix} n/2 & n \\ n & 2n^2 \end{pmatrix}^{-1}\begin{pmatrix} 1+b_\theta \\ b_{\mu\mu} \end{pmatrix} \le \frac{2}{n+1},$$

implies that $b(\theta) = -2\theta/(n+1)$. This inequality reduces to

$$\frac{b^2}{\theta^2} + 2\,\frac{(1+b_\theta)^2}{n-1} + \frac{b_{\mu\mu}^2}{2n(n-1)} - \frac{2(1+b_\theta)b_{\mu\mu}}{n(n-1)} - \frac{2}{n+1} \le 0. \tag{4.1}$$

Considering (4.1) as a quadratic in $b_{\mu\mu}$, the discriminant must be non-negative, i.e.,

$$\frac{(1+b_\theta)^2}{n^2(n-1)^2} - \frac{1}{2n(n-1)}\left\{\frac{b^2}{\theta^2} + 2\,\frac{(1+b_\theta)^2}{n-1} - \frac{2}{n+1}\right\} \ge 0, \tag{4.2}$$

or equivalently

$$2\,\frac{(1+b_\theta)^2}{n} + \frac{b^2}{\theta^2} \le \frac{2}{n+1}\ .$$

Similarly as a quadratic in $(1+b_\theta)^2$, we obtain

$$\frac{b_{\mu\mu}^2}{2n^2} + \frac{b^2}{\theta^2} \le \frac{2}{n+1}\ . \tag{4.3}$$

From (4.2), it is clear that $b_\theta < 0$.

Case (i). $b(\theta,\mu) = c(\mu)d(\theta)$. Suppose for some $\mu = \mu_0$, $b_{\mu\mu} \ne 0$. Then (4.3) reduces to

$$\frac{d^2(\theta)c_{\mu\mu}^2}{2n^2} + \frac{c^2(\mu)d^2(\theta)}{\theta^2} \le \frac{2}{n+1}\ .$$

This implies that $|d(\theta)| < k_1\theta$ for some $k_1 > 0$ and also $|d(\theta)| < k_2$ for some $k_2 > 0$. From the fact that $b_\theta < 0$, we have that $c(\mu)d_\theta < 0$. If $c(\mu_0) > 0$, then $d_\theta < 0$ for all θ. Consequently $d(\theta)$ decreases as θ increases and is bounded below by k_2. This is possible only if $d_\theta \to 0$ as $\theta \to \infty$, which would contradict (4.2). Therefore, $b_{\mu\mu} = 0$ for all θ,μ. The inequality (4.1) reduces to

$$(4.4) \qquad \frac{b^2}{\theta^2} + \frac{2(1+b_\theta)^2}{n-1} \leq \frac{2}{n+1} .$$

But (4.4) is exactly an inequality treated by Hodges and Lehmann [4], who show that (4.4) implies that $b(\theta,\mu) = -2\theta/(n+1)$.

Case (ii). $b(\theta,\mu) = c(\mu) + d(\theta)$. Inequalities (4.3) and (4.2) reduce to

$$(4.5) \qquad \frac{c_{\mu\mu}^2}{2n^2} + \frac{[c(\mu)+d(\theta)]^2}{\theta^2} \leq \frac{2}{n+1} ,$$

$$(4.6) \qquad \frac{2(1+d_\theta)^2}{n} + \frac{[c(\mu)+d(\theta)]^2}{\theta^2} \leq \frac{2}{n+1} .$$

It is clear from (4.5) that $c(\mu) = -\lim_{\theta\to 0} d(\theta)$; thus $c_{\mu\mu} = 0$. The inequality (4.1) reduces to

$$\frac{d^2(\theta)}{\theta^2} + \frac{2(1+d_\theta)^2}{n-1} \leq \frac{2}{n+1} \text{ for all } \theta,$$

which is identical to (4.4), from which it follows that $b(\theta,\mu) = -2\theta/(n+1)$. ||

Acknowledgment. The authors are grateful to J. Zidek for his comments and suggestions, and to L. Brown for providing substantial improvements in one of the proofs.

References

[1] Anderson, T. W. (1958). *An Introduction to Multivariate Statistical Analysis*, John Wiley and Sons, Inc., New York.

[2] Eaton, M. L. (1970). Some problems in covariance estimation. (Preliminary report) Technical report No. 49 NSF Grant GP17122, Dept. of Statistics, Stanford University.

[3] Efron, B. and Morris, C. (1974). Multivariate Empirical Bayes and estimation of covariance matrices. The Rand Paper Series P-5276, 1-21. Rand Corporation.

[4] Hodges, J. L. Jr., and Lehmann, E. L. (1951). Some applications of the Cramér-Rao inequality. *Proc. Second Berkeley Symp. Math. Statist. Prob.* 13-22. University of California Press.

[5] Karlin, S. (1958). Admissibility for estimation with quadratic loss. *Ann. Math. Statist.* 29, 406-436.

[6] Kiefer, J. (1957). Invariance, minimax sequential estimation and continuous time processes. *Ann. Math. Statist.* 28, 573-601.

[7] Kiefer, J. (1966). Multivariate optimality results. In P. R. Krishnaiah (Ed.) *Multivariate Analysis*, 255-274. Academic Press, New York.

[8] Shorrock, R. W. and Zidek, J. V. An improved estimator of the generalized variance. Technical Report, University of Montreal, Centre de Researche Mathematiques.

[9] Stein, C. (1964). Inadmissibility of the usual estimate for the variance of a normal distribution with unknown mean. *Ann. Inst. Statist. Math.* 16, 155-160.

[10] Stein, C., and James, W. (1960). Estimation with quadratic loss. *Proc. Fourth Berkeley Symp. Math. Statist. Prob.* 1, 424-443. University of California Press.

SIMULTANEOUS ESTIMATION OF PARAMETERS - A COMPOUND DECISION PROBLEM

By C. Radhakrishna Rao
Indian Statistical Institute, New Delhi

1. *Introduction.* The origin of simultaneous estimation of parameters can be traced to the papers by Fairfield Smith [8], Hazel [9] and Panse [16] on the construction of selection indices in genetics based on a method proposed by R. A. Fisher. The problem was as follows. Let $\underset{\sim}{\theta}$ and $\underset{\sim}{y}$ be vector variables representing the genotypic and phenotypic values of an individual, of which $\underset{\sim}{\theta}$ is unobservable and $\underset{\sim}{y}$ has the decomposition $\underset{\sim}{y} = \underset{\sim}{\theta} + \underset{\sim}{\varepsilon}$ with $E(\underset{\sim}{\varepsilon}) = 0$ and $\text{Cov}(\underset{\sim}{\theta},\underset{\sim}{\varepsilon}) = 0$. The genetic worth of an individual is measured by $\underset{\sim}{p}'\underset{\sim}{\theta}$ where $\underset{\sim}{p}$ is a given vector. Suppose we draw k individuals from a population with values $(\underset{\sim}{\theta}_1,\underset{\sim}{y}_1),\ldots,(\underset{\sim}{\theta}_k,\underset{\sim}{y}_k)$. What is the best way of estimating $\underset{\sim}{p}'\underset{\sim}{\theta}_1,\ldots,\underset{\sim}{p}'\underset{\sim}{\theta}_k$ for purposes of comparing and selecting individuals with the largest genetic values? Under the assumed model, $\underset{\sim}{p}'\underset{\sim}{y}_i$ is an unbiased estimate of $\underset{\sim}{p}'\underset{\sim}{\theta}_i$. However, Fisher suggested the regression estimate of $\underset{\sim}{p}'\underset{\sim}{\theta}_i$ given $\underset{\sim}{y}_i$ which involved the knowledge of the expectation of $\underset{\sim}{\theta}$ and covariance matrices of $\underset{\sim}{\theta}$ and $\underset{\sim}{\varepsilon}$. When repeated observations on $\underset{\sim}{y}$ are available for each $\underset{\sim}{\theta}$, all the unknown parameters can be estimated by analysis of variance and covariance, in which case an empirical regression estimate (ERE) of $\underset{\sim}{p}'\underset{\sim}{\theta}_i$ given $\underset{\sim}{y}_i$ (or $\bar{y}_i$ if more observations are available) can be obtained by substituting the estimates for the unknown parameters. The method of constructing such ERE's and associated inference problems are discussed in a paper by the author [17]. Further work in this direction considering more general situations was done by Henderson [10], [11]. It has been pointed out by Bartlett in the discussion on a paper by Stein [29] that regression estimators have also been suggested by Sir Godfrey Thompson in his book on The Factorial Analysis of Human Ability (e.g., 5th ed., pp. 353 and 360).

The empirical Bayes estimation procedure suggested by Robbins

[26] is analogous to Fisher's method used in a more general context. Let (θ,y) represent a parameter θ and an observation y whose distribution depends on θ and $T(F,y)$ be the Bayes estimator of θ for a given loss function and a given prior distribution F of θ. In some situations it may be possible to estimate F using only past data on the observable variable y. If $\hat{F}$ is an estimate of F, then the empirical Bayes estimate (EBE) of θ is $T(\hat{F},y)$. In many problems $T(F,y)$ depends on certain parameters of F in which case we need only estimate these parameters from past data for substitution in $T(F,y)$.

In a remarkable paper, Stein [28] showed that the sample mean of observations from a k (>2) variate normal distribution is not an admissible estimator of the population mean with respect to a quadratic loss function. For instance if $x_i \sim N(\mu_i,1)$, $i=1,\ldots,k$, and x_i are independent, then

$$t_i^* = \left(1 - \frac{k-2}{\Sigma x_i^2}\right) x_i, \quad i = 1,\ldots,k \ (>2) \tag{1.1}$$

has a smaller average mean square error as an estimator of $(\mu_1,\ldots,\mu_k)$ when compared with the usual estimator $(x_1,\ldots,x_k)$ whatever may be the true value of $(\mu_1,\ldots,\mu_k)$. From this result it was suggested that an estimator of the type (1.1) is preferable to $(x_1,\ldots,x_k)$.

There have been a number of contributions extending the result of Stein, searching for other classes of admissible estimators and advocating the use of Stein type estimators in practice. Principal contributors to this area are Baranchik [1], Bhattacharya [2], Bock [4], Efron and Morris [5], [6], [7], James and Stein [12], Shinozaki [27] and Strawderman [30].

The object of the present paper is to review the results of Stein and others in the framework of ERE and EBE of Fisher and Robbins, provide extensions to cases not considered by earlier writers, and examine their usefulness in practice in estimating several parameters (simultaneous estimation problem) and in ranking individuals with respect to unobservable intrinsic traits (selection problem).

The problem of simultaneous prediction of future observations in linear models was examined on similar lines in an earlier paper (Rao, [23]).

2. *The problem of simultaneous estimation.* We consider a (p+q)-vector random variable (θ, y) where θ is p dimensional and y is q dimensional. Let $(\theta_1, y_1), \ldots, (\theta_k, y_k)$ be k independent samples on (θ, y), but with only the y_i values observed. The problem is to estimate $\theta_1, \ldots, \theta_k$ or a subset $\theta_i, \theta_{i+1}, \ldots, \theta_k$.

There are various approaches to the problem depending on the purpose for which estimates are required and the information available on the joint distribution of (θ, y). Let $p(\theta, y | \phi)$ represent the density of (θ, y) depending on an unknown vector parameter ϕ.

Then we have the following decompositions

$$p(\theta, y | \phi) = p_1(\theta | \phi) p_2(y | \theta, \phi) = p_3(y | \phi) p_4(\theta | y, \phi). \tag{2.1}$$

The first decomposition in (2.1) gives rise to likelihoods of the following types

$$\prod_{i=1}^{k} p_1(\theta_i | \phi) p_2(y_i | \theta_i, \phi) \tag{2.2}$$

$$\prod_{i=1}^{k} p_2(y_i | \theta_i, \phi) \tag{2.3}$$

from any one of which the unknowns θ_i and ϕ may be estimated by maximum likelihood or some appropriate method. It may be noted that the alternative (2.3) does not depend on the marginal distribution of θ_i. The relative merits of using the likelihoods (2.2) and (2.3) will be examined in Section 3.

Another approach is provided by the second decomposition in (2.1). If ϕ is known we need only consider the conditional density

(2.4) $$\prod_{i=1}^{k} p_4(\underset{\sim}{\theta}_i | \underset{\sim}{y}_i, \underset{\sim}{\phi})$$

for drawing inferences on $\underset{\sim}{\theta}_i$. If $\underset{\sim}{\phi}$ is unknown we may estimate $\underset{\sim}{\phi}$ using the likelihood

(2.5) $$\prod_{i=1}^{k} p_3(\underset{\sim}{y}_i | \underset{\sim}{\phi})$$

and obtain the empirical conditional density

(2.6) $$\prod_{i=1}^{k} p_4(\theta_i | y_i, \hat{\phi})$$

substituting the estimate $\hat{\underset{\sim}{\phi}}$ for the unknown ϕ in (2.4). The conditional mean of $\underset{\sim}{\theta}_i$ computed from (2.5) provides an ERE of $\underset{\sim}{\theta}_i$ under quadratic loss function.

Faced with an unknown $\underset{\sim}{\phi}$, Bayesians would consider a prior distribution of $\underset{\sim}{\phi}$, say with density $p_0(\underset{\sim}{\phi})$ and use the decomposition

(2.7) $$p_0(\underset{\sim}{\phi}) \prod_{i=1}^{k} p(\underset{\sim}{\theta}_i, \underset{\sim}{y}_i | \underset{\sim}{\phi})$$
$$= p_5(\underset{\sim}{y}_1, \ldots, \underset{\sim}{y}_k) p_6(\underset{\sim}{\theta}_1, \ldots, \underset{\sim}{\theta}_k | \underset{\sim}{y}_1, \ldots, \underset{\sim}{y}_k) p_7(\underset{\sim}{\phi} | \underset{\sim}{\theta}_i, \underset{\sim}{y}_i, i=1, \ldots, k).$$

Then the conditional joint density of $\underset{\sim}{\theta}_1, \ldots, \underset{\sim}{\theta}_k$

(2.8) $$p_6(\underset{\sim}{\theta}_1, \ldots, \underset{\sim}{\theta}_k | \underset{\sim}{y}_1, \ldots, \underset{\sim}{y}_k)$$

which does not involve unknown elements, is used for drawing simultaneous inference on $\underset{\sim}{\theta}_1, \ldots, \underset{\sim}{\theta}_k$. We do not consider the Bayesian approach in the present paper but refer the readers to papers by Lindley [14] and Lindley and Smith [15] for details.

We shall consider various special cases and examine the relative merits of different methods of estimation.

3. *Direct and inverse regression estimates.* Let us consider a vector random variable

(3.1) $$(\theta, y_1, \ldots, y_n)$$

such that distribution of y_i given θ are independent and each has a normal distribution with mean θ and variance σ_1^2. Further let us suppose that the marginal distribution of θ is normal with mean μ and variance σ_2^2. We have a sample of k individuals with values

$$(3.2)\qquad \begin{array}{c}(\theta_1, y_{11}, \ldots, y_{n1})\\ \cdot \;\; \cdots \;\; \cdot\\ (\theta_k, y_{1k}, \ldots, y_{nk})\end{array}$$

but only with y_{ij} values observed (known) and the problem is to estimate (or predict) the unobserved values $\theta_1, \ldots, \theta_k$.

Case 1: $\mu, \sigma_1^2, \sigma_2^2$ *are known*. If μ, σ_1^2 and σ_2^2 are known, the regression of θ on $y_1, \ldots, y_n$ is

$$(3.3)\qquad \begin{aligned}\hat{\theta} &= \mu + \beta(\bar{y}-\mu)\\ &= (1-\beta)\mu + \beta\bar{y}\\ &= \bar{y} - (1-\beta)(\bar{y}-\mu)\end{aligned}$$

where $\beta = \sigma_2^2/(\sigma_2^2+\sigma_e^2)$, $\sigma_e^2 = \sigma_1^2/n$. The direct regression estimator (3.3) of θ is thus a linear combination of the general average μ and the observed average $\bar{y}$ for a given individual. On the other hand, the regression of $\bar{y}$ on θ is θ itself so that $\bar{y}$ is the inverse regression estimate of θ. Let us note that (3.3) is the best linear estimator under quadratic loss function without normality assumption.

Under the mean squared error loss function the risk associated with direct and inverse regression estimators are respectively

$$(3.4)\qquad E[(1-\beta)\mu + \beta\bar{y} - \theta]^2 = \sigma_e^2\sigma_2^2/(\sigma_e^2 + \sigma_2^2), \text{ and}$$

$$(3.5)\qquad E(\bar{Y} - \theta)^2 = \sigma_e^2.$$

Then the average risk in estimating θ_i by $(1-\beta)\mu + \beta\bar{y}_i$, $i=1,\ldots,k$ is (3.4) and in estimating θ_i by $\bar{y}_i$ is (3.5), and (3.4) is less than (3.5) unless σ_e is zero. Thus it appears that the knowledge that an individual belongs to a population with known characteristics enables us to obtain a better estimator for the individual.

The risks (3.4) and (3.5) give the average mean square error over all individuals of a population, i.e., in routine estimation for individuals drawn from a given population. In some problems it may be of interest to compute the risk for a given individual. We compute the following conditional risks, i.e., for an individual with given θ.

$$E(\bar{y}|\theta) = \theta \quad (3.6)$$

$$E[(\bar{y}-\theta)^2|\theta] = \sigma_e^2 \quad (3.7)$$

$$E[(1-\beta)\mu + \beta\bar{y}|\theta] = \theta - (1-\beta)(\theta-\mu) \quad (3.8)$$

$$E\{[(1-\beta)\mu+\beta\bar{y}-\theta]^2|\theta\}=\sigma_e^2[\beta^2+(1-\beta)^2\delta^2], \delta=(\mu-\theta)/\sigma_e. \quad (3.9)$$

It is seen from (3.8) that if $|\theta-\mu|$ is large there is a serious bias in the direct regression estimator resulting in over estimation if $\theta-\mu$ is positive and under estimation if $\theta-\mu$ is negative. Further (3.7) < (3.9) for large values of $|\theta-\mu|$ indicating that for some individuals inverse regression estimator $\bar{y}$ may be better than the direct regression estimator (3.3). The superiority of the latter over the former is realized only when the risk is averaged over all individuals of the population. We may call (3.4) statistician's loss and (3.9) individual's loss. In the usual decision theory we try to minimize (3.4) ignoring (3.9), i.e., minimize overall loss, which may be appropriate in some situations. But such an approach may impose heavy losses on certain individuals which may not be desirable (see Rao [21]). It appears that we should look for decision rules which achieve some sort of balance between individual's loss and statistician's loss. One method for achieving this is given in Rao [20].

Case 2: $\mu,\sigma_1^2,\sigma_2^2$ are unknown. If μ, σ_1^2, σ_2^2 are unknown, we can obtain their estimates from the observed values (y_{ij}) in (3.2) provided $n > 1$. For this purpose we consider the statistics

$$\bar{y}_{..} = \Sigma\Sigma y_{ij}/kn \quad , E(\bar{y}_{..}) = \mu \quad (3.10)$$

$$B = \Sigma(\bar{y}_{.i} - \bar{y}_{..})^2 \quad , E(B) = (\sigma_2^2 + \sigma_e^2)(k-1) \quad (3.11)$$

$$W = \Sigma\Sigma(y_{ij} - \bar{y}_{.i})^2 \qquad , \ E(W) = k(n-1)n\sigma_e^2$$

Under the normality assumption $B/(\sigma_2^2 + \sigma_e^2)$ and $W/n\sigma_e^2$ are distributed independently as chi-squares on k-1 and k(n-1) degrees of freedom respectively. In our analysis we shall assume that a statistic W is available such that it is distributed independently of B, and $W/n\sigma_e^2$ is a chi-square variable on s degrees of freedom.

The ERE (empirical regression estimator) of θ_i

$$\tilde{\theta}_i = \bar{y}_{.i} - \frac{cW}{B}(\bar{y}_{.i} - \bar{y}_{..}), \ i=1,\ldots,k \tag{3.12}$$

is obtained by substituting the estimates $\bar{y}_{..}$ and cW/B (for a suitable choice of c) for μ and $(1-\beta)$ respectively in the regression estimator (3.3). The estimator (3.12) is a special case of a vector estimator obtained by Fairfield Smith [8], Panse [16] and Rao [17]. In these papers c is chosen as (k-1)/nk(n-1). A slightly different choice of c seems to have some advantage in small samples. Rao [17] also considered the problem when the sample size n was not the same for all individuals.

Now under the assumptions made on the joint distribution of $(\theta, y_1, \ldots, y_n)$,

$$E\sum_1^k (\tilde{\theta}_i - \theta_i)^2 = k\sigma_e^2 + g\sigma_e^4/(\sigma_2^2 + \sigma_e^2) \tag{3.13}$$

where

$$g = \frac{c^2 s(s+2)n^2}{k-3} - 2c\,sn \tag{3.14}$$

where s is the degrees of freedom of W. The optimum choice of c, obtained by minimizing g, is (k-3)/[(s+2)n] in which case (3.12) can be written as

$$\tilde{\theta}_i = \bar{y}_{.i} - \frac{k-3}{n(s+2)}\frac{W}{B}(\bar{y}_{.i} - \bar{y}_{..}), \ i=1,\ldots,k \tag{3.15}$$

which is the ERE given in Rao [25]. The expression (3.15) differs from that of James and Stein [12]

$$\hat{\hat{\theta}}_i = \bar{y}_{.i}\left(1 - \frac{k-2}{n(s+2)}\frac{W}{\Sigma\bar{y}_{.i}^2}\right), \quad i = 1,\ldots,k. \tag{3.16}$$

Expressions of the type (3.15) have also been suggested by Efron and Morris [5], [6], [7] and by Lindley in the discussion on the paper by Stein [29]. But the performance characteristic of the ERE as defined in (3.15) compared with (3.16) and other estimators has not been worked out in detail. Some computations were reported in Rao [25].

Substituting the optimum value of c in (3.13) we obtain

$$(3.17)\qquad E\sum_{i}^{k}(\tilde{\theta}_i-\theta_i)^2 = k\sigma_e^2 - \frac{\sigma_e^4}{(\sigma_2^2+\sigma_e^2)}\,\frac{s(k-3)}{s+2}\,.$$

Let us recall that

$$(3.18)\qquad E\sum_{1}^{k}(\hat{\theta}_i-\theta_i)^2 = k\sigma_e^2\sigma_2^2/(\sigma_e^2+\sigma_2^2)$$

$$(3.19)\qquad E\sum_{1}^{k}(\bar{y}_{.i}-\theta_i)^2 = k\sigma_e^2$$

and observe that

$$(3.19) \geq (3.17) \geq (3.18)$$

so that under the loss function assumed, $\tilde{\theta}_i$ is better than $\bar{y}_{.i}$. The difference

$$(3.20)\qquad (3.17) - (3.18) = \frac{k(3n-1)}{s+2}\cdot\frac{\sigma_e^4}{\sigma_e^2+\sigma_2^2}$$

$$\rightarrow \frac{3n-1}{n-1}\,\frac{\sigma_e^4}{\sigma_e^2+\sigma_2^2}\ \text{as } k\rightarrow\infty\ [\text{choosing } s = k(n-1)]$$

is the additional loss due to using estimates of σ_e^2 and σ_2^2 in the place of true values in (3.3). The difference in the average loss

$$(3.21)\qquad \frac{(3.17) - (3.18)}{k} = \frac{(3n-1)}{s+2}\cdot\frac{\sigma_e^4}{\sigma_e^2+\sigma_2^2}$$

$$\rightarrow 0 \text{ as } k\rightarrow\infty\quad [\text{choosing } s = k(n-1)]$$

which is as expected. The difference in average loss when $\tilde{\theta}_i$ are compared with $\bar{y}_{.i}$ is

$$(3.22)\qquad \frac{(3.19)-(3.17)}{k} = \frac{(k-3)s}{k(s+2)} \frac{\sigma_e^4}{\sigma_e^2+\sigma_2^2}$$

$$\to \frac{\sigma_e^4}{\sigma_e^2+\sigma_2^2} \text{ as } k \to \infty \text{ [choosing } s = k(n-1)]$$

which is different from zero. The average loss in using James-Stein estimator (3.16) is, as $k \to \infty$,

$$(3.23)\qquad \sigma_e^2 - \frac{\sigma_e^4}{\sigma_e^2+\sigma_2^2+\mu^2}$$

which is greater than the corresponding expression for the ERE unless $\mu = 0$, where μ is as defined in (3.10). Thus as $k \to \infty$, the efficiency of James-Stein estimator compared to the regression estimator (3.3) is less than unity whereas it is unity for the ERE. Such a deficiency of James-Stein estimator is pointed out by Susarla [32] in a slightly different context, viz., of estimating the parameter for a (k+1)-th individual with an observed value y using the observed data on past k individuals.

We shall now evaluate the risks in $\tilde{\theta}_i$ and $\hat{\hat{\theta}}_i$ for fixed values of θ_i. Using the results of Lemma 2 in the appendix we find

$$(3.24)\qquad E(\tilde{\theta}_i|\theta_1,\ldots,\theta_k) = \theta_i - (\theta_i-\bar{\theta}_.)\frac{s}{s+2}E(\frac{k-3}{2K_1+k-1})$$

$$(3.25)\qquad E[\sum_1^k(\tilde{\theta}_i-\theta_i)^2|\theta_1,\ldots,\theta_k)] = \sigma_e^2[k - \frac{(k-3)^2 s}{s+2} E(\frac{1}{2K_1+k-3})]$$

where K_1 is a Poisson variable with $E(K_1) = \Sigma(\theta_i-\bar{\theta}_.)^2/2\sigma_e^2$. In the case of $\hat{\hat{\theta}}_i$, the corresponding expressions are

$$(3.26)\qquad E(\hat{\hat{\theta}}_i|\theta_1,\ldots,\theta_k) = \theta_i - \theta_i\frac{s}{s+2}E(\frac{k-2}{2K_2+k})$$

$$(3.27)\qquad E[\sum_1^k(\hat{\hat{\theta}}_i-\theta_i)^2|\theta_1,\ldots,\theta_k)] = \sigma_e^2[k - \frac{(k-2)^2 s}{s+2} E(\frac{1}{2K_2+k-2})]$$

where K_2 is a Poisson variable with $E(K_2) = \Sigma\theta_i^2/2\sigma_e^2$. Finally the corresponding results for $\bar{y}_{.i}$ are

(3.28) $$E[\bar{y}_{.i}|\theta_i] = \theta_i.$$

(3.29) $$E[\sum_1^k (\bar{y}_{.i} - \theta_i)^2|\theta_1,\ldots,\theta_k] = k\sigma_e^2.$$

Note that (3.29) $\geq$ (3.25) if $k > 3$ so that $\tilde{\theta}_i$ is better than $\bar{y}_{.i}$ for $k > 3$, whereas (3.29) $\geq$ (3.27) if $k > 2$ so that $\hat{\hat{\theta}}_i$ is better than $\bar{y}_{.i}$ for $k > 2$. However, the difference in the average losses for $\tilde{\theta}_i$ and $\hat{\hat{\theta}}_i$ is

(3.30) $$\frac{s}{s+2}\left[\frac{(k-2)^2}{k} E\left(\frac{1}{2K_2+k-2}\right) - \frac{(k-3)^2}{k} E\left(\frac{1}{2K_1+k-3}\right)\right]\sigma_e^2$$

which is of the order

(3.31) $$E\left(\frac{k-2}{2K_2+k-2}\right) - E\left(\frac{k-3}{2K_1+k-3}\right).$$

The expression (3.31) is negative if θ_i are not close to the origin indicating that pulling the observed averages towards the general average as in the case of $\tilde{\theta}_i$ is better than pulling the averages towards the origin as in the case of $\hat{\hat{\theta}}_i$ unless the θ_i's are all small.

Since (3.29) is larger than (3.25) or (3.27) for given values of $\theta_1,\ldots,\theta_k$, it follows that under the quadratic loss function the estimators $\tilde{\theta}_i$ and $\hat{\hat{\theta}}_i$ are preferable to $\bar{y}_{.i}$ whether θ_i are considered as a <u>random sample from a distribution</u> of θ's or individual parameters associated with possibly <u>unrelated</u> problems. There cannot be any controversy about such a statement once the quadratic loss function is accepted and minimizing the associated risk is chosen as the criterion.

However, the main drawback of the estimators $\tilde{\theta}_i$ and $\hat{\hat{\theta}}_i$ is the large bias that may be introduced in the estimation of parameters as pointed out by Bartlett and others (in the discussion on Efron and Morris [7]) and Rao [20], [21], [25]. The expressions (3.24) and (3.26) give the actual magnitude of bias in the estimates of individual parameters. The bias is large for parameters with high values resulting in underestimation which may not be desirable in many situations. Similarly, there is overestimation in the case of parameters with low values. Efron and Morris [6], [7] suggest some methods for reducing such a bias at the expense of increasing the overall risk. Rao [20] suggested another method. When such modifications are made, the net gain may be small. Further, the compound loss function of the type we are considering may not be appropriate if individual estimates are of interest or if we wish to have uniform loss for the estimation of each parameter. Thus, on the whole the improvement achieved in using estimators of the type (3.15) and (3.16) over the simple averages may not be of any practical value. Further the use of such adjusted values instead of simple averages may be misleading in some situations, as in the illustration below.

We shall consider a problem which I had the occasion to examine some time ago. The problem was to estimate the yields of a certain crop in six different regions of a state. For this purpose 100 random cuts of "crop in a unit area" were made in each region. If y_{ij} represents the yield of the ith cut in the jth region, $\bar{y}_{.j}$ is an unbiased estimator of θ_j the true yield per unit area in region j. Then the total yield in the jth region can be estimated by $\bar{y}_{.j} \times A_j$, where A_j is the known area under the crop in the jth region. Alternatively one may estimate θ_j by $\tilde{\theta}_j$, j = 1,...,6 simultaneously and estimate the yields of the regions by $\tilde{\theta}_j \times A_j$, j=1,...,6. Table 1 below gives the estimates of yields obtained by these two methods. The value of W/n(kn-k+2) with n= 100 and k=6 was 0.2311 and the value of $B/(k-3)= \sum_j(\bar{y}_{.j}-\bar{y}_{..})^2/(6-3)$ calculated from the values in column (2) of the above table is 2.8837. Then using the formula (3.15),

$$\tilde{\theta}_i = 0.0801\ \bar{y}_{..} + 0.9199\ \bar{y}_{.j}$$

$$= .5135 + 0.9199\ \bar{y}_{.j}$$

substituting $\bar{y}_{..} = 6.4117$. The values of $\tilde{\theta}_j$ are shown in column (3).

Table 1: Estimated yield of a crop in different regions

Region	yield per unit area		Area under crop (A)	total yield in 1000 maunds	
	$\bar{y}_{.}$	$\tilde{\theta}$	(in 1000 units)	$A \times \bar{y}_{.}$	$A \times \tilde{\theta}$
1	8.23	8.08	85.84	706.46	693.59
2	5.04	5.15	50.71	255.58	261.16
3	7.88	7.76	62.32	491.08	483.60
4	5.93	5.97	34.87	206.78	208.17
5	5.54	5.61	40.35	223.54	226.36
6	5.85	5.89	18.92	110.68	111.44
Total	38.47	38.46	293.01	1994.12	1984.32

The statistician's job was to estimate yield rates of different regions which were unknown, and the adjustment of the averages ($\bar{y}_{.j}$) to obtain improved estimates ($\tilde{\theta}_j$) was, perhaps, acceptable although the effect was to reduce the higher averages and increase the lower averages. However, the consequences of using such adjusted estimates in conjunction with known areas under crop to compute total yields do not seem to be desirable. It may be noted that certain regions of the state are more fertile than others and in such regions more land will be brought under cultivation. The adjustment has resulted in underestimation of total yield in more fertile regions and overestimation in less fertile regions, and what is more serious is that the yield for the state as a whole is underestimated by about 10,000 maunds. It appears that $\bar{y}_{.j}$, although inadmissible in some sense, provides a good insurance against gross errors involved in using them for a variety of purposes by customers.

Elsewhere (Rao [22]), I have given an example to show how the rate of increase in production of cereals in India over time is underestimated if trend is fitted to data adjusted to begin with by the formula (3.15) "to improve" on the original individual unbiased estimates.

These examples raise serious questions concerning

simultaneous estimation of parameters and compound loss functions.

In reporting results of several independent experiments should we give estimates of each parameter computed from the data of the relevant experiment or provide Stein type estimates using a formula like (3.15). If the latter procedure is recommended, what loss function should one use?

One may argue that, in the above example of estimating the yields in different regions, the parameters of interest are $A_i\theta_i$ and not θ_i, and therefore the loss function should be

$$\Sigma A_i^2 (t_i-\theta_i)^2. \tag{3.32}$$

The effect of minimizing the risk associated with (3.32) would be again to over estimate yields in poorer regions which may not be desirable. Thus when individual parameters are of interest, minimizing the expected value of a compound loss function may not be appropriate.

Further, while reporting the results of several experiments one should envisage a variety of uses of the individual estimates. There may not be a single loss function appropriate for different uses of the estimates. In such a case, it would appear best to report estimates of each parameter (together with a measure of precision) computed from the data of the relevant experiment, and allow the customer to make appropriate adjustment in the estimates, if necessary, for a particular use.

4. *Estimation of parameters with an underlying stochastic structure.* In a recent paper (Rao, [22]), an example was given to show how serious underestimation may occur in the parameter with the largest value by using an estimator of the type (3.15). It was however observed that θ_i refer to values which are expected to increase with i which represented time. The underlying structure of θ_i appeared to be a linear trend with a random component super-imposed on it. Can estimators of the type (3.15) be constructed taking this information into account?

We shall consider an underlying structure of a more general

type for $\theta' = (\theta_1,\ldots,\theta_k)$ in (3.2) and estimate θ using the observations y_{ij} in (3.2). We suppose that θ is a random variable with the structure

$$\theta = X\beta + \eta \tag{4.1}$$

where X is kxm known matrix of rank r, β is an m-vector of unknown parameters and η is a k-vector random variable such that $E(\eta) = 0$ $D(\eta) = \sigma_2^2 I$. In such a case we may write

$$\bar{y} = X\beta + \eta + \varepsilon \tag{4.2}$$

where $\bar{y}'_{.} = (\bar{y}_{.1},\ldots,\bar{y}_{.k})$ are averages of observations in (3.2), and ε is such that $E(\varepsilon) = 0$, $D(\varepsilon) = \sigma_e^2$ and $\text{Cov}(\eta,\varepsilon) = 0$. The model (4.2) was considered by Henderson [10], where η represented the true breeding ability of an individual and $X\beta$, the effect of age, etc. The problem was to predict η given $\bar{y}$.

Our problem is one of estimating $\theta = X\beta + \eta$ using the setup (4.2). If β, σ_2^2, σ_e^2 are known, then the regression of η on $\bar{y}_{.}$ is

$$\frac{\sigma_2^2}{\sigma_e^2 + \sigma_2^2}(\bar{y}_{.} - X\beta) \tag{4.3}$$

as shown by Henderson [10]. Then the estimate of $X\beta + \eta$ is

$$\hat{\theta} = X\beta + \frac{\sigma_2^2}{\sigma_e^2 + \sigma_2^2}(\bar{y}_{.} - X\beta). \tag{4.4}$$

If β, σ_2^2, σ_e^2 are unknown, we consider the following statistics, in addition to W defined in (3.11) for the estimation of σ_e^2:

$$X\hat{\beta} = X(X'X)^{-1}X'\bar{y}_{.}, \quad E(X\hat{\beta}) = X\beta \tag{4.5}$$

$$B = (\bar{y}_{.}-X\hat{\beta})'(\bar{y}_{.}-X\hat{\beta}), \quad E(B) = (k-r)(\sigma_e^2 + \sigma_2^2) \tag{4.6}$$

where r = Rank X. Substituting the estimates for unknowns in (4.4), we obtain the ERE

$$\tilde{\tilde{\theta}} = \bar{y}_{.} - \frac{cW}{B}(\bar{y}_{.} - X\hat{\beta}) \tag{4.7}$$

As in (3.15), the optimum value of c is found to be

(k-r-2)/n(s+2) in which case (4.7) becomes

$$\tilde{\tilde{\theta}} = \bar{y}_{.} - \frac{k-r-2}{n(s+2)} \frac{W}{B} (\bar{y}_{.} - X\hat{\beta}) \tag{4.8}$$

where $X\hat{\beta}$ and B are as defined in (4.5) and (4.6) respectively. Now

$$E[(\tilde{\tilde{\theta}}-\theta)'(\tilde{\tilde{\theta}}-\theta)] = \sigma_e^2[k - \frac{s(k-r-2)}{(s+2)} \frac{\sigma_e^2}{\sigma_2^2 + \sigma_e^2}] \tag{4.9}$$

so that the overall error is less than $k\sigma_e^2$ if $k > r+2$. We shall compute the bias and risk function for $\tilde{\tilde{\theta}}$ for given values of θ. Using the results of Lemma 2 of the appendix

$$E(\tilde{\tilde{\theta}}-\theta|\theta) = -\frac{s}{s+2} E(\frac{k-r-2}{2K_3+k-r}) (I-P)\theta \tag{4.10}$$

$$E[(\tilde{\tilde{\theta}}-\theta)'(\tilde{\tilde{\theta}}-\theta)|\theta] = \sigma_e^2[k - \frac{s(k-r-2)^2}{s+2} E(\frac{1}{2K_3+k-r-2})] \tag{4.11}$$

where K_3 is a Poisson variable with $E(K_3) = \theta'(I-P)\theta/2\sigma_e^2$ and P is the projection operator on the space spanned by the columns of X.

The estimator $\tilde{\tilde{\theta}}$ will have less bias and smaller mean square error than $\tilde{\theta}$ if θ has the decomposition $X\beta + \eta$ and the components of η are small provided $k-r-2 > 0$. Adjustment of the type (4.7) was also considered by Efron and Morris [5].

5. *Gauss-Markoff model.* Consider the Gauss-Markoff model

$$Y = X\beta + \varepsilon \tag{5.1}$$

where β is a random variable such that

$$\beta \sim N_m(i\gamma, \sigma_2^2 F), \quad \varepsilon \sim N_k(0, \sigma_0^2 V) \tag{5.2}$$

and β and ε are independent, where in (5.1), $i' = (1,1,\ldots,1)$. In such as case, when all the parameters σ_0^2, σ_2^2 and γ are known, the regression estimator of β is

$$\beta^{(b)} = \beta^{(\ell)} - \sigma_0^2 U(\sigma_2^2 F + \sigma_0^2 U)^{-1}(\beta^{(\ell)} - i\gamma) \tag{5.3}$$

where $U = (X'V^{-1}X)^{-1}$ and $\beta^{(\ell)} = (X'V^{-1}X)^{-1}X'V^{-1}Y$, the least squares estimator of $\underset{\sim}{\beta}$.

When the parameters are unknown we estimate them from the equations

$$(5.4)\qquad \hat{\gamma} = (\underset{\sim}{i}'\underset{\sim}{U}^{-1}\underset{\sim}{\beta}^{(\ell)}/\underset{\sim}{i}'\underset{\sim}{U}^{-1}\underset{\sim}{i}) \quad \text{where } i'=(1,1,\ldots,1)$$

$$(5.5)\qquad (k-m+2)\hat{\sigma}_0^2 = (\underset{\sim}{Y}-\underset{\sim}{X}\underset{\sim}{\beta}^{(\ell)})'V^{-1}(\underset{\sim}{Y}-\underset{\sim}{X}\underset{\sim}{\beta}^{(\ell)})$$

$$(5.6)\qquad (m-3)\left[\frac{\hat{\sigma}_2^2}{m-1}\left(\mathrm{tr}\underset{\sim}{U}^{-1}\underset{\sim}{F} - \frac{\underset{\sim}{i}'\underset{\sim}{U}^{-1}\underset{\sim}{F}\underset{\sim}{U}^{-1}\underset{\sim}{i}}{\underset{\sim}{i}'\underset{\sim}{U}^{-1}\underset{\sim}{i}}\right)+\hat{\sigma}_0^2\right]=(\underset{\sim}{\beta}^{(\ell)}-\underset{\sim}{i}\hat{\gamma})'\underset{\sim}{U}^{-1}(\underset{\sim}{\beta}^{(\ell)}-\underset{\sim}{i}\hat{\gamma}).$$

Note that the estimating equations (5.5) and (5.6) are not unbiased. Substituting the estimates for the unknowns in (5.3) we obtain the ERE

$$(5.7)\qquad \underset{\sim}{\beta}^{(e)} = \underset{\sim}{\beta}^{(\ell)} - \hat{\sigma}_0^2\underset{\sim}{U}(\hat{\sigma}_2^2\underset{\sim}{F}+\hat{\sigma}_0^2\underset{\sim}{U})^{-1}(\underset{\sim}{\beta}^{(\ell)}-\underset{\sim}{i}\hat{\gamma})$$

The estimator (5.7) differs from those given by Bhattacharya [2], Berger and Bock [3] and Shinozaki [27]. However, unlike the estimators of these authors, $\beta^{(e)}$ as defined in (5.7) need not dominate $\beta^{(\ell)}$. The exact computation of the mean square error for (5.7) is not easy. It would be of interest to compare the performance of (5.7) with those of Bhattacharya and other authors and also with ridge and shrunken estimators (see Rao [24] for details) through simulation studies.

6. *Simultaneous estimation of vector parameters - the problem of selection index.* Consider a 2p-variate normal random vector $(\underset{\sim}{\mu},\underset{\sim}{x})$ such that

$$(6.1)\qquad E(\underset{\sim}{\mu},\underset{\sim}{x}) = (\underset{\sim}{\alpha},\underset{\sim}{\alpha})$$

$$D(\underset{\sim}{\mu},\underset{\sim}{x}) = \begin{pmatrix} \underset{\sim}{\Gamma} & \underset{\sim}{\Gamma} \\ \underset{\sim}{\Gamma} & \underset{\sim}{\Gamma}+\underset{\sim}{\Sigma} \end{pmatrix}$$

Let $(\underset{\sim}{\mu}_1,\underset{\sim}{x}_1),\ldots,(\underset{\sim}{\mu}_k,\underset{\sim}{x}_k)$ represent the values of $(\underset{\sim}{\mu},\underset{\sim}{x})$ on k individuals. We have observations on $\underset{\sim}{x}_i$ only and the problem is to estimate (or predict) $\underset{\sim}{p}'\underset{\sim}{\mu}_i$, a linear function of unobserved $\underset{\sim}{\mu}_i$, defined as the genetic worth of individual $\underset{\sim}{i}$.

It is shown by Henderson [10] that if $c_0 + \underset{\sim}{c}'\underset{\sim}{x}_i$ is the regression of $\underset{\sim}{p}'\underset{\sim}{\mu}_i$ on $\underset{\sim}{x}_i$ then

$$\text{(6.2)} \qquad \text{Prob}(\underset{\sim}{c}'\underset{\sim}{x}_i > \underset{\sim}{c}'\underset{\sim}{x}_j | \underset{\sim}{p}'\underset{\sim}{\mu}_i - \underset{\sim}{p}'\underset{\sim}{\mu}_j = d > 0)$$

$$\geq \text{Prob}(\underset{\sim}{m}'\underset{\sim}{x}_i > \underset{\sim}{m}'\underset{\sim}{x}_j | \underset{\sim}{p}'\underset{\sim}{\mu}_i - \underset{\sim}{p}'\underset{\sim}{\mu}_j = d > 0)$$

for any vector $\underset{\sim}{m}$, so that $\underset{\sim}{c}'\underset{\sim}{x}$ is a good index for comparing and selecting best individuals with respect to the genetic value $\underset{\sim}{p}'\underset{\sim}{\mu}$.

The computational problem of estimating the regression of $\underset{\sim}{p}'\underset{\sim}{\mu}$ on $\underset{\sim}{x}$ (or $\bar{\underset{\sim}{x}}$ which is the mean of repeated observations on $\underset{\sim}{x}$ if available) has been discussed in papers by Fairfield Smith [8], Panse [16], Rao [17] and Henderson [11]. We shall review the previous results and study some optimum properties of these estimators.

Case 1: $\underset{\sim}{\alpha}$, $\underset{\sim}{\Gamma}$, $\underset{\sim}{\Sigma}$ *are known.* The regression of $\underset{\sim}{p}'\underset{\sim}{\mu}$ on $\underset{\sim}{x}$ is $\underset{\sim}{p}'\hat{\underset{\sim}{\mu}}$ where

$$\text{(6.3)} \qquad \hat{\underset{\sim}{\mu}} = \underset{\sim}{\alpha} + \underset{\sim}{\Gamma}(\underset{\sim}{\Gamma}+\underset{\sim}{\Sigma})^{-1}(\underset{\sim}{x}-\underset{\sim}{\alpha})$$

$$= \underset{\sim}{x} - \underset{\sim}{\Sigma}(\underset{\sim}{\Gamma}+\underset{\sim}{\Sigma})^{-1}(\underset{\sim}{x}-\underset{\sim}{\alpha})$$

Now

$$\text{(6.4)} \qquad D(\hat{\underset{\sim}{\mu}}-\underset{\sim}{\mu}) = \underset{\sim}{\Sigma} - \underset{\sim}{\Sigma}(\underset{\sim}{\Gamma}+\underset{\sim}{\Sigma})^{-1}\underset{\sim}{\Sigma},$$

$$\text{(6.5)} \qquad D(\underset{\sim}{x}-\underset{\sim}{\mu}) = \underset{\sim}{\Sigma}.$$

Since the difference of the matrices (6.4) and (6.5) is non-negative definite, $\hat{\underset{\sim}{\mu}}$ is uniformly better than $\underset{\sim}{x}$ as an estimator of $\underset{\sim}{\mu}$ with respect to mean dispersion error (MDE) matrix, and consequently for any given $\underset{\sim}{p}$, $\underset{\sim}{p}'\hat{\underset{\sim}{\mu}}$ is better than $\underset{\sim}{p}'\underset{\sim}{x}$ as an estimator of $\underset{\sim}{p}'\underset{\sim}{\mu}$.

Thus, if $\underset{\sim}{\alpha}$, $\underset{\sim}{\Gamma}$ and $\underset{\sim}{\Sigma}$ are known we may estimate $\underset{\sim}{p}'\underset{\sim}{\mu}_i$ by $\underset{\sim}{p}'\hat{\underset{\sim}{\mu}}_i$ where

$$\text{(6.6)} \qquad \hat{\underset{\sim}{\mu}}_i = \underset{\sim}{\alpha} + \underset{\sim}{\Gamma}(\underset{\sim}{\Gamma}+\underset{\sim}{\Sigma})^{-1}(\underset{\sim}{x}_i-\underset{\sim}{\alpha}), \; i=1,\ldots,k$$

and select individuals on the basis of large values of $\underset{\sim}{p}'\hat{\underset{\sim}{\mu}}_i$. As in the univariate case, $\hat{\underset{\sim}{\mu}}_i$ is a biased estimate of $\underset{\sim}{\mu}_i$ for given $\underset{\sim}{\mu}_i$ and, therefore, $\underset{\sim}{p}'\hat{\underset{\sim}{\mu}}_i$ is a biased estimate of $\underset{\sim}{p}'\underset{\sim}{\mu}_i$. However, bias does not detract the value of $p'\hat{\mu}_i$ as a selection index, although as an estimate of $p'\mu_i$ it may not be useful in certain situations.

Case 2: α, Γ *are unknown and* Σ *is known.* Let us estimate α and Γ by

$$\bar{x} = \Sigma x_i/k \tag{6.7}$$

$$\Sigma + \hat{\Gamma} = f\, \Sigma(x_i-\bar{x})(x_i-\bar{x})' = fB \tag{6.8}$$

where f is a suitably chosen constant. Substituting these estimates in (6.3) for the estimation of μ_i, we obtain

$$\hat{\mu}_i = x_i - c\, \Sigma B^{-1}(x_i-\bar{x}), \quad i = 1,\ldots,k \tag{6.9}$$

where c is a suitable constant. The total MDE matrix for the estimation of $\mu_1,\ldots,\mu_k$ is

$$E\sum_1^k[x_i-\mu_i-c\, \Sigma B^{-1}(x_i-\bar{x})][x_i-\mu_i-c\, \Sigma B^{-1}(x_i-\bar{x})]' \tag{6.10}$$

$$= E\sum_1^k[(x_i-\mu_i)(x_i-\mu_i)' + c^2\, \Sigma B^{-1}\Sigma-2c\Sigma+c\Sigma B^{-1}(x_i-\bar{x})(\mu_i-\bar{\mu})'$$

$$+ c(\mu_i-\bar{\mu})(x_i-\bar{x})'B^{-1}\Sigma]$$

$$= k\Sigma + c^2(k-p-2)^{-1}\Sigma(\Gamma+\Sigma)^{-1}\Sigma - 2c\, \Sigma(\Gamma+\Sigma)^{-1}\Sigma. \tag{6.11}$$

Choosing c = k-p-2 to minimize (6.11), the estimate (6.9) becomes

$$\hat{\mu}_i = x_i - (k-p-2)\, \Sigma B^{-1}(x_i-\bar{x}), \quad i = 1,\ldots,k. \tag{6.12}$$

The total MDE for the estimates (6.12) is

$$k\Sigma - (k-p-2)\Sigma(\Gamma+\Sigma)^{-1}\Sigma \tag{6.13}$$

while the total MDE when Γ and α are known is

$$k[\Sigma-\Sigma(\Gamma+\Sigma)^{-1}\Sigma] \tag{6.14}$$

and the total MDE if μ_i is estimated by x_i is

$$k\Sigma. \tag{6.15}$$

It may be seen that

$$(6.15) > (6.13) > (6.14)$$

so that both the regression and empirical regression estimators are better than the unbiased estimator $\underset{\sim}{x}_i$, i = 1,...,k. The additional loss due to estimation of $\underset{\sim}{\Gamma}$ in empirical regression is $(p+2)\underset{\sim}{\Sigma}(\underset{\sim}{\Gamma}+\underset{\sim}{\Sigma})^{-1}\underset{\sim}{\Sigma}$.

We shall compute the total MDE for (6.9) for given $\underset{\sim}{\mu}_1,\ldots,\underset{\sim}{\mu}_k$. Using the result of Lemma 4 of the appendix, the expectation in (6.10) for given $\underset{\sim}{\mu}_1,\ldots,\underset{\sim}{\mu}_k$ is

$$k\underset{\sim}{\Sigma} + c^2\underset{\sim}{\Sigma}E(\underset{\sim}{B}^{-1}|\underset{\sim}{M})\underset{\sim}{\Sigma} - 2c(k-p-2)\underset{\sim}{\Sigma}E(\underset{\sim}{B}^{-1}|\underset{\sim}{M})\underset{\sim}{\Sigma} \tag{6.16}$$

where $E(\underset{\sim}{B}^{-1}|\underset{\sim}{M})$ stands for expectation of $\underset{\sim}{B}^{-1}$ for given $\underset{\sim}{M}$ = $(\underset{\sim}{\mu}_1,\ldots,\underset{\sim}{\mu}_k)$. The optimum value of c is again k-p-2 so that the estimator is as in (6.12). The total MDE for given $\underset{\sim}{M}$ is

$$k\underset{\sim}{\Sigma} - (k-p-2)^2\underset{\sim}{\Sigma}E(\underset{\sim}{B}^{-1}|\underset{\sim}{M})\underset{\sim}{\Sigma} \tag{6.17}$$

which is smaller than $k\underset{\sim}{\Sigma}$, the total MDE for the estimator $\underset{\sim}{x}_i$, i = 1,...,k.

The results of Sections (6.2) and (6.3) are similar to those of Efron and Morris [5]. These authors were concerned with a quadratic loss function considering $\underset{\sim}{\mu}_1,\ldots,\underset{\sim}{\mu}_k$ as a set of pk parameters. In our problem we consider estimates of $\underset{\sim}{\mu}_i$ as such and the matrix loss function MDE. These estimates automatically provide improved estimates of any linear function of $\underset{\sim}{\mu}_i$.

Case 3: $\underset{\sim}{\alpha}$, $\underset{\sim}{\Gamma}$ and $\underset{\sim}{\Sigma}$ unknown. Let us suppose that we have a statistic $\underset{\sim}{S}$ which is independent of $\underset{\sim}{x}_i$ and

$$\underset{\sim}{S} \sim W_p(\underset{\sim}{\Sigma},s) \tag{6.18}$$

i.e., central Wishart on s degrees of freedom. In such a case the ERE is of the form

$$\hat{\underset{\sim}{\mu}}_i = \underset{\sim}{x}_i - c\underset{\sim}{S}\underset{\sim}{B}^{-1}(\underset{\sim}{x}_i-\bar{\underset{\sim}{x}}),\ i = 1,\ldots,k. \tag{6.19}$$

The total MDE (over variations of $\underset{\sim}{x}_i$, $\underset{\sim}{\mu}_i$, $\underset{\sim}{S}$) for such an estimator is

$$k\underset{\sim}{\Sigma}+(k-p-2)^{-1}c^2[(s^2+s)\underset{\sim}{\Sigma}(\underset{\sim}{\Gamma}+\underset{\sim}{\Sigma})^{-1}\underset{\sim}{\Sigma}+s[tr(\underset{\sim}{\Gamma}+\underset{\sim}{\Sigma})^{-1}\underset{\sim}{\Sigma}]\underset{\sim}{\Sigma}]- -2cs\underset{\sim}{\Sigma}(\underset{\sim}{\Gamma}+\underset{\sim}{\Sigma})^{-1}\underset{\sim}{\Sigma} \tag{6.20}$$

using the result (see Styan [31])

$$(6.21) \qquad E(\underset{\sim}{S}\underset{\sim}{A}\underset{\sim}{S}) = (s^2+s)\underset{\sim}{\Sigma}\underset{\sim}{A}\underset{\sim}{\Sigma} + s(\text{tr } \underset{\sim}{A}\underset{\sim}{\Sigma})\underset{\sim}{\Sigma}.$$

Unfortunately, there is no value of c independent of the unknown parameters which minimizes (6.20). However, we may choose $c = (k-p-2)/(s+p+1)$ and obtain the MDE

$$(6.22) \quad k\underset{\sim}{\Sigma} - \frac{(k-p-2)s}{s+p+1} \underset{\sim}{\Sigma}(\underset{\sim}{\Gamma}+\underset{\sim}{\Sigma})^{-1}\underset{\sim}{\Sigma} + \frac{(k-p-2)s}{(s+p+1)^2} [(\text{tr}(\underset{\sim}{\Gamma}+\underset{\sim}{\Sigma})^{-1}\underset{\sim}{\Sigma})\underset{\sim}{\Sigma} - p\underset{\sim}{\Sigma}(\underset{\sim}{\Gamma}+\underset{\sim}{\Sigma})^{-1}\underset{\sim}{\Sigma}].$$

The third term in (6.22) arises due to estimation of $\underset{\sim}{\Sigma}$ and is zero when p=1. Its magnitude is likely to be small and tends to zero as the degrees of freedom of S becomes large.

The total MDE for fixed $M = (\underset{\sim}{\mu}_1,\ldots,\underset{\sim}{\mu}_k)$ is

$$(6.23) \quad k\underset{\sim}{\Sigma} - \frac{(k-p-2)^2 s}{s+p+1} \underset{\sim}{\Sigma}E(\underset{\sim}{B}^{-1}|\underset{\sim}{M})\underset{\sim}{\Sigma} + \frac{(k-p-2)^2 s}{(s+p+1)^2}[(\text{tr } E(\underset{\sim}{B}^{-1}|\underset{\sim}{M})\underset{\sim}{\Sigma})\underset{\sim}{\Sigma} - p\underset{\sim}{\Sigma}E(\underset{\sim}{B}^{-1}|\underset{\sim}{M})\underset{\sim}{\Sigma}].$$

We note that if repeated observations are available on $\underset{\sim}{x}$ for each $\underset{\sim}{\mu}$ in the setup of (6.1), $\underset{\sim}{\Sigma}$ can be estimated by $\underset{\sim}{S}$, the dispersion matrix within individuals. In such a case we replace $\underset{\sim}{x}_i$ by $\bar{\underset{\sim}{x}}_i$ the mean of the observations corresponding to $\underset{\sim}{\mu}_i$ and $n^{-1}\underset{\sim}{S}$ for $\underset{\sim}{S}$ in all the formulae.

Appendix. The following lemma is given in Baranchik [1].

LEMMA 1. Let $y_i \sim N(\mu_i, \sigma^2)$, $i = 1,\ldots,p$, be independent and define $B = \Sigma y_i^2$. Then

$$(A.1) \qquad \sigma^2 E(B^{-1}) = E(2K + p-2)^{-1}$$

$$(A.2) \qquad \nu^2\sigma^2 E(y_i B^{-1}) = \mu_i E[2K/(2K + p-2)]$$

$$= \nu^2\mu_i E(2K+p)^{-1}$$

where K is a Poisson variable with $E(K) = \nu^2/2$ and $\nu^2 = (\Sigma\mu_i^2)/\sigma^2$.

From (A.1) and (A.2) we have the interesting result

$$(A.3) \qquad E(B^{-1}\Sigma\mu_i y_i) = 1 - \sigma^2(p-2)E(B^{-1}).$$

We give the following extension of Lemma 1 which can be

established on similar lines.

LEMMA 2. Let $\underset{\sim}{Y} \sim N_p(\underset{\sim}{\theta}, \sigma^2 \underset{\sim}{I})$, $\underset{\sim}{X}\hat{\underset{\sim}{\beta}} = \underset{\sim}{X}(\underset{\sim}{X}'\underset{\sim}{X})^{-1}\underset{\sim}{X}'\underset{\sim}{Y}$ and Rank $\underset{\sim}{X} = r$. Define $B = (\underset{\sim}{Y}-\underset{\sim}{X}\hat{\underset{\sim}{\beta}})'(\underset{\sim}{Y}-\underset{\sim}{X}\hat{\underset{\sim}{\beta}})$. Then

$$\sigma^2 E(B^{-1}) = E(2K_1 + p-r-2)^{-1} \tag{A.4}$$

$$E\left(\frac{\underset{\sim}{Y}-\underset{\sim}{X}\hat{\underset{\sim}{\beta}}}{B}\right) = \frac{(\underset{\sim}{I}-\underset{\sim}{P})\underset{\sim}{\theta}}{\sigma^2} E(2K_1+p-r)^{-1} = \frac{(\underset{\sim}{I}-\underset{\sim}{P})\underset{\sim}{\theta}}{\sigma^2\nu^2} E\left(\frac{2K_1}{2K_1+p-r-2}\right) \tag{A.5}$$

where K_1 is a Poisson variable with $E(K_1) = \nu^2/2$, $\nu^2 = \underset{\sim}{\theta}'(\underset{\sim}{I}-\underset{\sim}{P})\underset{\sim}{\theta}/\sigma^2$, and $\underset{\sim}{P}$ is the projection operator on the space spanned by the columns of $\underset{\sim}{X}$. Further

$$E\left(\frac{\underset{\sim}{\theta}'(\underset{\sim}{Y}-\underset{\sim}{X}\hat{\underset{\sim}{\beta}})}{B}\right) = 1-\sigma^2(p-r-2)E\left(\frac{1}{B}\right) \tag{A.6}$$

The multivariate extensions of Lemmas 1 and 2 are given in Lemmas 3 and 4. Lemma 3 is proved by following the arguments used by Efron and Morris [5] and an unpublished note by Khatri [13].

LEMMA 3. Let $\underset{\sim}{Y}_i \sim N_p(\underset{\sim}{\mu}_i, \underset{\sim}{\Sigma})$, $i = 1,\ldots,k$ be independent vector random variables, and $\underset{\sim}{B} = \Sigma \underset{\sim}{Y}_i \underset{\sim}{Y}_i'$. Then

$$E(\underset{\sim}{B}^{-1} \sum_1^k \underset{\sim}{Y}_i \underset{\sim}{\mu}_i') = I-(k-p-1)\Sigma E(\underset{\sim}{B}^{-1}) \tag{A.7}$$

where the expectations are taken for fixed $\underset{\sim}{\mu}_1,\ldots,\underset{\sim}{\mu}_k$.

LEMMA 4. Let $\underset{\sim}{Y}_i \sim N_p(\underset{\sim}{\mu}_i, \underset{\sim}{\Sigma})$, $i = 1,\ldots,k$ where $\underset{\sim}{M}=(\underset{\sim}{\mu}_1,\ldots,\underset{\sim}{\mu}_k)$ has the structure

$$\underset{\sim}{M}' = \underset{\sim}{X}(\underset{\sim}{\beta}_1,\ldots,\underset{\sim}{\beta}_p) = \underset{\sim}{X}\underset{\sim}{\beta} \tag{A.8}$$

Further let $B = (\underset{\sim}{Y}'-\underset{\sim}{X}\hat{\beta})'(\underset{\sim}{Y}'-\underset{\sim}{X}\hat{\underset{\sim}{\beta}})$ where $\underset{\sim}{Y} = (\underset{\sim}{Y}_1,\ldots,\underset{\sim}{Y}_k)$ and $\hat{\underset{\sim}{\beta}}$ is the least squares estimate of $\underset{\sim}{\beta}$ (see Rao [19] pp. 544-547). Then

$$E[\underset{\sim}{B}^{-1}(\underset{\sim}{Y}'\underset{\sim}{X}\hat{\underset{\sim}{\beta}})'\underset{\sim}{M}'] = \underset{\sim}{I}-(k-p-r-1)\underset{\sim}{\Sigma}E(\underset{\sim}{B}^{-1})$$

where r = Rank $\underset{\sim}{X}$.

References

[1] Baranchik, A. J. (1973). Inadmissibility of maximum likelihood estimators in some multiple regression problems with three or more independent variables. *Ann. Statist.* 1, 312-321.

[2] Bhattacharya, P. K. (1966). Estimating the mean of a multivariate normal population with general quadratic loss function. *Ann. Math. Statist.* 37, 1819-1824.

[3] Berger, J. and Bock, M. E. (1976). Improved minimax estimators of normal mean vectors for certain types of covariance matrices.

[4] Bock, M. E. (1975). Minimax estimators of the mean of a multivariate normal distribution. *Ann. Statist.* 3, 209-218.

[5] Efron, B. and Morris, C. (1972). Empirical Bayes on vector observations. *Biometrika* 59, 335-347.

[6] Efron, B. and Morris, C. (1973a). Stein's estimation rule and its competitors - An empirical Bayes approach. *J. Amer. Statist. Assoc.* 68, 117-130.

[7] Efron, B. and Morris, C. (1973b). Combining possibly related estimation problems. *J. Roy. Statist. Soc.* B, 35, 379-421.

[8] Fairfield Smith, H. (1936). A discriminant function for plant selection. *Ann. Eugen.* (London) 7, 240-260.

[9] Hazel, L. N. (1943). The genetic basis for constructing selection indexes. *Genetics* 28, 476.

[10] Henderson, C. R. (1963). Selection index and expected genetic advance. *Statistical Genetics and Plant Breeding* NAS-NRC 982, 141-163.

[11] Henderson, C. R. (1972). Sire evaluation and genetic trends. *Proc. Animal Breeding and Genetics Symp. in Honor of Dr. Lush,* pp. 10-41.

[12] James, W. and Stein, C. (1961). Estimation with quadratic loss. *Proc. Fourth Berkeley Symp. Math. Stat. Prob.* 1, 361-379. University of California Press, Berkeley.

[13] Khatri, C. G. (1976). Personal communication.

[14] Lindley, D. V. (1971). The estimation of many parameters. *Foundations of Statistical Inference,* Ed. Godambe and Sprott, pp. 435-454. Holt, Rinehard and Winston, New York.

[15] Lindley, D. V. and Smith, A. F. M. (1972). Bayes estimates for the linear model (with discussion). *J. Roy. Statist. Soc.* B, 34, 1-41.

[16] Panse, V. G. (1946). An application of the discriminant function for selection in poultry. *J. Genetics* (London) 47, 242-253.

[17] Rao, C. Radhakrishna (1953). Discriminant function for genetic differentiation and selection. *Sankhya* 12, 229-246.

[18] Rao, C. Radhakrishna (1952,1971). *Advanced Statistical Methods in Biometric Research.* Hafner, New York.

[19] Rao, C. Radhakrishna (1965,1973). *Linear Statistical Inference and Its Applications.* John Wiley, New York.

[20] Rao, C. Radhakrishna (1975a). Simultaneous estimation of parameters in different linear models and applications to biometric problems. *Biometrics* 31, 545-554.

[21] Rao, C. Radhakrishna (1975b). Some thoughts on regression and prediction. *Sankhya* C, 37, 102-120.

[22] Rao, C. Radhakrishna (1975c). Some problems of sample surveys. *Adv. Appl. Prob. Supp.* 7, 50-61.

[23] Rao, C. Radhakrishna (1976a). Prediction of future observations with special reference to linear models. *Multivariate Analysis - IV*, Ed. by P. R. Krishnaiah. Academic Press, New York.

[24] Rao, C. Radhakrishna (1976b). Estimation of parameters in a linear model - Wald Lecture 1. *Ann. Statist.* 4, to appear.

[25] Rao, C. Radhakrishna (1976c). Characterization of prior distributions and solution to a compound decision problem. *Ann. Statist.* 4, to appear.

[26] Robbins, H. (1955). An empirical Bayes approach to statistics. *Proc. Third Berkeley Symp. on Math. Statist. Prob.* 1, 157-163. University of California Press. Berkeley.

[27] Shinozaki, Nobuo (1974). A note on estimating the mean vector of a multivariate normal distribution with general quadratic loss function. *Keio Engineering Reports* 27, 105-112.

[28] Stein, C. (1955). Inadmissibility of the usual estimator for the mean of a multivariate normal distribution. *Proc. Third Berkeley Symp. Math. Statist. Prob.* 1, 197-206. University of California Press, Berkeley.

[29] Stein, C. (1962). Confidence sets for the mean of a normal distribution. *J. Roy. Statist. Soc.* B, 265-296.

[30] Strawderman, W. E. (1973). Proper Bayes minimax estimators of the multivariate normal mean vector for the case of common variances. *Ann. Statist.* 1, 1189-1194.

[31] Styan, G. P. H. (1973). Expectations of second degree expressions in a Wishart-matrix and its inverse. Abst. No. 141-9, IMS Bulletin.

[32] Susarla, V. (1976). A property of Stein's estimator for the mean of a multivariate normal distribution. *Statistica Neerlandica* 30, 1-5.

ROBUST BAYESIAN ESTIMATION

By Herman Rubin*
Purdue University

1. *Motivation.* There is a considerable body of literature expounding that proper statistical behavior is to take a convenient (but not necessarily true) prior, compute the posterior, and act accordingly. The advocates of this approach proclaim that asymptotically all is fine, so why worry? However, we shall show that all is not fine, and that much can be gained with little lost, at least in the case of estimating the mean of a univariate normal with known variance. The simple linear estimator has an infinite expected risk if the loss is squared error and the true prior has infinite variance. Even truncating the risk does not help too much. If we use an estimator based on a distribution with a flatter tail (such as double exponential, logistic, or Cauchy), the Bayes risk becomes finite and has reasonable robustness properties if the prior is not too concentrated. However, even this simple problem has its pitfalls and the results here are not completely satisfactory. In the higher, and especially infinite, dimensional cases (such as estimating a distribution function) the problem is likely to be much more difficult.

We shall approach the question of robustness from the behavioristic Bayes approach. Briefly, if one assumes certain axioms, which can be made much weaker than those in [6], the conclusion is that the utility under uncertainty as to the state of Nature is a positive linear functional of the utility considered as a function of the state of Nature, or alternatively

$$R(a) = \int \rho(\omega,a)d\xi(\omega),$$

where R is the risk of the action a, and $\rho(\omega,a)$ is the risk of the action a given the state ω. However, it is in general <u>impossible</u> [1], [3] to exactly evaluate <u>either</u> ρ or ξ. In some

*Research support in part by the National Science Foundation under grant #74-07836 at Purdue University, W. Lafayette, IN.

cases [4], [5], partial robustness is possible.

2. *The specific problem and procedures.* We assume that the observation X is normal with mean θ and variance 1, and that θ has a prior, which, for purposes of this paper, we take to be normal, double-exponential, logistic, or Cauchy, with mean 0. We investigate the behavior of Bayes estimates when the true prior and assumed prior are different members of this set of distributions. Since there is no clear scale parameter for different forms we have used the usual representation for each. We have taken the loss to be squared error; it can be argued that this is inappropriate since losses should be bounded, but if we take

$$L(\omega,a) = (\omega-a)^2/(1+\varepsilon^2(\omega-a)^2),$$

where ε is small, the result will change appreciably only when the true prior is Cauchy and the assumed prior is a distribution with small scale parameter or the normal. In this last case, if the scale parameter of the Cauchy is σ and that of the normal is τ, the risk is approximately

$$\left(\frac{\tau^2}{\tau^2+1}\right)^2 + \frac{\sigma}{\varepsilon(\sigma\varepsilon+\tau^2+1)} .$$

Since both ε and $\sigma\varepsilon$ are likely to be small, τ will have to be large enough that the other estimators being considered are already likely to be satisfactory without considering the bounding factor.

One of the major problems was computing the estimates. For the normal and double-exponential the procedures are straightforward. The logistic density can be written as

$$f(x) = \frac{e^{-|x|}}{(1+e^{-|x|})^2}$$

and $(1+e^{-|x|})^{-2}$ was approximated by $P(e^{-|x|})$ for a suitable polynomial P. (P was obtained by expanding $(1+x)^{-2}$ in a Taylor series about x = .5 to the desired accuracy and then expanding the powers of x - .5.)

This is the most expensive of the estimates to compute.

For the Cauchy,

$$\int_{-\infty}^{\infty} \frac{1}{\sqrt{2\pi}} \frac{e^{-\frac{1}{2}(\theta-x)^2}}{\sigma-i\theta}\, d\theta = \int_{-\infty}^{\infty} \frac{1}{\sqrt{2\pi}} \frac{\sigma+i\theta}{\sigma^2+\theta^2}\, e^{-\frac{1}{2}(\theta-x)^2} d\theta = A+iB,$$

so the Bayes estimate is $\sigma B/A$. But $A+iB$ is Mills' ratio [2, p. 137] at $\sigma-ix$.

The risks for the Bayes estimates for non-normal priors were computed by numerical integration for certain values of θ, followed by numerical integration over θ. Because of the need to use different scale factors, the values of θ were chosen so that a piecewise Simpson's rule could be used. Certain internal checks indicate that for small θ there is a relative error of .00013 in the computed Bayes risk. This can not seriously affect the conclusions. The risks were computed for scale parameters 10^{-2}, $10^{-1.9}, \ldots, 10^{1.9}, 10^{2}$, both for the true prior and the estimates, and also for Cauchy estimates with scale parameters $10^{-6}, 10^{-5.9}$, $\ldots, 10^{-2.1}$. The locations of the minimizing points and their Bayes risks was obtained by fitting the quotients of two quadratics in σ^2 to the points obtained. For the case where the prior distribution used for computing the risk was the same as that for computing the estimator, the location of the fitted minimum was surprisingly close to the known value. In addition, even a 10% error in the optimizing scale factor of the assumed prior increases the Bayes risk by less than 1% for all the combinations we considered except for the true prior Cauchy with $\sigma < .4$.

3. *Results.* The results of our study fall into three parts: σ small, σ large, and σ intermediate.

For σ small, it seems that not too much can be done either substituting one of the other priors for the Cauchy or vice-versa. If the variance of the prior is less than .1, using one of the other priors with the same variance increases the risk by less than .15%, and up to a variance of .6, matching variances is nearly optimal. It can be shown that this is true in general for small σ if the prior has a finite variance. It is possible to get reasonable estimators using Cauchy priors, but for variances less than .04, the Cauchy estimator is not much better than using the estimator 0. Why this is true can be seen by an examination of the Cauchy estimator for small σ.

For large σ, the bias for a prior of density $\sigma^{-1}e^{-\varphi(x/\sigma)}$ is approximately $-\sigma^{-1}\varphi'(x/\sigma)$, and the optimal alternative corresponds asymptotically to fitting this from the other distribution. The corresponding scale factors are

Table 1

	Normal	D.E.	Logistic	Cauchy
Normal	1	$(\sqrt{\pi/2})$1.2533	.6053	1.0650
D.E.	$(\sqrt{2})$1.4142	1	.6489	.7872
Logistic	$(\pi/\sqrt{3})$1.8138	2	1	1.6389
Cauchy	∞	$(\pi/2)$1.5708	.9677	1

The rate of approach to these limiting ratios is as σ^{-2}, except for the assumed prior double exponential where the discontinuity in φ' at 0 makes the rate of approach σ^{-1}. The estimators for double exponential, logistic, and Cauchy priors have an additional advantage in that if a prior with large σ is used the corresponding estimator is robust against all large tails. This is not true for linear estimators.

For intermediate values of σ, a transition from the "small σ" to the "large σ" case occured smoothly. We append a small table of the results. The figures may be slightly off in the last digit.

4. *Conclusions.* The main conclusions we can draw are

1. For distributions with not too extreme tails, it should be possible to get reasonable efficiency with a "wrong" prior.
2. If the prior is concentrated with respect to the likelihood, fitting the variance of the correct prior yields good results.
3. If the prior is diffuse with respect to the likelihood, the

form of the prior determines the bias, but good robustness is relatively easy to obtain.

4. Much more computational effort and ingenuity are required.

5. *Acknowledgment.* The author thanks Mr. J. K. Ghorai for invaluable assistance with all phases of the numerical work.

Table 2 Ratios of Bayes Risks

	Normal			Double-Exponential			Logistic			Cauchy	
σ	D	L	C	N	L	C	N	D	C	D	L
0	1.000	1.000	1.000	1.000	1.000	1.000	1.000	1.000	1.000	∞	∞
.1	1.000	1.000	1.010	1.000	1.000	1.002	1.000	1.000	1.033	1.42	1.65
.2	1.000	1.000	1.039	1.000	1.000	1.065	1.000	1.001	1.086	1.26	1.44
.5	1.006	1.001	1.111	1.024	1.006	1.060	1.009	1.007	1.071	1.11	1.22
1	1.028	1.007	1.097	1.056	1.011	1.021	1.012	1.011	1.030	1.043	1.116
2	1.030	1.007	1.045	1.042	1.011	1.005	1.006	1.008	1.009	1.013	1.042
5	1.010	1.002	1.010	1.013	1.005	1.001	1.003	1.004	1.004	1.002	1.009
10	1.003	1.000	1.002	1.003	1.001	1.000	1.000	1.000	1.000	1.001	1.002
∞	1.000	1.000	1.000	1.000	1.000	1.000	1.000	1.000	1.000	1.000	1.000

Table 3 $\hat{\sigma}$(assumed prior)/σ(true prior)

	Normal			Double-Exponential			Logistic			Cauchy	
σ	D	L	C	N	L	C	N	D	C	D	L
0	.71	.55	0	1.41	.78	0	1.81	1.28	0		
.1	.71	.55	.000047	1.41	.78	.00019	1.81	1.28	.0021		
.2	.71	.55	.003	1.41	.78	.006	1.81	1.27	.18	3.36	2.72
.5	.70	.55	.22	1.41	.78	.48	1.81	1.30	.85	2.12	1.66
1	.74	.56	.56	1.41	.75	.75	1.81	1.42	1.30	1.68	1.27
2	.87	.58	.86	1.41	.70	.85	1.81	1.60	1.53	1.48	1.08
5	1.06	.60	1.03	1.41	.66	.83	1.81	1.80	1.61	1.44	1.02
10	1.15	.60	1.06	1.41	.65	.80	1.81	1.90	1.64	1.49	.97
∞	1.25	.61	1.07	1.41	.65	.79	1.81	2.00	1.64	1.57	.97

References

[1] I. J. Good, "The probabilistic explication of information, evidence, surprise, causality, explanation, and utility. Appendix," pp. 123-127. *Foundations of Statistical Inference*, V. P. Godambe and D. A. Sprott, eds. Holt, Rinehart, and Winston of Canada, Ltd. Toronto, Montreal, 1970.

[2] M. G. Kendall and A. Stuart, *The Advanced Theory of Statistics*, vol. 1, 3rd ed, Charles Griffin & Company Ltd., London, 1969, pp. 136-137.

[3] H. Rubin, Comments on C. R. Rao's paper, p. 195, *Foundations of Statistical Inference*, V. P. Godambe and D. A. Sprott, eds. Holt, Rinehart, and Winston of Canada, Ltd. Toronto, Montreal, 1970.

[4] H. Rubin, "On large sample properties of certain nonparametric procedures," *Proc. Sixth Berkeley Symp. on Math. Statistics and Probability*, vol. 1, pp. 429-435, Univ. of California Press, Berkeley, 1972.

[5] H. Rubin and J. Sethuraman, "Bayes risk efficiency," *Sankhyā* Ser. A 27 (1965), pp. 347-356.

[6] L. J. Savage, *The Foundations of Statistics*, John Wiley & Sons, New York, 1954.

THE DIRICHLET-TYPE 1 INTEGRAL AND ITS APPLICATIONS TO MULTINOMIAL-RELATED PROBLEMS

By Milton Sobel
University of California, Santa Barbara

1. *Introduction.* In this paper it is desired to show the wide variety of problems in which it would be useful to apply tables of the incomplete Dirichlet integral

$$(1.1)\qquad I_p^{(b)}(r,n) = \frac{n!}{\Gamma^b(r)(n-br)!}\int_0^p \cdots \int_0^p (1-\sum_{i=1}^b x_i)^{n-br} \prod_{i=1}^b x_i^{r-1} dx_i$$

where we assume that $0 \leq p \leq 1/b$, $n \geq rb$ and n, b, r are all positive integers (in one exception we allow $b = 0$ with no integrals and define the value as 1 for all n,r and p, including $p = 0$). We emphasize the role of I as a generalization of the incomplete beta function by taking $b = 1$ in which case we have

$$(1.2)\qquad I_p^{(1)}(r,n) = I_p(r,n-r+1) = \frac{n!}{\Gamma(r)(n-r)!}\int_0^p x^{r-1}(1-x)^{n-r}dx,$$

where $I_p(a,b)$ is the usual incomplete beta function. As a tool for summing either tail of a binomial or negative binomial we also generalize the well known results (for $q = 1-p$)

$$(1.3)\qquad I_p^{(1)}(r,n) = \sum_{\alpha=r}^{n} \binom{n}{\alpha} p^\alpha q^{n-\alpha} = \sum_{\alpha=0}^{n-r} \binom{n}{\alpha} q^\alpha p^{n-\alpha} = 1 - I_q^{(1)}(n-r+1,n),$$

$$(1.4)\qquad I_p^{(1)}(r,n) = q^{n-r+1}\sum_{\alpha=r}^{\infty}\binom{n-r+\alpha}{\alpha}p^\alpha = p^r \sum_{\alpha=0}^{n-r}\binom{r+\alpha-1}{\alpha}q^\alpha$$

where the last result in (1.4) follows from (1.3) and the first result in (1.4).

To interpret (1.1) we consider a multinomial model based on n observations and b+1 cells, b of which have a common cell probability p and are called red cells. Let M denote the minimum frequency in the b red cells. Then I in (1.1) is the probability that $M \geq r$.

To generalize the result (1.3) we let S denote the sum $x_1 + x_2 + \ldots + x_b$, let $\underset{\sim}{x}$ denote the vector $(x_1, x_2, \ldots, x_b)$ and let

$[{n \atop \underset{\sim}{x},n-S}]$ denote the multinomial coefficient $n!/(x_1!\ldots x_b!(n-S)!)$. Then

$$(1.5) \qquad I_p^{(b)}(r,n) = \Sigma[{n \atop \underset{\sim}{x},n-S}]p^S(1-pb)^{n-S}$$

where the sum is over all b-tuples $(x_1,x_2,\ldots,x_b)$ with $x_i \geq r$ $(i = 1,2,\ldots,b)$ and $S \leq n$; for $n = S$ and $pb = 1$ we set $0^0 = 1$ in (1.5). Thus we see that the I function is closely related to the cdf of the minimum frequency among equi-probable cells in a multinomial distribution.

A slight generalization of the I function allows us to develop a recursive formula which is the main tool that was used for tabulating this function. Consider the probability for a particular subset of j of the b red cell frequencies that each one equals r and the remaining b-j red cell frequencies are all at least r; denote it by $I_p^{(b,j)}(r,n)$. Using (1.1) and the fact that the conditional distribution, given that j particular frequencies are equal to r, is again multinomial with common cell probability $p/(1-jp)$, it is easily shown that

$$(1.4) \qquad I_p^{(b,j)}(r,n) = \frac{n!(p^r/r)^j}{\Gamma^b(r)(n-br)!}\int_0^p \cdots \int_0^p (1-jp-\sum_1^{b-j} x_\alpha)^{n-br} \prod_1^{b-j} x_\alpha dx_\alpha;$$

the latter is a b-j fold integral, which for j=b is the multinomial probability

$$(1.5) \qquad I_p^{(b,b)}(r,n) = \frac{n!}{(r!)^b(n-br)!}\, p^{br}(1-bp)^{n-br}.$$

This furnishes a first boundary condition for our recurrence below.

2. *Basic recurrence relation.* The basic recurrence formula for (1.1) is

$$(2.1) \qquad I_p^{(b,j)}(r,n) = \frac{n(1-jp)}{n-jr}\, I_p^{(b,j)}(r,n-1) + \frac{r(b-j)}{n-jr}\, I_p^{(b,j+1)}(r,n),$$

which holds for any j $(0 \leq j \leq b-1)$ and is used only for $n > br$.

(Note that $n-jr > n-br \geq 0$ and hence $n-jr > 0$.)

We give a brief sketch of a derivation of (2.1). One factor $(1-jp-\Sigma x_\alpha)$ in the integrand in (1.4) can be separated and the result expanded as the difference of two integrals, one with coefficient $1-jp$ and the other containing $b-j$ equivalent integrals using a symmetry argument. After an integration-by-parts in the second term, we obtain the same integral as appears on the left side of (2.1) with the additional coefficient $-r(b-j)/(n-br)$; this then leads to the result given in (2.1) by a straightforward calculation.

There are two boundary conditions that are associated with the recurrence relation (2.1). One of these (for $j = b$) has already been given in (1.5). The other is the result for $n = br$, where it is easy to show by direct integration in (1.4) that for all j

$$I_p^{(b,j)}(r,br) = \frac{(br)!}{(r!)^b} p^{br}. \tag{2.2}$$

Note that when $p < 1/b$ this condition for $j = b$ is consistent with (1.5) for $n = br$; when $p = 1/b$ and $n = br$ we take the indefinite form 0^0 in (1.5) to be 1 and the consistency still holds.

It is of numerical as well as theoretical interest that the superscript j in (1.4) can be altered or brought to zero by a simple mathematical or a probabilistic argument (both of which are omitted), and we write for any i and j

$$I_p^{(b,i+j)}(r,n) = \frac{n!}{(r!)^j(n-jr)!} p^{rj}(1-jp)^{n-jr} I_{p/(1-jp)}^{(b-j,i)}(r,n-jr). \tag{2.3}$$

We refer to (2.3) as the reduction formula and use it mostly with $i = 0$ to reduce the second superscript to zero.

In calculating $I_p^{(b,j)}(r,n)$ for any j, the value for $p = 1/b$ is of particular interest since there is a simple summation formula that expresses the value of I_p for every p in terms of $I_{1/b}$-values. In fact we can condition on the total number of observations in the red cells and obtain for any p, b, j, r and n

$$I_p^{(b,j)}(r,n) = \sum_{\alpha=br}^{n} \binom{n}{\alpha}(bp)^{\alpha}(1-bp)^{n-\alpha} I_{1/b}^{(b,j)}(r,\alpha). \tag{2.4}$$

The last result is useful to obtain I function values for $p < 1/b$ and close to $1/b$ as a function of the tabled values for $p = 1/b$.

3. *Derivatives and differences.* A number of intersecting relations hold between the derivative of the I function with respect to p and differences with respect to the second argument, n. These relations are quite useful for interpolating in an I table for p values other than those used in the table. Moreover we can then use differences of the tabled values to evaluate higher derivatives and thus write down a Taylor expansion quite easily. Any such Taylor expansion must be finite and exact since the I function is a polynomial in p. However we may still be interested in the "speed of convergence" of such a finite series since we usually wish to deal with only a few terms of the series. In each case we sketch the development without giving detailed proof.

A straightforward differentiation of (1.1) gives

$$\frac{\partial}{\partial p} I_p^{(b)}(r,n) = \frac{n}{p} \Delta I_p^{(b)}(r,n-1) \quad (\text{for } p > 0), \tag{3.1}$$

where Δ operates on the second argument, n. To generalize (3.1) we first use induction to obtain a general Leibnitz-type result, namely that the sth difference of a product is given by

$$\Delta^s\{F(x)G(x)\} = \sum_{j=0}^{s} \binom{s}{j}[\Delta^j F(x)][\Delta^{s-j} G(x+j)]. \tag{3.2}$$

Then (3.2) can be used to generalize (3.1). Let $n^{[s]} = n(n-1)\ldots(n-s+1)$ then for $p > 0$

$$\frac{\partial^s}{\partial p^s} I_p^{(b)}(r,n) = \frac{n^{[s]}}{p^s} \Delta^s I_p^{(b)}(r,n-s); \tag{3.3}$$

both sides of (3.3) are identically zero unless $n \geq s + br$. Hence we can evaluate any derivative with respect to p by merely taking the appropriate difference in a table of I values.

Using (3.3) we can now write a Taylor expansion of (1.1)

about p_0 (assuming p is close to p_0 for fast convergence) as

$$(3.4) \qquad I_p^{(b)}(r,n) = \sum_{\alpha=0}^{n-br} \binom{n}{\alpha}\left(\frac{p-p_0}{p_0}\right)^{\alpha} \Delta^{\alpha} I_{p_0}^{(b)}(r,n-\alpha)$$

and this is exact if we use all n-br+1 terms. Furthermore if we write E-1 for Δ in (3.4) where $Ef(x) = f(x+1)$ then after a straightforward binomial summation we obtain an alternative exact form

$$(3.5) \qquad I_p^{(b)}(r,n) = \begin{cases} \sum_{\beta=0}^{n-br} B_{p_1}(n,\beta) I_{p_0}^{(b)}(r,n-\beta) & \text{for } p \le p_0 \\ \left(\frac{2p-p_0}{p_0}\right)^n \sum_{\beta=0}^{n-br} (-1)^{\beta} B_{p_2}(n,\beta) I_{p_0}^{(b)}(r,n-\beta) & \text{for } p \ge p_0 \end{cases}$$

where $B_p(n,r) = \binom{n}{r}p^r q^{n-r}$, $p_1=(p_0-p)/p_0$ and $p_2 = (p-p_0)/(2p-p_0)$. Although (3.4) and (3.5) are both exact, the use of 3 terms will not generally give the same result. Using the availability and common knowledge about binomial coefficients and the fact that I is increasing in the second argument, we can get better results with (3.5). Thus for $r = 2$, $n = 10$, $b = 3$, $p = .3$ and $p_0 = 1/3$ the first 3 terms of (3.5) for $p \le p_0$ give

$$(3.6) \qquad I_{.3}^{(3)}(2,10) \sim (3.4868)(.69349) + (.38742)(.58467) + (.19371)(.44810) = .55512$$

This is clearly a lower bound to the correct answer, which is .5730 in this case. If we replace the third coefficient, $B_{p_1}(10,2) = .19371$ by $B_{p_1}(10,2) + B_{p_1}(10,3) + B_{p_1}(10,4) = .26227$ then we obtain a closer upper bound, namely

$$(3.7) \quad I_{.3}^{(3)}(2,10) \sim (.34868)(.69349) + (.38742)(.58467) + (.26227)(.44810) = .58584.$$

The average of these bounds .57048 is an approximation to the correct I value. Thus if tables of I values are tabulated for p equal to the reciprocal of an integer like 1/3, we can get I for other values by appropriate interpolation.

4. *Dual quantities and generalizations.* The natural dual to the minimum frequency in a multinomial is the maximum frequency and we now consider for the same model as before the probability that j particular red cells have frequency exactly r and the remaining b-j red cells each have frequency less than r. This can, of course, be written in the form (1.5) with the appropriate summation and also as an upper tail integral

$$J_p^{(b)}(r,n) = \frac{n!}{\Gamma^b(r)(n-br)!} \int_p \cdots \int_p \left(1-\sum_1^b x_i\right)^{n-br} \prod_1^b x_i^{r-1} dx_i, \tag{4.1}$$

where the upper limits of integration are determined by the condition that the integrand is non-negative (even when n = br). An equivalent, perhaps more desirable, definition of the J-function is by the method of inclusion-exclusion. Stated with the extra parameter, this takes the form

$$J_p^{(b,j)}(r,n) = \sum_{\alpha=j}^{b} (-1)^{\alpha-j} \binom{b-j}{\alpha-j} I_p^{(\alpha,j)}(r,n). \tag{4.2}$$

The parameter j can be reduced by exactly the same reduction formula (2.3) with J replacing I. For j = 0 this deals with the probability that the maximum of the red cell frequencies is less than r (or $\leq$ r-1). This function was tabulated earlier by Steck [7] in an unpublished table. For small values of b and n this can also be obtained from an I table by using (4.2).

It should be noted that many generalizations of the I function in (1.1) are possible. For example we could ask for the joint probability that a certain set of c red cells each have frequency at least r and another disjoint set of d = b - c white cells each have frequency at least w let the associated common cell probabilities be p_r and p_w, respectively, so that $cp_r + dp_w \leq 1$. Letting R = cr + dw, this probability is equal to the c+d fold integral

(4.3) $$I^{(c,d)}_{p_r,p_w}(r,w,n) = \frac{n!}{\Gamma^c(r)\Gamma^d(w)(n-R)!} \cdot$$

$$\int_0^{p_r} \cdots \int_0^{p_r} \int_0^{p_w} \cdots \int_0^{p_w} (1-\sum_1^c x_\alpha - \sum_1^d y_\beta)^{n-R} \prod_{\alpha=1}^{c} x_\alpha^{r-1} dx_\alpha \prod_{\beta=1}^{d} y_\beta^{w-1} dy_\beta;$$

here we assume only that $n \geq R$ and the I function is defined to be zero if $n < R$. This is still a special case of a more general result proved in [5], since every cell C_α (or any subset of cells with common cell probability) can be assigned a color and a v value v_α such that we will call it crowded if the frequency $f_\alpha \geq v_\alpha$ and the joint probability that certain cells are all crowded is given by the corresponding integral. Similarly, cells can be called sparse if their frequency $f_\alpha < u_\alpha$ and the joint probability that certain cells are all sparse is given by an appropriate generalization of the J-function (or J-integral). The joint probability that a certain number of cells are crowded and a certain number are sparse is more involved and is considered in [6].

Moreover, if we add the appropriate extra parameter(s) then each new generalization has a corresponding recurrence and boundary conditions. Thus we add the parameters i and j to (4.3) to signify that i of the c red cells have frequency <u>exactly</u> r and j of the d white cells have frequency <u>exactly</u> w. The resulting I function satisfies the recurrence

(4.4) $$I^{(c,i;d,j)}_{p_r,p_w}(r,w,n) = \frac{n(1-ip_r-jp_w)}{n-ir-jw} I^{(c,i;d,j)}_{p_r,p_w}(r,w,n-1)$$

$$+ \frac{r(c-i)}{n-ir-jw} I^{(c,i+1;d,j)}_{p_r,p_w}(r,w,n) + \frac{w(d-j)}{n-ir-jw} I^{(c,i;d,j+1)}_{p_r,p_w}(r,w,n).$$

The boundary conditions are for $n = R$

(4.5) $$I^{(c,i;d,j)}_{p_r,p_w}(r,w,R) = \frac{R!}{(r!)^c(w!)^d} p_r^{cr} p_w^{dw} \text{ for all } i \text{ and } j,$$

for $i = c$

$$(4.6)\quad I_{p_r,p_w}^{(c,c;d,j)}(r,w,n) = \frac{n!}{(r!)^c(n-cr)!}\, p_r^{cr}(1-cp_r)^{n-cr} I_{p_w/(1-cp_r)}^{(d,j)}(w,n-cr),$$

and for j = d the boundary condition is similar to (4.6) with c,d, r,w,j replaced by d,c,w,r,i, respectively.

5. *Combinatorial aspects and generalized Stirling numbers.* A simple multiple of the I function yields Stirling numbers and generalized Stirling numbers that have a combinatorial interpretation. For $j = 0$ and $r > 0$ we obtain the number of partitions of n distinct objects into exactly b subsets (or parts), with a least r objects in each subset. Thus for $n = 4$, $b = 2$, $j = 0$, and $r = 1$ we have the seven partitions

$$(5.1)\quad \begin{matrix} (x_1;x_2,x_3.x_4) & (x_1,x_2;x_3,x_4) \\ (x_2;x_1,x_3,x_4) & (x_1,x_3;x_2,x_4) \\ (x_3;x_1,x_2,x_4) & (x_1,x_4;x_2,x_3) \\ (x_4;x_1,x_2,x_3) & \end{matrix}$$

For any b,r,n,j with $0 \leq j \leq b$, $n \geq br$ and b and n not both zero, we define

$$(5.2)\quad S_{n,r}^{(b,j)} = \frac{b^n}{(b-j)!} I_{1/b}^{(b,j)}(r,n), \qquad (S_{n,1}^{(b,0)} = S_n^{(b)}).$$

If $b = n = 0$ then $j = 0$ and we define the left side of (5.2) to be 1 for all r. If $b = 0 < n$ then (5.2) is zero and if $n = 0 < b$ then (5.2) is zero for $r > 0$ and δ_{0j} (Kronecker delta) for $r = 0$. For the illustration above (5.2) gives

$$(5.3)\quad S_4^{(2)} = \frac{16}{2} I^{(2)}(1,4) = 8(12)\int_0^{\frac{1}{2}}\int_0^{\frac{1}{2}}(1-x-y)^2 dxdy = 96\left(\frac{7}{96}\right) = 7,$$

and this is the number obtained by enumeration in (5.1).

Substituting (5.2) in (2.1) gives us the recursive formula for the generalized Stirling numbers

$$(5.4)\quad S_{n,r}^{(b,j)} = \frac{n(b-j)}{n-jr} S_{n-1,r}^{(b,j)} + \frac{r}{n-jr} S_{n,r}^{(b,j+1)}.$$

The two boundary conditions for this recurrence are

(5.5) $$S_{br,r}^{(b,j)} = \frac{(br)!}{(r!)^b(b-j)!} \quad \text{for } 0 \leq j \leq b,$$

(5.6) $$S_{n,r}^{(b,b)} = \frac{(br)!}{(r!)^b}\,\delta_{n,br},$$

and these are consistent when both $j = b$ and $n - br$ hold.

For $j = 0$ and $r = 1$ we used the notation $S_n^{(b)}$ above and these are the usual Stirling numbers of the second kind, most simply defined for $n \geq 1$ by the identity

(5.7) $$x^n = \sum_{\alpha=1}^{n} S_n^{(\alpha)} x^{[\alpha]}.$$

We note that $S_n^{(n)} = 1$ and from (5.2) that $S_n^{(1)} = 1$ for $n \geq 1$.

The combinatorial interpretation is easier for $j = 0$ and we have already mentioned that $S_{n,r}^{(b)}$ represents the number of partitions of n distinct objects into exactly b subsets, with each subset containing at least r objects. Since this is a partition there is no ordering of subsets, cf (5.1).

More generally $S_{n,r}^{(b,j)}$ represents the number of "compo-partitions" of n objects into exactly b subsets. The first j subsets each contain exactly r objects and there is an ordering between but not within the subsets; the last b-j subsets each contain at least r objects and these subsets are not ordered. Thus for b=4, $j = 2$, $n = 5$ and $r = 1$, $S_{5,1}^{(4,2)}$ is the number of ordered selections (i.e., permutations) of 2 things out of 5 multiplied by the number of partitions, $S_{3,1}^{(2)}$. Since the former is 20 and the latter is 3 the result is 60. More generally, (2.3) with $p = 1/b$ and (5.2) give the reduction formula

(5.8) $$S_{n,r}^{(b,i+j)} = \frac{n!}{(r!)^j(n-jr)!}\, S_{n-jr,r}^{(b-j,i)}.$$

Using this to remove the second superscript in (5.4), we obtain for $0 \leq j \leq b$

(5.9) $$S_{n-jr,r}^{(b-j)} = (b-j)S_{n-jr-1,r}^{(b-j)} + \binom{n-jr-1}{r-1} S_{n-jr-r,r}^{(b-j-1)}.$$

For $j = 0$ this reduces to

$$(5.10) \qquad S_{n,r}^{(b)} = b\, S_{n-1,r}^{(b)} + \binom{n-1}{r-1} S_{n-r,r}^{(b-1)};$$

this can be used with (5.5) alone to calculate $S_{n,r}^{(b)}$.

We now sketch a proof of (5.10). Letting x denote the one extra object in the set of size n on the left side of (5.10) that is not in the set of size n-1 in the first term on the right side of (5.10). We can put x into any one of the b subsets and this explains the multiplications by b. In addition, we can combine x with any subset of size r - 1 and use the result as one of the b subsets of size r. This number has to be multiplied by the number of partitions of the remaining n-r objects into exactly b-1 subsets, each with at least r objects. Moreover each such partition is a new one and all possible partitions are now included. This proves the combinatorial interpretations since the boundary conditions are easily seen to represent the appropriate number of partitions.

Using (4.4) we can now count "colored partitions" by defining a second-order Stirling number by

$$(5.11) \qquad S_{n,r,w}^{(c,i;d,j)} = \frac{b^n}{(c-i)!(d-j)!} I_{\frac{1}{b},\frac{1}{b}}^{(c,i;d,j)}(r,w,n)$$

where $b = c+d$, $0 \leq i \leq c$ and $0 \leq j \leq d$. Here we partition n distinct objects into exactly $b = c+d$ subsets of which c are red and d are white. In the red ones i particular subsets have exactly r and the remaining c-i each have at least r. In the white ones j particular subsets have exactly w and the remaining d-j each have at least w. The main point is that if we have 2 partitions that differ only in that we interchanged a red subset with a white subset, the present S-value counts them as different partitions (i.e., it is color conscious), whereas the interchange of 2 subsets of the same color, as usual, does not produce a new partition. This is what is meant by the above term "colored partitions".

Substituting (5.11) into (4.4) gives us the basis recursive relation for the second-order Stirling numbers

(5.12) $$S_{n,r,w}^{(c,i;d,j)} = \frac{n(b-i-j)S_{n-1,r,w}^{(c,i;d,j)} + rS_{n,r,w}^{(c,i+1;d,j)} + wS_{n,r,w}^{(c,i;d,j+1)}}{n-ir-jw}.$$

This has three boundary conditions: for n = R (= cr+dw)

(5.13) $$S_{R,r,w}^{(c,i;d,j)} = \frac{R!}{(c-i)!(d-j)!(r!)^c(w!)^d},$$

for c = i

(5.14) $$S_{n,r,w}^{(c,c;d,j)} = \frac{n!}{(r!)^c(n-cr)!} S_{n-cr,w}^{(d,j)}$$

and a similar result holds for d = j. The right side of (5.14) then utilizes (5.4), (5.5) and (5.6) to complete the recursion.

As an illustration suppose n = 5, c = d = 1, r = w = 2 and i = j = 0. By (5.12), (5.13) and (5.14)

(5.15) $$S_{5,2,2}^{(1,0;1,0)} = \frac{10\ S_{4,2,2}^{(1,0;1,0)} + 4\ S_{5,2,2}^{(1,1;1,0)}}{5} = \frac{10(6)+4(10)}{5} = 20$$

and a simple enumeration will confirm this result; here the partitions (1,2;3,4,5) and (3,4,5;1,2) are both counted since the two subsets (separated by a semicolon) in each case have different colors.

Clearly the extension to partitions with many different colors is now straightforward.

6. *Some applications of Dirichlet-type* 1. We categorize the applications as being

A. Related to Combinatorics and/or Number Theory.

B. Related to Multinomial Probabilities:

1. Sparse and Crowded Cell problems.
2. Goodness of Fit problems.
3. Sample Size problems.
4. Hypergeometric problems.

C. Related to problems of Ranking and Selection.

D. Related to Dirichlet-Type 2 Evaluation problems.

Although we cannot discuss all of these in great detail, we will try to illustrate the different types of applications.

A. How many numbers less than 10^9 have each odd digit

(i.e., 1, 3, 5, 7, 9) appearing at least once? Consider the associated multinomial probability with 9 observations (corresponding to the 9 digits in our numbers) with common cell probability $p = .1$ (since each digit has ten possible values from 0 to 9) with $b = 5$ (the 5 cells on which we put a restriction are the red cells) and with $r = 1$ (since we want to have at least 1 in each red cell). Since there are 10^9 possible numbers from .000000000 to .999999999 and each is equally likely under the multinomial model, it follows that the answer to the question above in 10^9 $I_{1/10}^{(5)}(1,9)$. Using (2.4), (5.2) and tables of S or I values, we have

$$(6.1)\quad 10^9 I_{1/10}^{(5)}(1,9) = 5^9 \sum_{\alpha=5}^{9} \binom{9}{\alpha} I_{1/5}^{(5)}(1,\alpha) = 5! \sum_{\alpha=5}^{9} \binom{9}{\alpha} 5^{9-\alpha} S_{\alpha}^{(5)}$$

$$= 120[78,750(1)+10,500(15)+900(140)+45(1050)+1(6951)]$$

$$= 49,974,120.$$

This type of problem shows an unusual tie-in between combinatorics, probability and the Dirichlet-type 1. It is not claimed that this is the only way of handling this type of problem; we are claiming that tables of the I values (or S values) would be useful in such problems.

B.3. Mice in a certain experiment can die of 5 different "relevant" causes with probability p_i for the ith cause. The p_i have a common value $p \leq 1/5$ (there may be other causes of death not relevant to the experiment), how many mice do we need to have probability at least $P^* = .95$ (say) that there will be at least r deaths in each of the 5 categories.

Here the answer is simply $I_p^{(5)}(r,n)$ so that for given p, r and P* we simply look up the answer in an I table. Thus for $p=1/5$ and $r = 5,10$ we find that $n = 55$ mice, and 89 mice respectively, are needed if randomization is not used (and at most 1 less if randomization is used). For $p = 1/10$ (so that only about half of the mice are dying from relevant causes) and the same r and P* we need $n = 113$ and 183, respectively; this is only slightly more than double the previous answers. This suggests the general inequality

$$I_p^{(b)}(r,n/bp) \leq I_{1/b}^{(b)}(r,n), \tag{6.2}$$

except for values of n less than or close to br; for $n < br$ the right side of (6.2) is zero but the left side could still be positive.

B.1. In the paper [6] a sparse cell is defined as one with frequency at most u whereas a crowded cell is defined as one with frequency at least v $(0 \leq u \leq v \leq n)$. The I-function was then used to evaluate the density, the cdf, the moments and the joint distribution of the number S of sparse red cells and the number C of crowded red cells in our multinomial model. For example, if the common cell probability is $p \leq 1/b$ for all the red cells then

$$P\{S=s\} = \binom{b}{s} \sum_{i=0}^{s} (-1)^i \binom{s}{i} I_p^{(b-s+i)}(u+1,n), \tag{6.3}$$

$$P\{C=c\} = \binom{b}{c} \sum_{j=0}^{b-c} (-1)^j \binom{b-c}{j} I_p^{(c+j)}(v,n), \tag{6.4}$$

$$E\{C^{[m]}\} = b^{[m]} I_p^{(m)}(v,n) \tag{6.5}$$

$$\sigma^2(C) = bI_p^{(1)}(v,n)[1-bI_p^{(1)}(v,n)]+b(b-1)I_p^{(2)}(v,n) \tag{6.6}$$

and similar results as (6.4), (6.5) and (6.6) hold for S if we replace v by u+1, I by J, and c by s.

B.2. Any goodness of fit test based on the number of empty cells when the hypothesized distribution is partitioned into b equally likely intervals can be modified by basing the criterion on the number of sparse cells or on some combination of sparse and crowded cells. This can only improve the test since the case u=0 corresponds to empty cells and we can usually do better. The details for efficiently carrying out such a test are in the process of being developed. The presence of tables of I values makes this a realistic project.

B. For a homogeneous multinomial with b cells (so that $p = 1/b$) and n observations the probability of getting no empty cells is

$$(6.7) \qquad I_{1/b}^{(b)}(1,n) = \sum_{\alpha=0}^{b} (-1)^{\alpha}\binom{b}{\alpha}\left(1 - \frac{\alpha}{b}\right)^{n} = \Delta^{b}\left(\frac{x}{b}\right)^{n}\Big]_{x=0} = \frac{b!}{b^{n}} S_{n}^{(b)}.$$

The fact that this probability is related to Stirling numbers of the second kind is well known, cf. [4].

B.3 Let $\nu_{b,r}$ denote the number of observations required to achieve (for the first time) a frequency of at least r in each red cell. Then

$$(6.8) \qquad P\{\nu_{b,r} \leq n\} = I_{p}^{(b)}(r,n)$$

and hence the 'density' of $\nu_{b,r}$ is given by

$$(6.9) \qquad P\{\nu_{b,r} = n\} = I_{p}^{(b)}(r,n) - I_{p}^{(b)}(r,n-1) = \Delta I_{p}^{(b)}(r,n-1)$$

and this is $(p/n) \frac{\partial}{\partial p} I_{p}^{(b)}(r,n)$ by (3.1) above.

D. What is the expectation of $\nu_{b,r}$? A straightforward calculation gives the result $(br/p)C_{1}^{(b-1)}(r,r+1)$ where

$$(6.10) \qquad C_{a}^{(b)}(r,M) = \frac{\Gamma(M+br)}{\Gamma(M)\Gamma^{b}(r)} \int_{0}^{a} \cdots \int_{0}^{a} \frac{\prod_{\alpha=1}^{b} x_{\alpha}^{r-1} dx_{\alpha}}{\left(1 + \sum_{\alpha=1}^{b} x_{\alpha}\right)^{M+br}}.$$

This is a type 2 Dirichlet integral. It arises in a very natural manner from the type 1 Dirichlet. In addition, it can be evaluated exactly (and/or be bounded) by Dirichlet type 1 expressions. We omit these results as they would take up too much space here.

B.4 An urn contains bM objects of b distinct types and exactly M of each type. If we draw a sample of size n without replacement, what is the probability that our sample contains at least r of each type; assume $n \geq br$. Strictly speaking, this is not a multinomial or Dirichlet problem, but it is the hypergeometric analogue. As $M \to \infty$ the exact answers approach those given by the corresponding multinomial or Dirichlet problem. In fact the latter answers are always conservative in the sense that the tabled I value is a lower bound and hence the resulting n-value is an upper bound on the number of observations needed. The following table illustrates this relationship for $b = 3$, $r = 1$, and $P^* = .95$

M	P{No empty cells | b=3,r=1,n=10}	Sample size n needed to reach P* = .95
50	.9556	10
100	.9519	10
500	.9488	11
1000	.9484	11
∞	.9480	11

C. Two different ranking problems have been considered in connection with the multinomial. One [1] deals with selecting the cell with the smallest cell probability; the other [2] deals with selecting the cell with the largest cell probability. In both cases the I-function can be used to help evaluate the probability of a correct selection (PCS). In addition, to properly assess the efficiency of the procedure we should evaluate the expected sample size needed to make the PCS $\geq P^*$ for all configurations outside a preassigned indifference zone. In both of these goals the problem of evaluating the effect of curtailment arises since we may terminate the experimenter before taking all n observations, if the winning cell is determined early. The most common example here is the baseball 'World Series' where $b = 2$ and, although we have preassigned n value equal to 7, we stop the series as soon as either team has won 4 games. Curtailment was <u>not</u> taken into account in either of the two references mentioned above, probably because the effect is generally small and the computations are difficult. With the help of the I function, reasonable expressions for the PCS and for the results of curtailment can be derived and tables can be calculated.

For the problem of selecting the cell with the smallest probability our procedure R is simply to select the cell with the smallest frequency, using randomization when there are ties for first place, and we use the difference $p_{[2]} - p_{[1]}$ of the ordered p values as a means of keeping one population separated from the others. In the so-called least favorable (LF) configuration we set $p_{[3]} = p_{[4]} = \ldots = p_{[k]} = p_{[2]} = p$ (say) and $p_{[1]} = \underline{p}$. Here we use a simpler model with k red cells and with $p_{[1]} + \ldots + p_{[k]} = 1$. Then it follows that in the configuration with $p - \underline{p} = \delta$

$$\underline{p} = \frac{1-b\delta}{k}, \quad p = \frac{1+\delta}{k}, \tag{6.11}$$

where $b = k-1$ and we assume that $0 \leq \delta \leq 1/b$, so that $\underline{p} \geq 0$. Let $p_1 = \underline{p}/(1-p)$ and $p_2 = p/(1-p)$. Let $\underset{\sim}{p}$ denote the vector with p_2 repeated b times; let $\underline{q} = 1-\underline{p}$, $q = 1-p$ and $q_1 = 1-p_1$. Let $P_m(CS)$ (resp., $P_m(IS)$) denote the probability of terminating our procedure after exactly m observations without ties for first place with a correct (resp., incorrect) selection. Then $P_m(S) = P_m(CS) + P_m(IS)$ is the probability of terminating after exactly m observations without such ties. Let $\{x\}$ denote the smallest integer not less than x, the possible values of m are from $m_0 = \{b(n+1)/(b+1)\}$ to n. Exact expressions for $P_m(CS)$ and $P_m(IS)$ are given in terms of I functions by

$$P_m(CS) = \underline{q} \sum_{y=m_0-1}^{m-1} B_{\underline{q}}(m-1,y)[I_{1/b}^{(b)}(n-y,y+1) - I_{1/b}^{(b)}(n-y+1,y)], \tag{6.12}$$

$$P_m(IS) = bq \sum_{y=m_0-1}^{m-1} B_q(m-1,y)[I_{p_1,\underset{\sim}{p}}^{(b)}(n-y,y+1) - I_{p_1,\underset{\sim}{p}}^{(b)}(n-y+1,y)], \tag{6.13}$$

where $B_p(n,x)$ is the usual binomial probability. In (6.13) we assumed that $b \geq 2$ in the notation; for $b = 1$ we disregard $\underset{\sim}{p}$ and note that $p_1 = 1$ and $I_1^{(1)}(r,s) = 1$ if $s \geq r$ and 0 otherwise. Then for $b = 1$ in (4.26) and (4.27) only 1 term is different from zero. For $b \geq 2$ we use the fact that

$$I_{p_1,\underset{\sim}{p}}^{(b)}(r,s) = \sum_{i=(b-1)r}^{s-r} B_{q_1}(s,i) I_{1/(b-1)}^{(b-1)}(r,i). \tag{6.14}$$

For $m < n$ we can only terminate without ties for first place but for $m = n$ we can terminate with or without them. Hence the desired expected number $E\{N\}$ of observations for procedure R for any configuration as in (6.11) is

$$E\{N\} = \sum_{m=m_0}^{n} mP_m(S)+n[1-\sum_{m=m_0}^{n} P_m(S)]. \tag{6.15}$$

For the LF configuration we simply set $\delta = \delta^*$ in (6.11) and this affects all the computations above.

The asymptotic percent saving due to curtailment is $(n-E\{N\})100/n$. As $n \to \infty$ the minimum number m of observations also $\to \infty$. For large values of m and n we can stop only if $x+n-m <$ the smallest of the b remaining cell frequencies, which tends to $(m-x)/b$ with probability one; here x is the frequency of the best cell (i.e., the one with cell probability $\underline{p}$) which must be close to $m\underline{p}$. Hence we obtain, using (6.11)

$$\text{Lim}\,\frac{n-\bar{E}\{N\}}{n} = \text{Lim}\,\frac{n-m}{n} = 1 - \frac{b}{(b+1)(1-\underline{p})} = 1 - \frac{1}{kp} = \frac{\delta}{1+\delta} \tag{6.16}$$

and hence $100\delta/(1+\delta)$ is the asymptotic percent saving for configurations with a common $p \geq \underline{p}$ and $\delta = p-\underline{p}$. For the LF configuration we simply set $\delta = \delta^*$. The percent saved is a maximum (50%) when $\underline{p} = 0$ and $b = 1$. Although the percentage saved is generally not large, it should be noted that this percentage does not vanish as $n \to \infty$ and, except for very special situations, there is always a positive saving. Hence the effect of curtailment should not be overlooked in treating the fixed sample size multinomial selection problem.

The latter remark also applies to papers like [3], where the values of n (needed if the LF configuration) to satisfy the condition that PCS $\geq$ P* are compared with the number of observations required by sequential methods to satisfy the same P* condition (in the same LF configuration); since there is no curtailment possible under the sequential procedure it would be more appropriate to use $E\{N\}$, rather than n, in assessing the fixed sample size procedure.

References

[1] Alam, K., and Thompson, J. R. (1973). On selecting the least probable multinomial event. *Ann. Math. Statist.* 43, 1981-1990.

[2] Bechhofer, R. E., Elmaghraby, S. A., and Morse, N. (1959). A single-sample multiple-decision procedure for selecting the multinomial event with the largest probability. *Ann. Math. Statist.* 30, 102-119.

[3] Cacoullos, T., and Sobel, M. (1966). An inverse sampling procedure for selecting the most probable event in a multinomial distribution. In *Multivariate Analysis* I (ed. by P. R. Krishnaiah) pp. 423-455, Academic Press, New York.

[4] Jordan, D. (1964). *Calculus of Finite Differences*, 2nd. Ed. Dover Publications, New York.

[5] Olkin, I., and Sobel, M. (1965). Integral expressions for tail probabilities of the multinomial and negative multinomial distributions. *Biometrika* 52, 167-179.

[6] Sobel, M., and Uppuluri, V.R.R. (1974). Sparse and crowded cells and Dirichlet distributions. *Ann. Statist.* 2, 977-987.

[7] Steck, G. P. (1960). Tables of the distribution and moments of the maximum of a homogeneous multinomial distribution. (Unpublished.)

OPTIMAL SEARCH DESIGNS, OR DESIGNS OPTIMAL UNDER BIAS FREE OPTIMALITY CRITERIA*

By J. N. Srivastava
Colorado State University

1. *Introduction.* In this paper, we raise the question whether the usual optimal designs are really optimal. The reason is that the underlying model may be biased. Earlier authors have talked of "minimizing" or "balancing" the bias. However, the theory of search models offers the possibility of "searching" the bias, and "correcting" for the same. In other words, one may try to search for the right model within a given family of models, it being assumed that exactly one model in the family holds good. Since the true model is rarely known, and since it is far easier to postulate a family of models, one of which is the true model, the idea of search should be inherent in the statistical aspects of most scientific experimentation. Thus, the problem of optimal designs is more a problem of obtaining a design which would be good from the viewpoint of a whole family of models rather than of any single model inside this family. In this paper, we study certain optimality criteria for designs, where optimality takes the above view point into account. First, we present a brief discussion of search linear models needed for the present development. We exemplify this particularly with respect to 2^m factorial designs. Next, the concept of optimal search designs is introduced and the criteria of AD-optimality and DD-optimality are introduced. The mathematical development leading to a simple-to-use form of the AD-optimality criterion is discussed. Some bounds connected with DD-optimality are considered. The problem of adding new assemblies to a given set of assemblies so as to make the total set into a search design is studied. A result useful in studying the rank of a (0,1) matrix over the real field

*This work was supported by NSF Grant No. MPS73-05086 A02.

is presented. Finally, a new class of row-column type of designs are considered; these designs seem to be better than the classical row-column designs in a certain sense.

Various criteria for optimality, like the A, D, E, G-optimality are well known. (See, for example, the papers of Kiefer, Wolfowitz, Elfving, Chernoff, Wynn, Karlin, Studden, etc., referenced at the end. This list of references is merely illustrative, and not exhaustive.)

It is clear that the above optimality criteria are related to the underlying model. Thus, let $M_j (j = 0,\ldots,\nu_2)$ denote the model $E(y(\underline{x})) = \sum_{i=1}^{\nu_1} \xi_{i1} f_i(\underline{X}) + \xi_{j2} g_j(\underline{X})$, where $\xi_{j2} \neq 0$ $(j = 1,2,\ldots,\nu_2)$, $\xi_{02} = 0$, $f_i(x)$ and $g_j(x)$, $(i = 1,\ldots,\nu_1;\ j = 1,\ldots,\nu_2)$ are known functions of $\underline{x} = (x_1,\ldots,x_m)'$, the functions g_j are distinct, and $(x_1,\ldots,x_m)$ ($\varepsilon \mathcal{X}$, say) denotes a combination of levels of m factors in which the rth factor occurs at level x_r. Let M denote the class of models $\{M_j,\ j = 1,\ldots,\nu_2\}$, and suppose that in some particular situation, the true model is M_{j_0} where $M_{j_0} \varepsilon M$, but j_0 is not known. Now suppose the model assumed is M_0, and D_0 is the optimal design under M_0. Since M_0 is biased, D_0 would generally be "non-optimal". On the other hand, since M_{j_0} is not known, the corresponding optimal design can not be obtained.

We can give a similar example from the field of discrete designs. For example, consider a situation where a 2^m factorial experiment is planned. Suppose the experimenter believes that 3-factor and higher effects are negligible. Then his model (say M_0^*) is of the form $E\ y(x_1,\ldots,x_m) = \mu + \sum_{i=1}^{m} \alpha_i F_i + \sum_{i<j=1}^{m} \alpha_i \alpha_j F_{ij}$,

where for $(r = 1,\ldots,m)$, $x_r = 0$ or 1, $\alpha_r = (+1)$ or (-1), according as $x_r = 1$ or 0, and μ, F_i and F_{ij} are respectively (in an obvious notation), the general mean, the main effects, and the 2-factor interactions. On the other hand, in an obvious extension of the above notation, the true model generally is $E\ y(x_1,\ldots,x_m) =$

$$\mu + \sum_{i=1}^{m} \alpha_i F_i + \sum_{i_1<i_2} \alpha_{i_1}\alpha_{i_2} F_{i_1 i_2} + \sum_{i_1<i_2<i_3} \alpha_{i_1}\alpha_{i_2}\alpha_{i_3} F_{i_1 i_2 i_3} + \ldots +$$

$\alpha_1\alpha_2\cdots\alpha_m F_{12\ldots m}$, where, (say) at least $(\nu_2 - k)$ of the 3-factor and higher interactions $F_{123}, F_{124},\ldots,F_{12\ldots m}$, are negligible, where $k \geq 0$, and $\nu_2 = 2^m - \nu_1$, $\nu_1 = 1 + m + \binom{m}{2}$. If $k > 0$, the model M_0^* is biased and has an unknown amount of bias. Thus, a design optimal under M_0^* would in general not be truly optimal.

Two points emerge from the above. Firstly, the situations noted above are quite representative of most scientific experimentation. Secondly, for any such situation, what would be generally possible to postulate (with relatively much greater chance of being correct) is a family of models (like M) rather than a particular model in the family. Whatever is "common" to all the models in the family would then correspond to M_0. Thus, one could symbolically represent the jth model by $M_0 + M_{j1}$. Note that experiments would, in general, supply information not only regarding M_0, but also regarding the "specific" part M_{j1}. Indeed, one could at least expect this to happen in case of "well-planned" or "optimum" experiments. Given the data of the experiment, one problem would then be to distinguish between the different M_{j1} (for various values of j). This aspect of the analysis of the data could be referred to as the "search" aspect. The problem of planning the experiment, in most situations, is thus to obtain information on M_0, to search for the correct value of j (say j_0) and to obtain information on the corresponding $M_{j_0 1}$. In the literature on statistical planning so far, the attempt was to obtain an optimal or good design under M_0, the remaining part (M_{j1}) being usually ignored.

An exception to the last remark is the work on "minimization" or "balancing" of the bias. Thus, Hedayat, Raktoe, and Federer [5] have suggested, in the context of general factorial designs, that the design should be made in such a way, that (briefly speaking) if for $(i = 1,\ldots,\nu_1)$, θ_i are parameters to be estimated, $\hat{\theta}_i$ is the best linear estimate of θ_i, and $E(\hat{\theta}_i) = \theta_i + \sum_{j=1}^{\nu_2} a_{ij}\phi_j$, where the ϕ_j are the remaining parameters (which are sometimes assumed zero), then $(\sum_{j=1}^{\nu_2} a_{ij}^2)$ be a constant for all i. Clearly, even though the bias will be balanced, it will still be very much existent.

Box and Draper [1] considered the minimization of bias in the following sense. Suppose a polynomial of degree d, in m continuous variables $x_1,\ldots,x_m$ is being fitted over a region of interest $\mathscr{X}$, while the true response function is of degree $d_2 > d_1$. then they consider the quantity B, where B is the average (over all points $x \in \mathscr{X}$) of $[b(\underline{x})]^2$, where $b(\underline{x})$ denotes the bias at $\underline{x}$. They discuss how to choose the design points in $\mathscr{X}$ so that B is minimized for all values of the coefficient (of terms of degree > d_1) occuring in the true response function. (However, this minimum value (say B_0) still depends heavily on these last mentioned (unknown) coefficient. The quantity B_0 (which could be very large) can not be further touched by the classical methods. "Search" is an obvious way out.)

2. *Search linear models.* The ordinary general linear model can be expressed in the form

$$E(\underline{y}) = A_1\underline{\xi}_1, \quad V(\underline{y}) = \sigma^2 I_N, \tag{2.1}$$

where $\underline{y}(N \times 1)$ is a vector of observations taken on N experimental units, $A_1(N \times \nu_1)$ is a known matrix (depending upon the structure or design of the experiment or investigation), $\sigma^2 (\geq 0)$ is either known or unknown, and $\underline{\xi}_1(\nu_1 \times 1)$ is an unknown vector of parameters. Clearly, in the above model, in order to be able to estimate $\underline{\xi}_1$, it is necessary (but not sufficient) that we have $N \geq \nu_1$.

In contrast with the above, a (relatively simple) search linear model is of the form

$$E(\underline{y}) = A_1\underline{\xi}_1 + A_2\underline{\xi}_2, \; V(\underline{y}) = \sigma^2 I_N, \tag{2.2}$$

where all symbols are as in (2.1) except that A_2 $(N \times \nu_2)$ is also a known matrix, and $\underline{\xi}_2(\nu_2 \times 1)$ is a vector of unknown parameters on which we have partial information. About $\underline{\xi}_2$, we know that at least ν_2 - k elements of $\underline{\xi}_2$ are negligible, where k is a non-negative integer which may or may not be known. (In applications in this paper, we will assume k known.) The problem is to search the nonzero elements of $\underline{\xi}_2$, and to draw inferences on these and on the elements of $\underline{\xi}_1$.

The search problem is clearly a multiple decision problem and is being studied by the author and his student D. G. Mallenby. We shall not go into this problem here.

It should, however, be important to point out some of the advantages which search linear models possess over ordinary linear models in real life situations. Very often, in such situations, when one attempts to fit a linear model, one comes to a point where one feels that he has included all parameters in his model which he could lay his fingers on. At the same time, he may feel that the model he has hypothesized may not fit well enough since he knows from his experience that there must be a few more parameters which are nonnegligible. However, he can not include these parameters in his model since these parameters could be any ones out of a large set of parameters, and he does not know exactly which parameters he should include. He may not include all of the remaining parameters because trying to estimate all of them using the ordinary linear model, and thus determining the nonnegligible parameters would involve too many observations. Thus, the situation in which he is in is most realistically describable by a search linear model.

In the context of the last section, the search linear model (2.2) corresponds to the family of models M, and the common part M_0 corresponds to the ordinary linear model as at (2.1). A model M_j in M then contains the common part M_0, and the remaining part M_{j1} corresponds to the part of (2.2) involving $\underline{\xi}_2$, the different possible sets of nonnegligible elements of $\underline{\xi}_2$ being in correspondence with the suffix j in M_j.

Notice that in general one could have a probability distribution $\{q_j\}$ over M where q_j is the probability that M_j is the true model. In the work done by the author so far, such probability distributions have not been considered, it being implicitly assumed that all values of j are on the same footing.

As in all statistical problems, two aspects arise here as well. The first is the inference aspect, which, in this case will correspond to obtaining efficient methods of searching nonnegligible parameters, and of drawing inferences regarding these and the elements of $\underline{\xi}_1$. We shall not be concerned with this aspect in the present paper. Here, we shall be concerned with the design aspect which deals with the question of how to obtain data so that the above inferences can be carried out in a reasonably good manner.

From the design point of view, the case when $\sigma^2 = 0$ (called the noiseless case) is of great importance. The reason is that if a design T does not work well for the noiseless case, it can not be expected to work in the noisy case ($\sigma^2 > 0$) either. Also, any inference procedure which works perfectly for the noiseless case would generally be expected to yield correct results in the noisy case with probability less than one, the actual probability depending upon the extent of noise. The following result is basic.

THEOREM 2.1. *Consider the model* (2.2) *and let* $\sigma^2 = 0$. *A necessary and sufficient condition that the search-cum-inference problem can be completely solved in the noiseless case, is that for every* ($N \times 2k$) *submatrix* A_{20} *of* A_2, *we have*

(2.3) $$\text{Rank}(A_1 : A_{20}) = \nu_1 + 2k.$$

By "completely solved", we mean that we will be able to search the nonzero elements of $\underline{\xi}_2$, without any error, and furthermore obtain estimators of the nonzero elements of $\underline{\xi}_2$ and the elements of $\underline{\xi}_1$ which have variance zero.

COROLLARY 2.1. *Under the conditions of Theorem 2.1 we must have*

(2.4) $$N \geq \nu_1 + 2k$$

A set of observations $\underline{y}$ for which condition (2.3) is satisfied, are said to form a "search design". When $\nu_1 = 0$, the model (2.2) is called the pure search model. Usually, ν_1 is relatively large, and hence, condition (2.3) is cumbersome to check, particularly since the checking has to be done for every possible submatrix A_{20} of A_2. (When k is fixed, the number of matrices A_{20} is $\binom{\nu_2}{2k}$, which is usually extremely large.) The following result shows that in a sense the model (2.2) can be reduced to a pure search model.

THEOREM 2.2. *Consider the model (2.2) where* $\underline{y}$ *etc., are partitioned as below:*

(2.5) $$\underline{y} = \left[\frac{\underline{y}_1}{\underline{y}_2}\right], \quad A_1 = \left[\frac{A_{11}}{A_{21}}\right], \quad A_2 = \left[\frac{A_{12}}{A_{22}}\right]$$

where $\underline{y}_1$, A_{11} *and* A_{12} *have* N_1 *rows each and* $N = N_1 + N_2$. *Then a sufficient condition that (2.3) holds for every* A_{20} *is that the following two conditions hold:*

(2.6) (i) $A_{11}(N_1 \times \nu_1)$ *has rank* ν_1, *and*

(2.7) (ii) $Q = A_{22} - A_{21}(A_{11}'A_{11})^{-1}A_{11}'A_{12}$, *has property* P_{2k}.

(*A matrix is said to have property* P_t *if every set of t columns are linearly independent.*)

THEOREM 2.3. *Under the conditions of Theorem 2.2, let*

(2.8) $$\underline{y}_0 = \underline{y}_2 - A_{21}(A_{11}'A_{11})^{-1}A_{11}'\underline{y}_1,$$

$K(N_2 \times N_2)$ be a matrix such that

$$KK' = I_{N_2} + A_{21}(A_{11}'A_{11})^{-1}A_{21}', \tag{2.9}$$

$$\underset{\sim}{y}_0^* = K^{-1}\underset{\sim}{y}_0, \quad Q^* = K^{-1}Q. \tag{2.10}$$

Then we have

$$E[\underset{\sim}{y}_0^*] = Q^*\underset{\sim}{\xi}_2, \quad V(\underset{\sim}{y}_0^*) = \sigma^2 I_{N_2}, \tag{2.11}$$

so that (2.11) is a pure search model.

Notice that since Q^* and $\underset{\sim}{y}_0^*$ are known, the model (2.11) can be used as a pure search model for searching the nonnegligible elements of $\underset{\sim}{\xi}_2$.

3. *Factorial designs.* The new optimality criteria to be developed in this paper will here be applied mostly to the theory of factorial designs of the 2^m type. We shall, therefore, consider here some of the details which will be needed in the sequal.

Without loss of generality, we shall denote the two levels of each factor by -1 and 1 respectively. The treatments can then be denoted by $(x_1,\ldots,x_m)$ where $x_r = \pm 1$, $r = 1,2,\ldots,m$. The "true effect" of the treatment $(x_1,\ldots,x_m)$ will be denoted by $\phi(x_1,\ldots,x_m)$. The model can then be written as

$$\phi(x_1,\ldots,x_m) = \mu + \sum_{i=1}^{m} x_i F_i + \sum_{i_1<i_2=1}^{m} x_{i_1}x_{i_2}F_{i_1i_2} + \sum_{i_1<i_2<i_3=1} x_{i_1}x_{i_2}x_{i_3}F_{i_1i_2i_3} + \ldots + x_1x_2\ldots x_m F_{1,2\ldots m}, \tag{3.1}$$

where μ denotes the general mean F_i the main effect of the ith factor, $F_{i_1i_2}$ the two factor interaction between the factors i_1 and i_2 and so on. We shall also sometimes use the dual notation

$$F_{i_1i_2\ldots i_k} = F_{i_1}F_{i_2}\ldots F_{i_k}, \tag{3.2}$$

where for $1 \leq k \leq m$, $(i_1, i_2,\ldots,i_k)$ is a subset of k distinct integers out of the set $(1,2,\ldots,m)$. The above notation will be used in the usual group theoretic sense. In other words, the set of elements $\{\mu; F_1, F_2, \ldots, F_m; F_1F_2, F_1F_3, \ldots, F_{m-1}F_m; F_1F_2F_3, F_1F_2F_4, \ldots, F_{m-2}F_{m-1}F_m; F_1F_2F_3F_4, F_1F_2F_3F_5, \ldots, \ldots; \ldots; F_1F_2F_3\ldots F_m\}$, constitute a commutative group of order 2^m in which μ is the identity element, $F_1, \ldots, F_m$ can be regarded as the generators, each element is idempotent (i.e., $F_i^2 = \mu$, $i = 1, \ldots, m$), multiplication being defined as in ordinary algebra. (Thus, for example, $(F_1F_2F_3)(F_1F_3F_5) = F_2F_5$).

Now we can define the two parameter vector $\underline{\xi}_1(\nu_1 \times 1)$, $\underline{\xi}_2(\nu_2 \times 1)$, where

$$\nu_1 = 1 + \binom{m}{1} + \binom{m}{2}, \quad \nu_2 = 2^m - \nu_1, \tag{3.3}$$

$$\underline{\xi}_1' = (\mu; F_1, F_2, F_3, \ldots, F_m; F_{12}, F_{13}, \ldots, F_{1m}, F_{23}, F_{24}, \ldots, F_{m-1,m}) \tag{3.4}$$

$$\underline{\xi}_2' = (F_{123}, F_{124}, \ldots, F_{m-2,m-1,m}; F_{1234}, F_{1235}, \ldots, ; \ldots; F_{12\ldots m})$$

Let T be a design, i.e., T is an arbitrary set of N treatment combinations which are not necessarily distinct. The design T is usually written as a matrix of size $N \times m$ such that the rows of T denote the treatments, and the columns denote the factors. Let the ith row of T be $(x_{i1}, \ldots, x_{im})$, $i = 1, 2, .., N$. Suppose that there are N experimental units to which these N treatments are applied. We shall assume that there are no blocks. (The main thrust of the paper is in the conceptually new direction of optimal search designs. The ideas of this paper could be developed in a similar way when blocks are present.) Let $y(x_1, \ldots, x_m)$ denote the observed yield, for the treatment $(x_1, \ldots, x_m)$, so that

$$E[y(x_1, x_2, \ldots, x_m)] = \phi(x_1, x_2, \ldots, x_m), \tag{3.5}$$

where ϕ is as defined in (3.1). Let $\underline{y}(N \times 1)$ denote the vector of the N observations $y(x_{i1}, \ldots, x_{im})$ $i = 1, 2, \ldots, N$, the treatments of $\underline{y}$ being arranged in the same order as in T. Let $A_1(N \times \nu_1)$ and $A_2(N \times \nu_2)$ be real matrices whose columns correspond respectively to the elements of $\underline{\xi}_1$ and $\underline{\xi}_2$, and whose rows correspond respectively to the treatments in T. Thus, for example, the ith row of A_1 is $(1, x_{i1}, x_{i2}, \ldots, x_{im}, x_{i1}x_{i2}, \ldots, x_{im-1} x_{im})$, and the ith row of A_2 is $(x_{i1}x_{i2}x_{i3}, \ldots, x_{i1}x_{i2}\ldots x_{im})$. Let

(3.6) $$A = [A_1 \vdots A_2],\ \underline{\xi}' = (\underline{\xi}_1', \underline{\xi}_2'),$$

where the columns of A and $\underline{\xi}$ correspond to the total set of 2^m effects. Then

(3.7) $$E[\underline{y}] = A\underline{\xi} = A_1\underline{\xi}_1 + A_2\underline{\xi}_2.$$

In the theory of designs of resolution $\underline{\bar{V}}$, one assumes that $\underline{\xi}_2$ is null. Optimum balanced designs of resolution $\underline{\bar{V}}$ for 2^m factorials for every value of N in a practical range have been obtained in various papers of the author and his former student D. V. Chopra, a few of which are referenced at the end for illustration. (Briefly speaking, a balanced design of resolution $\underline{\bar{V}}$ is one for which $V(\hat{F}_i)$, $V(\hat{F}_{ij})$, $Cov(\hat{\mu}, \hat{F}_i)$, $Cov(\hat{\mu}, \hat{F}_{ij})$, $Cov(\hat{F}_i, \hat{F}_j)$, $Cov(\hat{F}_i, \hat{F}_{ij})$, $Cov(\hat{F}_i, \hat{F}_{k\ell})$, $Cov(\hat{F}_{ij}, \hat{F}_{ik})$, $Cov(\hat{F}_{ij}, \hat{F}_{k\ell})$ are independent of the suffixes i, j, k and ℓ. In other words, the dispersion matrix V of the estimates of the parameters is symmetric with respect to "factors". Also, the designs are optimum with respect to the trace criterion, which means that among all possible balanced designs with a given value of N, our design minimizes trace V. It is shown in unpublished work of the author that for almost all such designs, det V is also minimized. Thus, they are both A- and D-optimal in the class of balanced designs. For many designs the efficiency is high enough to suspect that they are optimal or near optimal in the class of all designs, balanced or unbalanced. It so turns out that all of these optimum balanced designs of Srivastava and Chopra are balanced arrays of full strength. (A design T is said to be a balanced array (B-array) of full strength if it is such that for every treatment $(x_1, x_2, \ldots, x_m)$, the number of times $(x_1, x_2, \ldots, x_m)$ occurs in T is the same as the number of times any permutation of $(x_1\ x_2 \ldots x_m)$ occurs in T. Indeed, let Ω_{mi}, $(i = 0, 1, \ldots, m)$ denote the set of $\binom{m}{i}$ distinct treatments $(x_1\ x_2 \ldots x_m)$, which are of weight i, i.e., which have exactly i 1's in them. Then it is easy to check that T is a B-array of full strength if and only if it contains the set Ω_{mi} $(i = 0,1,\ldots,m)$ exactly λ_i times, where λ_i is a nonnegative integer. The vector $\underline{\lambda}' = (\lambda_0,\lambda_1,\ldots,\lambda_m)$ is called the index set of the array.)

EXAMPLE 1. In Table 1 we present, in the column headed T, a B-array of full strength with index set (1 1 1 0 1 0). For ease of reference, the other columns in this table are headed μ, 1, 2, ..., which correspond to the effects μ, F_1, $F_2,\ldots$, as they occur in $\underline{\xi}'$. Under the model (3.7), the 16 columns in table 1 under $\{\mu, 1,\ldots,45\}$ correspond to A_1 and the remaining 16 columns to A_2. (Note that throughout our discussion + and - stand for +1 and -1 respectively.)

We now consider search designs for 2^m factorials. Suppose that atmost k elements of $\underline{\xi}_2$ are nonzero, but we do not know these elements. A design T is said to be a search design of resolution 5.k if it satisfies condition (2.3) of Theorem 2.1 where A_1 and A_2 have interpretation as in (3.7). In this paper we shall restrict our attention usually to case $k = 1$. Srivastava and Ghosh [13] have obtained a class of resolution 5.1 designs for various values of N when $4 \leq m \leq 8$, and have presented [14] an infinite series of such designs for all m. These designs are all B-arrays of full strength. Also, in these papers, the question of their optimality as search designs has not been discussed. Our purpose of this paper is to make some headway in this direction.

4. *Optimum search designs*. We now consider designs suitable under the search linear model (2.2). In view of Theorem 2.1, one could consider the estimation of $\underline{\xi}_1$ and $\underline{\xi}_{20}$, where $\underline{\xi}_{20}$ is a vector having an arbitrary set of 2k elements of $\underline{\xi}_2$. Let $H(\underline{\eta})$ denote the information matrix (occuring in the normal equations) for the estimation of a vector $\underline{\eta}$. Now, for fixed $\underline{\xi}_{20}$, the determinant criterion leads us to consider $|H(\underline{\xi}_1,\underline{\xi}_{20})|$, which is the

Table 1: Example of a Resolution $\underline{\bar{V}}$ Plus One Plan with m=5 and N=21

T	μ	1	2	3	4	5	12	13	14	15	23	24	25	34	35	45
-----	+	-	-	-	-	-	+	+	+	+	+	+	+	+	+	+
+----	+	+	-	-	-	-	-	-	-	-	+	+	+	+	+	+
-+---	+	-	+	-	-	-	-	+	+	+	-	-	-	+	+	+
--+--	+	-	-	+	-	-	+	-	+	+	-	+	+	-	-	+
---+-	+	-	-	-	+	-	+	+	-	+	+	-	+	-	+	-
----+	+	-	-	-	-	+	+	+	+	-	+	+	-	+	-	-
++---	+	+	+	-	-	-	+	-	-	-	-	-	-	+	+	+
+-+--	+	+	-	+	-	-	-	+	-	-	-	+	+	-	-	+
+--+-	+	+	-	-	+	-	-	-	+	-	+	-	+	-	+	-
+---+	+	+	-	-	-	+	-	-	-	+	+	+	-	+	-	-
-++--	+	-	+	+	-	-	-	-	+	+	+	-	-	-	-	+
-+-+-	+	-	+	-	+	-	-	+	-	+	-	+	-	-	+	-
-+--+	+	-	+	-	-	+	-	+	-	+	-	+	-	+	-	-
--++-	+	-	-	+	+	-	+	-	-	+	-	-	+	+	-	-
--+-+	+	-	-	+	-	+	+	-	+	-	-	+	-	-	+	-
---++	+	-	-	-	+	+	+	+	-	-	+	-	-	-	-	+
++++-	+	+	+	+	+	-	+	+	+	-	+	+	-	+	-	-
+++-+	+	+	+	+	-	+	+	+	-	+	+	-	+	-	+	-
++-++	+	+	+	-	+	+	+	-	+	+	-	+	+	-	-	+
+-+++	+	+	-	+	+	+	-	+	+	+	-	-	-	+	+	+
-++++	+	-	+	+	+	+	-	-	-	-	+	+	+	+	+	+

Table 1 (continued)

123	124	125	134	135	145	234	235	245	345	1234	1235	1245	1345	2345	12345
-	-	-	-	-	-	-	-	-	-	+	+	+	+	+	+
+	+	+	+	+	+	-	-	-	-	-	-	-	-	+	+
+	+	+	-	-	-	+	+	+	-	-	-	-	+	-	+
+	-	-	+	+	-	+	+	-	+	-	-	+	-	-	+
-	+	-	+	-	+	+	-	+	+	-	+	-	-	-	+
-	-	+	-	+	+	-	+	+	+	+	-	-	-	-	+
-	-	-	+	+	+	+	+	+	-	+	+	+	-	-	-
-	+	+	-	-	+	+	+	-	+	+	+	-	+	-	-
+	-	+	-	+	-	+	-	+	+	+	-	+	+	-	-
+	+	-	+	-	-	-	+	+	+	-	+	+	+	-	-
-	+	+	+	+	-	-	-	+	+	+	+	-	-	+	-
+	-	+	+	-	+	-	+	-	+	+	-	+	-	+	-
+	+	-	-	+	+	+	-	-	+	-	+	+	-	+	-
+	+	-	-	+	+	-	+	+	-	+	-	-	+	+	-
+	-	+	+	-	+	+	-	+	-	-	+	-	+	+	-
-	+	+	+	+	-	+	+	-	-	-	-	+	+	+	-
+	+	+	+	+	+	+	+	+	+	+	-	-	-	-	+
+	+	+	+	+	+	+	+	+	+	-	+	-	-	-	+
+	+	+	+	+	+	+	+	+	+	-	-	+	-	-	+
+	+	+	+	+	+	+	+	+	+	-	-	-	+	-	+
+	+	+	+	+	+	+	+	+	+	-	-	-	-	+	+

quantity to be minimized (under the variation of the underlying design) when we are looking for D-optimality. Let

$$K_A(\underline{\xi}_1, \underline{\xi}_2) = \sum_{\underline{\xi}_{20} \varepsilon \underline{\xi}_2} |H(\underline{\xi}_1, \underline{\xi}_{20})| \tag{4.1}$$

$$K_D(\underline{\xi}_1, \underline{\xi}_2) = \prod_{\underline{\xi}_{20} \varepsilon \underline{\xi}_2} |H(\underline{\xi}_1, \underline{\xi}_{20})| \tag{4.2}$$

$$K_A^*(\underline{\xi}_1, \underline{\xi}_2) = [K_A(\underline{\xi}_1, \underline{\xi}_2)] \cdot \frac{1}{\binom{\nu_2}{2k}} \tag{4.3}$$

$$K_D^*(\underline{\xi}_1, \underline{\xi}_2) = [K_D(\underline{\xi}_1, \underline{\xi}_2)]^{1/\binom{\nu_2}{\nu_k}} \tag{4.4}$$

The quantities K_A^* and K_D^* will be referred to respectively as the criteria for AD-optimality, and DD-optimality. Here, AD means average (or arithmetic mean) of the determinant, and DD, the geometric mean. (Note that D-optimality itself corresponds to the geometric mean of the roots of $H(\underline{\xi}_1)$; hence the notation DD.). We may also consider a probability distribution $[q(\underline{\xi}_{20}), \underline{\xi}_{20} \varepsilon \underline{\xi}_2]$, where $\sum_{\underline{\xi}_{20} \varepsilon \underline{\xi}_2} q(\underline{\xi}_{20}) = 1$, and $q(\underline{\xi}_{20})$ represents the probability (or degree of belief) that $\underline{\xi}_{20}$ represents the nonnegligible subset. Then we may define the criteria (the suffix "w" standing for "weighted")

$$K_{wA}^* = \sum_{\underline{\xi}_{20} \varepsilon \underline{\xi}_2} q(\underline{\xi}_{20}) |H(\underline{\xi}_1, \underline{\xi}_{20})| \tag{4.5}$$

$$K_{wD}^* = \prod_{\underline{\xi}_{20} \varepsilon \underline{\xi}_2} |H(\underline{\xi}_1, \underline{\xi}_{20})|^{q(\underline{\xi}_{20})} . \tag{4.6}$$

Analogous criteria based on the average of trace $[H(\underline{\xi}_1, \underline{\xi}_{20})]^{-1}$ or the smallest root of $H(\underline{\xi}_1, \underline{\xi}_{20})$, etc. can be defined. These will be extensions of the notions of A-, or E-optimality of Kiefer. For lack of space, we shall study these elsewhere.

Consider now $H(\underline{\xi}_1, \underline{\xi}_{20})$. We have

(4.7) $$H(\underline{\xi}_1, \underline{\xi}_{20}) = \left[\begin{array}{c|c} A_1'A_1 & A_1'A_{20} \\ \hline A_{20}'A_1 & A_{20}'A_{20} \end{array}\right]$$

(4.8) $$|H(\underline{\xi}_1, \underline{\xi}_{20})| = C_1 |H_{1,20}|,$$

where

(4.9) $$H_{1,20} = A_{20}'GA_{20},\ C_1 = |A_1'A_1|$$

(4.10) $$G = I_N - A_1(A_1'A_1)^{-1}A_1',$$

and where we are assuming that $(A_1'A_1)$ is nonsingular, since otherwise even $\underline{\xi}_1$ would not be estimable. Notice that G (N×N) is idempotent. Clearly, in order to study the criteria K_A^* etc., we need to study $H_{1,20}$. We shall look at a somewhat more general case, and consider H_{1t}, where

(4.11) $$H_{1t} = B_t'GB_t,$$

and where B_t contains an arbitrary, but fixed set of t columns of A_2, so that

(4.12) $$B_t = [\underline{b}_1, \ldots, \underline{b}_t],$$

where $\underline{b}_i$ (i = 1,...,t) are a set of t <u>distinct</u> columns of A_2. We shall sometimes write $H_{1t} = H_{1t}(B_t) = H_{1t}(\underline{b}_1, \ldots, \underline{b}_t)$ to emphasize its dependence upon the $\underline{b}_i$, which will be denoted dually as $\underline{b}(i)$. We have

$$H_{1t}(B_t) = \begin{bmatrix} \underline{b}_1' \\ \vdots \\ \underline{b}_t' \end{bmatrix} G[\underline{b}_1, \ldots, \underline{b}_t]$$

$$= \begin{bmatrix} \underline{b}_1'G\underline{b}_1 & \underline{b}_1'G\underline{b}_2 & \cdots & \underline{b}_1'G\underline{b}_t \\ \underline{b}_2'G\underline{b}_1 & \underline{b}_2'G\underline{b}_2 & \cdots & \underline{b}_2'G\underline{b}_t \\ \cdots & \cdots & \cdots & \cdots \\ \underline{b}_t'G\underline{b}_1 & \underline{b}_t'G\underline{b}_2 & \cdots & \underline{b}_t'G\underline{b}_t \end{bmatrix} .$$

Let $h(i,j) = \underline{b}_i'G\underline{b}_j$, i,j = 1,...,t. Let P_t denote the symmetric group of permutations of the set $I_t^* = \{1,2,\ldots,t\}$. If $p \in P_t$, then p can be decomposed as a product of disjoint cycles, say n_p in number. Let p have the cycle structure $[\theta_{p1}, \ldots, \theta_{pn_p}]$. If $i \in I_t^*$, let $p(i)$ denote the image of i under p. Also, define sign $p =$ $(+1)$ or (-1), according as p is an even or odd permutation. Then, we have,

$$(4.14)\qquad |H_{1t}(B_t)| = \sum_{p\varepsilon P_t} (\text{sign } p)[\prod_{i=1}^{t} h(i,p(i))].$$

For any $p\varepsilon P_t$, let I_t^* be decomposed into the disjoint sets I_{pj} $(j = 1,\ldots,n_p)$, such that I_{pj} has θ_{pj} elements, and that if $i\varepsilon I_{pj}$, then the elements of I_{pj} can be written as $i = p^0(i)$, $p^1(i)$, $p^2(i)$, $\ldots$, $p^{\theta_{pj}-1}(i)$, since $p^{\theta_{pj}}(i) = i$. Let $i_{pj}\varepsilon I_{pj}$, for $j = 1,\ldots,n_p$, and all $p\varepsilon P_t$. Then the following is clear.

LEMMA 4.1. *We have*

$$(4.15)\qquad |H_{1t}(B_t)| = \sum_{p\varepsilon P_t} (\text{sign } p) \prod_{j=1}^{n_p} [\prod_{u=1}^{\theta_{pj}} h(p^{u-1}(i_{pj}),p^u(i_{pj}))].$$

The parameter set $\underline{\xi}_2$ will be said to be a "k-independent" set if for all columns $\underline{b}$ in A_2, there exist positive numbers $q(\underline{b})$, such that

$$(4.16)\qquad q(\underline{\xi}_{20}) = \prod_{\underline{b}\varepsilon A_{20}} q(\underline{b}),$$

for all $\underline{\xi}_{20}\varepsilon\underline{\xi}_2$. Roughly speaking, "k-independence" of $\underline{\xi}_2$ will mean that we believe with positive probability that k (but no more) elements of $\underline{\xi}_2$ are nonnegligible, and that the degree of belief of one parameter being nonnegligible does not depend upon that of another parameter being so. Let

$$(4.17)\qquad B = \sum_{\underline{b}\varepsilon A_2} (\underline{b}\underline{b}')q(\underline{b}),\ B_0 = \sum_{\underline{b}\varepsilon A_2} (\underline{b}\underline{b}')$$

the summation running over all columns $\underline{b}$ of A_2. Then, if $\underline{\xi}_2$ is k-independent, we have,

$$K_{wA}^* = c_1 \sum_{\underline{\xi}_{20}\varepsilon\underline{\xi}_2} \prod_{\underline{b}\varepsilon A_{20}} q(\underline{b})|A_{20}'GA_{20}|$$

$$= c_1 \sum_{B_t} q(\underline{b}_1)\ldots q(\underline{b}_t)|H_{1t}(B_t)|.$$

Now, B_t consists of a set of t distinct columns of A_2, and $H_{1t}(B_t)$ is constructed using these. From (4.13), it is clear that if (say) $\underline{b}_1 = \underline{b}_2$, then $H_{1t}(B_t) = 0$. Hence we obtain

LEMMA 4.2. *We have*

$$(4.19)\qquad K^*_{wA} = \frac{c_1}{t!} \sum_{\underline{b}_1} \sum_{\underline{b}_2} \cdots \sum_{\underline{b}_t} q(\underline{b}_1)\ldots q(\underline{b}_t)\,|H_{1t}(\underline{b}_1,\ldots,\underline{b}_t)|,$$

where each summation runs over the set of all possible (ν_2) *columns of* A_2.

The reason for the term (1/t!) is that there are (t!) permutations of rows of $H_{1t}(B_t)$. From (4.15) and (4.19) we obtain

$$(4.20)\qquad K^*_{wA} = c_1 \sum_{\underline{b}_1} \sum_{\underline{b}_2} \cdots \sum_{\underline{b}_t} q(\underline{b}_1)\ldots q(\underline{b}_t) \sum_{p\varepsilon P_t} (\text{sign } p) \prod_{j=1}^{n_p} H^*(p,j),$$

where

$$(4.21)\qquad H^*(p,j) = \prod_{u=1}^{\theta_{pj}} (\underline{b}(p^{u-1}(i_{pj})))'G(\underline{b}(p^{u}(i_{pj})))$$

$$= [\underline{b}(i_{pj})]'G[\underline{b}(p(i_{pj}))][\underline{b}(p(i_{pj}))]'G[\underline{b}(p^2(i_{pj}))]$$

$$\times [\underline{b}(p^2(i_{pj}))]'G[\underline{b}(p^3(i_{pj}))]\ldots[\underline{b}(p^{\theta_{pj}-1}(i_{pj}))[G]\ \underline{b}(i_{pj})]$$

Hence,

$$(4.22)\qquad K^*_{wA} = \frac{c_1}{t!} \sum_{p\varepsilon P_t} (\text{sign } p) \prod_{j=1}^{n_p} [\text{trace } G^*(p,G,j)],$$

where $G^*(p,G,j)$ is an $(N\times N)$ matrix depending on p, G and j, and is given by

$$G^*(p,G,j) = G\{\sum_{\underline{b}(i_{pj})} q(p(i_{pj}))[\underline{b}(p(i_{pj}))][\underline{b}(p(i_{pj}))]'\} \times$$

$$\times\ G\{\sum_{\underline{b}(p^2(i_{pj}))} q(p^2(i_{pj}))[\underline{b}(p^2(i_{pj}))][\underline{b}(p^2(i_{pj}))]'\}\times$$

$$\times \ldots \times$$

$$\times\ G\{\sum_{\underline{b}(p^{\theta_{pj}}(i_{pj}))} q(p^{\theta_{pj}}(i_{pj}))[\underline{b}(p^{\theta_{pj}}(i_{pj}))][\underline{b}(p^{\theta_{pj}}(i_{pj}))]'$$

$$= \prod_{u=1}^{\theta_{pj}} [G\{\sum_{\underline{b}(p^{u}(i_{pj}))} q(p^{u}(i_{pj}))[\underline{b}(p^{u}(i_{pj}))][\underline{b}(p^{u}(i_{pj}))]'\}]$$

$$= (GB)^{\theta_{pj}}.$$

LEMMA 4.3. <u>We have</u>

$$K^*_{wA} = c_1 c_2, \tag{4.23}$$

$$c_2 = \frac{1}{t!} \sum_{p \varepsilon P_t} (\text{sign } p) \prod_{j=1}^{n_p} [\text{trace}(GB)^{\theta_{pj}}] \tag{4.24}$$

Now, let $n(\ell_1,\ldots,\ell_t)$ denote the number of distinct permutations p^* in P_t such that p^* has exactly ℓ_i cycles of length i ($i = 1,\ldots,t$). Then, clearly,

$$\ell_1 + 2\ell_2 + \ldots + t\ell_t = t, \quad \ell_i \geq 0 \ (i = 1,\ldots,t). \tag{4.25}$$

Also, it is well known (for example, De Bruijn [3]) that

$$n(\ell_1,\ldots,\ell_t) = \frac{t!}{(\ell_1!)(2^{\ell_2}\ell_2!)(3^{\ell_3}\ell_3!)\ldots(t^{\ell_t}\ell_t!)}. \tag{4.26}$$

Now let $p^* \varepsilon P_t$, and suppose p^* can be decomposed as a product of exactly ℓ_i cycles of length i ($i = 1,\ldots,t$). Now, every cycle of length i (≥ 2) can be expressed as a product of $(i - 1)$ cycles of length 2. Also, a cycle of length 1 can be trivially expressed as a product of 2 cycles of length 2 each. On the other hand, p^* is an odd or even permutation according as p^* can be expressed as the product of (respectively) an odd or even number of permutations. It thus follows that we shall have

$$(\text{sign } p^*) = \prod_{i=2}^{t} [(-1)^{i-1}]^{\ell_i} = (-1)^t(-1)^{\ell_1+\ldots+\ell_t}, \tag{4.27}$$

using (4.22). Hence, from (4.24) - (4.27), we get

$$(4.28)\quad c_2=(-1)^t \sum_{\ell_1,\ldots,\ell_t} \frac{(-1)^{\ell_1+\ldots+\ell_t}}{(\ell_1!)(2^{\ell_2}\ell_2!)\ldots(t^{\ell_t}\ell_t!)} \prod_{i=1}^{t} [\mathrm{trace}(GB)^i]^{\ell_i},$$

where the summation is overall values of $(\ell_1,\ldots,\ell_t)$ satisfying (4.25). We summarize the result in

THEOREM 4.1. _We have_

$$(4.29)\qquad K^*_{wA} = c_1c_2$$

where c_1 _and_ c_2 _are given by_ (4.9) _and_ (4.28). (_Note that_ t=1 _corresponds to_ $\ell_1 = 1$.)

COROLLARY 4.1. _Let_ $c_2(t)$ _denote the value of_ c_2 _as a function of_ t. _Then_

$$(4.30)\qquad c_2(1) = tr(GB),\ c_2(2) = \frac{1}{2}[\{tr(GB)\}^2 - tr(GB)^2].$$

We now consider an important special case. The search model (2.2) is said to be orthogonal if for some real number a_0, we have

$$(4.31)\qquad [A_1 \vdots A_2][A_1 \vdots A_2]' = a_0I_N.$$

Suppose (4.31) holds. Then

$$B_0 = \sum_{\underline{b}\varepsilon A_2} \underline{bb}' = A_2A_2' = a_0I_N - A_1A_1',\ \text{and}$$

$$GB_0 = [I_N-A_1(A_1'A_1)^{-1}A_1'][a_0I_N-A_1A_1']=a_0[I_N-A_1(A_1'A_1)^{-1}A_1']=a_0G.$$

Since G is idempotent, we obtain (for all positive integers i),

$$tr(GB_0)^i=a_0^i trG^i=a_0^i trG=a_0^i[N-trA_1(A_1'A_1)^{-1}A_1']=a_0^i[N-trI_r]=a_0^i(N-\nu_1).$$

This leads to the following result.

THEOREM 4.2. _Under_ (4.31), c_2 _is a function of_ a_0, t, N, _and_ ν_1 _alone, and is given by_

$$(4.32)\qquad c_2 = (-a_0)^t \sum_{\ell_1,\ldots,\ell_t} \frac{(\nu_1-N)^{\ell_1+\ldots+\ell_t}}{(\ell_1!)(2^{\ell_2}\ell_2!)\ldots(t^{\ell_t}\ell_t!)},$$

where the sum runs over all $(\ell_1,\ldots,\ell_t)$ _satisfying_ (4.25). _In_ this case

$$(4.33) \qquad c_2(1) = a_0(N-\nu_1),\ c_2(2) = \frac{1}{2}\, a_0^2(N-\nu_1)(N-\nu_1-1)$$

One important application of the above is to the theory of general $s_1 \times \ldots \times s_m$ factorial experiments. Thus, it is well known that if $(\underline{\xi}_1', \underline{\xi}_2')$ contains all the $s_1 \times \ldots \times s_m$ effects (in some order), and if $\underline{y}(N\times 1)$ corresponds to observations on N distinct treatment-combinations, then (2.2) holds, and, furthermore, the model is orthogonal (provided the effects are defined in the normalized form). In particular, the 2^m factorial model discussed in Section 3, is orthogonal, and Theorem 4.2 is applicable to it with $a_0 = 2^m$. Note that the above remarks hold for every subset of parameters $\underline{\xi}_1$ (out of the total set of $\nu_1 + \nu_2$ parameters).

5. *DD-optimality*. The results on AD-optimality in Theorem 4.2 indicate that at least in the case of orthogonal search models, the AD-optimality criterion K_A does not depend on A_2, and is proportional to $|A_1'A_1|$. Thus, for such models, a design which is optimal for estimating $\underline{\xi}_1$ (when $\underline{\xi}_2$ is assumed zero) is also AD-optimal. In other words, the AD criterion is not "sensitive" to the "search" aspect. This fact is also revealed by a close examination of K_{wA}^* at (4.5). Notice that if a design T minimizes K_{wA}^*, it could still be such that $|H(\underline{\xi}_1, \underline{\xi}_{20})|$ is zero, for some $\underline{\xi}_{20} \varepsilon \underline{\xi}_{20}$. Thus, T could minimize K_{wA}^* although it may not even be a search design (i.e., not satisfy (2.3)). On the other hand, k_{wD}^* will be nonzero if and only if (2.3) is satisfied. Hence, from this angle, the DD-optimality seems preferable. Nevertheless, K_{wA}^* is also useful. Knowing its value, we know an arithmetic mean, which gives some idea of the geometric mean K_{wD}^*. Also, we have found it mathematically tractable, a property which is not enjoyed by K_{wD}^*.

Indeed, a look at (4.20), (4.21) and (4.6) shows what problems we run into. Below, we obtain a bound on K_D when t = 1. Since G is idempotent of rank $(N - \nu_1)$, we can write

$G = \sum_{i=1}^{N-\nu_1} \underset{\sim}{g}_i \underset{\sim}{g}_i'$, where $\underset{\sim}{g}_i$ $(i = 1,\ldots,N-\nu_1)$ are mutually orthogonal normalized vectors of size $(N\times 1)$ each. Hence, $\underset{\sim}{b}'G\underset{\sim}{b}=\underset{\sim}{b}'(\sum_{i=1}^{N-\nu_1} \underset{\sim}{g}_i\underset{\sim}{g}_i')\underset{\sim}{b}$

$= \sum_{i=1}^{N-\nu_1} (\underset{\sim}{b}'\underset{\sim}{g}_i)^2 \geq (N-\nu_1)^{-1}[\sum_{i=1}^{N-\nu_1} \underset{\sim}{b}'\underset{\sim}{g}_i]^2 = (N-\nu_1)^{-1}(\underset{\sim}{b}'\underset{\sim}{g})^2$, where $\underset{\sim}{g} = \sum_{i=1}^{N-\nu_i} \underset{\sim}{g}_i$. Thus

$$K^*_{wD} = \prod_{\underset{\sim}{b}\varepsilon A_2} [c_1(\underset{\sim}{b}'G\underset{\sim}{b})]^{q(\underset{\sim}{b})} \geq c_1(N-\nu_1)^{-1} \prod_{\underset{\sim}{b}\varepsilon A_2} (\underset{\sim}{b}'\underset{\sim}{g})^{2q(\underset{\sim}{b})}, \tag{5.1}$$

where $q(\underset{\sim}{b}) = q(\xi_{20})$, ξ_{20} being the element of $\underset{\sim}{\xi}_2$ which corresponds to the column $\underset{\sim}{b}$ of A_2. Notice that in (5.1), we have used the fact that we must have $\sum_{\underset{\sim}{b}\varepsilon A_2} q(\underset{\sim}{b}) = 1$. This leads to

THEOREM 5.1. In the notation introduced above, and for t=1, the bound (5.1) on K^*_{wD} holds.

The bound (5.1), and particularly the inequality

$$\underset{\sim}{b}'G\underset{\sim}{b} \geq (N-\nu_1)^{-1}(\underset{\sim}{b}'\underset{\sim}{g})^2, \tag{5.2}$$

should be useful in investigations on optimal search designs. To test whether a design is a search design, we need to test whether $|A_{20}'GA_{20}| > 0$, for all $A_{20}(2k\times N) \subset A_2$. A necessary condition for this to happen is that we have $\underset{\sim}{b}'G\underset{\sim}{b} > 0$ for all columns $\underset{\sim}{b}\varepsilon A_2$. Now $\underset{\sim}{b}'\underset{\sim}{g} > 0$ implies (by (5.2)) that $\underset{\sim}{b}'G\underset{\sim}{b} > 0$. Since, often it would be much easier to obtain the value of $\underset{\sim}{b}'\underset{\sim}{g}$ than of $\underset{\sim}{b}'G\underset{\sim}{b}$, the above helps in verifying an important necessary condition for a design to be a search design.

It is clear that both in AD and DD-optimality, we need to check whether the matrix $H_{1t}(B_t)$ is nonsingular for all sets of t columns B_t contained in A_2, which of course is equivalent to verifying (2.3). The nature of the expression for K^*_{wD} is such that it seems (at this stage of the development of the theory) that it will be necessary to verify the nonsingularity of $H_{1t}(B_t)$ separately for each of the $\binom{\nu_2}{t}$ values of B_t. (Recall that for a given value of k in the search linear model (2.2), the value of t in the above discussion will be 2k.) On the other hand, if for a design T, the verification of (2.3) has been conducted, and T is found to be a search design then both K^*_{wA} and K^*_{wD} become meaningful. This shows the importance of K^*_{wA}. It also shows the importance of developing tools for checking (2.3) for a tremendously large number of cases. In the next section we present results, which seem to be very useful in the context of 2^m factorials.

Because of the above, the problems associated with the construction of designs which are optimal with respect to K^*_{wA} or K^*_{wD} will be considered in future publications.

It would be instructive to make some remarks here on the quadratic forms $\underline{b}'G\underline{b}(\underline{b}\in A_2)$ for factorial designs. Consider, for example, a design of the type given by Srivastava and Ghosh in Table II, Sec. 8. Actually, to evaluate K^*_D, we did not have to calculate $|H_{12}(B_2)|$ for all B_2 in A_2. The reason is that $|H_{12}(B_2)|$ takes on only a few distinct values (under variation of B_2) because of the fact that the designs considered are all B-arrays of full strength. It will take much space and will require the theory of association schemes to elaborate this fully and prove exactly worded theorems for the general case. However, some intuitive hints will be offered. If a design T is a B-array of full strength, it is clear that it is symmetric with respect to factors. In other words, for $1 \leq \ell \leq m$, any subset of ℓ factors occurs in T in exactly the same way as any other subset of ℓ-factors. This is reflected in the structure of G (see Srivastava [19], Bose and Srivastava [17], Srivastava and Chopra [18]). Thus, suppose $\underline{\xi}_1$ contains μ, the main effects and the two factor interactions, and T consists of two sets Ω_{mu} and Ω_{mv} $(1 \leq u < v \leq m)$. Then it is intuitively clear that G will be of the form $[\frac{G_{11} | G_{12}}{G_{21} | G_{22}}]$, where G_{11}, G_{22} are of the form aI + bJ, and G_{12} and G_{21} are of the form cJ where a, b, c are real numbers, I is the identity matrix, and J the matrix all of whose elements are unity and where I and J are of appropriate orders. Also, suppose $\underline{b}_1$ and $\underline{b}_2$ are two columns of A_2, which respectively correspond to two distinct ℓ-factor $(1 \leq \ell \leq m)$ interactions, then it is clear that we shall have $\underline{b}_1 = [\frac{\underline{b}_{11}}{\underline{b}_{12}}]$, $\underline{b}_2 = [\frac{\underline{b}_{21}}{\underline{b}_{22}}]$, where the order of $\underline{b}_{11}$ and $\underline{b}_{12}$ are the same as those of G_{11}, and where $\underline{b}_{11}$ shall be a permutation of $\underline{b}_{12}$, and $\underline{b}_{21}$ a permutation of $\underline{b}_{22}$. From this, it is not difficult to check that $\underline{b}_1' G \underline{b}_1$ and $\underline{b}_2' G \underline{b}_2$ are equal. Thus, it is clear that $\underline{b}' G \underline{b}$ will take the same value for all columns $\underline{b}$ which correspond to ℓ-factor interactions. Thus, $\underline{b}' G \underline{b}$ will, take atmost (m-2) distinct values, as $\underline{b}$ ranges over A_2. Similar remarks apply to $|H_{1t}(B_t)|$ for larger t.

In passing, some general comments on optimality criteria for search designs seem to be called for. In a decision theory set up, one could consider vector loss risk functions. For example, one could be explicitly interested in maximizing (i) K^*_{WA}, and (ii) q^*, where q^* is the probability of correct search. (In other words, if $\underline{\xi}_{21}$ denotes the subset of parameters $\underline{\xi}_2$ which is nonnegligible, then q^* is the probability that our search procedure (for searching the nonnegligible set of parameters) will deliver $\underline{\xi}_{21}$). Since the study of q^* depends upon the search procedure selected, it will be considered elsewhere in conjunction with studies on the latter.

6. *"Main effect plus plans for 2^m factorials.* Consider the search linear model (2.2) corresponding to a 2^m factorial when $\underline{\xi}_1 = \{\mu; F_1, \ldots, F_m\}$, and $\underline{\xi}_2$ is the set of remaining effects out of which some (unknown) effects are nonnegligible and are to be searched. A design T suitable for this purpose is said to be a "main effect plus plan" (MEPP). Such plans will be discussed in

in the sequel in the frame work of Theorem 2.2. Thus, let $T = T_1 + T_2$, i.e., T $(N \times m)$ consists of 2 disjoint sets T_1 $(N_1 \times m)$ and T_2 $(N_2 \times m)$, when $N_1 + N_2 = N$, and where T_1 and T_2 correspond respectively to $\underline{y}_1$ and $\underline{y}_2$.

In this approach, we take T_1 to be some "good" design for estimating $\underline{\xi}_1$ when $\underline{\xi}_2 = \underline{0}$ (i.e., some "good" main effect plan (MEP)). For a fixed T_1, we then ask the question of how to select T_2 so that $T_1 + T_2$ would be a "good" MEPP. In Srivastava and Gupta (1974), T_1 was taken to be the set of $(m + 2)$ treatments consisting of Ω_{m0}, Ω_{m1}, and Ω_{mm}, which is a B-array of full strength. For this T_1, necessary and sufficient conditions on T_2 were obtained so that T_1 and T_2 would allow the search and estimation of one extra effect in $\underline{\xi}_2$.

The problem with taking B-arrays of full strength is that the choice of T_1 gets severely constrained. In particular, usually the available T_1 is not good from the point of view of variance-optimality. In view of this, we shall develop a new approach in this section.

We shall consider designs T_1 such that A_{11} has a "group structure" with respect to columns, a property which is often possessed by the classical "optimal" designs. (A matrix L is said to have a group structure with respect to columns if the Schur (or Hadamard) product of any two or more (not necessarily distinct) columns of L is also a column of L.) Notice that in the present case, the columns of A_{11} (and also of A_{21}) correspond respectively to $\mu, F_1 \ldots F_m$. Thus, the elements of A_{11} are ± 1, the first column having +1 everywhere. Thus if $m = 2^n - 1$, then the columns of A_{11} corresponding to $F_1 \ldots F_h$ can be chosen arbitrarily (subject to the restriction that the Hadamard product of any ℓ of these $(1 \leq \ell \leq h)$ does not equal $(1,1,\ldots,1)'$), and the other columns of A_{11} being filled by multiplying any two or more of these. In view of this, we shall name the columns of A_{11} respectively as $(\Pi_0, \Pi_1, \ldots, \Pi_h, \Pi_{12}, \ldots, \Pi_{h-1,h}, \Pi_{123}, \ldots, \Pi_{123\,h})$, where $\Pi_0, \Pi_1, \ldots, \Pi_h$ are the first $(h+1)$ columns corresponding to $\mu, F_1, \ldots, F_h$ chosen as above. Also, for $2 \leq \ell \leq h$, and any subset $(j_1,\ldots,j_h)$ of ℓ distinct integers from the set, the column of A_{11} which is the Hadamard product of the columns with names $\Pi_{j_1},\ldots,\Pi_{j_\ell}$ will be given the name $\Pi_{j_1\ldots j_\ell}$. The columns of A_1 and A_{11} will be given the same names. Also, we shall treat the Π's as the set of symbols of effects of a 2^h factorial, with $\Pi_{j_1\ldots j_\ell} = \Pi_{j_1}\ldots\Pi_{j_\ell}$, and the Π's forming a group of order 2^h as explained in Section 3. (For example, $\Pi_{12}\Pi_1 = \Pi_2$, etc.). Thus, the columns of A_1 (and hence of A_{11}) have dual names, these being the F_j's $(j = 0,1,\ldots,m)$, and the Π's. For ease of correspondence between the two notations, we shall define, for $j = 0,1,\ldots,m$, the function $\gamma(j)$ as being such that $\Pi_{\gamma(j)}$ and F_j are names for the <u>same</u> columns of A_{11}. (For example, when $h = 3$, these names are respectively $F_0 \equiv \Pi_0$, $F \equiv \Pi_1$, $F_2 \equiv \Pi_2$, $F_3 \equiv \Pi_3$, $F_4 \equiv \Pi_{12} \equiv \Pi_1\Pi_2$, $F_5 \equiv \Pi_{13} \equiv \Pi_1\Pi_3$, $F_6 \equiv \Pi_{23} \equiv \Pi_2\Pi_3$, and $F_7 \equiv \Pi_{123} \equiv \Pi_1\Pi_2\Pi_3$. Then $\gamma(1) = 1$, $\gamma(5) = 13$, $\gamma(7) = 123$, etc.)

Consider now a design T_2 to go with the above T_1. We shall obtain a sufficient condition so that

(6.1) $$\text{rank}(A_{11},\underline{b}_1,\underline{b}_2) = m + 3$$

for all distinct $\underline{b}_1$ and $\underline{b}_2$ in A_2, or equivalently, so that (2.6), (2.7) are satisfied for $k = 1$. We shall assume throughout that T_1 is so chosen that (2.6) holds.

Let

(6.2) $$A_1 = \left[\frac{A_{11}}{A_{21}}\right] = \left[\frac{\underline{a}_0,\underline{a}_1,\ldots,\underline{a}_m}{\underline{\alpha}_0,\underline{\alpha}_1,\ldots,\underline{\alpha}_m}\right];\ \underline{b}_j = \left[\frac{\underline{d}_j}{\underline{\delta}_j}\right],\quad j = 1,2;$$

where $\underline{a}_0$ $(N_1 \times 1)$ and $\underline{\alpha}_0$ $(N_2 \times 1)$ have 1 everywhere, and $\underline{a}_i$ and $\underline{d}_j$ are $(N_1 \times 1)$ and $\underline{\alpha}_i$ and $\underline{\delta}_j$ are $(N_2 \times 1)$. Now, consider Q in (2.7), given by $Q = A_{22} - A_{21}(A_{11}'A_{11})^{-1}A_{11}'A_{12}$. A sufficient condition for (6.1) is that Q has property P_2. Let

(6.3) $$\underline{Q}_j = \underline{\delta}_j - A_{21}(A_{11}'A_{11})^{-1}A_{11}'\underline{d}_j;\ j = 1,2.$$

We need to obtain a sufficient condition that $\underline{Q}_1$ and $\underline{Q}_2$ are linearly independent. Now $(A'_{11}A_{11})^{-1}(A'_{11}A_{11}) = I_{\nu_1}$. Hence,

(6.4) $$(A'_{11}A_{11})^{-1}A'_{11}\underline{a}_j = \underline{\rho}_j,$$

where $\underline{\rho}_j(\nu_1 \times 1)$ is a vector having 1 in the jth position and 0 elsewhere. This gives

(6.5) $$A_{21}(A'_{11}A_{11})^{-1}A'_{11}\underline{a}_j = \underline{\alpha}_j;\ j = 0,\ldots,m.$$

Consider an (arbitrary) interaction (say, Θ) involving the factors $i_1,\ldots,i_\ell$, where $\ell \geq 2$, and $(i_1,\ldots,i_\ell)$ is a subset of distinct integers out of the set $(1,2,\ldots,m)$. Then the column of A_2 corresponding to Θ is the Hadamard product of the columns $F_{i_1},\ldots,F_{i_\ell}$ of A_1. These ℓ columns are also respectively named $\Pi_{\gamma(i_u)}$, $u = 1,\ldots,\ell$. Since the Π's form a group, there exists an integer $\varepsilon = \varepsilon(i_1,\ldots,i_u) = \varepsilon(\Theta)$, in the set $\{0,1,\ldots,m\}$, such that $\Pi_{\gamma(\varepsilon)} = \Pi_{\gamma(i_1)}\Pi_{\gamma(i_2)}\cdots\Pi_{\gamma(i_\ell)}$. Then, clearly, the column of A_2 corresponding to Θ will be the same as the column of A_1 corresponding to F_ε, i.e., the column $\left[\frac{\underline{a}_\varepsilon}{\underline{\alpha}_\varepsilon}\right]$. Now, suppose $\underline{b}_1$ and $\underline{b}_2$ of (6.2) are two columns of A_2 corresponding to the two distinct effects Θ_1 and Θ_2. Let

(6.6) $$\varepsilon_1 = \varepsilon(\Theta_1),\ \varepsilon_2 = \varepsilon(\Theta_2).$$

Using (6.3) - (6.6), we get

(6.7) $$\underline{Q}_j = \underline{\delta}_j - \underline{\alpha}_{\varepsilon_j};\ j = 1,2.$$

Since we want Q to have property P_2, we get

LEMMA 6.1. *Under the set up of the above discussion, a necessary and sufficient condition that Q has property P_2 is that for every pair of distinct Θ_1 and $\Theta_2 \in \xi_2$, we have*

(6.8) $$\underline{0} \neq (\underline{\delta}_1 - \underline{\alpha}_{\varepsilon_1}) \neq \pm(\underline{\delta}_2 - \underline{\alpha}_{\varepsilon_2}) \neq \underline{0}.$$

PROOF. The result follows from the fact that $\underline{Q}_1$ and $\underline{Q}_2$ can have only three distinct elements 0, 2, and (-2).

THEOREM 6.1. *Under the above set up, a necessary condition* that Q has P_2 is

(6.9) $$N_2 \geq \log_3(1 + 2\nu_2)$$

PROOF. The number of nonnull ($N_2 \times 1$) vectors with elements (0,2,-2) such that none is (± 1) times the other is clearly $\frac{1}{2}(3^{N_2} - 1)$. Hence, from Lemma 6.1, we have

(6.10) $$\frac{1}{2}(3^{N_2} - 1) \geq \nu_2,$$

leading to (6.9).

A subset of ℓ ($\leq m$) distinct factors $F_{i_1}, \ldots, F_{i_\ell}$ will be said to be a "complimentary" set, if and only if we have

(6.11) $$\varepsilon(i_1, \ldots, i_\ell) = 0, \text{ or } \Pi_{\gamma(i_1)}\Pi_{\gamma(i_2)} \cdots \Pi_{\gamma(i_\ell)} = \Pi_0.$$

Also, the set of ℓ distinct integers $(i_1, \ldots, i_\ell)$ in $\{1,2,\ldots,m\}$ is said to be a "complimentary" set if and only if $F_{i_1}, \ldots, F_{i_\ell}$ are so.

THEOREM 6.2. *A sufficient condition that Q has P_2 is that for every pair of distinct complimentary sets $(i_1, \ldots, i_u)$ and $(j_1, \ldots, j_v)$ in $\{1,2,\ldots,m\}$, where u, $v \leq m$, there exists a treatment $x = (x_1, \ldots, x_m)$ in T_2, where x may depend on $(i, \ldots, i_u)$, and $(j_1, \ldots, j_v)$ such that*

(6.12) $$x_{i_1}x_{i_2}\cdots x_{i_u}x_{j_1}\cdots x_{j_v} = -1.$$

(Recall that factor levels are denoted by ± 1.)

PROOF. Take two elements Θ_1 and Θ_2 of $\underline{\xi}_2$ as in Lemma 6.1, and let $(x_1^*, \ldots, x_m^*) = x^*$ be a treatment in T_2. Let Θ_1 involve the distinct factors $i_1, \ldots, i_{\ell_1}$, and Θ_2 involve $j_1, \ldots, j_{\ell_2}$. Let $\varepsilon_1 = \varepsilon(\Theta_1) = \varepsilon(i_1, \ldots, i_{\ell_1})$, and $\varepsilon_2 = \varepsilon(\Theta_2) = \varepsilon(j_1, \ldots, j_{\ell_2})$. Then the element of $(\underline{\delta}_1 - \underline{\alpha}_{\varepsilon_1})$ corresponding to x^* is $(x_{i_1}^* \cdots x_{i_{\ell_1}}^* - x_{\varepsilon_1}^*)$ ($= \eta_1$, say), and that of $(\underline{\delta}_2 - \underline{\alpha}_{\varepsilon_2})$ is $(x_{j_1}^* \cdots x_{j_{\ell_2}}^* - x_{\varepsilon_2}^*)$($= \eta_2$, say). Clearly, (6.8) will hold if there exists an $x^* \varepsilon T_2$ such that one of the expressions η_1 and η_2 is zero, and the other is non-zero. Now, $\eta_1 = \eta_2$ is equivalent to $x^*_{\varepsilon_1}(x^*_{i_1}\ldots x^*_{i_{\ell_1}} x^*_{\varepsilon_1} - 1) =$ $x^*_{\varepsilon_2}(x^*_{j_1}\ldots x^*_{j_{\ell_2}} x^*_{\varepsilon_2} - 1)$. Now $x^*_{i_1}\ldots x^*_{i_{\ell_1}} x^*_{\varepsilon_1}$ and $x^*_{j_1}\ldots x^*_{j_{\ell_2}} x^*_{\varepsilon_2}$ are both ± 1; hence, if one of them is 1, and the other is (-1), then (6.8) will hold. But $(i_1,\ldots,i_{\ell_1},\varepsilon_1)$ and $(j_1,\ldots,j_{\ell_2},\varepsilon_2)$ are both complimentary sets. Hence, letting $u = \ell_1 + 1$, $v = \ell_2 + 1$, we find from (6.12) that there exists a treatment $x \varepsilon T_2$ such that $x_{i_1}\ldots x_{i_{\ell_1}} x_{\varepsilon_i} = -x_{j_1}\ldots x_{j_{\ell_2}} x_{\varepsilon_2}$. Hence, in view of the earlier remarks, (6.8) holds. This completes the proof.

The last theorem is very useful for considering a candidate T_2 to go with a given T_1 (for which A_{11} has the group structure). The reason is that the number of complimentary sets in the set $\{1,\ldots,m\}$ are relatively few, and the condition (6.12) can be simply and quickly checked.

We now show the inadequacy of the classical (nonsequential) fractionally replicated factorial designs, from the search view point. Consider, in the usual notation, an s^m factorial experiment, where s is a prime number or a power of a prime number, so that the finite field with s elements GF(s) exists. Consider the finite Euclidean geometry EG(m,s) of m dimensions based on GF(s). A "proper" fractionally replicated design T_0 is a set of points lying on some (m-h)dimensional $(h > D)$, hyperplane of EG(m,s); such designs are usually termed s^{m-h} fractional replicates. For such designs, it is well known that we can estimate exactly s^{m-h} $(< s^m)$ linear combinations of factorial effects, each combination containing exactly s^h distinct factorial effects which are called aliases of each other. Now if two elements of $\underline{\xi}_2$ happen to belong to the same alias set, a situation which would mostly arise except in pathological cases, then they are confounded. Thus, if used as single stage designs, the classical designs are inadequate.

This leads us back to the problem of obtaining good search

designs. Notice that the basic problem is to check statements like (6.1). We close this section with a new theorem on (0,1) matrices which (though easy to prove) has been found very valuable in this regard. Notice that in (6.1), we are essentially working with (0,1) matrices since the first column of A_{11} (which has 1 everywhere) can be added to all columns of A_1 and A_2, and the result divided by 2, to give us a (0,1) matrix on which the same rank condition needs to be checked.

Now let R $(g \times h)$, $g \geq h$, be a (0,1) matrix over the real field, and R_0 the same matrix over GF(2). Let $\underline{R}$ $(g \times 1)$ denote some column of R, and let $\underline{R}^*$ denote the sum of $\underline{R}$ <u>and</u> several <u>other</u> columns of R. Let c be the largest integer $(c \geq 0)$ such that 2^c divides all elements of $\underline{R}^*$, and let $\underline{R}_0 = 2^{-c}\underline{R}^*$. Let R_1 be the matrix obtained from R by replacing the column $\underline{R}$ by $\underline{R}_0$ and keeping the other columns as they are. Let R_{10}^* be the matrix obtained from R_1 by replacing all even and odd numbers in R_1 respectively by 0 and 1. Then R_{10}^* is a (0,1) matrix over the real field; let R_{10} be the same matrix over GF(2).

THEOREM 6.3. With these definitions, (i) <u>if the columns of</u> R_0 <u>are linearly independent, so are those of</u> R, (ii) Rank R = Rank R_{10}, (iii) <u>if the columns of</u> R_{10} <u>are linearly independent, so are those of</u> R_1, <u>and hence, also those of</u> R.

PROOF. Clearly, it is sufficient to consider the case when $g = h$. Consider $|R|$, where for any square matrix K, $|K|$ denotes the determinant of K. Note that $|R|$ is an integer, which is odd if and only if $|R_0| \neq 0$. This proves (i). The result (ii) is obvious from the theory of determinants. Again, from (i), it follows that $|R_{10}| \neq 0$ implies $|R_{10}^*| \neq 0$. On the other hand, since R_{10}^* is obtained from R_1 by reducing each element (mod 2), it is easy to see that $(|R_{10}^*| - |R_1|)$ is an even integer. Now if $|R_{10}| \neq 0$, then $|R_{10}^*|$ must be an odd integer. Hence, from the proceeding remark $|R_1|$ must be an odd integer, and hence $|R_1| \neq 0$. This completes the proof.

The utility of this theorem lies in the fact that it gives us a very fast method of checking the rank of a (0,1) matrix R over the real field. We first check the rank of R_0. The rank of R_0 is relatively far easier to check because of the ease of operations over GF(2). If R_0 turns out to be of full rank, the job is done. On the other hand, it is possible that R be of full rank, while R_0 may not be of full rank. In this case, we further study the rank of R by considering R_{10}, which is easily obtainable. In practice, for checking conditions like (6.1), this theorem has, in general, been found most useful among all results known to the author on (0,1) matrices.

7. *"Bypassing" search: Block designs with multi-dimensional blocks.* We have discussed the importance of search, and the criteria for optimal search designs. We have also considered applications to factorial designs without blocks. One important situation where the search concept may be gainfully employed is in the search for interaction in two-way or multi-way cross-classification.

One might be tempted to think that the above would also be true of large Latin squares, and other similar designs, like the rwo-column designs for eliminating heterogeneity in two directions. Such designs, sometimes called multi-dimensional designs (Bose and Srivastava [17], Srivastava and Anderson [16]), are, of course, potential areas of application of the search concept. However, in large designs of this type, like, for example, large row-column designs, the author is of the opinion that instead of using such designs and searching for interactions, we could almost bypass the latter by simply not using such designs. The reason is that a new class of designs, whose concept we now introduce, seems to be by far much more efficient in an over all sense.

We shall briefly consider a new type of row-column design, which would be parallel to the idea of the row-column designs for the two-way elimination of heterogeneity. The concept can be easily extended to the multi-way case.

We call the new designs, the block-treatment designs with

two-dimensional blocks. In other words, we have, say, v treatments, and b blocks such that in each block one can eliminate heterogeneity in two directions. Of course, heterogeneity between blocks is also eliminated as usual. Thus, the ith block ($i = 1,\ldots,b$) can be represented as a rectangle, say of size $m_i \times n_i$, heterogeneity being eliminated both between the rows of the rectangle, and the columns of the rectangle. Observations could be assumed (with much more certainty than usual) to be independent from one experimental unit to another, and to have constant variance σ^2. An important special case would be when all rectangles are of the same size, i.e., $m_i = m$, $n_i = n$, for all i. Designs with smaller values of m and n are likely to be more useful.

We illustrate by a simple example with $v = 5$, $b = 5$, the blocks being as in (7.1) below:

(7.1)

I		II		III		IV		V	
1	2	2	3	3	4	4	0	0	1
3	4	4	0	0	1	1	2	2	3

Notice that, in the general case of blocks of size $m \times n$, if any block has four treatments a, b, c, d, in the configuration

a	b
c	d

, i.e., a and b occur in the same row, c and d occur in the same row (but a row different from that in which a and b occur), a and c occur in the same column, and b and d occur together in another column, then the contrast (a-b-c+d) would be estimable. The model assumed is

$$y_{i\alpha\beta u} = g_{i\alpha} + h_{i\beta} + \tau_u + \varepsilon_{i\alpha\beta u}, \tag{7.2}$$

where $y_{i\alpha\beta u}$ denotes the observed yield when treatment u is applied to the experimental unit occurring in the αth row and βth column inside the ith block, $g_{i\alpha}$ is the effect of the αth row of the ith block, $h_{i\beta}$ is the effect of the βth column of the ith block, τ_u is the effect of the uth treatment, and $\varepsilon_{i\alpha\beta u}$ denotes the error, the errors being independent with mean 0, and variance σ^2. Thus, in the design of (7.1), the contrasts $(\tau_1-\tau_2-\tau_3+\tau_4)$, $(\tau_2-\tau_3-\tau_4+\tau_0)$,

$(\tau_3-\tau_4-\tau_0+\tau_1)$, $(\tau_4-\tau_0-\tau_1+\tau_2)$, and $(\tau_0-\tau_1-\tau_2+\tau_3)$ are estimable, implying that $(\tau_i-\tau_j)$, $i,j = 0,\ldots,4$, are all estimable. (Hence, this design is connected.) The analysis of the data can be conducted as usual using the theory of linear models. It is intuitively clear that there is much less chance of encountering row-column interactions in such designs than in ordinary large row-column designs involving the same number (bmn) of units. Also, it seems that heterogeneity will be eliminated to a much greater extent. The "negative" effect (on the power of tests of linear hypothesis) of the existence of nuisance parameters in the form of block effects, can be taken care of by using the theory of inter-block information.

8. *Some numerical results on resolution* 5.1 *designs*. We present here the values of $c_1^{-1}K_A^*$ and $c_1^{-1}K_D^*$ for balanced designs of resolution 5.1 for 2^m factorials obtained in Srivastava and Ghosh [13]. Note that the value of K_A^* is obtainable directly using (4.32), for designs in which all treatments are distinct. (Such designs have an index set $\underline{\lambda}'$ in which all elements are 0 or 1.) The value of a_0 in this case is 2^m. The equation (4.32) gives $c_2(2) = K_A c_1^{-1}$, while Table 2 presents $c_1^{-1}K_A^*$. Thus, to obtain the entries for K_A^* from $c_2(2)$, we must divide the latter by $\binom{\nu_2}{2}$, where $\nu_2 = 2^m - 1 - \binom{m}{1} - \binom{m}{2}$.

It is interesting to observe that the arithmetic and geometric mean are quite close to each other when m = 5. Also as N increases, for any fixed m, both means increase.

Table 2: Value of $c_1^{-1}K_A^*$ and $c_1^{-1}K_D^*$ for some resolution 5.1 designs

$\underline{\lambda}'$	$c_1^{-1}K_A^*$	$c_1^{-1}K_D^*$	$\underline{\lambda}'$	$c_1^{-1}K_A^*$	$c_1^{-1}K_D^*$
m = 5					
111010	85.33	83.86	11100011	94.56	66.62
111011	128.00	128.00	11100100	709.25	480.40
211011	139.41	139.14	11010010	709.23	461.16
211012	155.94	155.50	01100101	709.25	471.19
011101	384.00	384.00	01010011	709.27	455.95
111100	384.00	381.81	11100101	780.20	518.71
111101	469.34	469.09	11010011	780.18	569.66
111102	476.18	475.91	01100110	1185.48	923.84
211102	507.38	506.92	11100110	1276.66	1012.12
011110	776.53	773.54	11100111	1371.22	1085.97
111110	896.00	896.00	11110000	2009.61	1884.05
			01110001	2009.62	1901.42
m = 6			11101000	2009.57	1551.01
			01101001	2009.57	1513.47
1110010	71.36	59.18	11001100	2009.56	1950.87
1100110	71.28	64.82	01001101	2009.61	1953.81
1110011	99.91	85.16	11110001	2127.80	2010.48
1110100	499.52	410.64	11101001	2127.79	1661.65
0110101	499.51	380.31	11001101	2127.80	2080.00
1110101	570.87	452.17			
0110102	507.65	389.28			
1110102	584.68	460.04	m = 8		
1110103	590.55	463.37	111000010	76.90	46.96
			011000011	76.89	50.60
m = 7			111000011	98.81	61.24
11100010	70.93	49.16			

References

[1] Box, G. E. P. and Draper, N. R. (1959). A basis for the selection of a response surface design. *J. Amer. Statist. Assoc.* 54, 622-654.

[2] Chernoff, H. (1953). Locally optimum designs for estimating parameters. *Ann. Math. Statist.* 24, 586-602.

[3] DeBruijn, N. G. (1964). Polya's theory of counting. In E. F. Beckenbach (Ed.) *Applied Combinatorial Mathematics*, 144-184. Wiley, New York.

[4] Elfving, G. (1952). Optimum allocation in linear regression theory. *Ann. Math. Statist.* 23, 255-262.

[5] Hedayat, A., Raktoe, B. L. and Federer, W. T. (1974). On a measure of aliasing due to fitting an incomplete model. *Ann. Statist.* 2, 650-660.

[6] Karlin, S. and Studden, W. J. (1966). Optimal experimental designs. *Ann. Math. Statist.* 37, 1439-1888.

[7] Kiefer, J. (1959). Optimum experimental designs. *J. Roy. Statist. Soc. Ser. B*, 21, 272-319.

[8] Kiefer, J. (1961). Optimum designs in regression problems, II. *Ann. Math. Statist.* 32, 298-325.

[9] Kiefer, J. (1961). Optimum experimental designs V, with applications to systematic and rotatable designs. *Proc. Fourth Berkeley Symp. Math. Stat. Prob.* 1, 381-405. Univ. of California Press.

[10] Kiefer, J. (1962). Two more criteria equivalent to D-optimality of designs. *Ann. Math. Statist.* 33, 792-796.

[11] Kiefer, J. (1975). Construction and optimality of generalized Youden designs. In J. N. Srivastava (Ed.) *A Survey of Statistical Design and Linear Models*, 333-353. North-Holland Publishing Company, Inc., New York.

[12] Kiefer, J. and Wolfowitz, J. (1959). Optimum designs in regression problems. *Ann. Math. Statist.* 30, 271-294.

[13] Srivastava, J. N. and Ghosh, S. (1975). Balanced 2^m factorial designs of resolution V which allow search and estimation of one extra unknown effect $4 \leq m \leq 8$. *Communications in Statistics*, to appear.

[14] Srivastava, J. N. and Ghosh, S. (1976). Infinite series of designs. Unpublished.

[15] Srivastava, J. N. and Chopra, D. V. (1974). Trace-optimal 2^7 fractional factorial designs of resolution V, with 56 to 68 runs. *Utilitas Mathematicas*, 5, 263-279.

[16] Srivastava, J. N. and Anderson, D. A. (1971). Factorial subassembly association scheme and multidimensional partially balanced designs. *Ann. Math. Statist.* 42, 1167-1181.

[17] Srivastava, J. N. and Bose, R. C. (1964). Multidimensional partially balanced designs and their analysis, with applications to partially balanced factorial fractions. *Sankhya*, Ser. A, 26, 145-168.

[18] Srivastava, J. N. and Chopra, D. V. (1971). On the characteristic roots of the information matrix of balanced fractional 2^m factorial designs of resolution V, with applications. *Ann. Math. Statist.* 42, 722-734.

[19] Srivastava, J. N. (1961). Contribution to the construction and analysis of designs. Ph.D. thesis. Univ. of North Carolina at Chapel Hill.

[20] Srivastava, J. N. and Chopra, D. V. (1971). Optimal balanced 2^m factorial designs of resolution V: $m \leq 6$. *Technometrics* 13, 257-269.

[21] Srivastava, J. N. (1975). Designs for searching non-negligible effects. In J. N. Srivastava (Ed.) *A Survey of Statistical Design and Linear Models*, 507-519. North-Holland Publishing Company, Inc., New York.

[22] Srivastava, J. N. (1976). Some further theory of search linear models. Contribution to *Applied Statistics*. Published by the Swiss-Australian Region of Biometry Society. 249-256.

[23] Wynn, H. P. (1975). Simple conditions for optimum design algorithms. In J. N. Srivastava (Ed.) *A Survey of Statistical Design and Linear Models*, 571-579. North-Holland Publishing Company, Inc., New York.

OPTIMAL DESIGNS FOR INTEGRATED VARIANCE IN POLYNOMIAL REGRESSION

By W. J. Studden
Purdue University

1. *Introduction.* For each $x \in [-1,1]$ an experiment can be performed with a random outcome with mean value $\sum_{i=0}^{n} \theta_i x^i$ and constant variance σ^2. The parameters θ_i and σ^2 are assumed to be unknown. A design is a probability measure μ on $[-1,1]$. Observations are to be taken at the points $x \in [-1,1]$ proportional in number to the measure μ. If N observations are allowed and μ concentrates mass p_ν at points x_ν, $\nu = 1,2,\ldots,r$, and $Np_\nu = n_\nu$ are integers then the experimenter takes n_ν observation at x_ν. All observations are assumed to be uncorrelated. The information matrix of a design μ is the matrix $M(\mu) = (m_{ij})_{i,j=0}^{n}$ where $m_{ij} = \int_{-1}^{1} x^{i+j} d\mu(x)$. From the standard least squares theory we know that the covariance matrix of the least squares estimates of the parameters θ_i is given by $(\sigma^2/N)M^{-1}(\mu)$. The variance of the estimate of an arbitrary linear combination $\sum d_i\theta_i$ is proportional to $d'M^{-1}(\mu)d$ where $d'=(d_0,d_1,\ldots,d_n)$. In particular if $d'=(1,x,\ldots,x^n)$ for a fixed x value then $\sum d_i\theta_i = \sum \theta_i x^i$ is the regression curve at x. If $f'(x)=(1,x,\ldots,x^n)$ then the variance is proportional to $V(\mu,x) = f'(x)M^{-1}(\mu)f(x)$. The problem discussed here is the minimization of this variance integrated with respect to some measure σ on $[-1,1]$. That is, we wish to minimize

$$V(\mu,\sigma) = \int f'(x)M^{-1}(\mu)f(x)d\sigma(x) \tag{1.1}$$

with respect to the design μ for a fixed probability measure σ. A design μ^* which minimizes $V(\mu,\sigma)$ will be called an I_σ-optimal design. If trA denotes the trace of A then (1.1) can be written as

*Research supported by National Science Foundation Grant MP575-08294.

$$(1.2) \qquad V(\mu,\sigma) = \mathrm{tr} M^{-1}(\mu) M(\sigma)$$

The results presented here were stimulated by a preprint of Fedorov and Malyutov [2] that indicated that for σ = constant, the appropriate μ was supported on the same set of points as the D-optimal design which maximizes the determinant of $M(\mu)$. The proof given there seems to be in error. The error seems to be that the term $(-1)^i$ in equation (3.8) below is missing. The resulting design is extremely close to the minimizing design μ^*.

In Section 2 we consider a general procedure for constructing an approximate I_σ-optimal design and illustrate the procedure with an example. Although the minimization of (1.1) can be done fairly readily on a computer the procedure outlined here can be done by hand for low values of n and the results give further insight into the design problem. In Section 3 we consider bounds for our error in calculating (1.1); indicate a class of σ where the procedure is exact and give a fairly accurate simple expression for the bound for the case of Lebesque measure. In Section 4 we consider the asymptotic behavior of $V(\mu,\sigma)$ as n becomes large. This resulting asymptotic expression can be minimized with respect to μ when σ has a density. In the asymptotic case, as usual, the relationship between μ and σ in $V(\mu,\sigma)$ becomes considerable simplified.

2. *General procedure.* Optimal designs for minimizing a number of expressions involving the information matrix $M(\mu)$ have been considered. The design minimizing $\sup_{x\in[-1,1]} V(\mu,x)$ or equivalently maximizing the determinant of $M(\mu)$ is called a D-optimal design. The extrapolation problem involves minimizing $V(\mu,x_0) = f'(x_0)M^{-1}(\mu)f(x_0)$ for some $x_0 \notin [-1,1]$. In both of these as well as minimizing $V(\mu,\sigma)$, or anything else involving $M^{-1}(\mu)$ in a positive definite sense, it is known that one can restrict oneself to designs μ concentrating on at most n-1 points on the interior (-1,1). An explanation for this will be indicated below. We

therefore restrict ourselves to designs μ concentrating on $x_0 = -1 < x_1 <\ldots< x_{n-1} < x_n = 1$ with corresponding weights $p_0, p_1, \ldots, p_n$ some of which are possibly zero. Throughout the paper we assume that σ concentrates on at least n+1 points. In this case any μ minimizing $V(\mu,\sigma)$ must be such that $M(\mu)$ is non-singular. In this case all the p_i values must be positive.

Our problem reduces to finding the points x_i and the weights p_i. The difficult part seems to be the points x_i. For a given set of points x_i the appropriate weights can readily be found. This is done by noting that $V(\mu,x)$ is invariant with respect to a basis change for the powers of x. Thus we use, instead of $1, x, \ldots, x^n$, the n+1 Lagrange polynomials on the points $x_0, x_1, \ldots, x_n$. These are the polynomials $\ell_i(x)$ which are zero at all the points x_j, $j \neq i$ and are one at x_i. That is $\ell_i(x_j) = \delta_{ij}$, $i,j = 1,2,\ldots,n$. In terms of these polynomials we find that the information matrix is diagonal with elements $p_0, p_1, \ldots, p_n$ so that

$$V(\mu,x) = \sum_{i=0}^{n} \frac{\ell_i^2(x)}{p_i} \tag{2.1}$$

and

$$V(\mu,\sigma) = \sum_{i=0}^{n} \frac{k_i}{p_i}, \quad k_i = \int \ell_i^2(x)\,d\sigma(x) \tag{2.2}$$

Schwarz's inequality can then be used to minimize this expression. For the integrated variance $V(\mu,\sigma)$ in (2.2) the p_i should be chosen proportional to $\sqrt{k_i}$. That is

$$p_i = \frac{\sqrt{k_i}}{\sum_{n=0}^{n} \sqrt{k_j}} \quad \text{where } k_i = \int \ell_i^2(x)\,d\sigma(x). \tag{2.3}$$

With this choice of weights we have

$$V(\mu,\sigma) = \Big(\sum_{i=0}^{n} \big(\int \ell_i(x)\,d\sigma(x)\big)^{1/2}\Big)^2 \tag{2.4}$$

This quantity must then be minimized with respect to the points $x_1,\ldots,x_{n-1}$.

For the case n = 1 this gives $V(\mu,\sigma)$ as a function of one variable x_1. The quantity $V(\mu,\sigma)$ can be differentiated with respect to x_1 and set equal to zero. The resulting equation is given in Murty and Studden [6] and seems to be rather cumbersome. One or two iterations of Newton's method starting at $x_1 = 0$ give good approximations.

The reason that the design measures can be restricted to n-1 interior points and the basis for our procedure for finding I_σ-optimal designs rests on some well known moment theory. Note that the matrix $M(\sigma)$ involves the moments $c_i=\int x^i d\sigma(x)$, $i=0,1,\ldots$, 2n. Consider the class of measures with the same moments c_i as σ for $i = 0,1,\ldots,2n-1$. We omit the last moment c_{2n}. It is known that in this class there is a measure $\bar{\sigma}$ concentrating mass on at most n-1 interior points which maximizes the last moment c_{2n} to give $\bar{c}_{2n}$. The n-1 interior points are the roots of the polynomial Q_{n-1} where $Q_0,Q_1,Q_2,\ldots$ are the polynomials orthogonal with respect to the measure $(1-x^2)d\sigma(x)$. The appropriate weights on these points can be found, if needed, by setting up the linear equations matching up the moments $c_0,c_1,\ldots,c_n$.

If we apply this to the design measure μ we see that $\bar{\mu}$ is always at least as good a design as μ if our criterion involves $M(\mu)$ in an appropriate manner.

Our general procedure for finding an I_σ-optimal design is to start with the weight measure σ, pass to the measure $\bar{\sigma}$ maximizing the 2nth moment and then restrict our designs to having mass on the support points of $\bar{\sigma}$. Let $\bar{\sigma}$ concentrate mass $q_0,q_1,\ldots,q_{n-1},q_n$ on the points $x_0 = -1 < x_1 <\ldots< x_{n-1} < x_n = 1$. Then our approximate I_σ-optimal design will use this same set of points with weights given by (2.3). If we write $\sigma = \bar{\sigma} - (\bar{\sigma}-\sigma)$ then the integrals k_i can be expressed by

$$k_i = \int \ell_i^2(x)d\sigma = [q_i - (\bar{c}_{2n}-c_{2n})h_i^2] \tag{2.5}$$

where h_i is the coefficient of x^{2n} in the polynomial $\ell_i^2(x)$. If the x_i and q_i are readily available and the quantities $(\bar{c}_{2n}-c_{2n})h_i^2$ are not then a further approximation would be to use weights proportional to $\sqrt{q_i}$.

For the case n = 1 the two end points are the only points used. The appropriate weights can be calculated from (2.3). When n = 2 and σ is symmetric the resulting I_σ-optimum design is also symmetric so that $x_1 = 0$. The weights can again be calculated readily from (2.3).

We consider a nonsymmetrical case where σ has density $(1+x)/2$. The orthogonal polynomials can easily be written down in terms of the moments $c_0, c_1, c_2, \ldots$ of σ. These are given by

$$Q_{n-1}(x) = \begin{vmatrix} c_0-c_2 & \cdots & c_{n-2}-c_{n-1} & 1 \\ c_1-c_3 & & & x \\ \vdots & & \vdots & \vdots \\ c_{n-1}-c_{n+1} & \cdots & c_{2n-3}-c_{2n-1} & x^{n-1} \end{vmatrix}$$

For the case n = 2 we have

$$Q_1(x) = (c_1-c_3)-(c_0-c_2)x$$

so that the single interior support point is

$$x_1 = \frac{c_1-c_3}{1-c_2} \tag{2.6}$$

If σ has density $(1+x)/2$ this gives $x_1 = 1/5$. The weight's on -1, x_1 and 1 can be calculated by (2.3) to give approximately .1365, .5397 and .3238. The resulting value for $V(\mu,\sigma) = 1.987$. The correct $x_1 = .0642$ is closer to zero and gives a value $V(\mu,\sigma) = 1.936$.

We should remark that in practice it would be hard to gain anything using the exact or the approximate optimal design. The design using -1, 0, 1 with weights 1/8, 4/8, 3/8 would probably work just as well.

The D-optimal design which minimizes $\sup_{x\in[-1,1]} V(\mu,x)$ concentrates equal mass at the zeros of $(1-x^2)P_n'(x) = 0$ where P_n is the nth Legendre polynomial orthogonal with respect to the constant weight or Lebesque measure. If the measure already concentrates its mass on the zeros of $(1-x^2)P_n'(x) = 0$ then the procedure gives the exact I_σ optimal design. If the weight measure is constant the approximate procedure again gives the D-optimal points and the procedure seems to be extremely accurate although not exact. These results are given in the next section.

3. *Bounds on error and σ = constant.*

LEMMA 3.1. Let μ^* denote the design minimizing $V(\mu,\sigma)$. Then

$$V(\mu,\sigma) - V(\mu^*,\sigma) \leq B(\mu,\sigma) \tag{3.1}$$

where

$$B(\mu,\sigma) = \sup_{x\in[-1,1]} \varphi(x,\mu,\sigma) - V(\mu,\sigma) \tag{3.2}$$

and

$$\varphi(x,\mu,\sigma) = f'(x)M^{-1}(\mu)M(\sigma)M^{-1}(\mu)f(x). \tag{3.3}$$

PROOF. The inequality (3.1) can be obtained using the convexity of $V(\mu,\sigma)$ in the design variable μ. The proof follows in outline the similar result contained in Kiefer [4], page 388 for the D-optimal design. For the general case see Kiefer [5] p. 877.

COROLLARY. If σ has mass $q_0,\ldots,q_n$ on the points $x_0,x_1,\ldots,x_n$ and we take the design μ concentrating on $x_0,x_1,\ldots,x_n$ with weights proportional to $\sqrt{q_i}$, $i = 0,1,\ldots,n$ then

$$B(\mu,\sigma) = (\sum\sqrt{q_i})^2(\sup_x \sum_{i=0}^n \ell_i^2(x) - 1) \tag{3.4}$$

COROLLARY. If σ concentrates all its mass on the zeros of $(1-x^2)P_n'(x) = 0$ where $P_n(x)$ is the Legendre polynomial, then the approximate I_σ-design gives the exact I_σ-design.

PROOF. In this case $\sup_x \sum_{i=0}^n \ell_i^2(x) = 1$ and the polynomials

$P_n'(x)$ are orthogonal to $(1-x^2)dx$.

LEMMA 3.2. <u>Let</u> σ = <u>constant</u> = <u>Lebesque measure</u>. <u>Then the approximate</u> I_σ <u>design</u> μ_0 <u>concentrates on the zeros</u> $x_0, x_1, \ldots, x_n$ <u>of</u> $(1-x^2)P_n'(x) = 0$ <u>with weight proportional to</u> $|P_n(x_i)|^{-1}$. <u>Moreover</u>

$$V(\mu_0,\sigma) = \frac{2n}{n(n+1)(2n+1}\left(\sum_{i=0}^{n}|P_n(x_i)|^{-1}\right)^2 \tag{3.5}$$

and

$$B(\mu_0,\sigma) \leq \frac{\left(\sum_{i=0}^{n}|P_n(x_i)|^{-1}\right)^2}{n(n+1)(2n+1)} \tag{3.6}$$

PROOF. The approximate I_σ-optimal design concentrates on the two end points and the zeros of the (n-1)<u>st</u> polynomial orthogonal to $(1-x^2)dx$. This polynomial is $P_n'(x)$ where P_n is the Legendre polynomial.

The calculation of the weights and the evaluation of (3.5) and (3.6) are based on the formula

$$\int_{-1}^{1} \ell_i(x)\ell_j(x)dx = \frac{\left(\delta_{ij} - \frac{1}{2n+1}\right)}{n(n+1)P_n(x_i)P_n(x_j)} \tag{3.7}$$

This calculation was also made in [2]. The evaluation of (3.7) can also be made using quadrature formula from Ghizzetti and Ossicini [3], page 110.

Letting i = j in (3.7) we readily see that the weights for the approximate design are proportional to $|P_n(x_i)|^{-1}$. The value of $V(\mu_0,\sigma)$ given in (3.5) now follows from (2.4) and (3.7) with $i = j$.

In order to calculate the bound on $B(\mu_0,\sigma)$ in (3.6) we calculate the expression $\varphi(x,\mu_0,\sigma)$ used in Lemma 3.1. The leads to

$$\begin{aligned} &\varphi(x,\mu_0,\sigma) \\ &= \frac{\left(\sum_{i=0}^{n}|P_n(x_i)|^{-1}\right)^2}{n(n+1)}\left[\sum_{i=0}^{n}\ell_i^2(x) - \frac{\left(\sum_{i=0}^{n}(-1)^i\ell_i(x)\right)^2}{2n+1}\right] \end{aligned} \tag{3.8}$$

Since for the roots of $(1-x^2)P_n'(x) = 0$ we have $\sum \ell_i^2(x) \le 1$ for $x \in [-1,1]$ it follows that

$$\sup_x \varphi(x,\mu_0,\sigma) \le \frac{(\sum |P_n(x_i)|^{-1})^2}{n(n+1)} \tag{3.8}$$

The bound on $B(\mu_0,\sigma)$ given in (3.6) now follows from (3.8) and Lemma 3.1.

The calculations given below in Table 1 indicate that the value $V(\mu_0,\sigma)$ is extremely close to the actual minimum $V(\mu^*,\sigma)$. Moreover from (3.8), by inserting the value x_i, we see that very little is lost in estimating the bound $B(\mu_0,\sigma)$ by estimating $\sup_x \varphi(x,\mu_0,\sigma)$.

Table 1

	n = 3	n = 4	n = 5
$V(\mu^*,\sigma)$	2.9898	3.8676	4.7562
$V(\mu_0,\sigma)$	2.9920	3.8716	4.7615
$B(\mu_0,\sigma)$	.4987	.4840	.4762

4. *Asymptotic results.* In this section we assume that the measures μ and σ have densities $h(x)$ and $\omega(x)$ with respect to Lebesque measure. In this case we write $V_n(\mu,\sigma) = V_n(h,\omega)$. We have added a subscript n since we are interested in the dependence on n. Under mild restrictions on h and ω it can be shown that

$$V_n(h,\omega) = \operatorname{tr} M^{-1}(h)M(\omega) \simeq \frac{n}{\pi}\int_{-1}^{1} \frac{\omega(x)}{h(x)(1-x^2)^{1/2}}\,dx. \tag{4.1}$$

This is asymptotic in the sense that the ratio goes to one. A proof of this result will not be given here. We note that $\sup_x V_n(\mu,x) \ge n+1$ with equality when μ is the D-optimal design. This indicates that $V_n(\mu,\sigma)$ is of order n. It appears that, roughly speaking, the matrices $M(h)$ and $M(\omega)$ can be diagonalized by approximately the same orthogonal matrix and have eigenvalues $h(x_i)$ and $\omega(x_i)$, $i = 0,1,\ldots,n$, respectively where the x_i values become distributed like the density $\pi^{-1}(1-x^2)^{-1/2}$. This gives, roughly speaking,

$$\mathrm{tr}M^{-1}(h)M(\omega) \simeq \sum_{i=0}^{n} \frac{h(x_i)}{\omega(x_i)}$$

$$\simeq \frac{n}{\pi}\int_{-1}^{1} \frac{h(x)}{\omega(x)(1-x^2)^{1/2}}\, dx.$$

If we now fix ω, the integral in (4.1) can be minimized by applying Schwarz's inequality. The minimizing h is

$$h_0(x) = c\,\frac{\omega^{1/2}(x)}{(1-x^2)^{1/4}} \tag{4.2}$$

where c is a normalizing constant making h_0 have integral one.

The minimum value then gives

$$V_n(h,\omega) \simeq \frac{n}{\pi}\int_{-1}^{1} \frac{\omega(x)}{h(x)(1-x^2)^{1/2}}\, dx$$

$$\geq \frac{n}{\pi}\left(\int_{-1}^{1} \frac{\omega^{1/2}(x)}{(1-x^2)^{1/4}}\, dx\right)^2$$

We further note (again by Schwarz's inequality) that

$$\frac{1}{\pi}\left(\int_{-1}^{1} \frac{\omega^{1/2}(x)}{(1-x^2)^{1/4}}\, dx\right)^2 \leq 1$$

and equality occurs for $\omega_0(x) = \dfrac{1}{\pi(1-x^2)^{1/2}}$. Thus the density that gives the worst possible asymptotic integrated variance is $\omega_0(x)$. For $\omega(x) = 1/2$ we get

$$V_n(h_0,\omega) \simeq \frac{n}{2}\left(\int_{-1}^{1} \frac{1}{(1-x^2)^{1/4}}\, dx\right)^2$$

$$= n(.91388)$$

These values happen to give a fairly accurate estimate of the minimum values $\lambda_n^* = V_n(\mu^*,\omega)$ for $\omega(x) = 1/2$. Letting $\lambda_n = n(.91388)$ we get the following, correct to three figures.

n	2	3	4	5	6	7	8	9
λ_n^*	2.13	2.99	3.87	4.76	5.65	6.55	7.45	8.35
λ_n	1.83	2.74	3.66	4.57	5.48	6.40	7.31	8.22

We note that the design minimizing the asymptotic expression for $V(h,\omega)$ is the density h_0 given in (4.2). Since for a fixed n we should be using only n+1 points of support in our design, we may, replace h_0 by $\bar{h}_0$ in the same way we calculated $\bar{\sigma}$ from σ. This gives an alternate approximate design. For $\omega(x) = 1/2$ this seems to be slightly better than the D-optimal points considered in Section 3. For the example $\omega(x) = (1+x)/2$ considered in Section 2 we get, from (2.6), the value $x_1 = 1/8$ for $n = 2$. This gives $V(\mu,\omega) = 1.945$ as compared with 1.987 for our original procedure and the minimum value 1.936.

References

[1] Fedorov, V. V. (1972). *Theory of Optimal Experiments*. Academic Press, New York.

[2] Fedorov, V. V. and Malyutov, M. B. (1969). On the designs for certain weighted polynomial regression minimizing the average variance. Preprint No. 8, Moscow State University, Laboratory of Statistical Research.

[3] Ghizzetti, A. and Ossicini, A. (1970). *Quadrature Formulae*. Birkhäuser Verlag Basel.

[4] Kiefer, J. (1960). Optimum experimental designs V, with applications to systematic and rotatable designs. *Proc. Fourth Berkeley Symp. Math. Statist. Prob.* 1, 381-405.

[5] Kiefer, J. (1974). General equivalence theory for optimum designs (approximate theory). *Ann. Statist.* 2, 849-879.

[6] Murty, V. N. and Studden, W. J. (1972). Optimal designs for estimating the slope of a polynomial regression. *Jour. Amer. Statist. Assoc.* 67, 869-873.

ASYMPTOTIC EXPANSIONS FOR THE DISTRIBUTION FUNCTIONS OF LINEAR COMBINATIONS OF ORDER STATISTICS

By W. R. van Zwet
University of Leiden

1. *Introduction.* Let $U_{1:N} < U_{2:N} < \ldots < U_{N:N}$ be the order statistics of a sample of size N from the uniform distribution on (0,1), let a_{jN} be real numbers and let h be a real valued function on (0,1). In this paper it is shown that under certain smoothness conditions, $E \exp\{itN^{-\frac{1}{2}}\Sigma_{j=1}^{N} a_{jN}\, h(U_{j:N})\} = O(|t|^{-r} + e^{-\gamma N})$ for every positive integer r and $\gamma > 0$ depending on r (Theorem 4.1).

Without further motivation this result will seem utterly meaningless and the simple statement that it is a crucial step in obtaining an asymptotic expansion for the distribution function of a linear combination of order statistics will not satisfy most readers either. The topic of asymptotic expansions is enjoying a remarkable revival in recent years. However, many of the results that have been obtained are as yet unpublished and the extreme technicality inherent in the subject does not make the published results easily accessible. We shall therefore try to put the result of the present paper in its proper perspective by reviewing some of the relevant aspects of the recent developments.

In Section 2 we begin by explaining very briefly the need for asymptotic expansions. For a complete and much more eloquent account the reader should consult Hodges and Lehmann [17]. Next, we review the classical theory of Edgeworth expansions for sums of independent and identically distributed random variables and indicate the two main techniques for extending this theory to more general statistics. For additional information the reader is referred to the literature quoted in Section 2 and to an excellent review paper by Bickel [2] on the developments in nonparametric statistics.

In Section 3 we give an account of as yet unpublished

results of Bjerve [5] and Helmers [15] who establish Berry-Esseen type bounds for linear combinations of order statistics. At the same time we try to make it clear that the result of the present paper removes a major stumbling block in going from Berry-Esseen bounds to asymptotic expansions.

In Section 4 the result is proved. In Section 5 we note that on the basis of this result, Helmers [16] has indeed obtained an Edgeworth expansion for linear combinations of order statistics with smooth weights.

2. *Edgeworth expansions.* For many years mathematical statisticians have spent a great deal of effort and ingenuity towards applying the central limit theorem in statistics. The estimators and test statistics that interest statisticians are as a rule not sums of independent random variables (r.v.'s) and much work went into showing that they can often be approximated sufficiently well by such sums to ensure asymptotic normality. This work can be traced throughout the development of mathematical statistics from the proof of the asymptotic normality of the maximum likelihood estimator to much of the recent work in nonparametric statistics.

In the last decade it has become increasingly clear to many statisticians that it would be useful to go beyond limit theorems and investigate higher order terms of the distribution functions (d.f.'s) of asymptotically normal estimators and test statistics. The case for moving in this direction was stated with admirable clarity by Hodges and Lehmann [17]. Quite apart from a demand for better numerical approximations, one wishes to distinguish between the multitude of asymptotically efficient statistical procedures that limit theory has provided us with. To make such a distinction one needs asymptotic expansions for the d.f.'s of the statistics involved with remainder term $o(N^{-1})$, as the sample size $N \to \infty$. The problem is not just an academic one because there is ample room for the suspicion that certain asymptotically efficient procedures may converge to their optimal limiting behavior very slowly.

Here again probability theory provides us with a well developed theory for the case of sums of independent r.v.'s, the theory of Edgeworth expansions, of which many excellent accounts are available, e.g. Cramér [11], Gnedenko and Kolmogorov [14], Feller [13] or Petrov [18]. For the purposes of the present paper it will suffice to indicate briefly the techniques employed by showing how one obtains an expansion with remainder $o(N^{-1})$ for the d.f. of a sum of independent and identically distributed (i.i.d.) r.v.'s. The starting point is a result proved by Esseen [12].

LEMMA 2.1 (Smoothing lemma). _Let_ M _be a positive number_, F _a_ d.f. _on_ R^1 _and_ $\tilde{F}$ _a differentiable function of bounded variation on_ R^1 _with_ $\tilde{F}(-\infty) = 0$, $\tilde{F}(\infty) = 1$ _and_ $|\tilde{F}'| \leq M$. _Define the Fourier-Stieltjes transforms_ $\psi(t) = \int \exp\{itx\}dF(x)$ _and_ $\tilde{\psi}(t) = \int \exp\{itx\}d\tilde{F}(x)$. _Then there exists a constant_ C _such that for every_ T

$$(2.1) \qquad \sup_x |F(x)-\tilde{F}(x)| \leq \frac{1}{\pi} \int_{-T}^{T} \left|\frac{\psi(t)-\tilde{\psi}(t)}{t}\right| dt + \frac{CM}{T} .$$

Now let $X_1, X_2, \ldots$ be i.i.d. r.v.'s with $EX_1 = 0$, $EX_1^2 = 1$, $EX_1^3 = \mu_3$ and $EX_1^4 = \mu_4 < \infty$, and let F_N and ψ_N denote the d.f. and the characteristic function (c.f.) of $N^{-\frac{1}{2}} \Sigma_{j=1}^N X_j$. For $N \to \infty$ and $|t| = o(N^{\frac{1}{2}})$ we have

$$(2.2) \quad \log \psi_N(t) = -\frac{1}{2} t^2 - \frac{i}{6} \mu_3 N^{-\frac{1}{2}} t^3 + \frac{1}{24}(\mu_4 - 3)N^{-1}t^4 + o(N^{-1}t^4)$$

and this expansion is easily converted into

$$(2.3) \qquad \psi_N(t) = \tilde{\psi}_N(t) + o(N^{-1}|t|e^{-\frac{1}{4}t^2}),$$

where

$$(2.4) \quad \tilde{\psi}_N(t) = e^{-\frac{1}{2}t^2}[1 - \frac{i}{6} \mu_3 N^{-\frac{1}{2}} t^3 + \frac{1}{24}(\mu_4 - 3)N^{-1}t^4 - \frac{1}{72} \mu_3^2 N^{-1} t^6].$$

For a sufficiently small $\delta > 0$ this expansion for $\psi_N(t)$ remains valid for $|t| \leq \delta N^{\frac{1}{2}}$ because $|\psi_N(t)| \leq (1-t^2/3N)^N \leq \exp\{-\frac{1}{3}t^2\}$ for $|t| \leq \delta N^{\frac{1}{2}}$. It follows that

$$(2.5) \qquad \int_{-\delta N^{\frac{1}{2}}}^{\delta N^{\frac{1}{2}}} \left|\frac{\psi_N(t)-\tilde{\psi}_N(t)}{t}\right| dt = o(N^{-1}),$$

$$(2.6) \qquad \int_{|t|\geq\delta N^{\frac{1}{2}}} \left|\frac{\tilde{\psi}_N(t)}{t}\right| dt = o(N^{-1}).$$

If we now assume that Cramér's condition (C) holds, i.e. that the c.f. ψ_1 of X_1 satisfies

$$(2.7) \qquad \limsup_{|t|\to\infty} |\psi_1(t)| < 1,$$

we immediately find

$$(2.8) \qquad \int_{N^{3/2}\geq|t|\geq\delta N^{\frac{1}{2}}} \left|\frac{\psi_N(t)}{t}\right| dt = o(N^{-1}).$$

Finally we note that $\tilde{\psi}_N$ is the Fourier-Stieltjes transform of

$$(2.9) \quad \tilde{F}_N(x) = \Phi(x)-\phi(x)\left[\frac{\mu_3}{6N^{\frac{1}{2}}}(x^2-1)+\frac{\mu_4-3}{24N}(x^3-3x)+\frac{\mu_3^2}{72N}(x^5-10x^3+15x)\right],$$

where Φ and ϕ denote the d.f. and density of the standard normal distribution. Combining (2.5), (2.6), (2.8) and Lemma 2.1 we find

$$\sup_x |F_N(x)-\tilde{F}_N(x)| = o(N^{-1})$$

which establishes the Edgeworth expansion $\tilde{F}_N$ of F_N.

The problem of extending the theory of Edgeworth expansions from sums of independent r.v.'s to the r.v.'s of interest to statisticians is obviously more complicated - and certainly more laborious - than the corresponding problem for the central limit theorem. So far, only a limited number of specific cases have been solved. Although one might say that in each of these cases the solution depends on artificially creating sums of independent r.v.'s where none exist, two rather different techniques appear to have evolved.

One of these techniques is based on expansion of the r.v. itself. Suppose we wish to obtain an expansion for the d.f. of a statistic T_N which admits an expansion

$$T_N = N^{-\frac{1}{2}}S_{1N} + N^{-1}S_{2N} + N^{-3/2}S_{3N} + R_N, \tag{2.10}$$

where $S_N = (S_{1N}, S_{2N}, S_{3N})$ is a sum of i.i.d. random vectors and $R_N = o(N^{-1})$ with probability $1 - o(N^{-1})$. Then R_N can be neglected and we are basically back in the case of a sum of i.i.d. r.v.'s. Statistics admitting more complicated expansions in terms of S_N than (2.10) can be handled by appealing to the theory of Edgeworth expansions for sums of i.i.d. random vectors that was developed by Bikjalis (e.g. [4]). Such approximation techniques were used successfully by Chibisov and Pfanzagl [19]-[21], to obtain Edgeworth expansions for a number of statistics occurring in parametric models, such as maximum likelihood estimators and, more generally, minimum contrast estimators. Some version of Cramér's condition (C) is obviously required in each of these results.

A different technique was used by Albers, Bickel and van Zwet [1] and Bickel and van Zwet [3] to obtain Edgeworth expansions for linear rank statistics for the one- and two-sample problems under the hypothesis and under contiguous alternatives. Here the approach is based on conditioning rather than approximation of the statistic. Given the right conditioning, the conditional distribution of the linear rank statistic T_N is that of a sum of independent r.v.'s in the one-sample problem, or that of such a sum conditional on a similar sum in the two-sample problem. In both cases this yields an explicit representation for the conditional c.f. of T_N with enough product structure present to enable one to expand it and thus obtain an Edgeworth expansion for the conditional d.f. of T_N. An expansion for the unconditional d.f. of T_N is then obtained by taking the expected value. A complicating factor is that we are dealing with sums of lattice r.v.'s so that Cramér's condition (C) obviously does not hold.

3. *Linear combinations of order statistics.* Having outlined the results obtained and the techniques developed so far, let us now turn to the problem of obtaining an expansion for the d.f. of a linear combination of order statistics. Let $X_1, X_2, \ldots$ be i.i.d. r.v.'s with a common d.f. F and let $X_{1:N} \leq X_{2:N} \leq \cdots \leq X_{N:N}$ denote the order statistics corresponding to $X_1, \ldots, X_N$. For real numbers a_{jN}, $j = 1, \ldots, N$, $N = 1, 2, \ldots$, define

$$T_N = \frac{1}{\sigma_N} \left(\sum_{j=1}^{N} a_{jN} X_{j:N} - \mu_N \right) \tag{3.1}$$

where μ_N and σ_N^2 denote the expected value and the - supposedly finite - variance of $\Sigma a_{jN} X_{j:N}$. Let F_N and ψ_N denote the d.f. and c.f. of T_N.

The problem of obtaining an asymptotic expansion for F_N is, of course, trivial if F is the exponential distribution. In this case $n X_{1:N}$, $(n-1)(X_{2:N} - X_{1:N}), \ldots, (X_{N:N} - X_{N-1:N})$ have independent exponential distributions and T_N is distributed as a normalized weighted sum of i.i.d. exponentially distributed r.v.'s. If F is the uniform distribution on (0,1) the expansion for F_N can also be obtained in a straightforward manner by using the fact that the vector of uniform order statistics is distributed like $(Y_1 + \ldots + Y_{N+1})^{-1} \cdot (Y_1, Y_1 + Y_2, \ldots, Y_1 + \ldots + Y_N)$, where $Y_1, \ldots, Y_{N+1}$ are i.i.d. exponentially distributed r.v.'s.

Another comparatively simple case is that of the α - trimmed mean

$$\bar{X}_\alpha = (N-2m)^{-1} \sum_{j=m+1}^{N-m} X_{j:N}, \quad m = [\alpha N].$$

Given $X_{m:N}$ and $X_{N-m+1:N}$, the α - trimmed mean is distributed as a mean of i.i.d. r.v.'s. Exploiting this fact Bjerve [5] established an Edgeworth expansion for this case.

In general, however, the problem is far from trivial. No satisfactory representation for the c.f. ψ_N of T_N is known so the only available line of attack is expansion of the statistic itself. A first attempt in this direction was made by Bjerve [5] who obtained a Berry-Esseen type bound, i.e.

$$\sup_x |F_N(x)-\Phi(x)| = O(N^{-\frac{1}{2}}). \tag{3.2}$$

His proof proceeds as follows. First $X_{j:N}$ is replaced by $g(Y_{j:N})$, where $g(y) = F^{-1}(1-e^{-y})$ and $Y_{1:N} <\ldots< Y_{N:N}$ are order statistics from an exponential distribution. The function g is assumed to be smooth and Taylor expansion of $g(Y_{j:N})$ about the point $\nu_{jN} = EY_{j:N}$ yields

$$T_N = S_N + Q_N + R_N, \tag{3.3}$$

where

$$S_N = \frac{1}{\sigma_N} \sum a_{jN} g'(\nu_{jN})(Y_{j:N}-\nu_{jN}),$$

$$Q_N = \frac{1}{2\sigma_N} \sum a_{jN} g''(\nu_{jN})\{(Y_{j:N}-\nu_{jN})^2 - E(Y_{j:N}-\nu_{jN})^2\}.$$

At this point the first problem arises, which is to show that $E|R_N| = O(N^{-1})$. More generally, one needs to control the order of magnitude of each term in the expansion and the terms with j near 1 and N create difficulties. Bjerve therefore assumes that there are no weights in the tails, i.e. $a_{jN} = 0$ for $j \leq \alpha N$ or $j \geq \beta N$, $0 < \alpha < \beta < 1$. This gives the required control under reasonable conditions and ensures that $E|R_N| = O(N^{-1})$. Now $Y_{j:N}$ is replaced by $\Sigma_{i=1}^{j}(N-i+1)^{-1} Y_i$, where $Y_1,\ldots,Y_N$ are i.i.d. exponentially distributed r.v.'s. Then S_N becomes a weighted sum of $(Y_1-1),\ldots,(Y_N-1)$ and Q_N becomes a weighted sum of terms (Y_iY_j-1), $i \neq j$, and (Y_i^2-2). Taking minor liberties with Bjerve's proof - he concludes that $P(|R_N| \geq N^{-\frac{1}{2}}) = O(N^{-\frac{1}{2}})$ and proceeds to establish a Berry-Esseen bound for $S_N + Q_N$ but we shall find this inconvenient when we turn to Edgeworth expansions - we write

$$(3.4)\qquad \psi_N(t) = Ee^{itT_N} = Ee^{itS_N}(1+itQ_N+\ldots+\frac{(itQ_N)^{m-1}}{(m-1)!})$$

$$+ O(\frac{|t|^m}{m!}E|Q_N|^m + |t|E|R_N|).$$

Since S_N is a sum of independent r.v.'s it is easy to show that $E\exp\{itS_N\} = \exp\{-\frac{1}{2}t^2\} + O(N^{-\frac{1}{2}}|t|\exp\{-\frac{1}{4}t^2\})$ which is a truncated version of (2.2). Because of the special structure of Q_N, $|E\exp\{itS_N\}Q_N| = O(N^{-\frac{1}{2}}\exp\{-\frac{1}{4}t^2\})$ where the factor $N^{-\frac{1}{2}}$ occurs because $E|Q_N| = O(N^{-\frac{1}{2}})$ and the factor $\exp\{-\frac{1}{4}t^2\}$ reflects the fact that S_N is almost independent of every single term in Q_N. Finally, $E\,Q_N^2 = O(N^{-1})$ and $E|R_N| = O(N^{-1})$, so that (3.4) with $m = 2$ yields

$$(3.5)\qquad \psi_N(t) = e^{-\frac{1}{2}t^2} + O(N^{-\frac{1}{2}}|t|e^{-\frac{1}{4}t^2} + N^{-1}(|t|+t^2)),$$

$$(3.6)\qquad \int_{-N^\varepsilon}^{N^\varepsilon} \left|\frac{\psi_N(t)-e^{-\frac{1}{2}t^2}}{t}\right| dt = O(N^{-\frac{1}{2}})$$

for every $0 < \varepsilon \leq \frac{1}{4}$. To prove (3.2) it remains to consider values of $|t|$ in $(N^\varepsilon, \delta N^{\frac{1}{2}})$ for some $\delta > 0$. This gives rise to the second major problem in the proof which is solved by an ingenious device due to Bickel [2]. First it is shown that $|E\exp\{itS_N\}Q_N^k| = O((kN)^k \exp\{-\eta t^2\})$ for $\eta > 0$ and $k = o(N)$, where the factor $\exp\{-\eta t^2\}$ again arises because S_N is almost independent of each term in Q_N^k. Then the bound $|E\,Q_N^m| = O((cm)^{m}N^{-m/2})$ for some constant c and $m < N$ is established. It follows from (3.4) that for $m = [\log N]$ and $0 < \delta \leq (ce^2)^{-1}$

$$\text{(3.7)} \qquad \int_{N^{\varepsilon}\leq|t|\leq\delta N^{\frac{1}{2}}} \left|\frac{\psi_N(t)-e^{-\frac{1}{2}t^2}}{t}\right| dt = O\left(\frac{\delta^m N^{\frac{1}{2}m}}{(m+1)!}(cm)^m N^{-m/2}+N^{-\frac{1}{2}}\right)$$

$$= O((\delta ce)^m+N^{-\frac{1}{2}}) = O(N^{-\frac{1}{2}}).$$

Together with (3.6) and Lemma 2.1 this proves (3.2).

As a general rule, to every asymptotic result for linear combinations of order statistics for smooth F, there corresponds a similar result for smooth weights a_{jN}. In this case, the Berry-Esseen bound (3.2) for smooth weights was established by Helmers [15]. He takes $a_{jN} = J(j/(N+1))$ for a smooth and bounded function J, shows that for his purpose the weights may be replaced by

$$\text{(3.8)} \qquad a^*_{jN} = N \int_{\frac{j-1}{N}}^{\frac{j}{N}} J(s)ds,$$

and writes

$$\text{(3.9)} \qquad \sum_{j=1}^{N} a^*_{jN}X_{j:N} = N\int_{-\infty}^{\infty}\psi(F_N(x))dx + \int J(s)ds\sum_{i=1}^{N}X_i,$$

where F_N is the empirical d.f. of $X_1,\ldots,X_N$ and

$$\text{(3.10)} \qquad \psi(u) = \int_u^1 J(s)ds - (1-u)\int_0^1 J(s)ds.$$

Now in (3.9), $\psi(F_N)$ is expanded about $\psi(F)$ and this yields

$$\text{(3.11)} \qquad T_N = N^{-\frac{1}{2}}\sum_{i=1}^{N}h_1(X_i)+N^{-3/2}\sum_{i=1}^{N}\sum_{j=1}^{i-1}h_2(X_i,X_j)+R_N$$

for certain functions h_1 and h_2, where $E|R_N| = O(N^{-1})$. In fact Helmers proves $P(|R_N| \geq N^{-\frac{1}{2}}) = O(N^{-\frac{1}{2}})$ but the stronger statement can be proved under only slightly stronger conditions. From here on the proof works in much the same way as Bjerve's proof. There is no need for the assumption that there are no weights in the tails, but the use of Bickel's device at the end of the proof entails the rather unpleasant condition $\int_0^1 |J'(s)|dF^{-1}(s) < \infty$.

The aim of this paper is to open a way to go from these Berry-Esseen bounds to an asymptotic expansion for the d.f. of a linear combination of order statistics. The first step towards an expansion is, of course, to include more terms in the Taylor expansions (3.3) or (3.11). This will not give rise to essentially new difficulties in the analysis of $\psi_N(t)$ for $|t| \leq N^{\varepsilon}$ for sufficiently small $\varepsilon > 0$. Surely this analysis will be much more laborious and technical than before and stronger regularity conditions will have to be imposed. Obtaining an expansion with remainder $o(N^{-1})$, for instance, will involve functions of four r.v.'s instead of two and one will actually have to compute things like

$$E \exp\{itN^{-\frac{1}{2}} \sum_{i=1}^{N} h_1(X_i)\}h_3(X_1,X_2,X_3) =$$

$$= E \exp\{itN^{-\frac{1}{2}} \sum_{i=4}^{N} h_1(X_i)\}E\ h_3(X_1,X_2,X_3)\{1+itN^{-\frac{1}{2}} \sum_{i=1}^{3} h_1(X_i)+\ldots\}.$$

However, after such computations one does arrive at an expansion ψ_N^* of ψ_N satisfying

$$(3.12) \qquad \int_{-N^{\varepsilon}}^{N^{\varepsilon}} \left|\frac{\psi_N(t)-\psi_N^*(t)}{t}\right| dt = o(N^{-1}),$$

or even $o(N^{-\frac{1}{2}k})$ if one takes sufficiently many terms and perseveres. Also, $\psi_N^*(t) = O(\exp\{-\frac{1}{4}t^2\})$ and hence

$$(3.13) \qquad \int_{|t|\geq N^{\varepsilon}} \left|\frac{\psi_N^*(t)}{t}\right| dt = o(\exp\{-\frac{1}{4} N^{2\varepsilon}\}).$$

It follows that if we can show that for some sufficiently small $\varepsilon > 0$ and some sequence $\eta_N \to \infty$,

(3.14) $$\int_{N^{\varepsilon}\leq|t|\leq\eta_N N} \left|\frac{\psi_N(t)}{t}\right| dt = o(N^{-1}),$$

we can establish an asymptotic expansion for F_N with remainder $o(N^{-1})$. If we can prove the stronger assertion that for every positive ε and r,

(3.15) $$\int_{N^{\varepsilon}\leq|t|\leq N^r} \left|\frac{\psi_N(t)}{t}\right| dt = O(N^{-r}),$$

then for every positive integer k, we can -in principle- establish an expansion for F_N with remainder $o(N^{-\frac{1}{2}k})$.

Even in the case of a sum of i.i.d. r.v.'s that was discussed in Section 2, we can't obtain (3.14) or (3.15) from the expansion for $\psi_N(t)$ because this expansion simply breaks down for $|t| \geq CN^{\frac{1}{2}}$ for some $C > 0$. This is true a fortiori in the present case and it is easily checked that Bickel's device cannot produce anything more than a Berry-Esseen type bound. What is needed is obviously something that ensures sufficient smoothness of F_N, which is what Cramér's condition (C) does in the case of sums of i.i.d. r.v.'s. In Section 4 we shall formulate such a set of conditions and show that they ensure that for every positive integer r,

(3.16) $$|\psi_N(t)| = O(|t|^{-r} + e^{-\gamma N}),$$

where $\gamma > 0$ depends on r. This obviously implies (3.15) for every positive ε and r and thus opens a way to obtain asymptotic expansions for the d.f. of a linear combination of order statistics.

4. *A bound on the characteristic function.* Throughout this section h is a real valued function on (0,1) and for N = 1,2,..., $U_{1:N} < U_{2:N} <\ldots< U_{N:N}$ denote the order statistics of a sample of size N from the uniform distribution on (0,1). We begin by showing that for smooth and bounded h and j/N bounded away from 0 and 1, $E \exp\{itN^{\frac{1}{2}}h(U_{j:N})\} = O(|t|^{-1})$ uniformly in N. In view of

Feller [13], Lemma XV.4.4, this is a rather obvious result.

LEMMA 4.1. _Let_ h _be twice differentiable on_ $(0,1)$ _and suppose that positive numbers_ c, C _and_ ε _exist such that_ $h' \geq c$, $|h''| \leq C$ _and_ $\varepsilon N \leq j \leq (1-\varepsilon)N$. _Then there exists a positive number_ M _depending only on_ c, C _and_ ε, _such that_

$$|E \exp\{itN^{\frac{1}{2}}h(U_{j:N})\}| \leq M|t|^{-1} \quad \text{for } t \neq 0.$$

PROOF. For the density $b_{j:N}$ of $U_{j:N}$, Stirling's formula ensures

$$(4.1) \quad b_{j:N}(u) = \frac{N!}{(j-1)!(N-j)!} u^{j-1}(1-u)^{N-j} \leq b_{j:N}\left(\frac{j-1}{N-1}\right) \leq M_\varepsilon N^{\frac{1}{2}},$$

and $b_{j:N}$ is monotone on $(0, \frac{j-1}{N-1})$ and on $(\frac{j-1}{N-1}, 1)$. For $t \neq 0$ partial integration yields

$$(4.2) \quad E \exp\{itN^{\frac{1}{2}}h(U_{j:N})\} = \int_0^1 \exp\{itN^{\frac{1}{2}}h(u)\}h'(u) \frac{b_{j:N}(u)}{h'(u)} du$$

$$= \frac{1}{itN^{\frac{1}{2}}} \exp\{itN^{\frac{1}{2}}h(u)\}\frac{b_{j:N}(u)}{h'(u)} \Big|_0^1 - \frac{1}{itN^{\frac{1}{2}}} \int_0^1 \exp\{itN^{\frac{1}{2}}h(u)\}\left(\frac{b_{j:N}(u)}{h'(u)}\right)' du.$$

If $2 \leq j \leq n-1$, the first term on the right in (4.2) is zero, so

$$|E \exp\{itN^{\frac{1}{2}}h(U_{j:N})\}| \leq \frac{1}{|t|N^{\frac{1}{2}}} \left\{\int_0^1 \frac{|b'_{j:N}(u)|}{c} du + \int_0^1 \frac{C}{c^2} b_{j:N}(u) du\right\}$$

$$\leq \frac{1}{|t|N^{\frac{1}{2}}} \left\{\frac{2}{c} b_{j:N}\left(\frac{j-1}{N-1}\right) + \frac{C}{c^2}\right\} \leq \frac{1}{|t|} \left\{\frac{2M_\varepsilon}{c} + \frac{C}{c^2} N^{-\frac{1}{2}}\right\}. \qquad \square$$

Let a_{jN}, $j = 1,\ldots,N$, $N = 1,2,\ldots$, be real numbers and define

$$(4.3) \qquad L_N = N^{-\frac{1}{2}} \sum_{j=1}^{N} a_{jN} h(U_{j:N}).$$

We shall show that $E \exp\{itL_N\} = O(|t|^{-1} + e^{-\gamma N})$ for smooth and bounded h and for rather severely restricted weights a_{jN}. The proof depends on a conditioning argument and shows in effect that the total influence on L_N of a single order statistic is already sufficient to obtain the required bound.

LEMMA 4.2. _Let_ h _be as in Lemma 4.1 and suppose that_

positive numbers a and A exist such that $a \le a_{jN} \le A$ for $j = 1,\ldots,N$. Then there exist a positive number M depending only on c, C, a and A, and also a positive constant γ such that

$$|E \exp\{itL_N\}| \le M|t|^{-1} + e^{-\gamma N} \quad \text{for } t \neq 0.$$

PROOF. Take $m = [N/2]$. Let $V = (V_{1:m-1},\ldots,V_{m-1:m-1})$ and $W = (W_{1:N-m},\ldots,W_{N-m:N-m})$ be vectors of order statistics from the uniform distribution on (0,1) for sample sizes (m-1) and (N-m) and let $U_{m:N}$, V and W be independent. Then the joint distribution of $U_{1:N},\ldots,U_{N:N}$ is the same as that of $U_{m:N}V_{1:m-1},\ldots,U_{m:N}V_{m-1:m-1}$, $U_{m:N}$, $U_{m:N} + (1-U_{m:N})W_{1:N-m},\ldots,U_{m:N} + (1-U_{m:N})W_{N-m:N-m}$. It follows that

$$(4.4) \qquad E(\exp\{itL_N\}|V=v,W=w) = E \exp\{itN^{\frac{1}{2}}\tilde{h}(U_{m:N})\},$$

where

$$\tilde{h}(u) = N^{-1}[\sum_{j=1}^{m-1} a_{jN}h(v_{j:m-1}u) + a_{mN}h(u) + \\ + \sum_{j=1}^{N-m} a_{m+j,N}h(w_{j:N-m} + (1-w_{j:N-m})u)],$$

$v = (v_{1:m-1},\ldots,v_{m-1:m-1})$ and $w = (w_{1:N-m},\ldots,w_{N-m:N-m})$. Since $h' \ge c > 0$, $|h''| \le C$ and $0 < a \le a_{jN} \le A$, we have for all u

$$(4.5) \qquad \tilde{h}'(u) \ge \frac{ac}{N}[\sum_{j=1}^{m-1} v_{j:m-1}+1+\sum_{j=1}^{N-m}(1-w_{j:N-m})]$$

$$(4.6) \qquad |\tilde{h}''(u)| \le \frac{AC}{N}[\sum_{j=1}^{m-1} v^2_{j:m-1}+1+\sum_{j=1}^{N-m}(1-w_{j:N-m})^2] \le AC.$$

Now there exists a positive constant γ such that for every N

$$(4.7) \qquad P(\sum_{j=1}^{m-1} V_{j:m-1}+1+\sum_{j=1}^{N-m}(1-W_{j:N-m}) \ge \frac{N}{4}) \ge 1-e^{-\gamma N},$$

and hence $\tilde{h}' \ge \frac{ac}{4}$ on a set with probability $1-e^{-\gamma N}$. On this set we can apply Lemma 4.1 with h and j replaced by $\tilde{h}$ and m and the proof is complete. □

Lemma 4.2 certainly looks like a poor result under rather

severe conditions but this appearance is somewhat deceptive. With the aid of one more conditioning argument we shall show that Lemma 4.2 already contains the essence of an apparently much stronger statement under weaker conditions.

THEOREM 4.1. *Suppose that there exist* $0 \leq \alpha < \beta \leq 1$ *and positive numbers* c, C, a *and* A *such that*

(i) h *is twice differentiable on* (α,β) *with* $h' \geq c$ *and* $|h''| \leq C$ *on* (α,β);

(ii) $a \leq a_{jN} \leq A$ *for all* j *with* $\alpha < jN^{-1} < \beta$.

Then for every positive integer r *there exists a positive number* M_1 *depending only on* α, β, c, C, a, A *and* r, *and positive numbers* M_2 *and* γ *depending only on* α,β *and* r *such that*

$$|E\ \exp\{itL_N\}| \leq M_1|t|^{-r} + M_2\ e^{-\gamma N} \quad \text{for} \quad t \neq 0.$$

PROOF. Choose a positive integer r and take $n=[(\beta-\alpha)N/(3r)]$. There exists an integer $N_0 = N_0(\alpha,\beta,r)$ such that for $N \geq N_0$, we have $n \geq 1$ and we can choose integers $k_1,k_2,\ldots,k_{r+1}$ with

$$(4.8) \quad \frac{k_1}{N} \geq \alpha+\tfrac{1}{4}(\beta-\alpha),\ \frac{k_{r+1}}{N} \leq \beta-\tfrac{1}{4}(\beta-\alpha),k_{\nu+1}-k_\nu=n+1,\ \nu=1,\ldots,r.$$

For $N < N_0$ the result of the theorem is trivially true for an appropriate choice of M_2 and γ. Suppose therefore that $N \geq N_0$ and that (4.8) holds with $n \geq 1$. Define the event

$$B = \{U_{k_1:N} > \alpha, U_{k_{r+1}:N} < \beta, U_{k_{\nu+1}:N}-U_{k_\nu:N} > \frac{\beta-\alpha}{4r},\ \nu=1,\ldots,r\}.$$

By applying an exponential bound for the tails of the binomial distribution we find positive numbers M_2 and γ depending on α, β and r, such that

$$(4.9) \qquad P(B) \geq 1 - M_2\ e^{-\gamma N}.$$

Conditional on $U_{k_\nu:N} = s_\nu$, $\nu = 1,\ldots,r+1$, the vectors $(U_{1:N},\ldots,U_{k_1-1:N})$, $(U_{k_\nu+1:N},\ldots,U_{k_{\nu+1}-1:N})$, $\nu=1,\ldots,r$, and $(U_{k_{r+1}+1:N},\ldots,U_{N:N})$ are independent and $(U_{k_\nu+1:N},\ldots,U_{k_{\nu+1}-1:N})$

is distributed as $(s_\nu + (s_{\nu+1}-s_\nu) V_{1:n},\ldots,s_\nu + (s_{\nu+1}-s_\nu)V_{n:n})$ where $V_{1:n} <\ldots< V_{n:n}$ are order statistics from the uniform distribution on $(0,1)$ for sample size n. Taking $k_0 = s_0 = 0$, $k_{r+2} = N + 1$ and $s_{r+2} = 1$, we see that

$$(4.10) \qquad |E(\exp\{itL_N\}|U_{k_\nu:N} = s_\nu,\ \nu = 1,\ldots,r+1)|$$

$$= \prod_{\nu=0}^{r+1} |E(\exp\{itN^{-\frac{1}{2}} \sum_{j=k_\nu+1}^{k_{\nu+1}-1} a_{jN}h(U_{j:N})\}|U_{k_\nu:N}=s_\nu,\ \nu=1,\ldots,r+1)|$$

$$\leq \prod_{\nu=1}^{r} |E(\exp\{itN^{-\frac{1}{2}} \sum_{j=1}^{n} a_{k_\nu+j,N}h(s_\nu+(s_{\nu+1}-s_\nu)V_{j:n})\})|,$$

where the final result is obtained by simply bounding the factors with $\nu = 0$ and $\nu = r+1$ by 1. Defining

$$h_\nu(v) = h(s_\nu+(s_{\nu+1}-s_\nu)v),\ \nu = 1,\ldots,r,$$

we find that if $\{U_{k_\nu:N} = s_\nu,\ \nu=1,\ldots,r+1\} \subset B$,

$$(4.11) \qquad h'_\nu(v) = (s_{\nu+1}-s_\nu)h'(s_\nu+(s_{\nu+1}-s_\nu)v) \geq \frac{(\beta-\alpha)c}{4r}$$

$$(4.12) \qquad |h''_\nu(v)| = (s_{\nu+1}-s_\nu)^2 h''(s_\nu+(s_{\nu+1}-s_\nu)v)| \leq (\beta-\alpha)^2\ C$$

for $\nu = 1,\ldots,r$. Because of (4.8) we also have

$$(4.13) \qquad a \leq a_{k_\nu+j,N} \leq A,\ \nu = 1,\ldots,r,\ j = 1,\ldots,n.$$

It follows that on the set B we can apply Lemma 4.2 to every factor on the right in (4.10) with h_ν, $a_{k_\nu+j,N}$ and n playing the part of h, a_{jN} and N in Lemma 4.2. Noting that $(x+y)^r \leq 2^{r-1}(|x|^r+|y|^r)$, that $n/N \geq (\beta-\alpha)N/(6r)$ for $n \geq 1$ and using (4.9) the proof is completed. □

Theorem 4.1 is certainly an improvement over Lemma 4.2 but condition (ii) on the weights is of course not very satisfactory. Its only function is to ensure that in the proof of Lemma 4.2 we can give a positive lower bound for $\tilde{h}'$ as well as an upper bound for $|\tilde{h}''|$ except on a set of exponentially small probability. This will obviously be possible in many cases where condition

(ii) does not hold. On the other hand, Theorem 4.1 would seem to cover most cases of interest.

To apply Theorem 4.1 to the situation of Section 3 and obtain (3.16), all one has to do is to replace h by F^{-1} in Theorem 4.1 and add the requirement that $0 < b \leq N^{-\frac{1}{2}} \sigma_N \leq B$. The number M_1 will then also depend on b and B.

5. *Edgeworth expansions for linear combinations of order statistics.* In Section 3 we indicated that a result like (3.16) would make it possible to obtain an asymptotic expansion for the d.f. of a linear combination of order statistics. Using Theorem 4.1, Helmers [16] has indeed obtained an Edgeworth expansion with remainder $o(N^{-1})$ for the case of smooth weights.

References

1. Albers, W., Bickel, P. J. and van Zwet, W. R. (1976). Asymptotic expansions for the power of distributionfree tests in the one-sample problem. *Ann. Statist.* 4, 108-156.

2. Bickel, P. J. (1974). Edgeworth expansions in nonparametric statistics. *Ann. Statist.* 2, 1-20.

3. Bickel, P. J. and van Zwet, W. R. (1976). Asymptotic expansions for the power of distributionfree tests in the two-sample problem. Report SW 38/76, Mathematisch Centrum, Amsterdam.

4. Bikjalis, A. (1968). Asymptotic expansions for densities and distributions of sums of identically distributed independent random vectors. *Litovsk. Mat. Sb.* 8, 405-422 (in Russian); *Selected Translations in Math. Statist. and Prob.* 13 (1973), 213-234.

5. Bjerve, S. (1974). Error bounds and asymptotic expansions for linear combinations of order statistics. Ph.D. thesis, University of California, Berkeley.

6. Chibisov, D. M. (1972). An asymptotic expansion for the distribution of a statistic admitting an asymptotic expansion. *Theor. Prob. & Appl.* 17, 620-630.

7. Chibisov, D. M. (1973a). An asymptotic expansion for a class of estimators containing maximum likelihood estimators. *Theor. Prob. & Appl.* 18, 295-303.

8. Chibisov, D. M. (1973b). An asymptotic expansion for the distribution of sums of a special form with an application to minimum contrast estimates. *Theor. Prob. & Appl.* 18, 649-661.

9. Chibisov, D. M. (1973c). Asymptotic expansions for Neyman's C(α) tests. Lecture notes in mathematics 330, Springer, 16-45.

10. Chibisov, D. M. (1974). Asymptotic expansions for some asymptotically optimal tests. Proc. Prague Symp. on Asymptotic Statistics, Vol. II, 37-68.

11. Cramér, H. (1962). *Random Variables and Probability Distributions*, 2nd Ed. Cambridge University Press.

12. Esseen, C. F. (1945). Fourier analysis of distribution functions. A mathematical study of the Laplace-Gaussian law. *Acta Math.* 77, 1-125.

13. Feller, W. (1971). *An Introduction to Probability Theory and Its Applications*, Vol. 2, 2nd Ed. Wiley, New York.

14. Gnedenko, B. V. and Kolmogorov, A. N. (1954). *Limit Distributions for Sums of Independent Random Variables.* Addison-Wesley, Reading.

15. Helmers, R. (1975). The order of the normal approximation for linear combinations of order statistics with smooth weight functions. Report SW 41/75, Mathematisch Centrum, Amsterdam.

16. Helmers, R. (1976). Edgeworth expansions for linear combinations of order statistics with smooth weight functions. Report SW 44/76, Mathematisch Centrum, Amsterdam. In preparation.

17. Hodges, J. L. and Lehmann, E. L. (1970). Deficiency. *Ann. Math. Statist.* 41, 783-801.

18. Petrov, V. V. (1975). *Sums of Independent Random Variables.* Springer, Berlin.

19. Pfanzagl, J. (1973). Asymptotic expansions related to minimum contrast estimators. *Ann. Statist.* 1, 993-1026.

20. Pfanzagl, J. (1974a). Asymptotically optimum estimation and test procedures. Proc. Prague Symp. on Asymptotic Statistics, Vol. I, 201-272.

21. Pfanzagl, J. (1974b). Nonexistence of tests with deficiency zero. Preprints in Statistics 8, Univ. of Köln.

ASYMPTOTIC PROPERTIES OF BAYES TESTS OF NONPARAMETRIC HYPOTHESES*

By Lionel Weiss
Cornell University

1. *Introduction.* This paper is concerned with a continuation of the investigation of asymptotic properties of tests of fit carried out in [4]. First we summarize the techniques and results from [4] that we will need.

A basic idea used in [4] is the concept of the asymptotic equivalence of two sequences of distributions. Suppose that for each postive integer m, $F_m(x(1),\ldots,x(k(m)))$ and $G_m(x(1),\ldots, x(k(m)))$ are joint cumulative distribution functions for a k(m)-dimensional random vector. For any Borel measurable region R(k(m)) in k(m)-dimensional space, $P_{F_m}(R(k(m))$, $P_{G_m}(R(k(m))$ denote the probabilities assigned to R(k(m)) by F_m and G_m respectively. We say that the sequences $\{F_m\}$ and $\{G_m\}$ are "asymptotically equivalent" if $\lim_{m\to\infty} \sup_{R(k(m))} |P_{F_m}(R(k(m)))-P_{G_m}(R(k(m)))| = 0$. If f_m, g_m are the densities corresponding to F_m, G_m respectively, then either of the following two conditions is necessary and sufficient for $\{F_m\}$ and $\{G_m\}$ to be asymptotically equivalent:

(1) $\log \dfrac{f_m(X_m(1),\ldots,X_m(k(m)))}{g_m(X_m(1),\ldots,X_m(k(m)))}$ converges stochastically to zero as m increases, when the joint cdf for $X_m(1),\ldots,X_m(k(m))$ is F_m.

(2) Same, except that the joint cdf is G_m. The sufficiency was proved in [2]. The necessity (which we do not need for our present purposes) is easily shown. The following special case will be used below: Suppose F_m is a k(m)-variate normal cdf with unit variances, zero covariances, and means $A_m(1),\ldots,A_m(k(m))$, and suppose G_m is a k(m)-variate normal cdf with unit variances, zero covariances, and means $B_m(1),\ldots,B_m(k(m))$. Then $\{F_m\}$ and $\{G_m\}$

*Research supported by National Science Foundation Grant No. MPS74-24270.

are asymptotically equivalent if and only if $\sum_{i=1}^{k(m)} [A_m(i)-B_m(i)]^2$ approaches zero as m increases.

The following standard problem of testing fit was discussed in [4], using somewhat different notation. For each positive integer n, we observe $X_n(1),\ldots,X_n(n)$, which are independent and identically distributed continuous random variables, $P[0 < X_n(i) < 1] = 1$. The common probability density function is $1 + r_n(x)$ for $0 \leq x \leq 1$, where $r_n(x) \geq -1$ for all x in [0,1], and $\int_0^1 r_n(x)dx = 0$. (Below we will sometimes use a function $r_n(x)$ of the special form $c(n)r(x)$, where $c(n)$ approaches zero as n increases, such that for small n and certain values of x, $c(n)r(x)$ may be below -1. This will not concern us, since we are interested in what happens as $n \to \infty$). The problem is to test the hypothesis of uniformity: $r_n(x) = 0$ for all x in [0,1].

For the time being, we assume that there exist fixed values ε,δ satisfying the inequalities $\frac{1}{4} < \varepsilon < \frac{1}{2}$, $\frac{3}{4} < \delta < \frac{1}{3}(2+\varepsilon)$, $\delta < 2\varepsilon$, such that

$$\lim_{n\to\infty} n^{\varepsilon} \max_{0\leq x\leq 1} |r_n(x)| = 0$$

$$\lim_{n\to\infty} n^{\varepsilon} \sup_{0<x<1} \left|\frac{d^s}{d_x^s} r_n(x)\right| = 0, \quad s = 1,2,3.$$

For typographical simplicity, whenever we write n^{γ} for a positive γ, we interpret it as the largest integer no greater than n^{γ}. Let $k(n)$ denote the largest integer such that $k(n)n^{\delta} < n$, and let $Y_n(1) <\ldots< Y_n(n)$ denote the ordered values of $X_n(1),\ldots,X_n(n)$. It was shown in [4] that the random variables $\{Y_n(n^{\delta}),Y_n(2n^{\delta}),\ldots,Y_n(k(n)n^{\delta})\}$ are asymptotically sufficient for our hypothesis testing problem.

Define $f_n(x)$ as $1 + r_n(x)$, $F_n(x)$ as $\int_0^x f_n(t)dt$ for $0 \leq x \leq 1$, and $Z_n'(j)$ as $n^{\frac{1}{2}}[Y_n(jn^{\delta}) - F_n^{-1}(\frac{jn^{\delta}}{n})]$ for $j = 1,\ldots,k(n)$. It follows from the results of [3] that if $n-k(n)n^{\delta}$ approaches infinity as n increases, then the sequence of joint asymptotic distributions of $\{Z_n'(1),\ldots,Z_n'))\}$ is asymptotically equivalent to a sequence of joint normal distributions given explicitly in [3]. (For this result, the conditions on the second and third derivatives of $r_n(x)$ are not required.) These joint normal distributions take a particularly simple form if we consider only values of n and δ for which n^{δ} and n/n^{δ} are integers, as in [2], so that $k(n)=n/n^{\delta}-1$. We make this restriction on the values of n considered. This is for convenience, and does not affect the asymptotic results. It was made implicitly in [4].

Define

$$Z_n(j) = n^{\frac{1}{2}}\,[Y_n(jn^{\delta}) - \frac{jn^{\delta}}{n}] \qquad j = 1,\ldots,k(n)$$

$$W_n(1) = [\frac{n(n^{\delta}-1)}{n^{2\delta}}]^{\frac{1}{2}}[Z_n(1) - \frac{1+(1+k(n))^{\frac{1}{2}}}{k(n)}\,Z_n(k(n))]$$

$$W_n(j) = [\frac{n(n^{\delta}-1)}{n^{2\delta}}]^{\frac{1}{2}}[Z_n(j)-Z_n(j-1)-\frac{1+(1+k(n))^{\frac{1}{2}}}{k(n)}\,Z_n(k(n))]$$

$$j = 2,\ldots,k(n).$$

It was shown in [4] that the sequence of joint distributions of $\{W_n(1),\ldots,W_n(k(n))\}$ is asymptotically equivalent to a sequence of joint normal distributions with unit variances, zero covariances, and

$$E\{W_n(j)\} = -[\frac{n^2(n^{\delta}-1)}{n^{2\delta}}]^{\frac{1}{2}}[n^{\delta}-F_n^{-1}(\frac{jn^{\delta}}{n})+F_n^{-1}(\frac{(j-1)n^{\delta}}{n})$$
$$-\,(\frac{1+(1+k(n))^{\frac{1}{2}}}{k(n)})(\frac{k(n)n^{\delta}}{n} - F_n^{-1}(\frac{k(n)n^{\delta}}{n}))]$$

for $j = 1,\ldots,k(n)$. Using the sufficient condition for asymptotic equivalence for the special case described above, a straightforward calculation shows that this sequence of joint normal distributions is asymptotically equivalent to a sequence of joint nor-

normal distributions with unit variances, zero covariances, and

$E\{W_n(j)\} = -n^{\delta/2}\, r_n(\frac{jn^\delta}{n})$ for $j = 1,\ldots,k(n)$. For the special case where $r_n(x) = \frac{r(x)}{\sqrt{n}}$ this was proved in [4], but this special case is too restrictive for our present purposes.

2. *Some tests of fit.* In Section 1 we have defined, for each n, observable random variables $W_n(1),\ldots,W_n(k(n))$ which are asymptotically sufficient for our hypothesis testing problem (since there is a one to one correspondence between them and $Y_n(n^\delta),\ldots,Y_n(k(n)n^\delta))$. Also, for all asymptotic probability calculations we can (and will) assume that $W_n(1),\ldots,W_n(k(n))$ are independent normal random variables with unit variances and $E\{W_n(j)\} = -n^{\delta/2}\, r_n(\frac{jn^\delta}{n})$. Thus the original hypothesis testing problem has become the problem of testing the hypothesis that $E\{W_n(j)\} = 0$ for $j = 1,\ldots,k(n)$. We use this form of the problem from now on.

It was pointed out in [4] that a natural measure of distance between the uniform density over (0,1) and the density $1+r_n(x)$ is $\int_0^1 r_n^2(x)dx$, but that no test procedure has asymptotic power staying above the level of significance against all alternatives with $\int_0^1 r_n^2(x)dx = \frac{c}{n}$, $c > 0$. (The term "test procedure" refers to a sequence of tests, one for each n). For example, suppose $r_n(x) = \frac{A}{\sqrt{n}} \cos t\pi x$, where A is arbitrary and t is a positive integer, so that $\int_0^1 r_n^2(x)dx = \frac{A^2}{2n}$, and $n^{1/2} \max_{0\leqq x\leqq 1} |F_n(x)-x| \leqq \frac{A}{\pi t}$. By taking t large enough, we can make $n^{1/2} \max_{0\leqq x\leqq 1} |F_n(x)-x|$ arbitrarily close to zero, and against such an alternative the Kolmogorov-Smirnov test procedure has power arbitrarily close to the level of significance.

Now suppose that for each n, we have a set of s(n) functions $r_n(x;1),\ldots,r_n(x;s(n))$, and our set of alternative distributions to the uniform distribution consists of all distributions with densities of the form $1 + \sum_{i=1}^{s(n)} \theta(i) r_n(x;i)$ for some unknown $\theta(1),\ldots,\theta(s(n))$. That is, we are testing the hypothesis that $E\{W_n(j)\} = 0$ against the alternative that

$$E\{W_n(j)\} = -n^{\delta/2} \sum_{i=1}^{s(n)} \theta(i) r_n\left(\frac{jn^\delta}{n} ; i\right), \quad j = 1,\ldots,k(n).$$

Let the t(n) vectors $v_n(i)$ $(i = 1,\ldots,t(n))$, where $v_n(i) = (v_n(1;i),\ldots,v_n(k(n);i))$ be a normal orthogonal basis for the vector space generated by the s(n) vectors $\{(r_n(\frac{n^\delta}{n} ; i),\ldots, r_n(\frac{k(n)n^\delta}{n} ; i)); i = 1,\ldots,s(n)\}$. We have $t(n) \leqq s(n)$, with strict inequality possible even if the s(n) functions $r_n(x;1),\ldots, r_n(x;s(n))$ are linearly independent. Our alternative can now be stated as follows:

$$E\{W_n(j)\} = -n^{\delta/2} \sum_{i=1}^{t(n)} \beta(i) v_n(j;i),$$

for $\beta(1),\ldots,\beta(t(n))$ unknown. We want to test the hypothesis that $\sum_{i=1}^{t(n)} \beta^2(i) = 0$. This is a classical regression problem with the common variance known to be unity. Using a Bayes decision rule with respect to an a priori distribution which assigns a probability in the open interval (0,1) to the point $\Sigma\beta^2(i) = 0$, and distributes the remaining probability uniformly over the sphere $\Sigma\beta^2(i) = c$ $(c > 0)$, we find that the rule rejects the hypothesis that $\Sigma\beta^2(i) = 0$ when $n^\delta \sum_{i=1}^{t(n)} \hat{\beta}_n^2(i)$ is "too large," where

$$\hat{\beta}_n(i) = -n^{-\delta/2} \sum_{j=1}^{k(n)} W_n(j) v_n(j;i).$$

When the hypothesis is true, $n^{\delta} \sum_{i=1}^{t(n)} \hat{\beta}_n^2(i)$ has a central chi-square distribution with t(n) degrees of freedom, and so a test of level of significance α is given by rejecting the hypothesis if $n^{\delta} \sum_{i=1}^{t(n)} \hat{\beta}_n^2(i)$ is greater than $C(\alpha;t(n))$, the appropriate value from the chi-square table with t(n) degrees of freedom. This is the test of level of significance α which is minimax against the class of alternatives with $\Sigma\beta^2(i) = c$, for any positive c.

To investigate the power of the test just developed, suppose that the true density is $1 + \bar{r}_n(x)$. We write the vector $(\bar{r}_n(\frac{n^{\delta}}{n}),\ldots,\bar{r}_n(\frac{k(n)n^{\delta}}{n}))$ as $\sum_{i=1}^{t(n)} \bar{\beta}_n(i)v_n(i) + \beta_n^* v_n^*$, where the vector v_n^* is orthogonal to $v_n(1),\ldots,v_n(t(n))$. Then $n^{\delta} \sum_{i=1}^{t(n)} \hat{\beta}_n^2(i)$ has a noncentral chi-square distribution with t(n) degrees of freedom and noncentrality parameter $n^{\delta} \sum_{i=1}^{t(n)} \bar{\beta}_n^2(i)$. If t(n) is large, then the distribution of

$$\frac{n^{\delta} \sum_{i=1}^{t(n)} \hat{\beta}_n^2(i) - t(n) - n^{\delta} \sum_{i=1}^{t(n)} \bar{\beta}_n^2(i)}{[2t(n)+4n^{\delta}\Sigma\bar{\beta}_n^2(i)]^{\frac{1}{2}}}$$

is approximately standard normal. From this, it follows that in order to have the asymptotic power of the test against the alternative $1 + \bar{r}_n(x)$ stay above the level of significance, we must have $\frac{n^{\delta}}{t(n)^{\frac{1}{2}}} \sum_{i=1}^{t(n)} \bar{\beta}_n^2(i)$ bounded away from zero. Roughly speaking, this means that $\frac{1}{t(n)^{\frac{1}{2}}} [n \int_0^1 \bar{r}_n^2(x)dx]$ must be bounded away from zero. If $\bar{r}_n(x)$ is of the form $n^{-\frac{1}{2}}\,\bar{r}(x)$ and t(n) gets larger as n gets larger, this boundedness away from zero cannot occur. It can occur if $\bar{r}_n(x)$ is of the form $\frac{\bar{r}(x)}{\Delta(n)}$, where $\Delta(n)$ approaches infinity more slowly than $n^{\frac{1}{2}}$.

The Bayes test just decribed is not an "all purpose" test, since it depends on a choice of a special set of alternatives. Now we try to construct a Bayes test which will be more of an all-purpose test. It is based on the fact that if a density over (0,1) goes above one in one neighborhood, it must go below one somewhere else to compensate. To incorporate this idea into our decision problem, we construct a Bayes decision rule relative to an a priori distribution which assigns a probability p in the open interval (0,1) to the point $E\{W_n(j)\} = 0$ for $j = 1,\ldots,k(n)$, and assigns probability $\frac{(1-p)}{k(n)[k(n)-1]}$ to each of the $k(n)[k(n)-1]$ points with $E\{W_n(i)\} = -\Delta_n$, $E\{W_n(j)\} = \Delta_n$, $E\{W_n(g)\} = 0$ for $g \neq i,j$, obtained as i,j vary over $1,\ldots,k(n)$ with $i \neq j$. This is not an entirely reasonable a priori distribution, because it allows $E\{W_n(i)\}$ and $E\{W_n(i+1)\}$ to differ by a relatively large amount, whereas in the original model $-n^{\delta/2}\, r_n(\frac{in^\delta}{n})$ and $-n^{\delta/2}\, r_n(\frac{(i+1)n^\delta}{n})$ should be relatively close, since r_n is differentiable. It is possible to assign a priori distributions which take this smoothness into account, but the analysis is more complicated, so we use the a priori distribution described as a first approximation to a more reasonable one. The usual simple calculation shows that the Bayes decision rule relative to the a priori distribution rejects the hypothesis when $T_n(\Delta_n)$ is "too large," where

$$T_n(\Delta_n) = \sum_{i \neq j}\sum e^{\Delta_n(W_n(i)-W_n(j))} .$$

To analyze the properties of the test based on $T_n(\Delta_n)$, we expand

$$e^{\Delta_n(W_n(i)-W_n(j))} = 1+\Delta_n(W_n(i)-W_n(j))+ \frac{\Delta_n^2}{2}(W_n(i)-W_n(j))^2+\ldots,$$

use the fact that $\sum_{i \neq j}\sum (W_n(i)-W_n(j))^s$ is zero if s is an odd integer, and find that if Δ_n approaches zero as n increases, the asymptotic properties of the test based on $T_n(\Delta_n)$ are the same as the test which rejects when $\sum_{i=1}^{k(n)} W_n^2(i) = T_n'$, say, is "too large." If the true density is $1+\bar{r}_n(x)$, T_n' has a noncentral chi-square distribution with $k(n)$ degrees of freedom and noncentrality parameter $n^\delta \sum_{i=1}^{k(n)} \bar{r}_n^2(\frac{in^\delta}{n})$, and thus the asymptotic distribution of

$$\frac{T_n' - k(n) - n^\delta \sum_{i=1}^{k(n)} \bar{r}_n^2(\frac{in^\delta}{n})}{[2k(n) + 4n^\delta \Sigma \bar{r}_n^2(\frac{in^\delta}{n})]^{\frac{1}{2}}}$$

is standard normal.

For the same level of significance, the asymptotic power of the test based on T_n' is greater than the asymptotic power of the test based on $\Sigma\hat{\beta}_n^2(i)$ if

$$\frac{n^\delta \Sigma \bar{r}_n^2(\frac{in^\delta}{n})}{\sqrt{k(n)}} > \frac{n^\delta \Sigma \bar{\beta}_n^2(i)}{\sqrt{t(n)}} .$$

Now $n^\delta \Sigma \bar{r}_n^2(\frac{in^\delta}{n})$ is asymptotically $n \int_0^1 \bar{r}_n^2(x)dx$, but if β_n^* is nonzero, then $n^\delta \Sigma \bar{\beta}_n^2(i)$ is less than $n \int_0^1 \bar{r}_n^2(x)dx$. On the other hand, $t(n)$ will usually be less than $k(n)$. This illustrates the fact that the test based on T_n' is an all-purpose test, but the test based on $\Sigma\hat{\beta}_n^2(i)$ will be better against the alternatives used in its construction.

Next we construct an all-purpose test using a priori distributions over smooth functions $r_n(x)$. We start by writing $r_n''(x)$ as the Fourier cosine series $A_n(0) + \sum_{j=1}^{\infty} A_n(j)\sqrt{2} \cos j\pi x$ for $0 \leq x \leq 1$. Then

$$r_n(x) = A_n(0)\,\frac{x^2}{2} - \sum_{j=1}^{\infty} A_n(j)\sqrt{2}\,\frac{\cos j\pi x}{(j\pi)^2} + \sum_{j=1}^{\infty} \frac{A_n(j)\sqrt{2}}{(j\pi)^2}\,.$$

Since $\int_0^1 r_n(x)dx = 0$, we find

$$\frac{A_n(0)}{6} + \sum_{j=1}^{\infty} \frac{\sqrt{2}\,A_n(j)}{(j\pi)^2} = 0,$$

so

$$r_n(x) = A_n(0)\left(\frac{x^2}{2} - \frac{1}{6}\right) - \sum_{j=1}^{\infty} A_n(j)\sqrt{2}\,\frac{\cos j\pi x}{(j\pi)^2}\,.$$

Using the Fourier expansion

$$x^2 = \frac{1}{3} - \frac{4}{\pi^2}\sum_{j=1}^{\infty}(-1)^{j+1}\,\frac{\cos j\pi x}{j^2}\,,$$

we obtain finally that

$$r_n(x) = \sum_{j=1}^{\infty} B_n(j)\sqrt{2}\,\frac{\cos j\pi x}{j^2}\,,$$

$$B_n(j) = \frac{1}{\pi^2}\,[-A_n(j) \pm (-1)^j\sqrt{2}\,A_n(0)].$$

Our hypothesis is that $B_n(j) = 0$ for $j = 1,2,\ldots$. We construct a Bayes decision rule relative to a priori distributions of the following type: Choose a value b_n in the open interval $(0,1)$, and two infinite sequences of positive values, $\{p_n(j);\ j=1,2,\ldots\}$, $\{\Delta_n(j);\ j=1,2,\ldots\}$, with $\sum_{j=1}^{\infty} p_n(j) = 1$. Assign a priori probability b_n to $\{B_n(j) = 0,\ j=1,2,\ldots\}$, and for each positive integer j, assign a priori probability $\frac{1}{2}(1-b_n)p_n(j)$ to $\{B_n(j) = -\Delta_n(j);\ B_n(\gamma) = 0 \text{ if } \gamma \neq j\}$, and a priori probability $\frac{1}{2}(1-b_n)p_n(j)$ to $\{B_n(j) = \Delta_n(j); B_n(\gamma) = 0 \text{ if } \gamma \neq j\}$. Denote

$$H_n(j) = \frac{-n^{\delta}\Delta_n^2(j)}{j^4}\sum_{i=1}^{k(n)} \cos^2\left(\frac{j\pi i n^{\delta}}{n}\right)$$

$$J_n(j) = \frac{n^{\delta/2}\Delta_n(j)\sqrt{2}}{j^2}\sum_{i=1}^{k(n)} W_n(i)\cos\left(\frac{j\pi i n^{\delta}}{n}\right).$$

Then a Bayes decision rule relative to this a priori distribution rejects the hypothesis if S_n is above 1, where

$$S_n = \frac{1-b_n}{2b_n} \sum_{j=1}^{\infty} p_n(j) e^{H_n(j)} (e^{J_n(j)} + e^{-J_n(j)})$$

The asymptotic distribution of S_n is not tabled (in most cases it is not normal), so at present we cannot compare the properties of S_n to the properties of the two tests developed above. However, as the following example shows, the test based on S_n seems to be quite good. We take $\Delta_n(j) = n^{-\frac{1}{2}}$ for all j and all n, and $p_n(j) = p(j) > 0$ for all n and j. We assume the true density is $1+n^{-\frac{1}{2}} L_n \sqrt{2}\, \frac{\cos \gamma\pi x}{\gamma^2}$ for some positive integer γ, where $|L_n|$ approaches infinity as n increases, but slowly enough so that $n^{-\frac{1}{2}+\varepsilon} L_n$ and $n^{\delta-1} L_n$ both approach zero as n increases. Define

$$\bar{W}_n(i) = W_n(i) + \frac{n^{\delta/2}}{n^{\frac{1}{2}}} \sqrt{2}\, \frac{L_n}{\gamma^2} \cos \gamma\pi \left(\frac{in^{\delta}}{n}\right),$$

and $\bar{J}_n(j)$ as the same function of $\{\bar{W}_n(i)\}$ as $J_n(j)$ is of $\{W_n(i)\}$. Thus, under our present density, $\{\bar{W}_n(i)\}$ have the same joint distribution as $\{W_n(i)\}$ have when the true density is uniform over (0,1).

$$\bar{J}_n(j) = \left[\frac{n^{\delta}}{n}\right]^{\frac{1}{2}} \frac{\sqrt{2}}{j^2} \sum_{i=1}^{k(n)} \left[W_n(i) + \left[\frac{n^{\delta}}{n}\right]^{\frac{1}{2}} \frac{\sqrt{2} L_n}{\gamma^2} \cos \gamma\pi\left(\frac{in^{\delta}}{n}\right)\right] \cos\left(\frac{j\pi in^{\delta}}{n}\right)$$

$$= J_n(j) + 2\left(\frac{n^{\delta}}{n}\right) L_n \sum_{i=1}^{k(n)} \left(\frac{\cos \gamma\pi\left(\frac{in^{\delta}}{n}\right)}{\gamma^2}\right)\left(\frac{\cos j\pi\left(\frac{in^{\delta}}{n}\right)}{j^2}\right)$$

$$= J_n(j) + \frac{L_n}{\gamma^4} D(j,\gamma) + \left(\frac{n^{\delta}}{n}\right) L_n \bar{B}_n(j)$$

where $D(j,\gamma) = 0$ if $j \neq \gamma$, $D(\gamma,\gamma) = 1$, and $|\bar{B}_n(j)| < \bar{B} < \infty$ for all j,n. Now in the expression for S_n, we replace $J_n(j)$ by the function of $\bar{J}_n(j)$ just developed, and look at the term containing $\bar{J}_n(\gamma)$. We have

$$S_n \geqq \frac{1-b_n}{2b_n} p(\gamma)e^{H_n(\gamma)}\{\exp[\bar{J}_n(\gamma)- \frac{L_n}{\gamma^4} -(\frac{n^\delta}{n})L_n\bar{B}_n(\gamma)] + \exp[-\bar{J}_n(\gamma)+ \frac{L_n}{\gamma^4} + (\frac{n^\delta}{n})L_n\bar{B}_n(\gamma)].\}$$

It follows that if b_n is bounded away from one, S_n approaches ∞ in probability as n increases, because $H_n(\gamma)$ approaches $-\frac{1}{2\gamma^4}$ as n increases, and $\bar{J}_n(\gamma)$ remains finite with probability approaching one. If we want our level of significance to be a fixed value in the open interval (0,1), b_n will be bounded away from one. Thus the test based on S_n will be consistent against the density $1 + n^{-\frac{1}{2}} L_n\sqrt{2}\,\frac{\cos \gamma\pi x}{\gamma^2}$. Things are even better if other values of $B_n(j)$ besides $B_n(\gamma)$ are nonzero.

The simple form of the asymptotic distribution of $\{W_n(i);\ i = 1,\ldots,k(n)\}$ suggests other tests which may not be Bayes, but have some intuitive appeal. We briefly describe some of these.

For a given $\Delta_n > 0$, let $U_n(\Delta_n)$ denote the number of $\{W_n(i)\}$ satisfying $|W_n(i)| > \Delta_n$. A possible test is to reject if $U_n(\Delta_n)$ is "too large." To analyze the properties of this test, let Φ denote the standard normal cdf. Assume that $1 + \bar{r}_n(x)$ is the true density, and define

$$A_n(i) = \Phi(\Delta_n + n^{\delta/2}\,\bar{r}_n(\frac{in^\delta}{n}))-\Phi(-\Delta_n + n^{\delta/2}\,\bar{r}_n(\frac{in^\delta}{n})).$$

Then the asymptotic distribution of

$$\frac{U_n(\Delta_n)-\sum_{i=1}^{k(n)} (1-A_n(i))}{[\sum_{i=1}^{k(n)} (1-A_n(i))A_n(i)]^{\frac{1}{2}}}$$

is standard normal.

Define $V_n = \max\{|W_n(1)|,\ldots,|W_n(k(n))|\}$. Another possible test is to reject if V_n is "too large." For any given $v > 0$,

$$P[V_n \leqq v]= \prod_{i=1}^{k(n)} [\Phi(v+n^{\delta/2}\,\bar{r}_n(\frac{in^\delta}{n}))-\Phi(-v+n^{\delta/2}\,\bar{r}_n(\frac{in^\delta}{n}))]$$

assuming that $1 + \bar{r}_n(x)$ is the true density.

A different type of test can be based on runs. A run is defined as a sequence of consecutive $\{W_n(i)\}$ all of the same sign. One possible test is to reject if there are too few runs. Another possible test is to reject if the longest run is too long.

3. *Concluding remarks.* In a paper by Reiss [1], a bound is given on the error made by assuming that the joint distribution of $\{W_n(i);\ i = 1,\ldots,k(n)\}$ is normal.

It is shown in [5] that the test procedures developed above also apply to the two-sample problem.

References

[1] Reiss, R. D. (1974). The asymptotic normality and asymptotic expansions for the joint distribution of several order statistics. Preprints in statistics, no. 11, University of Cologne.

[2] Weiss, L. (1969). The asymptotic joint distribution of an increasing number of sample quantiles. *Ann. Inst. Statist. Math.* 21, 257-263.

[3] Weiss, L. (1973). Statistical procedures based on a gradually increasing number of order statistics. *Comm. Statist.* 2, 95-114.

[4] Weiss, L. (1974). The asymptotic sufficiency of a relatively small number of order statistics in tests of fit. *Ann. Statist.* 2, 795-802.

[5] Weiss, L. Two-sample tests and tests of fit. *Comm. Statist.* 5, to appear.

OBSTRUCTIVE DISTRIBUTIONS IN A SEQUENTIAL RANK-ORDER TEST BASED ON LEHMANN ALTERNATIVES

By R. A. Wijsman*

University of Illinois at Urbana-Champaign

1. *Introduction.* In [2] Savage and Sethuraman considered the stopping time N of a sequential probability ratio test (SPRT) based on ranks for the following problem. Let X and Y be real valued and independent random variables, possessing distribution functions F and G, respectively. The problem is to test sequentially the hypothesis $G = F$ against the alternative $G = F^A$ for some given $A \neq 1$. To this end independent observations (X_1, Y_1), $(X_2, Y_2) \ldots$ on (X, Y) are taken and at each sampling stage all information in the sample is discarded except the ranks of the Y's among the X's and Y's. This reduces the composite hypotheses to simple ones. Let L_n be the log probability ratio at the nth sampling stage (an explicit expression is given in (2.1)), then the SPRT continues sampling as long as $\ell_1 < L_n < \ell_2$ for some chosen stopping bounds ℓ_1, ℓ_2. Let N be the stopping time, i.e. the smallest $n \geq 1$ at which $L_n \notin (\ell_1, \ell_2)$. It is desired to study the distribution of N when the true joint distribution of (X, Y) is P. Here P may belong to the model, i.e. be such that X and Y are independent and either $G = F$ or $G = F^A$, or P may be something quite different. Whatever P is, though, the (X_i, Y_i) will always be assumed to be iid, with common distribution P. N is called exponentially bounded under P if for every pair ℓ_1, ℓ_2 it is possible to choose $c > 0$ and $\rho < 1$ such that

$$(1.1) \qquad P(N > n) < c\,\rho^n, \quad n = 1, 2, \ldots .$$

Savage and Sethuraman [2] proved that N is exponentially bounded under every P except possibly if P is such that

*Research supported by the National Science Foundation under grants GP 28154 and MPS75-07978.

$$(1.2) \qquad E_P\, f(X,Y) = 0$$

where f is given in (2.2). Sethuraman [3] strengthened this result considerably by proving that N is exponentially bounded under P except possibly if

$$(1.3) \qquad P\{f(X,Y) = 0\} = 1.$$

(Another proof of this appears in [5], Section 4.2.)
The property of exponential boundedness of N is a desirable one. In the contrary case, when N is not exponentially bounded under P, we shall say that P is obstructive.

Sethuraman [3], Section 7, wondered whether there are any distributions P for which (1.3) holds. He showed that there are none if X and Y are required to be independent under P, and conjectured that perhaps there are none even without this requirement. However, it turns out that there are indeed distributions P satisfying (1.3), as will be shown in this paper. In fact, under the additional assumption that F and G are continuous, a description of the totality of all P satisfying (1.3) will be given. In addition, it will be shown that each such P is obstructive. Thus, for continuous F and G a complete classification of P's into those under which N is exponentially bounded and those that are obstructive has been obtained. The situation when F and G are unrestricted seems much more difficult, and no attempt will be made here to obtain results of any generality. However, one example of a discrete distribution P that is obstructive will be exhibited in Section 6. Questions beyond mere obstructiveness, such as which functions of N possess what moments, will not be considered in this paper.

Complete or partial results on exponential boundedness of N and obstructive distributions have been obtained in several other testing situations (for examples and references see [5]). This author believes that such results, besides their intrinsic interest, may be an important preliminary to the study of the best possible value of ρ in (1.1). Intuitively one expects distributions

P very close to an obstructive one to have "bad" values of ρ, i.e. values very close to 1. One could imagine the obstructive distributions among all distributions to play a role somewhat similar to the singularities of an analytic function.

2. *The main results.* It will be assumed in this section that F and G are continuous. Let F_n and G_n be the empirical distribution functions of $X_1,\ldots,X_n$ and $Y_1,\ldots,Y_n$, respectively, i.e. $F_n(z) = (1/n)(\text{number of X's} \le z)$, and similarly G_n. Following the notation in [3] denote $W = F + AG$ and $W_n = F_n + AG_n$. Also, if S is a set in some space, let I_S denote its indicator. In terms of this notation one obtains from [2][3] the following expressions for the log probability ratio L_n and for the function f in (1.2) and (1.3):

$$(2.1)\qquad L_n = \sum_1^n [-\log W_n(X_i)W_n(Y_i) + \log 4A-2] + \frac{1}{2}\log n,$$

omitting a quantity that is uniformly (in n) bounded, and

$$(2.2)\qquad f(x,y) = -\log W(x)W(y) + \log 4A$$
$$- \int (W(z))^{-1}[I_{(-\infty,z]}(x) + AI_{(-\infty,z]}(y)](F(dz) + G(dz)).$$

The integral on the right hand side in (2.2) splits into two integrals. The first one is

$\int_x^\infty (W(z))^{-1}(F(dz) + G(dz))$, or $\int_x^\infty W^{-1}(dF + dG)$ for short. It can be reduced to $\int_x^\infty W^{-1}dW + (1-A)\int_x^\infty W^{-1}dG = \log(1+A) - \log W(x) + (1-A)\int_x^\infty W^{-1}dG$. Analogously, the second can be reduced to $\log(1+A) - \log W(y) + (A-1)\int_y^\infty W^{-1}dF$. Substitution into (2.2) gives $f(x,y) = (A-1)\int_x^\infty W^{-1}dG - (A-1)\int_y^\infty W^{-1}dF + \log 4A - 2\log(A+1)$. Introducing

$$(2.3)\qquad c(A) = \frac{1}{A-1}\log\frac{(A+1)^2}{4A}$$

the equation (1.3) becomes

$$(2.4) \qquad P\{\int_X^\infty \frac{G(dz)}{W(z)} - \int_Y^\infty \frac{F(dz)}{W(z)} = c(A)\} = 1.$$

It suffices to treat the case $A > 1$ since the case $A < 1$ can be reduced to the former by interchanging $A \longleftrightarrow A^{-1}$, $X \longleftrightarrow Y$, $F \longleftrightarrow G$. For $A > 1$ it can easily be checked that $c(A) > 0$, and $c(A) \longrightarrow 0$ as $A \longrightarrow 1$.

Before stating the general results a special P satisfying (2.4) will be exhibited. There is a value of A and a value of v_1 for which

$$(2.5) \qquad \int_{v_1}^1 \frac{dv}{1+Av} = c(A) \text{ and } \int_0^{v_1} \frac{dv}{v_1+Av} = \int_{v_1}^1 \frac{dv}{v+Av_1},$$

(to 8 place accuracy $A = 8.2041497$, $v_1 = .25909257$). Choose any three points $a_1 < b_1 < a_2$ and take any P such that

$$P(X < a_1 < Y < b_1) = v_1, \; P(b_1 < X < a_2 < Y) = 1 - v_1,$$

and such that X and Y have continuous distributions. Then in the interval $(-\infty, a_1)$ only F rises, from 0 to v_1, while G remains 0; in (a_1, b_1) only G rises, from 0 to v_1, while F remains constant $(= v_1)$; in (b_1, a_2) only F rises, from v_1 to 1, while G stays at the value v_1; finally, in (a_2, ∞) G rises from v_1 to 1. Thus, F and G have a plateau at the same height v_1, and each distribution function is constant where the other increases. To verify (2.4), take first (X,Y) such that $b_1 < X < a_2 < Y$. Then since F does not change beyond a_2, $\int_Y^\infty W^{-1} dF = 0$. Furthermore, $\int_X^\infty W^{-1} dG = \int_{a_2}^\infty (1+AG)^{-1} dG$ since G does not change in (b_1, a_2), and on (a_2, ∞) $W(z) = 1+AG(z)$ since $F(z) = 1$ there. Making the substitution $G(z) = v$ the integral becomes $\int_{v_1}^1 (1+Av)^{-1} dv$, which equals $c(A)$ by the first of equations (2.5). Therefore, with this (X,Y) the event in (2.4) happens. Now taking $X < a_1 < Y < b_1$ it can be again verified that the event in (2.4) happens by a reasoning similar to the one above and using (2.5). Therefore, this P satisfies (2.4).

In this example there are two pairs of intervals. These intervals are arbitrary, but the values of A and v_1 turn out to be unique. The example can be extended to any number k of pairs of intervals.

PROPOSITION 2.1. For each $k = 2,3,\ldots$ there exists a unique $A > 1$ and a unique sequence $0 = v_0 < v_1 < \ldots < v_k = 1$ such that

$$(2.6)\quad \int_{v_{k-1}}^{1} \frac{dv}{1+av} = c(A),\ \int_{v_{i-1}}^{v_i} \frac{dv}{v_i+Av} = \int_{v_i}^{v_{i+1}} \frac{dv}{v+Av_i},\ i=1,\ldots,k-1,$$

in which c(A) is defined in (2.3).

In Proposition 2.1 the dependence of A and the v_i on k has been suppressed in the notation. Note that for $k = 2$, (2.6) reduces to (2.5).

PROPOSITION 2.2. For any integer $k \geq 2$ choose $2k-1$ points $a_1 < b_1 < \ldots < a_k$ arbitrarily. Let A and the v_i be as in Proposition 2.1 and define $p_i = v_i - v_{i-1}$, $i = 1,\ldots,k$. Let the joint distribution P of (X,Y) have continuous marginals and be such that

$$(2.7)\quad P(b_{i-1} < X < a_i < Y < b_i) = p_i, \quad i = 1,\ldots,k$$

in which $b_0 = -\infty$, $b_k = \infty$. Then P satisfies (2.4).

PROPOSITION 2.3 (converse of Proposition 2.2). Let P be such that F and G are continuous and (2.4) is satisfied, with $A > 1$. Then P has to be of the form given in Proposition 2.2 for some integer $k \geq 2$ and some sequence $a_1, b_1,\ldots,a_k$.

PROPOSITION 2.4. Every distribution P described in Proposition 2.2 is obstructive.

The proof of Proposition 2.2 will be omitted since it is essentially the same as in the special case $k = 2$. The other propositions will be proved in the next three sections.

This section concludes with a table of values of A and the p_i of Proposition 2.2, for $k = 2,\ldots,5$. These values were computed with help of some of the formulas of Section 3.

$k = 2$	$A = 8.204149713$	$p_1 = .259$
		$p_2 = .741$
$k = 3$	$A = 3.712349169$	$p_1 = .134$
		$p_2 = .328$
		$p_3 = .538$
$k = 4$	$A = 2.634668342$	$p_1 = .083$
		$p_2 = .191$
		$p_3 = .305$
		$p_4 = .421$
$k = 5$	$A = 2.164327986$	$p_1 = .057$
		$p_2 = .126$
		$p_3 = .199$
		$p_4 = .272$
		$p_5 = .346$

3. *Proof of Proposition* 2.1. Make the substitution $u_i = v_i/v_{i+1}$, $i = 0,\ldots,k-1$ (so that $u_0 = 0$), then (2.6) for $i = 1$ takes the form

$$\int_0^1 \frac{dx}{1+Ax} = \int_{u_1}^1 \frac{dx}{x+Au_1} \tag{3.1}$$

which determines u_1 as a function of A. Then (2.6) for $i = 2$ determines u_2, etc. Finally, having determined u_{k-1}, the integral $\int_{u_{k-1}}^1 (1 + Ax)^{-1}dx$ should equal $c(A)$ by the first of equations (2.6). It turns out more convenient to multiply all integrals and $c(A)$ by A. Therefore, introduce

$$f_i(A) = A\int_{u_i}^1 \frac{dx}{1+Ax}, \quad i = 0,1,\ldots, \tag{3.2}$$

$$g(A) = \frac{A}{A-1} \log \frac{(A+1)^2}{4A}, \; A > 1; \; g(1) = 0, \tag{3.3}$$

then the equation $\int_{u_{k-1}}^{1} (1+Ax)^{-1}dx = c(A)$ is the same as $f_{k-1}(A) = c(A)$. Thus, Proposition 2.1 takes the following form. Given the sequence of equations

$$f_i(A) = A\int_{u_{i+1}}^{1} \frac{dx}{x+Au_{i+1}}, \qquad i = 0,1,... \tag{3.4}$$

with f_i defined by (3.2). To show that for every $i = 1,2,...$ there exists a unique solution for $A > 1$ of the equation

$$f_i(A) = g(A). \tag{3.5}$$

Write definition (3.2) down for $i + 1$:

$$f_{i+1}(A) = A\int_{u_{i+1}}^{1} \frac{dx}{1+Ax}, \tag{3.6}$$

then comparison of the integrands in (3.4) and (3.6) shows that

$$f_{i+1}(A) < f_i(A). \tag{3.7}$$

(It also follows easily from a comparison of the integrands in (3.2) and (3.4) that $u_{i+1} > u_i$, but this fact will not be used.) Solving for u_{i+1} from (3.4) one obtains

$$u_{i+1}^{-1} = (1+A)\exp(A^{-1}f_i(A))-A \tag{3.8}$$

and substituting this into (3.6) yields the recursion relation

$$f_{i+1}(A) = \log[1+A-A\exp(-A^{-1}f_i(A))]. \tag{3.9}$$

By (3.2) for $i = 0$

$$f_0(A) = \log(1+A) \tag{3.10}$$

and it follows then from (3.9) by induction that all functions f_i are continuous. It also follows by induction that $f_i(1) > 0$ for all i. (In fact, it is easy to evaluate $f_i(1)=\log((i+2)/(i+1))$.) On the other hand $g(1) = 0$, so that

$$f_i(1) - g(1) > 0, \qquad i = 0,1,... . \tag{3.11}$$

The function f_1 follows from (3.9) and (3.10) as

$$f_1(A) = \log[1+A-A\exp(-A^{-1}\log(1+A))]. \tag{3.12}$$

Study of the asymptotic behavior of f_1 reveals that $f_1(A) \sim \log \log A$ as $A \longrightarrow \infty$, whereas $g(A) \sim \log A$. This implies, using (3.7), that

$$(3.13) \qquad f_i(A) - g(A) \longrightarrow -\infty \text{ as } A \longrightarrow \infty, \; i = 1,2,\ldots .$$

(Note: (3.13) does not hold for $i = 0$.) From (3.11), (3.13), and the continuity of $f_i - g$ it follows that for each $i = 1,2,\ldots$ there is a value of $A > 1$ such that (3.5) is satisfied. This takes care of the existence part of the proof.

The uniqueness part is much harder and rather lengthy. The main steps will be outlined here; most of the details are relegated to the Appendix. For $i = 0,1,\ldots$ the functions f_i are defined for $A \geq 0$, $f_i(0) = 0$ (using (3.10) and (3.9)), and $f_i(A) > 0$ if $A > 0$. The function f_0 is obviously strictly increasing and strictly concave. By induction, using (3.9), it can easily be established that each f_i enjoys the same properties; i.e.

$$(3.14) \qquad f_i'(A) > 0, \; f_i''(A) < 0, \; A \geq 0, \; i = 0,1,\ldots .$$

Uniqueness of a solution A of (3.5) follows if it is true that for every solution $g'(A) - f_i'(A) > 0$. This would trivially be the case if $g'(A) - f_i'(A)$ were > 0 for every $A > 1$, $i = 1,2,\ldots$ However, this turns out to be false for $i = 1$ (although true for $i \geq 2$). The argument is therefore of necessity slightly more involved. The following properties will be shown in the Appendix:

(i) $f_i'(A) - f_{i+1}'(A) > 0$, $A > 0$, $i = 0,1,\ldots$;

(ii) there exists $A_0 > 1$ such that $g'(A) - f_1'(A) <$ or > 0 according as $A <$ or $> A_0$, and $g(A) - f_1(A) < 0$ for $A < A_0$;

(iii) $g'(A) - f_2'(A) > 0$ for all $A > 1$.

Combining (i) and (iii) it follows that $g'(A) - f_i'(A) > 0$ for $i = 2,3,\ldots$, so that a solution of (3.5) is unique for those i. By (ii), a solution of (3.5) with $i = 1$ cannot be $< A_0$ since $g - f_1$ is negative there. Since for any $A > A_0$, $g'(A) - f_1'(A) > 0$, uniqueness of a solution is also guaranteed for $i = 1$.

4. *Proof of Proposition* 2.3. Let P be a distribution of (X,Y) such that (2.4) holds. From (2.4) it follows that

$$P\{\int_X^\infty W^{-1}dG \geq c(A)\} = 1. \tag{4.1}$$

The integral $\int_x^\infty W^{-1}dG$ is continuous as a function of x and $\longrightarrow 0$ as $x \longrightarrow \infty$. It follows from (4.1) that there is a value of x for which the integral equals c(A). Define

$$x_0 = \max\{x: \int_x^\infty W^{-1}dG = c(A)\}, \tag{4.2}$$

and define also

$$x_1 = \min\{x: F(x) = 1\}. \tag{4.3}$$

It will now be shown that $x_1 \leq x_0$. Suppose the contrary: $x_1 > x_0$. Take x_2 such that $x_0 < x_2 < x_1$, then, by the definition of x_1, $F(x_2) < 1$ so $P(X > x_2) > 0$, and by the definition of x_0,

$\int_{x_2}^\infty W^{-1}dG < c(A)$. But this would imply $P\{\int_X^\infty W^{-1}dG < c(A)\} > 0$, contradicting (1). Hence $x_1 \leq x_0$, so that $F(x_0) = 1$.

Put $G(x_0) = u_1$, then $\int_{x_0}^\infty W^{-1}dG = \int_{u_1}^1 (1+Au)^{-1}du$, using the change of variable $G(z) = u$ and observing $F(z) = 1$ if $z \geq x_0$. By (4.2):

$$\int_{u_1}^1 (1+Au)^{-1}\, du = c(A). \tag{4.4}$$

Here u_1 must be < 1 since $c(A) > 0$. AlsO, u_1 must be > 0 since $\int_0^1(1+Au)^{-1}du = A^{-1}\log(1+A)$ can be shown to be $> c(A)$ for all $A \geq 1$. Define

$$y_0 = \max\{x: F(x) = u_1\}, \tag{4.5}$$

so that $y_0 < x_1$. Furthermore, define

$$(4.6) \qquad S = \{(x,y) \in R^2: \int_x^\infty W^{-1}dG - \int_y^\infty W^{-1}dF = c(A)\}$$

and

$$(4.7) \qquad S_1 = \{(x,y) \in S: y > x_0\},$$

then, using (2.4),

$$(4.8) \qquad P\{(X,Y) \in S\} = 1,$$

$$(4.9) \qquad P\{(X,Y) \in S_1\} = 1-u_1 > 0.$$

It will be shown now that $G(y_0) = u_1$. Suppose not, so that $G(y_0) < u_1$. Choose $\varepsilon > 0$ so small that $y_0 + \varepsilon < x_0$ and $G(y_0 + \varepsilon) < u_1$. Since $G(x_0) = u_1$ it follows that

$$(4.10) \qquad \int_{y_0+\varepsilon}^{x_0} W^{-1}dG > 0.$$

Furthermore, by the definition (4.5) of y_0, $F(y_0 + \varepsilon) > u_1$. Then $P\{(X,Y) \in S_1, X > y_0 + \varepsilon\} \leq P(X > y_0 + \varepsilon) = 1-F(y_0 + \varepsilon) < 1-u_1$. Combining this with (4.9) we get

$$(4.11) \qquad P\{(X,Y) \in S_1, X \leq y_0 + \varepsilon\} > 0.$$

Take any $x \in \{x: (x,y) \in S_1, x \leq y_0 + \varepsilon\}$, then

$$\int_x^\infty W^{-1}dG \geq \int_{y_0+\varepsilon}^\infty W^{-1}dG = \int_{y_0+\varepsilon}^{x_0} W^{-1}dG + \int_{x_0}^1 W^{-1}dG > c(A),$$

using (4.10) and (4.4). Also, if $(x,y) \in S_1$, then $y > x_0$ so that $\int_y^\infty W^{-1}dF = 0$. Combining these two results, if $(x,y) \in S_1$ with $x \leq y_0 + \varepsilon$, then $\int_x^\infty W^{-1}dG - \int_y^\infty W^{-1}dF > c(A)$. But, using (4.11), this means that (2.4) is violated. Hence the conclusion $G(y_0) = u_1$.

Define

$$(4.12) \qquad c_1(A) = \int_{u_1}^1 (u + Au_1)^{-1}du.$$

The integral on the right hand side in (4.12) is $\int_{y_0}^{\infty} W^{-1}dF$. Since $G(y_0) = G(x_0)$ and since $F(x_0) = 1$, it may be assumed (after removing a set of probability 0 if necessary) that on S $y \notin (y_0,x_0)$ and $x \leq x_0$. On the set $S-S_1$, $y \leq x_0$ so $y \leq y_0$. For any such y,

$\int_y^{\infty} W^{-1}dF \geq \int_{y_0}^{\infty} W^{-1}dF = c_1(A) > 0$, so that, by (4.6), on $S-S_1$ we have

$\int_x^{\infty} W^{-1}dG > c(A)$. This rules out $x > y_0$, since for such x,

$\int_x^{\infty} W^{-1}dG = \int_{x_0}^{\infty} W^{-1}dG = c(A)$. Hence, on $S-S_1$ both x and y are $\leq y_0$.

Then $u_1 = F(y_0) = P\{(X,Y) \in S_1, X \leq y_0\} + P\{(X,Y) \in S-S_1, X \leq y_0\} = P\{(X,Y) \in S_1, X \leq y_0\} + P\{(X,Y) \in S-S_1\} = P\{(X,Y) \in S_1, X \leq y_0\} + u_1$, by (4.9). It follows that

$$(4.13) \qquad P\{(X,Y) \in S_1, X \leq y_0\} = 0.$$

Therefore, it may be assumed that on S_1, $y_0 < x \leq x_0$ as well as $y > x_0$. This proves Proposition (2.3) for the kth pair of intervals by identifying $x_0 = a_k$, $y_0 = b_{k-1}$, $u_1 = v_1$.

For $(x,y) \in S-S_1$ (where $x \leq y_0$, $y \leq y_0$),

$\int_x^{\infty} W^{-1}dG - \int_y^{\infty} W^{-1}dF = (\int_x^{y_0} W^{-1}dG - \int_y^{y_0} W^{-1}dF) + (\int_{y_0}^{\infty} W^{-1}dG - \int_{y_0}^{\infty} W^{-1}dF)$. The expression on the left hand side equals c(A), by (2.4), and the second expression in parentheses on the right hand side equals $c(A) - c_1(A)$, by (4.4) and (4.12). It follows that

$$(4.14) \qquad P\{\int_X^{y_0} W^{-1}dG - \int_Y^{y_0} W^{-1}dF = c_1(A) \mid (X,Y) \in S-S_1\} = 1.$$

The conditional problem (4.14) is identical to (2.4), after replacing c(A) by $c_1(A)$.

Continuing this process we obtain the sequence (u_1,c_1), $(u_2,c_2),\ldots$ (the dependence of the u's and c's on A has been suppressed in the notation), in which

$$\int_{u_i}^{1} \frac{du}{1+Au} = c_{i-1}, \int_{u_i}^{1} \frac{du}{u+Au_i} = c_i, \ (c_0=c). \tag{4.15}$$

(These u's are numerically the same as the u's in Section 3, but their subscripting is different.) Remembering that the u_i for $i \geq 2$ are conditional probabilities, their relation to the v_i of Proposition 2.2 is through $v_{k-i} = \Pi_{j \leq i} u_j$, $1 \leq i \leq k$. Proposition 2.3 will have been proved if we show that the sequence (u_i,v_i) is necessarily finite, say of length k. If that is the case, then u_k must be $= 0$ and $k \geq 2$ (since $u_1 > 0$). To show this, suppose the contrary, i.e. suppose there is an infinite sequence (u_1,c_1) $(u_2,c_2),\ldots$ satisfying (4.15) with all $u_i \geq 0$. From the first of the equations (4.15) it follows that each $c_i \leq 1$. Comparison of the two integrands in (4.15) makes it clear that $c_{i-1} < c_i$. Thus, c_i converges upwards to a limit $c_\infty \leq 1$. Note that all c_i are > 0 since $c_0 = c > 0$, so all u_i are < 1. From the first of the equations (4.15) it follows that the u_i are decreasing, say to $u_\infty \geq 0$. Taking limits in (4.15) there results

$$\int_{u_\infty}^{1} \frac{du}{1+Au} = \int_{u_\infty}^{1} \frac{du}{u+Au_\infty}. \tag{4.16}$$

However, (4.16) is impossible since the left hand side if finite, and the integrand on the left hand side is strictly less than the integrand on the right hand side. Therefore, the sequence is finite and Proposition 2.3 is proved.

5. *Proof of Proposition* 2.4. The distribution P of Proposition 2.2 is supported on k pairs of intervals, say I_j, J_j, $j = 1,\ldots,k$, such that $I_1 \leq J_1 \leq I_2 \leq \cdots \leq J_k$, and $P(X \in I_j, Y \in J_j) = p_j$, $j = 1,\ldots,k$. Therefore, for each $X_i \in I_j$ the corresponding Y_i is $\in J_j$. The value of L_n can then be expressed completely in terms of the numbers n_j of X's falling in I_j, $j = 1,\ldots,k$ (with

$\sum_1^k n_j = n$). For instance, in I_1 the value of G_n is identically 0 so that $W_n(X_i) = F_n(X_i)$ for any $X_i \in I_1$ and the contribution of I_1 to $\sum \log W_n(X_i)$ is $\sum_{i=1}^{n_1} \log (i/n)$. In J_1 the value of F_n is identically n_1/n so that $W_n(Y_i) = (n_1/n) + AG_n(Y_i)$ for $Y_i \in J_1$. The contribution of J_1 to $\sum \log W_n(Y_i)$ is therefore $\sum_{i=1}^{n_1} \log((n_1/n)$ $+Ai/n)$. For each j, n_j is a binomial variable and E $n_j/n = p_j$. It is readily shown, e.g. using Stirling's approximation, that $\sum_{i=1}^{n_1} \log((n_1/n) + Ai/n)$ differs from $n \int_0^{n_1/n} \log((n_1/n) + Ax)\, dx$ by an amount bounded uniformly in n and uniformly for n_1/n in some neighborhood of p_1. The resulting integral is an analytic function of n_1/n. The contributions of the I_j and J_j for $j = 2,\ldots,k$ can be treated analogously. Only the contribution of I_1 requires a little care. Here the result is that $\sum_{i=1}^{n_1} \log(i/n) - \frac{1}{2} \log n$ differs from $n \int_0^{n_1/n} \log x\, dx$ by a uniformly bounded constant. The term $\frac{1}{2} \log n$ that appears here cancels the term $\frac{1}{2} \log n$ in the expression (2.1) for L_n. Therefore, there is an analytic function Φ of k arguments and a constant $B > 0$ such that

$$|L_n - n\, \Phi\, (\frac{n_1}{n}, \ldots, \frac{n_k}{n})| < B \tag{5.1}$$

for $n = 1,2,\ldots$ and the n_j/n in some neighborhoods of the p_j, $j = 1,\ldots,k$.

The fact that L_n only depends on n and the n_j/n enables replacing the random vector (X,Y) by Z defined as the k-vector whose jth component equals 1 if $X \in I_j$ and 0 otherwise.

Analogously, replace (X_i, Y_i) by Z_i, $i = 1,2,\ldots$. Then Z, Z_1, $Z_2, \ldots$ are iid with $EZ = p = (p_1,\ldots,p_k)'$ and $E||Z||^2 < \infty$ (in fact, $||Z||$ has finite moment generating function). The vector $(n_1/n,\ldots,n_k/n)'$ can then be written as the sample mean $\bar{Z}_n = (1/n) \sum_1^n Z_i$, so that (5.1) takes the form

(5.2) $$|L_n - n\,\Phi\,(\bar{Z}_n)| < B,$$

valid for $n = 1,2,\ldots$ and $\bar{Z}_n$ in a neighborhood of p. Furthermore, (1.3) implies that

(5.3) $$\Phi\,(p) = 0 \text{ and } P\{\Delta'(Z-p) = 0\} = 1,$$

in which Δ equals grad Φ evaluated at p (see [5] Proposition 3.2.1 and Remark 2 following that proposition; equations (5.3) can also be verified directly, with a good deal of work). The obstructiveness of P now follows from Theorem 2.1 in [4].

6. *An example of an obstructive discrete* P. Ties between observations can now occur with positive probability. If that happens it is customary to replace ranks by midranks. This can be handled conveniently by redefining distribution functions: $n\,F_n(z) = \frac{1}{2}$ (number of X's $\leq z$ + number of X's $< z$), $F(z) = \frac{1}{2}\,(P(X \leq z) + P(X < z))$, and similarly G_n and G. Moreover, in (2.2) the indicator $I_{(-\infty,z]}$ should be replaced by $\frac{1}{2}\,(I_{(-\infty,z]} + I_{(-\infty,z)})$. Strictly speaking, in the definition (2.1) of L_n, $F_n(X_i)$ should be replaced by $F_n(X_i) + 1/(2n)$ (with the new definition of F_n) and $G_n(Y_i)$ by $G_n(Y_i) + 1/(2n)$. This replaces $W_n(X_i)$ by $W_n(X_i) + 1/(2n)$, $W_n(Y_i)$ by $W_n(Y_i) + A/(2n)$. However, these additions of $1/(2n)$ and $A/(2n)$ to $W_n(X_i)$ and $W_n(Y_i)$, respectively, make only a uniformly bounded difference in the value of L_n and can therefore be ignored for the purpose of proving exponential boundedness or obstructiveness.

In analogy to the type of obstructive distributions described in Proposition 2.2, with $k = 2$, let us try a discrete distribution P of the following form. Choose four points $a_1 < b_1 < a_2 < b_2$ arbitrarily, and for $0 < p < 1$, to be determined later, let

$P(X = a_1, Y = b_1) = p$, $P(X = a_2, Y = b_2) = 1-p$. For P to be obstructive it is necessary (but not sufficient) that (1.3) be satisfied, where f is given in (2.2) (with the new definitions of distribution functions and indicator). Equation (1.3) is equivalent to $f(a_1,b_1) = 0$, $f(a_2,b_2) = 0$, which furnishes two equations for the pair A, p. Instead of reproducing these rather messy equations another method will be presented yielding an equivalent set of equations. In the sample $(X_1,Y_1),\ldots,(X_n,Y_n)$ at the nth sampling stage let there be n_1 of the X's equal to a_1, so $n-n_1$ equal to a_2. Then one computes

$$L_n = n\,\Phi_A\left(\frac{n_1}{n}\right) + \frac{1}{2}\log n \tag{6.1}$$

in which for $0 \leq v \leq 1$

$$\Phi_A(v) = -\,v\log\left(\frac{v}{2}\right) - v\log\left(v\left(1 + \frac{1}{2}A\right)\right) \tag{6.2}$$

$$-\,(1-v)\log\left(\frac{1}{2} + \left(A + \frac{1}{2}\right)v\right) - (1-v)\log\left(1 + \frac{1}{2}A(1+v)\right) + \log 4A-2.$$

Note that in contrast to the situation with F and G continuous the $\frac{1}{2}\log n$ term in (2.1) is not absorbed by $\sum \log W_n(X_i)$ as it was in Section 5 (leading to (5.2)). It turns out that the equations $f(a_1,b_1) = 0$ are equivalent to the two equations $\Phi_A(p) = 0$, $\Phi_A'(p)$, where prime denotes differentiation. There are several solutions to these equations. For two of these solutions Φ_A has a local minimum at p. Then for n_1/n close to p both terms on the right hand side in (6.1) have the same sign. Since for the asymptotic behavior of L_n only the local properties of Φ_A near p matter, it is easy to see that N is exponentially bounded. Only one of the solutions has $\Phi''(p) < 0$. The values of A and p (rounded to 10 places) are $A = 1.320015126$, $p = .1401865276$, and $\Phi''(p) < -3.57666$. Now for n_1/n close to p the two terms in (6.1) have opposite sign and it is a nontrivial problem to decide whether N is exponentially bounded or not. The answer is given in [1] where it is proved that under the above circumstances P is obstructive.

Appendix. Proof of (i) *of Section* 3. It will first be shown that $u_i'(A) < 0$ for $i = 1,2,\ldots$, where u_i is defined in (3.8). Equating the right hand sides of (3.2) and (3.4) yields the following relation between u_i and u_{i+1}:

$$\text{(A1)} \quad A \log(A+u_{i+1}^{-1}) + \log(1+Au_i) = (1+A)\log(1+A).$$

Differentiate (A1):

$$\text{(A2)} \qquad \frac{-Au_{i+1}'}{u_{i+1}(1+Au_{i+1})} = \frac{-Au_i'}{1+Au_i} + \varphi(A)$$

in which

$$\text{(A3)} \qquad \varphi(A) = \log(1+A) - \log(A+u_{i+1}^{-1}) + (1+Au_{i+1})^{-1} - u_i(1+Au_i)^{-1}.$$

It will be shown now that

$$\text{(A4)} \qquad \varphi(A) > 0 \text{ for all } A > 0.$$

Using (A1) to eliminate u_{i+1} from the right hand side of (A3), and replacing u_i by u, gives

$$\text{(A5)} \qquad \varphi = \varphi(u,A) = 1 - \frac{u}{1+Au} + \frac{1}{A}\log\frac{1+Au}{1+A} - \frac{A(1+Au)^{1/A}}{(1+A)^{1+1/A}}.$$

The dependence of u on A will be ignored and it will be shown that $\varphi(u,A) > 0$ for all $0 \leq u < 1$, $A > 0$. Put first $x = (1+Au)/(1+A)$ and then $y = x^{-1/A}$, and put $A(1+A)\varphi(u,A) = g(y,A)$. The result is

$$\text{(A6)} \qquad g(y,A) = A^2 - 1 + y^A - A(1+A)\log y - A^2y^{-1}.$$

Note that $y > 1$ if $u < 1$. From (A6) it follows that $g(1,A) = 0$. By taking the derivative with respect to y several times it can be established that $\partial g/\partial y > 0$ for $y > 1$. It follows that $g(y,A) > 0$ for $y > 1$ so that (A4) has been shown.

Going back to (A2), and using (A4), it follows that $u_i' \leq 0$ implies $u_{i+1}' < 0$. Since $u_0' = 0$ it follows by induction that $u_i' < 0$ for $i = 1,2,\ldots$, as was to be shown.

Now put $h(A) = f_i(A) - f_{i+1}(A)$ (for any $i = 0,1,\ldots$) then to show (i) means to show $h'(A) > 0$ for $A > 0$. For f_i take the expression on the right hand side of (3.4) and for f_i+1 the definition (3.2) with i replaced by i+1. The result is

$$\text{(A7)}\quad h(A) = A \log(A+u_{i+1}^{-1}) + \log(1+Au_{i+1})-(1+A)\log(1+A).$$

Differentiate: $h'(A) = \partial h/\partial A + u_{i+1}' \; \partial h/\partial u_{i+1}$. Both terms on the right hand side of the latter equation will be shown to be > 0. The first term, with u_{i+1}^{-1} replaced by $v(> 1)$ is $\partial h/\partial A = \log(A+v) + (1+A)(A+v)^{-1} - \log(1+A) - 1$ which is easily shown to be > 0 for $v > 1$. The second term is $(-u_{i+1}')A(1+Au_{i+1})^{-1}(u_{i+1}^{-1}-1) > 0$ since $-u_{i+1}' > 0$ as shown before. This finishes the proof of (i).

Proof of (ii). Put $h(A) = A(g'(A)-f_1'(A))$. It will first be shown that $h'(A) > 0$. Differentiation of (3.3) and (3.12) yields

$$\text{(A8)}\qquad h(A) = \frac{\log\,(1+A)}{(1+A)^{1+1/A}-A} - \frac{A}{(A-1)^2}\log\frac{(A+1)^2}{4A},$$
$$= h_1(A) + h_2(A), \text{ say.}$$

It will be shown that both h_1 and h_2 have a positive derivative. Compute $h_2'(A) = (A+1)(A-1)^{-3}\log((A+1)^2/4A)-(A-1)^{-1}(A+1)^{-1}$. Making the substitution $A = \exp(2\theta)$, the claim follows from the inequality $2 \log\cosh\theta > (\tanh\theta)^2$ if $\theta \neq 0$. In h_1 put $x = A^{-1}\log(1+A)$. Since $\log(1+A)$ is concave and $= 0$ at $A = 0$, x is a decreasing function of A, decreasing from 1 to 0 as A increases from 0 to ∞. Consider $1 + A^{-1} = k(x)$ as a function of x, then $h_1(A)$ can be written in the form $x(k(x)e^x-1)^{-1}$. To show that this is an increasing function of A is equivalent to showing that $x^{-1}(k(x)e^x-1)$ is an increasing function of x. Since $k(x)e^x-1 = 0$ at $x = 0$, it suffices to show that $k(x)e^x-1$ is convex. Using the fact that k and k' are positive, it suffices to show that k is convex. Differentiating k twice, and writing x' for dx/dA, we compute $k''(x) = A^{-2}(x')^{-2}(2A^{-1}+x''/x')$. In order to show that the expression in parentheses is > 0 write

(A9) $$x = x(A) = \int_0^1 \frac{dy}{1+Ay}$$

and differentiate twice with respect to A on the right hand side of (A9) under the integral. The claim then follows from the inequality

(A10) $$\int_0^1 \frac{y}{(1+Ay)^2}\,dy > \int_0^1 \frac{Ay^2}{(1+Ay)^3}\,dy,$$

which is immediate from a comparison of the integrands on both sides. This concludes the demonstration that $h' > 0$.

In order to evaluate $h(A)$ at $A = 1$ compute from (A8) $h_1(1) = (1/3)\log 2$ and $\lim_{A \to 1} h_2(A) = -1/4$. Therfore, $h(1) = (1/3)\log 2 - 1/4 < 0$. Next, investigate $h(A)$ as $A \to \infty$. Clearly, $h_2(A) \to 0$ as $A \to \infty$. On the other hand, h_1 and h_1' are > 0 for all $A > 0$. Therefore, h is eventually positive.

It has been shown now that h is strictly increasing, negative at $A = 1$, and eventually positive. Since h is continuous there exists a unique A_0 such that $h <$ or > 0 according as $A <$ or $> A_0$. The same is then true for $g' - f_1'$, as was to be shown. Furthermore, $g(1) = 0$, $f_1(1)>0$, so $g-f < 0$ at $A = 1$. The function $g - f$ has to stay < 0 for $A < A_0$ since its derivative is negative there. This concludes the proof of (ii).

Proof of (iii). To show $g'(A) - f_2'(A) > 0$ for all $A > 1$. This is certainly true for $A > A_0$, because then $g'(A) - f_1'(A) > 0$ by (ii), and $f_1'(A) - f_2'(A) > 0$ for all $A > 0$ by (i). It remains to be shown that $g' - f_2' > 0$ for $A < A_0$. This is certainly true if shown to be true for $A < A_1$ if $A_1 > A_0$. Take $A_1 = 1.2$ and verify numerically that $g'-f_1' > 0$ at $A = 1.2$ so that $A_1 > A_0$. From definition (3.3) by twice differentiation it can be established that g is concave. The concavity of f_2 is implied by (3.14). It follows that

(A11) $$g'(A) - f_2'(A) \geq g'(A_1) - f_2'(1) \quad \text{if} \quad 1 \leq A \leq A_1.$$

Numerical computation of the two terms on the right hand side in

(A11) yields $g'(A_1) = g'(1.2) > .24$, and $f_2'(1) = 1/2 + (1/3) \log 2 - (1/2) \log 3 < .19$. Therefore, the right hand side in (A11) is $> .05$ so that certainly the left hand side is positive. This concludes the proof of (iii).

References

[1] Lai, T. L., and Wijsman, R. A. (   ). On the first exit time of a random walk from the stopping bounds $f(n) \pm cg(n)$ with application to obstructive distributions in sequential tests. *IMS Bull.* to appear.

[2] Savage, I. Richard, and Sethuraman, J. (1966). Stopping time of a rank-order sequential probability ratio test based on Lehmann alternatives. *Ann. Math. Statist.* 37, 1154-1160.

[3] Sethuraman, J. (1970). Stopping time of a rank-order sequential probability ratio test based on Lehmann alternatives II. *Ann. Math. Statist.* 41, 1322-1333.

[4] Wijsman, R. A. (1972). A theorem on obstructive distributions. *Ann. Math. Statist.* 43, 1709-1715.

[5] Wijsman, R. A. (   ). A general theorem with applications on exponentially bounded stopping time, without moment conditions. *Ann. Statist.* to appear.

OPTIMUM DESIGNS FOR FINITE POPULATIONS SAMPLING

By H. P. Wynn

University of California, Berkeley and Imperical College, London

1. *Statistical motivation.* The author has been trying for some while to see whether optimum design theory could be used in the superpopulation approach to survey sampling developed by Royall [2], [3] and Royall and Herson [4]. A beginning is made in an earlier paper [5] in which a G-optimum minimax approach is taken. In this paper we show further than an equivalence theory can exist similar in conception to and conceivably more general than the now classical D-optimality of Kiefer and Wolfowitz ([1], etc.). Although the motivation was genuinely from finite populations sampling it is possible that there will be further applications to more standard regression design situations.

There is a population S of N units labelled $i = 1,\ldots,N$. Each unit has a known attribute x_i. We can consider the x_i to lie in a general topological space as in optimum design theory but it is convenient to think of x as lying in a compact set of R^m. This is certainly in tune with classical survey sampling in which x_i may be a vector of stratification variables. Each unit i has a characteristic Y_i unknown before the sampling. We assume a superpopulation model in which each Y_i is a random variable and

$$E(Y_i) = f(x_i)^T\theta$$

$\mathrm{cov}(Y_i,Y_j) = 0$ $(i \neq j)$, $\mathrm{var}(Y_i) = \sigma^2$ $(i = 1,\ldots,N)$, $f(x) = (f_1(x),\ldots,f_k(x))^T$ is a vector of linearly independent continuous functions on H and $\theta = (\theta_1,\ldots,\theta_k')^T$ are unknown parameters.

On the basis of a purposive selection of a sample s of n units from S for which we observe the Y_i, $i \in s$ we seek to estimate some linear function $\tau = \sum c_i Y_i$ of the Y_i. Even after sampling τ is still random since it is a function of unobserved quantities also. We seek an estimator T for τ which is unbiased in the sense that $E(T) = E(\tau)$ and minimizes the m.s.e. $E(T-\tau)^2$. The solution turns out to be $T = \sum_{i\in s} c_i Y_i + \sum_{i\in\bar{s}} c_i \hat{Y}_i$ where $\bar{s} = S - s$ the unsampled units and the $\hat{Y}_i$ are the least squares estimates for the $E(Y_i)$ for $i \in \bar{s}$ based on Y_i, $i \in s$. The case $\tau = \sum Y_i$ has been studied at length by Royall and Herson in the above papers.

We shall consider the case when $\tau = Y_i$ for some $i \in \bar{s}$. Then the best linear unbiased estimator for Y_i in the above sense is just $\hat{Y}_i$. One of our design criteria is to minimize over all choices of s

$$\max_{i\in\bar{s}} E(Y_i-\hat{Y}_i)^2.$$

That is to minimize the m.s.e. of "prediction" over the individual unsampled units. The quantity

$$E(Y_i-\hat{Y}_i)^2 = \sigma^2(1 + f(x_i)^T(X_s^TX_s)^{-1}f(x_i))$$

where $X_s = [f(x_{(1)}),\ldots,f(x_{(n)})]^T$ and $s = \{(1),\ldots,(n)\}$. Thus because of homoscedascity we need only consider the maximum of the variance function $f(x_i)^T(X_s^TX_s)^{-1}f(x_i)$. Although this is close to the G-optimality criterion, the difference is that the region of maximization is itself design dependent. Since we have the sampled units in the sample we do not need to predict for them!

2. *Continuous generalization.* The exact theory described above has been developed briefly by the author in [5] in which he shows that

$$\min_{s\subset S}\ \max_{i\in\bar{s}} \{E(Y_i-\hat{Y}_i)^2\} \le \frac{n+1}{n-k+1}, \tag{2.1}$$

that is, an upper bound can be placed on the minimax value independent of the population size N. A more general theory is now available. We consider the population S to be replaced by a measure ξ_0 on the compact space $\mathcal{X}$ and the sample by a measure ξ on $\mathcal{X}$. The measure ξ_0 is a probability measure, integrating to unity, while ξ has all the properties of a probability measure except that

$$\int_{\mathcal{X}} \xi(dx) = v, \quad (0 < v < 1). \tag{2.2}$$

Also we assume that ξ_0 dominates ξ_1 in the strict sense that

$$\xi_0(A) \geq \xi(A) \tag{2.3}$$

for every measurable set $A \subset \mathcal{X}$. The moment matrix of ξ is defined in the usual way as

$$M(\xi) = \int_{\mathcal{X}} f(x)f(x)^T \xi(dx)$$

where f(x) is continuous on $\mathcal{X}$ as before. We assume also that at least one non-singular $M(\xi)$ exists.

An important special case will play the major role in this paper, that is when ξ_0 and ξ have finite support. Denoting by

$$\left\{\begin{matrix} p_1,\dots,p_L \\ x_1,\dots,x_L \end{matrix}\right\}$$

the measure placing mass $p_1,\dots,p_L$ at $x_1,\dots,x_L$ respectively the conditions (2.1) and (2.2) reduce to

$$\xi_0 = \left\{\begin{matrix} p_1,\dots,p_L \\ \\ x_1,\dots,x_L \end{matrix}\right\}$$

$$\xi = \left\{\begin{matrix} q_1,\dots,q_L \\ \\ x_1,\dots,x_L \end{matrix}\right\}$$

$\sum p_i = 1$, $\sum q_i = v$ and $0 \leq q_i \leq p_i$ $(i = 1,\dots,L)$.

For a given ξ_0 and v let $\mathcal{D}(\xi_0,v)$, sometimes abbreviated to $\mathcal{D}$, be the set of measures satisfying (2.2) and (2.3). We define $D(\xi_0,v)$-optimality as the maximizing of det $M(\xi)$ for ξ in $\mathcal{D}$. The definition of G-optimality requires some further notation to cope with the idea of maximization over the unsampled units.

Let $\underline{B}(\xi)$ denote the support, $\text{supp}(\xi_0-\xi)$, of $\xi_0-\xi$. This can be taken as a closed set. It represents the points where $\xi_0 > \xi$.

Similarly define $\bar{B}(\xi) = \text{supp}(\xi_0) \cap \text{supp}(\xi)$ also closed, those points where $\xi_0 \geq \xi > 0$. Define $B(\xi) = \underline{B}(\xi) \cap \bar{B}(\xi)$, that is where $\xi_0 > \xi > 0$. The variance function is defined as $d(x,\xi) = f(x)^T M^{-1}(\xi) f(x)$ in the usual way.

THEOREM 2.1. *The following two statements about a measure* ξ^* *in* $\mathcal{D}(\xi_0, v)$ *are equivalent:*

(i) $\sup_{\xi \in \mathcal{D}} \det M(\xi)$ *is achieved for* ξ^*

(ii) $d(x_1,\xi^*) \geq d(x_2,\xi^*)$ *for* $x_1 \in \bar{B}(\xi^*)$, $x_2 \in \underline{B}(\xi^*)$ *with equality when* x_1 *and* $x_2 \in B(\xi)$.

PROOF. We shall only give the complete proof in this paper for measures with finite support; the extension to arbitrary measures is purely technical.

Let $\xi_0 = \{\begin{smallmatrix} p_1,\ldots,p_L \\ x_1,\ldots,x_L \end{smallmatrix}\}$, $\xi = \{\begin{smallmatrix} q_1,\ldots,q_L \\ x_1,\ldots,x_L \end{smallmatrix}\}$, $p_i > 0$, $i = 1,\ldots,L$, $\sum p_i = 1$, $\sum q_i = v$ $(0 < v < 1)$ and $0 \leq q_i \leq p_i$ $(i = 1,\ldots,L)$. In this case,

$$\bar{B}(\xi) = \{x_i | p_i \geq q_i > 0\}$$

$$\underline{B}(\xi) = \{x_i | p_i > q_i \geq 0\}$$

and $$B(\xi) = \{x_i | p_i > q_i > 0\}.$$

(i) $\Rightarrow$ (ii). Suppose that (i) holds but (ii) does not. Then there exists points $x_i \in \bar{B}(\xi^*)$ and $x_j \in \underline{B}(\xi^*)$ with

$$d(x_j,\xi^*) > d(x_i,\xi^*).$$

For α sufficiently small $(0 < \alpha < 1)$

$$\xi^+ = \xi^* - \alpha[x_i] + \alpha[x_j]$$

belongs to $\mathcal{D}$, where $[x]$ denotes the measure with mass unity at x. But

$$\frac{\partial}{\partial\alpha} \log \det M(\xi^+)\Big|_{\alpha=0} = d(\xi^*,x_j) - d(\xi^*,x_i) > 0$$

contradicting that ξ^* satisfies (i). By putting x_i and x_j both in $B(\xi^*)$ and repeating the argument we obtain the equality of $d(x,\xi^*)$ on $B(\xi^*)$.

(ii) $\Rightarrow$ (i). By concavity of log det $M(\xi)$ for $\xi \in \mathcal{D}$ we need only show that

$$\frac{\partial}{\partial\alpha} \log \det((1-\alpha)M(\xi^*) + \alpha M(\xi'))\Big|_{\alpha=0} \leq 0,$$

for all $\xi' \in \mathcal{D}$. But this inequality is equivalent to

$$-\int_{\mathcal{X}} d(x,\xi^*)\xi^*(dx) + \int_{\mathcal{X}} d(x,\xi^*)\xi'(dx) \leq 0.$$

This certainly holds from the assumptions in (ii) and the fact that $\xi^* \geq \xi'$ when $d(x,\xi^*)$ takes its largest values. This completes the proof.

More can be said about the set of possible solutions. Since log det $M(\xi)$ is in fact strictly concave the set of solutions ξ^* to (i) (and (ii)) give a unique $M(\xi^*)$. Let ξ_1^* and ξ_2^* be two solutions. Then so is $\xi^* = \alpha\xi_1^* + (1-\alpha)\xi_2^*$ $(0 < \alpha < 1)$. Clearly

$$B(\xi^*) \supset (B(\xi_1^*) \cup B(\xi_2^*)).$$

Thus since (ii) says that $d(x,\xi^*)$ is a constant on $B(\xi^*)$ and all solutions have the same $d(x,\xi^*)$, all the $B(\xi^*)$ must have images under f, $f(B(\xi^*))$ on the same hyper-ellipsoid given by $d(x,\xi^*) =$ constant, in R^k. Thus the only allowable difference is for the x_i whose $f(x_i)$ vectors are on this ellipsoid. All x_i with $f(x_i)$ inside the ellipsoid are given zero measure by any ξ^*. This goes over naturally to more general measures.

The main difficulty with the theory is in giving conditions under which (i) and (ii) are equivalent to the following analogue of G-optimality:

(iii) $\inf_{\xi \in \mathcal{D}} \sup_{x \in \underline{B}(\xi)} d(x,\xi)$ is achieved by ξ^*.

We may call ξ^* $G(\xi_0,v)$-optimum in this case. It was thought initially that equivalence always held, particularly since it does for two important examples (see Section 3). However counterexamples can be found. Consider the behavior of $d(x,\xi^*)$ for fixed x when ξ^* is changed slightly. Let $\bar{\xi}=\xi^*+\alpha\xi_2-\alpha\xi_1$ $(0\leq\alpha<1)$.

It can easily be verified that

$$\frac{\partial}{\partial\alpha} d(x,\bar{\xi})\Big|_{\alpha=0} = \text{trace}\{M^{-1}(\xi^*)f(x)f(x)^T M^{-1}(\xi^*)(M(\xi_1)-M(\xi_2))\}$$

$$= \int_{\mathcal{X}} (f(x)^T M^{-1}(\xi^*)f(z))^2 \xi_1(dz)$$

$$- \int_{\mathcal{X}} (f(x)^T M^{-1}(\xi^*)f(z))^2 \xi_2(dz). \tag{2.4}$$

To obtain a counterexample we try to find probability measures ξ_1 and ξ_2 with supports in $\underline{B}(\xi^*)$ such that (2.4) is negative for all x in $B(\xi^*)$. This would imply that for small α and ξ_0 with finite support

$$\sup_{x\in \underline{B}(\bar{\xi})} d(x,\bar{\xi}) < \sup_{x\in \underline{B}(\xi^*)} d(x,\xi^*).$$

This follows since for small α, $\underline{B}(\bar{\xi}) = \underline{B}(\xi^*)$ and $\bar{\xi} \subset \mathcal{D}(\xi_0,v)$.

It is simpler to take $M(\xi^*) = cI$ a multiple of the $k \times k$ identity, in looking for counterexamples. Suppose we can find such an example in which $k = 2$. Let $B(\xi^*)$ consist of two points x_1 and x_2 for which

$$f(x_1) = \left(\sqrt{\tfrac{3}{4}}, \sqrt{\tfrac{1}{4}}\right),\ f(x_2) = \left(\sqrt{\tfrac{2}{3}}, \sqrt{\tfrac{1}{3}}\right).$$

Let $\underline{B}(\xi^*)$ contain one extra point beside x_1 and x_2 namely x_3 for which

$$f(x_3) = (\delta,0) \quad (0 < \delta < 1).$$

Let

$$\xi_2 = \begin{Bmatrix} \frac{1}{2} & \frac{1}{2} \\ x_1 & x_2 \end{Bmatrix} \qquad \xi_1 = \begin{Bmatrix} 1 \\ x_3 \end{Bmatrix} .$$

Evaluation of the right hand side of (2.4) at $x = x_1$ and $x = x_2$ respectively yields

$$\frac{1}{c^2}\left(\frac{1}{2} + \frac{1}{2}\left(\frac{7}{12} - \frac{1}{\sqrt{6}}\right) - \delta\sqrt{\tfrac{3}{4}}\right) \text{ and } \frac{1}{c^2}\left(\frac{1}{2} + \frac{1}{2}\left(\frac{7}{12} - \frac{1}{\sqrt{6}}\right) - \delta\sqrt{\tfrac{2}{3}}\right).$$

For δ sufficiently close to 1 both these quantities are negative. It is not hard to set up masses at outlying points, $\bar{B}(\xi^*)$, so that

$M(\xi^*) = cI$, and ξ^* is $D(\xi_0,v)$ optimum.

Although equivalence cannot be achieved for arbitrary measures a bound can be placed on inf sup $d(x,\xi)$ analogous to the value k in the usual D-optimality theory. This is the generalization of the exact theory bound (2.1).

THEOREM 2.2. <u>The</u> $D(\xi_0,v)$ <u>optimum measure</u> ξ_2^* <u>and</u> $G(\xi_0,v)$ <u>optimum measure</u> ξ_1^* <u>satisfy</u>

$$\sup_{x \in \underline{B}(\xi_1^*)} d(x,\xi_1^*) \le \sup_{x \in \underline{B}(\xi_2^*)} d(x,\xi_2^*) \le \frac{k}{v} .$$

PROOF. This is given only for discrete measures, as in Theorem 2.1. The first inequality follows by definition. Let $\xi = (1-\alpha)\xi_2^* + \alpha[x]_v$ where x is a point in $\underline{B}(\xi_2^*)$ at which $\sup d(x,\xi_2^*)$ is achieved, and $[x]_v$ denotes the measure consisting of mass v at x. For small α, ξ is in $\mathcal{D}(\xi_0,v)$. But

$$\left.\frac{\partial \log \det M(\xi)}{\partial \alpha}\right|_{\alpha=0} = vd(x,\xi_2^*) - k.$$

By the optimality of ξ_2^*, this cannot be positive.

3. *Examples*. There must certainly be a rich theory of $D(\xi_0,v)$-optimality complicated by the problem of when we have equivalence with $G(\xi_0,v)$ optimality. We present two simple examples where full equivalence holds.

(1) STRAIGHT LINE REGRESSION. Let ξ_0 be an arbitrary probability measure on an interval [a,b]. The problem reduces to finding the measure ξ^* with maximum variance subject to the constraints that $\xi_0 \ge \xi^*$ and $\int_a^b \xi^*(dx) = v < 1$. It is not hard to show that in this case the $D(\xi_0,v)$ and $G(\xi_0,v)$ optimum designs are unique and identical. The author has not come across such a problem before. The solution is clearly different in every case but can be given nicely in words. There are points c, d with $a < c < d < b$ such that $\xi^* \equiv \xi_0$ on [a,c] and [d,b] and $\xi^* \equiv 0$ on (c,d) and $\frac{c+d}{2} = \mu$ the mean of ξ^*. We can see this immediately

since the variance function must be equal at c and d.

(2) STRATIFIED SAMPLING. Let $x_1,\ldots,x_k$ be k stratification values. We give each stratum j its own mean θ_j. Thus we may take $f(x_1) = (1,0,0,\ldots,0)^T, f(x_2) = (0,1,0,\ldots,0)^T$, etc. Let

$$\xi_0 = \begin{Bmatrix} p_1,\ldots,p_k \\ x_1,\ldots,x_k \end{Bmatrix} ,$$

$$\xi^* = \begin{Bmatrix} q_1,\ldots,q_k \\ x_1,\ldots,x_k \end{Bmatrix}$$

$\sum p_i = 1$, $\sum q_i = v$, $0 \le q_i \le p_i$ $(i = 1,\ldots,k)$. Suppose also that $p_1 \le p_2 \le \cdots \le p_k$. The D and $G(\xi_0,v)$ optimum designs coincide. The solution is to put $q_i = p_i$ for $i = 1,\ldots,s$ and put all the remaining q_i equal. The integer s is determined uniquely by v.

The solution differs from proportional allocation or other designs. It can be interpreted as treating every unsampled unit equally.

References

[1] Kiefer, J. and Wolfowitz, J. (1959). Optimum designs in regression problems. *Ann. Math. Statist.* 30, 271-294.

[2] Royall, R. M. (1970). On finite population sampling theory under certain linear regression models. *Biometrika* 57, 377-87.

[3] Royall, R. M. (1971). Linear regression models in finite population theory. In Godambe, V. P. and Sprott, D. A. (Ed.) *Foundations of Statistical Inference*. Holt, Rinehart & Winston, Toronto.

[4] Royall, R. M. and Herson, J. (1973). Robust estimation in finite populations, I, II. *J. Amer. Statist. Soc.* 68, 880-893.

[5] Wynn, H. P. (1976). Minimax purposive survey sampling design. *J. Amer. Statist. Soc.* (to appear).

A
B 7
C 8
D 9
E 0
F 1
G 2
H 3
I 4
J 5